# Advanced Structured Materials

Volume 69

More information about this series at http://www.springer.com/series/8611

Francesco dell'Isola · Mircea Sofonea
David Steigmann
Editors

# Mathematical Modelling in Solid Mechanics

Springer

*Editors*
Francesco dell'Isola
Università di Roma "La Sapienza"
Roma
Italy

David Steigmann
Department of Mechanical Engineering
UC Berkeley
Berkeley, CA
USA

Mircea Sofonea
LAboratoire de Mathématiques et PhySique
Université de Perpignan Via Domitia
Perpignan
France

ISSN 1869-8433 ISSN 1869-8441 (electronic)
Advanced Structured Materials
ISBN 978-981-10-9961-8 ISBN 978-981-10-3764-1 (eBook)
DOI 10.1007/978-981-10-3764-1



Printed on acid-free paper

This Springer imprint is published by Springer Nature
The registered company is Springer Nature Singapore Pte Ltd.
The registered company address is: 152 Beach Road, #21-01/04 Gateway East, Singapore 189721, Singapore

# Preface

Mechanical process involving deformable solids are abundant in industry and everyday life and play an important role in engineering structures and systems. They include a large variety of phenomena and, therefore, the need of using mathematical models (based on fundamental physical principles) that can predict reliably the evolution of deformable bodies under the action of external loads was recognised long time ago. Mathematical models used in Solid Mechanics are usually expressed in terms of partial differential equations, associated to various boundary and initial conditions. Their validity depends on how well the theoretical results they predict agree with the results of repeatable experiments. Indeed, the crucial modelling criterion is that a mathematical model should be able to describe those mechanical properties of the real system that are under the consideration with the desired accuracy. In some special cases, the solution of the established mathematical model may be found analytically, where the obtained solution is 'exact' under the assumptions made during mechanical and mathematical modelling. Nevertheless, in most of the cases computational procedures have to be utilised to find the numerical solutions.

Mathematical modelling in Solid Mechanics requires, generally, three steps. The first one is construction of mathematical models based on constitutive assumptions and thermodynamic principles. It needs a lot of engineering experience since proper analogies between the reality and the model, dependent on the goals of the analysis, have to be established. The second step is given by the mathematical and numerical analysis of the models. It is devoted to existence, uniqueness and convergence results for the continuous or discrete solutions and error estimates for approximation schemes, among others. Finally, the third step consists in construction of reliable and efficient algorithms for the numerical approximations of the models, to provide numerical simulations and comparison of the numerical results with the experimental ones.

The aim of this edited volume is to present new and original results of mathematical modelling in Solid Mechanics. It contains a large variety of results which cover the path starting from the construction of the models and ending with the numerical solution and engineering applications. It is structured into 18 contributions that have not been published before, have been obtained by recognised scholars in the area and have gone through a rigorous refereeing process. Most of the contributions made the subject of oral presentations to the international conference *Emerging Trends in Applied Mathematics and Mechanics* (ETAMM 2016), organised by the *Laboratory of Mathematics and Physics* (LAMPS) of the University of Perpignan Via Domitia, France. This successful meeting held in Perpignan during May 30–June 3 2016 and gathered more than two hundred mathematicians, interdisciplinary scientists and engineers from all over the world. It was organised, together with this volume, with the collaboration of the Scientific Committee of the *International Research Centre for Mathematics and Mechanics of Complex Systems* (MEMOCS) of the Università dell'Aquila, Italy. The quest of a deeper interdisciplinary collaboration among mathematicians and applied scientists has been shared between the organisers of ETAMM 2016 and the MEMOCS researchers and was the main leitmotif of all talks presented.

This volume addresses engineers, applied mathematicians and scientists. Advanced graduate students can also benefit from the material presented in this book. Generally, the reader is expected to have background knowledge on mechanics of continua, partial differential equations, nonlinear analysis, numerical analysis and computer sciences.

As editors, we wish to express our gratitude to all authors for their valuable contributions to this special issue. We really appreciate their professional job in the achievement of this volume. We extend our thanks to the reviewers for their very helpful comments that have improved the final versions of some papers. We would like to express our sincere thanks to Springer staff for inviting us as editors of this volume and for their help in bringing it in your hand.

Roma, Italy
Perpignan, France
Berkeley, CA, USA
January 2017

Francesco dell'Isola
Mircea Sofonea
David Steigmann

# Emerging Trends in Applied Mathematics and Mechanics (ETAMM 2016)

This conference held in Perpignan (France) during May 30–June 3 2016, organised by the *Laboratory of Mathematics and Physics* (LAMPS) of the University of Perpignan Via Domitia. Its aim was to bring together mathematicians, engineers and interdisciplinary scientists from around the world, in order to identify new trends in the domains of applied mathematics and mechanics, and their applications in physics, engineering sciences, biology and economy, and to further promote collaborations among researchers from different disciplines. The programme contained seven invited plenary talks, 12 keynotes lectures and 14 minisymposia. The conference was rendered possible by the generous financial support of the Université de Perpignan Via Domitia, Conseil Général des Pyrénées Orientales and Région Languedoc-Roussillon.

## Organizing Committee

**Mircea Sofonea**, Université de Perpignan Via Domitia, France (**Chair**)
Mikaël Barboteu, Université de Perpignan Via Domitia, France
Robert Brouzet, Université de Perpignan Via Domitia, France
David Danan, Université de Perpignan Via Domitia, France
Sylvia Muñoz, Université de Perpignan Via Domitia, France
Joëlle Sulian, Université de Perpignan Via Domitia, France

## International Scientific Committee

Sanda Cleja-Ţigoiu, University of Bucharest, Romania
Victor Eremeyev, Institute of Mechanical Engineering, St. Petersburg, Russia
Xiqiao Feng, Tsinghua University, Beijing, China

Yibin Fu, Keele University, Staffordshire, UK and Tianjin University, China
Weimin Han, University of Iowa, Iowa City, USA
Francesco dell'Isola, Università di Roma La Sapienza, Italy
Yuri P. Kalmykov, Université de Perpignan Via Domitia, France
Amaury Lambert, Université de Paris 6, France
Ming Li, Zhejiang University, China
José Merodio, Universidad Politécnica de Madrid, Spain
Stanisław Migórski, Jagiellonian University, Krakow, Poland
Dumitru Motreanu, Université de Perpignan Via Domitia, France
Patrizio Neff, Universität Duisburg-Essen, Germany
Ray Ogden, University of Glasgow, UK
Qinqhua Qin, Australian National University, Australia
Tudor Raţiu, Shanghai Jiao Tong University, China
B. Daya Reddy, University of Cape Town, South Africa
Hugo A.F.A. Santos, Instituto Superior de Engenharia de Lisboa, Portugal
Meir Shillor, Oakland University, Rochester MI, USA
Mircea Sofonea, Université de Perpignan Via Domitia, France
Kostas P. Soldatos, University of Nottingham, UK
David Steigmann, University of California, Berkeley, USA
Tianhai Tian, Monash University, Australia
Juan. M. Viaño, Universidade da Santiago de Compostela, Spain
Jie Wang, Zhejiang University, Hangzhou, China
Qi Wang, University of South Carolina Columbia, USA
Zong-Ben Xu, Xi-an Jiaotong University, China
Song-Ping Zhu, University of Wollongong, Australia

## Plenary Lectures

Philippe G. Ciarlet, City University of Hong Kong, China
José M. Mazón, Universitat de Valencia, Spain
Ray Ogden, University of Glasgow, UK
Tudor Raţiu, Shanghai Jiao Tong University, China
B. Daya Reddy, University of Cape Town, South Africa
David J. Steigmann, University of California, USA
Zongben Xu, Xi-an Jiaotong University, China

## Keynotes Lectures

Marianne Akian, INRIA Saclay–Ile-de-France and CMAP, France
Peter Butkovic, University of Birmingham, UK
Weimin Han, University of Iowa, USA

Francesco dell'Isola, Università di Roma La Sapienza, Italy
Stanisław Migórski, Jagiellonian University in Krakow, Poland
Patrizio Neff, Universität Duisburg-Essen, Germany
Martin Ostoja-Starzewski, University of Illinois at Urbana-Champaign, USA
Luca Placidi, International Telematic University Uninettuno, Italy
Marc Quincampoix, Université de Bretagne Occidentale, France
Pierre Seppecher, Université de Toulon, France
Meir Shillor, Oakland University, USA
Mykhailo Zarichnyi, Ivan Franko Lviv National University, Ukraine

# Contents

# Contributors

**Jean-Jacques Alibert** IMATH, Université de Toulon, Toulon, France

**Gabriele Barbagallo** LaMCoS-CNRS \& LGCIE, INSA-Lyon, Universitité de Lyon, Villeurbanne Cedex, France

**Mikaël Barboteu** Laboratoire de Mathématiques et Physique, Université de Perpignan Via Domitia, Perpignan, France

**Emilio Barchiesi** Dipartimento di Ingegneria Meccanica e Aerospaziale, Universitá di Roma La Sapienza, Rome, Italy

**Antonio Battista** Laboratory of Science for Environmental Engineering, Université de la Rochelle, La Rochelle, France

**Mircea Bîrsan** Lehrstuhl für Nichtlineare Analysis und Modellierung, Fakultät für Mathematik, Universität Duisburg-Essen, Essen, Germany; Alexandru Ioan Cuza University of Iaşi, Department of Mathematics, Iaşi, Romania

**Carsten Carstensen** Humboldt-Universität Zu Berlin, Berlin, Germany

**Sanda Cleja-Ţigoiu** Faculty of Mathematics and Computer Science, University of Bucharest, Bucharest, Romania; Institute of Solid Mechanics, Romanian Academy, Bucharest, Romania

**Alessandro Della Corte** Dipartimento di Ingegneria Meccanica e Aerospaziale, Università di Roma La Sapienza, Rome, Italy

**Marco Valerio d'Agostino** LGCIE, INSA-Lyon, Université de Lyon, Villeurbanne Cedex, France

**Francesco dell'Isola** MeMoCS, International Research Center for the Mathematics & Mechanics of Complex Systems, Università dell'Aquila, L'Aquila, Italy

**François Ebobisse** University of Cape Town, Cape Town, South Africa

**Bernhard Eidel** Chair of Computational Mechanics, Universität Siegen, Siegen, Germany

**Ionel-Dumitrel Ghiba** Lehrstuhl Für Nichtlineare Analysis und Modellierung, Fakultät für Mathematik, Universität Duisburg-Essen, Essen, Germany; Department of Mathematics, Alexandru Ioan Cuza University of Iaşi, Iaşi, Romania; Octav Mayer Institute of Mathematics of the Romanian Academy, Iaşi Branch, Iaşi, Romania

**Ivan Giorgio** MeMoCS, International Research Center for the Mathematics & Mechanics of Complex Systems, Università dell'Aquila, L'Aquila, Italy

**Maciej Golaszewski** DCEBM, Warsaw University of Technology, Warsaw, Poland

**Weimin Han** Department of Mathematics, University of Iowa, Iowa City, IA, USA

**Jaroslav Haslinger** Charles University, Prague, Czech Republic; Institute of Geonics of the Czech Academy of Sciences, Ostrava, Czech Republic

**Loïc Le Marrec** IRMAR UMR 6625 CNRS, Université de Rennes 1, Rennes, France

**Angela Madeo** LGCIE, INSA-Lyon, Université de Lyon, Villeurbanne Cedex, France; IUF, Institut Universitaire de France, Paris Cedex 05, France

**Robert J. Martin** Lehrstuhl Für Nichtlineare Analysis und Modellierung, Fakultät für Mathematik, Universität Duisburg-Essen, Essen, Germany

**Andrew T. McBride** The University of Glasgow, Glasgow, UK

**Stanisław Migórski** Faculty of Mathematics and Computer Science, Jagiellonian University in Krakow, Krakow, Poland

**Cornel Marius Murea** Laboratoire de Mathématiques, Informatique et Applications, Université de Haute Alsace, Mulhouse, France

**Andrzej Myśliński** Systems Research Institute, Warsaw, Poland; Faculty of Manufacturing Engineering, Warsaw University of Technology, Warsaw, Poland

**Patrizio Neff** Lehrstuhl für Nichtlineare Analysis und Modellierung, Fakultät für Mathematik, Universität Duisburg-Essen, Mathematik-Carrée, Essen, Germany

**Martin Ostoja-Starzewski** Department of Mechanical Science & Engineering, Institute for Condensed Matter Theory, University of Illinois at Urbana-Champaign, Urbana, IL, USA; Department of Mechanical Science & Engineering, Beckman Institute, University of Illinois at Urbana-Champaign, Urbana, IL, USA

**Luca Placidi** Faculty of Engineering, International Telematic University Uninettuno, Rome, Italy

**Lalaonirina R. Rakotomanana** IRMAR UMR 6625 CNRS, Université de Rennes 1, Rennes, France

**B. Daya Reddy** University of Cape Town, Cape Town, South Africa

**Pierre Seppecher** IMATH, Université de Toulon, Toulon, France; L.M.A CNRS AMU Marseille, Marseille, France

**Meir Shillor** Department of Mathematics and Statistics, Oakland University, Rochester, MI, USA

**Mircea Sofonea** Laboratoire de Mathématiques et Physique, Université de Perpignan Via Domitia, Perpignan, France

**Paul Steinmann** University of Erlangen-Nuremberg, Erlangen, Germany

**Stanislav Sysala** Institute of Geonics of the Czech Academy of Sciences, Ostrava, Czech Republic

**Emilio Turco** Department of Architecture, Design and Urban Planning, Alghero, Italy

# Convergence of Hencky-Type Discrete Beam Model to Euler Inextensible Elastica in Large Deformation: Rigorous Proof

**Jean-Jacques Alibert, Alessandro Della Corte and Pierre Seppecher**

**Abstract** The present chapter concerns rigorous homogenization of a Hencky-type discrete beam model, which is useful for the numerical study of complex fibrous systems as pantographic sheets as well as woven fabrics. $\Gamma$-convergence of the discrete model towards the inextensible Euler's beam model is proven and the result is established for placements in $\mathbb{R}^d$ in large deformation regime.

## 1 Introduction

Rigorous results on homogenization are very important for today's theoretical and applied mechanics. This is especially true for the numerical investigation of very complex systems, as even with today's computational tools they may require a long computation time, and thus the *a priori* reliability of the results is of course desirable. The investigation of metamaterials (see [1] for a review of recent results) is among the topical research directions in which one often deals with very expensive numerical simulations, as the implementation of the desired (often exotic) properties at the macro-scale are usually realized by means of a very complex microstructure [2, 3]. The theory of microstructured/micromorphic continua is by now well developed, with several sound and interesting results (see e.g. [4, 5] as general references on Cosserat continua, [6–8] for related results and [9–16] for different kinds of applications of microstructured models). Still, it is necessary to develop suitable convergence arguments if one wants to solidly rely on the numerical simulations based on the solu-

J.-J. Alibert · P. Seppecher (✉)
IMATH, Université de Toulon, Toulon, France
e-mail: alibert@univ-tln.fr

P. Seppecher
L.M.A CNRS AMU Marseille, Marseille, France
e-mail: seppecher@univ-tln.fr

A. Della Corte
Dipartimento di Ingegneria Meccanica e Aerospaziale,
Università di Roma La Sapienza, Via Eudossiana 18, 00184 Rome, Italy
e-mail: alessandro.dellacorte@uniroma1.it



F. dell'Isola et al. (eds.), *Mathematical Modelling in Solid Mechanics*,
Advanced Structured Materials 69, DOI 10.1007/978-981-10-3764-1_1

tion of the simplified equations coming from the micromorphic/generalized continuum model used for the description of the metamaterial.

In the present chapter we focus on special micro-structured systems which can be described as discrete systems. In this case, the reliability of the homogenization has to be intended in two ways:

1. Real world micro-structured systems with suitably small characteristic lengths have to be well described by the homogenized continuum model;
2. The numerical simulation of the equations coming from the homogenized model (that are usually way simpler than the ones coming from the discrete model) has to converge in a suitable sense.

In principle, there is no reason to believe that the ordinary assumptions made for classical (Cauchy) continuum models are suitable for models describing objects that are so different from the phenomenology originally motivating them. Indeed, very often generalized continuum models are called for, and in particular higher gradient theories (see e.g. [17–22]) are being successfully employed in a number of cases for the homogenization of systems with complex geometry at the micro-scale [23–27]. In the present contribution we address this kind of question for Euler's beam model (also known as *Elastica*), which is the elementary constituent for a large class of complex fibrous systems, including the promising case of pantographic sheets (see [28–31] for theoretical and numerical results and [32] for experimental ones in this direction). Specifically, we want to provide a rigorous justification for the discrete approximation by Heinrich Hencky (1885–1951) [33] of Euler's beam model in large deformation, which is becoming increasingly topical in today's research in structural and computational mechanics [34–36] and metamaterials [37]. In particular, we address here the ideal case in which the beam is perfectly inextensible, while future investigation will be devoted to the more general extensible case.

## 2 Convergence of Measure Functionals

Before setting the mechanical problem we are interested in, we need to recall some (well known) mathematical tools for describing the placement and the energy of the discrete beam model and define a suitable convergence for the sequence of the discrete energy functionals.

Let $(C[0,1])^d$ be the space of vector valued continuous functions on $[0,1]$ endowed with the uniform norm $\|\varphi\|_\infty := \sup\{\|\varphi(t)\| : t \in [0,1]\}$ and $(\mathscr{M}[0,1])^d$ the set of vector valued bounded measures on $[0,1]$ endowed with the norm

$$\|\mu\|_{\mathscr{M}} := \sup\{\langle \mu, \varphi\rangle : \varphi \in (C[0,1])^d, \|\varphi\|_\infty = 1\}$$

where $\langle .,. \rangle$ stands for the duality bracket between $(\mathscr{M}[0,1])^d$ and $(C[0,1])^d$. Recall that if a sequence of vector valued bounded measures $(\mu_n)$ satisfies $\sup_n \|\mu_n\|_{\mathscr{M}} <$

$+\infty$ then there exists a vector valued bounded measure $\mu$ and a subsequence $(\mu_{n_k})$ which converges to $\mu$ with respect to the weak*−topology of $(\mathcal{M}([0,1])^d$ i.e.

$$\lim_{k\to\infty} \langle \mu_{n_k}, \varphi \rangle = \langle \mu, \varphi \rangle$$

for every $\varphi \in (C([0,1])^d$.

Let $(F_n)$ and $F$ be functionals on $(\mathcal{M}[0,1])^d$ with values in $\mathbb{R} \cup \{+\infty\}$. We say that $F_n$ $\Gamma-$ converges to $F$ if the following holds [38]:

i. *Upper bound inequality*. For every $\mu \in (\mathcal{M}([0,1])^d$, there exists a sequence $(\mu_n)$ in $(\mathcal{M}[0,1])^d$ weak*− converging to $\mu$ for which

$$\limsup_{n\to\infty} F_n(\mu_n) \leq F(\mu).$$

ii. *Lower bound inequality*. For every $\mu \in (\mathcal{M}[0,1])^d$ and every sequence $(\mu_n)$ in $(\mathcal{M}[0,1])^d)$ weak*−converging to $\mu$,

$$\liminf_{n\to\infty} F_n(\mu_n) \geq F(\mu).$$

Such a $\Gamma$-convergence result is efficient when the following property of the sequence $(F_n)$ holds:

iii. *Relative compactness*. For every sequence $(\mu_n)$ in $(\mathcal{M}[0,1])^d$

$$\sup_n F_n(\mu_n) < +\infty \quad \Longrightarrow \quad \sup_n \|\mu_n\|_{\mathcal{M}} < +\infty.$$

Informally speaking, relative compactness ensures that controlling the deformation energy is enough to control, with the help of boundary conditions, the norm of the measure employed for the description of the current configuration of the discrete model.

# 3 Micro-Model for Non-Linear Beams

## 3.1 *Discrete Configurations and Operators*

Let $\delta_t$ denote the Dirac measure at the point $t \in [0,1]$. The reference configuration of the discrete micro-system is constituted by $n+1$ nodes placed at the points $\frac{i}{n}$, $i = 0, \ldots, n$. Therefore it can be identified with a measure concentrated at the points $\frac{i}{n}$ where $i = 0, 1, ..., n$, more precisely with the positive Radon measure on [0, 1]

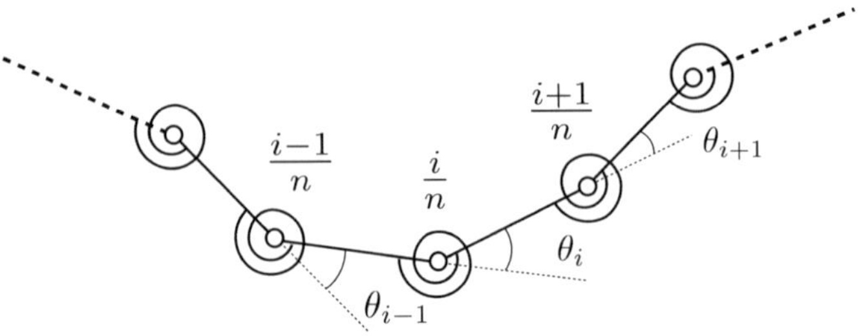


**Fig. 1** Graphical representation of Hencky discrete model consisting of inextensible bars and rotational springs. In the graph $\theta_i := \theta_n(u)(\frac{i}{n})$

$$\overline{\nu}_n := \frac{1}{n}\sum_{i=0}^{n}\delta_{\frac{i}{n}} \tag{1}$$

We assume that the reference (unstressed) configuration of the beam is straight, has unitary length and lays parallel to $e_1$, i.e. the first vector of the canonical base of $\mathbb{R}^d$. The current configuration of the beam can be described by a vector bounded measure $\mu$ on $[0, 1]$ of the form $\mu(dt) := u(t)\overline{\nu}_n(dt)$ where the placement function $u : [0, 1] \to \mathbf{R}^d$ is defined $\overline{\nu}_n$-almost everywhere i.e. at the points $\frac{i}{n}$ where $i = 0, 1, \ldots, n$ (see Fig. 1 for a graphical representation of the discrete model).

In what follows, we will use the following notations:

$$\nu_n^+ := \frac{1}{n}\sum_{i=0}^{n-1}\delta_{\frac{i}{n}}, \qquad \nu_n^- := \frac{1}{n}\sum_{i=1}^{n}\delta_{\frac{i}{n}}, \qquad \nu_n := \frac{1}{n}\sum_{i=1}^{n-1}\delta_{\frac{i}{n}}, \tag{2}$$

$$D_n^+u(t) := n\big(u(t+\tfrac{1}{n}) - u(t)\big), \qquad D_n^-u(t) := n\big(u(t) - u(t-\tfrac{1}{n})\big), \tag{3}$$

$$D_n^2 u := n(D_n^+u - D_n^-u). \tag{4}$$

Note that, if $u$ is a placement function, $D_n^+u$ is defined $\nu_n^+$—almost everywhere, $D_n^-u$ is defined $\nu_n^-$—almost everywhere and $D_n^2u$ is defined $\nu_n$—almost everywhere.

## 3.2 *Left Hand Side Clamped Inextensible Beam*

A placement function $u$ is said to be admissible for a left hand side clamped beam if the following condition holds:

$$u(0) = 0 \quad \text{and} \quad D_n^+u(0) = e_1. \tag{5}$$

It is said to be admissible for an inextensible beam if the following condition holds: $\|u(\frac{i+1}{n}) - u(\frac{i}{n})\| = \frac{1}{n}$. for $i = 0, 1, ..., n-1$. This condition can be written

$$\|D_n^+ u\| = 1 \quad \nu_n^+ - \text{almost everywhere} \tag{6}$$

### 3.3 *Deformation Energy Associated with Three Points Interactions*

At each node $\frac{i}{n}$, for $i = 1, \ldots, n-1$, a rotational spring is placed, whose deformation energy depends on the angle $\theta_n(u)(\frac{i}{n}) \in (-\pi, +\pi)$ formed by the vectors $u(\frac{i+1}{n}) - u(\frac{i}{n})$ and $u(\frac{i}{n}) - u(\frac{i-1}{n})$. This energy must vanish when the angle is zero. We assume, following [39, 40], that this energy is proportional to $1 - \cos(\theta_n(u)(\frac{i}{n}))$. Hence, when the discrete system is in the configuration described by the bounded measure $\mu(dt) = u(t)\overline{\nu}_n(dt)$, its energy is given by

$$E_n^3(\mu) := \frac{1}{n}\sum_{i=1}^{n-1} n^2(1 - \cos\theta_n(u)(\tfrac{i}{n})) \quad \text{where} \quad \cos\theta_n(u)(\tfrac{i}{n}) = \frac{D_n^+ u(\frac{i}{n}) \cdot D_n^- u(\frac{i}{n})}{\|D_n^+ u(\frac{i}{n})\|\|D_n^- u(\frac{i}{n})\|}$$

or, equivalently,

$$E_n^3(\mu) = \frac{1}{2}\int \left\| n\left(\frac{D_n^+ u(t)}{\|D_n^+ u(t)\|} - \frac{D_n^- u(t)}{\|D_n^- u(t)\|}\right)\right\|^2 \nu_n(dt).$$

The above energy is well defined if the placement function $u$ is such that $D_n^+ u \neq 0$ $\nu_n^+$–almost everywhere. This is clearly the case when $u$ is admissible for an inextensible beam. In this case, the discrete energy has the reduced form

$$E_n^3(\mu) = \frac{1}{2}\int \|D_n^2 u(t)\|^2 \, \nu_n(dt). \tag{7}$$

## 4 From Micro to Macro Model: *Γ*-Convergence Result

This section is devoted to left hand side clamped inextensible beams.

### 4.1 *Functionals Associated to the Micro Model*

Let $\mathscr{M}_n$ denote the set of those vector bounded measures of the form $\mu(dt) = u(t)\overline{\nu}_n(dt) \in (\mathscr{M}[0, 1])^d$ such that

$$u(0) = 0, \quad D_n^+ u(0) = \mathrm{e}_1 \quad \text{and} \quad \|D_n^+ u\| = 1 \quad \nu_n^+ - \text{almost everywhere.} \tag{8}$$

The total energy functional (associated to the discrete model) is given by

$$E_n(\mu) := \begin{cases} \frac{1}{2} \int \|D_n^2 u(t)\|^2 \, \nu_n(dt) & \text{if } \mu(dt) = u(t)\overline{\nu}_n(dt) \in \mathscr{M}_n, \\ +\infty & \text{otherwise.} \end{cases} \tag{9}$$

## 4.2 *Functional Associated to the Macro Model*

Let $H^2(0, 1)$ denote the usual Sobolev space. Relying on well-known embedding theorems, any function $u \in H^2(0, 1)$ will be considered as a $C^1[0, 1]$-function. Let $\mathscr{M}$ be the set of those vector bounded measures of the form $\mu(dt) = u(t)dt \in (\mathscr{M}[0, 1])^d$ with $u \in (H^2((0, 1))^d$ and such that

$$u(0) = 0, \quad u'(0) = \mathrm{e}_1 \quad \text{and} \quad \|u'(t)\| = 1 \quad \text{for every} \quad t \in [0, 1]. \tag{10}$$

The total energy functional (associated to the continuous model) is given by

$$E(\mu) := \begin{cases} \frac{1}{2} \int_0^1 \|u''(t)\|^2 \, dt & \text{if } \mu(dt) = u(t)dt \in \mathscr{M}, \\ +\infty & \text{otherwise.} \end{cases} \tag{11}$$

## 4.3 *$\Gamma$−Convergence result*

Our main result is the following:

**Theorem 1** *The sequence $(E_n)$ satisfies the relative compactness property and $\Gamma$-converges to the functional $E$.*

If we compare Theorem 4.1 with the results proved in [41], the difficulty consists in the fact that the beam is inextensible, which corresponds to a nonlinear constraint.

# 5 Proof of the Main Result

## 5.1 *Approximation of a Sequence with Bounded Energy*

Let $(\mu_n)$ be a sequence in $(\mathscr{M}(0, 1])^d$ with bounded energy. This means that there exists some positive real number $M$ such that

$$\mu_n(dt) = u_n(t)\overline{\nu}_n(dt) \in \mathscr{M}_n \quad \text{and} \quad \int \|D_n^2 u_n(t)\|^2 \, \nu_n(dt) \le M. \tag{12}$$

for every integer $n$. Let us define the sequence $(\overline{\mu}_n)$ by setting $\overline{\mu}_n(dt) = \overline{u}_n(t)dt$, with $\bar{u}_n$ piecewise $C^2$ in (0, 1) satisfying:

$$\overline{u}_n(0) = 0 \quad , \quad \overline{u}_n'(0) = \mathrm{e}_1$$
$$\overline{u}_n''(t) = D^2 u_n(\tfrac{i}{n}) \qquad \text{as soon as } t \in (\tfrac{i}{n} - \tfrac{1}{2n}, \tfrac{i}{n} + \tfrac{1}{2n}).$$

Notice that $\overline{u}_n \in (H^2(0,1))^d$ but in general $\overline{u}_n \notin \mathscr{M}$ because $\|u_n'(t)\|$ is not necessarily equal to 1. The following result will be used to establish the lower bound inequality.

**Lemma 1** *Let $(\mu_n)$ be a sequence in $(\mathscr{M}(0,1])^d$ with bounded energy. Then, the sequence $(\overline{u}_n)$ defined above is bounded with respect to the usual $H^2-$norm and satisfies the following properties.*

$$\int_0^1 \|\overline{u}_n''(t)\|^2 \, dt = \int \|D_n^2 u_n(t)\|^2 \, \nu_n(dt) \qquad \textit{for every } n, \tag{13}$$

$$\lim_{n\to\infty} \|\overline{u}_n'(t)\| = 1 \qquad \textit{for every } t \in [0,1], \tag{14}$$

$$\overline{\mu}_n - \mu_n \textit{ converges to } 0 \textit{ with respect to the weak}^* \textit{topology.} \tag{15}$$

*Proof* One has $\overline{u}_n(0) = 0$, $\overline{u}_n'(0) = \mathrm{e}_1$ and

$$\int_0^1 \|\overline{u}_n''(t)\|^2 \, dt = \sum_{i=1}^{n-1} \int_{\frac{i}{n}-\frac{1}{2n}}^{\frac{i}{n}+\frac{1}{2n}} \|\overline{u}_n''(t)\|^2 dt = \int \|D_n^2 u_n(t)\|^2 \nu_n(dt) \le M$$

which implies that the sequence $(\overline{u}_n)$ is bounded with respect to the usual $H^2-$ norm. Hence, the two sequences $(\overline{u}_n')$ and $(\overline{u}_n)$ are equicontinuous on [0, 1] and uniformly bounded on [0, 1]. More precisely, for any $s, t \in [0, 1]$,

$$\|\overline{u}_n'(t) - \overline{u}_n'(s)\| \le \sqrt{M}\sqrt{|t-s|} \quad \text{and} \quad \|\overline{u}_n'(t)\| \le 1 + \sqrt{M}, \tag{16}$$

$$\|\overline{u}_n(t) - \overline{u}_n(s)\| \le (1 + \sqrt{M})|t-s| \quad \text{and} \quad \|\overline{u}_n(t)\| \le 1 + \sqrt{M}. \tag{17}$$

On the other hand, a first computation gives that for any $i = 1, \ldots, n-1$,

$$\overline{u}_n'(\tfrac{i}{n} + \tfrac{1}{2n}) = \mathrm{e}_1 + \int_0^{\frac{i}{n}+\frac{1}{2n}} \overline{u}_n''(t)\, dt = \mathrm{e}_1 + \frac{1}{n}\sum_{k=1}^{i} D_n^2 u_n(\tfrac{k}{n}) = D_n^+ u_n(\tfrac{i}{n}).$$

Since $\|D_n^+ u_n\| = 1$ $\nu_n^+-$almost everywhere and the sequence $(\overline{u}_n')$ is equicontinuous on [0, 1], we obtain (14).

A second computation gives $\overline{u}_n(1) = u_n(1)$ and

$$\begin{aligned}
\overline{u}_n(\tfrac{i}{n}) &= \tfrac{i}{n}\mathrm{e}_1 + \int_0^{\frac{i}{n}} (\tfrac{i}{n} - s)\overline{u}_n''(s)ds \\
&= \tfrac{i}{n}\mathrm{e}_1 + \frac{1}{n}\sum_{k=1}^{i-1}\left(n\int_{\frac{k}{n}-\frac{1}{2n}}^{\frac{k}{n}+\frac{1}{2n}} (\tfrac{i}{n} - s)\, ds\right) D_n^2 u_n(\tfrac{k}{n}) + \left(\int_{\frac{i}{n}-\frac{1}{2n}}^{\frac{i}{n}} (\tfrac{i}{n} - s)\, ds\right) D_n^2 u_n(\tfrac{i}{n}) \\
&= \tfrac{i}{n}\mathrm{e}_1 + \frac{1}{n}\sum_{k=1}^{i-1} (\tfrac{i-k}{n}) D_n^2 u_n(\tfrac{k}{n}) + \frac{1}{8n^2} D_n^2 u_n(\tfrac{i}{n}) \\
&= u_n(\tfrac{i}{n}) + \frac{1}{8n^2} D_n^2 u_n(\tfrac{i}{n})
\end{aligned}$$

for every $i = 1, ..., n-1$. As a consequence, the inequality $\|\overline{u}_n - u_n\| \le \frac{\sqrt{M}}{8n}$ holds $\overline{\nu}_n$−almost everywhere.

Let $\varphi \in C([0,1])^2$. A third computation gives

$$\begin{aligned}
|\langle \overline{\mu}_n - \mu_n, \varphi\rangle| &= \left|\sum_{i=1}^{n}\int_{\frac{i-1}{n}}^{\frac{i}{n}} \left(\varphi(t)\cdot\overline{u}_n(t) - \varphi(\tfrac{i}{n})\cdot u_n(\tfrac{i}{n})\right) dt\right| \\
&\le \sum_{i=1}^{n}\int_{\frac{i-1}{n}}^{\frac{i}{n}} \|\varphi(t)\|\|\overline{u}_n(t) - \overline{u}_n(\tfrac{i}{n})\|\, dt + \sum_{i=1}^{n}\int_{\frac{i-1}{n}}^{\frac{i}{n}} \|\varphi(t) - \varphi(\tfrac{i}{n})\|\|\overline{u}_n(\tfrac{i}{n})\|\, dt \\
&\quad + \int \|\varphi(t)\|\|\overline{u}_n(t) - u_n(t)\|\nu_n(dt) \\
&\le \frac{1+\sqrt{M}}{n}\int_0^1 \|\varphi(t)\|\, dt + (1+\sqrt{M})\sum_{i=1}^{n}\int_{\frac{i-1}{n}}^{\frac{i}{n}} \|\varphi(t) - \varphi(\tfrac{i}{n})\|\, dt \\
&\quad + \frac{\sqrt{M}}{8n}\int \|\varphi(t)\|\nu_n(dt).
\end{aligned}$$

Since

$$\lim_{n\to\infty}\sum_{i=1}^{n}\int_{\frac{i-1}{n}}^{\frac{i}{n}} \|\varphi(t) - \varphi(\tfrac{i}{n})\|\, dt = 0 \quad \text{and} \quad \lim_{n\to\infty}\int \|\varphi(t)\|\,\nu_n(dt) = \int_0^1 \|\varphi(t)\|\, dt.$$

we conclude that the sequence $(\overline{\mu}_n - \mu_n)$ converges to 0 with respect to the weak*−topology of $(\mathscr{M}[0,1])^d$. The proof is complete.

## 5.2 *The Proof of Theorem 4.1*

We divide this proof in three steps.

Step 1. (*Relative compactness*). Let $\mu(dt) := u(t)\overline{\nu}_n(dt) \in \mathscr{M}_n$. Since $\|u(0)\| = 0$ and $\|D_n^+ u\| = 1$ $\nu_n^+-$ almost everywhere, one has $\|u\| \leq 1$ $\overline{\nu}_n-$almost everywhere, hence

$$\|\mu\|_{\mathscr{M}} = \int \|u(t)\| \,\overline{\nu}_n(dt) \leq 1.$$

Step 2. (*Upper bound inequality*). Let $\mu(dt) := u(t)dt \in \mathscr{M}$. Since $u \in (C^1[0,1])^d$, we define $\mu_n(dt) = u_n(t)\overline{\nu}_n(dt)$ by setting

$$u_n(0) = 0 \quad \text{and} \quad u_n(\tfrac{i}{n}) = \frac{1}{n}\sum_{k=0}^{i-1} u'(\tfrac{k}{n}) \quad (\text{for } i = 1, ..., n).$$

Note that $D_n^+ u_n(\frac{i}{n}) = u'(\frac{i}{n})$. Then $D_n^+ u_n(0) = \mathrm{e}_1$ and $\|D_n^+ u_n\| = 1$ $\nu_n^+$-almost everywhere. Hence one has $\mu_n \in \mathscr{M}_n$ and

$$D_n^2 u_n(\tfrac{i}{n}) = n\big(D_n^+ u_n(\tfrac{i}{n}) - D_n^+ u_n(\tfrac{i-1}{n})\big) = n\big(u'(\tfrac{i}{n}) - u'(\tfrac{i-1}{n})\big) = n\int_{\frac{i-1}{n}}^{\frac{i}{n}} u''(t)\,dt$$

then, using Jensen inequality we obtain

$$\limsup_{n\to\infty} \int \|D_n^2 u_n\|^2 d\nu_n = \limsup_{n\to\infty} \frac{1}{n}\sum_{k=1}^{n-1} \left\| n\int_{\frac{i-1}{n}}^{\frac{i}{n}} u''(t)\,dt \right\|^2 \leq \int_0^1 \|u''(t)\|^2\,dt.$$

Let $\varphi \in (C[0,1])^d$. Since $u'$ is continuous on $[0, 1]$ we obtain

$$\begin{aligned}
\lim_{n\to\infty} \langle \mu_n, \varphi\rangle &:= \lim_{n\to\infty} \int \varphi(t)\cdot u_n(t)\,\overline{\nu}_n(dt)\\
&= \lim_{n\to\infty} \frac{1}{n}\sum_{i=1}^{n} \varphi(\tfrac{i}{n})\cdot\left(\frac{1}{n}\sum_{k=0}^{i-1} u'(\tfrac{k}{n})\right)\\
&= \lim_{n\to\infty} \frac{1}{n}\sum_{i=1}^{n} \varphi(\tfrac{i}{n})\cdot\left(u(\tfrac{i}{n}) + \sum_{k=0}^{i-1}\int_{\frac{k}{n}}^{\frac{k+1}{n}} \big(u'(\tfrac{k}{n}) - u'(t)\big)\,dt\right)\\
&= \lim_{n\to\infty} \int \varphi(t)\cdot u(t)\overline{\nu}_n(dt)
\end{aligned}$$

Hence, Riemann's Theorem implies that the sequence $(\mu_n)$ converges to $\mu$ with respect to the weak*−topology of $(\mathscr{M}[0,1])^d$.

Step 3. (*Lower bound inequality*). Let $\mu, \mu_n \in (\mathscr{M}[0,1])^d$ such that $(\mu_n)$ converges to $\mu$ with respect to the weak*−topology of $(\mathscr{M}[0,1])^d$. Without loss of generality we may assume that $\mu_n(dt) = u_n(t)\overline{\nu}_n(dt) \in \mathscr{M}_n$ and there exists a nonnegative real number $M$ such that for every $n$

$$\int \|D_n^2 u_n(t)\|^2 \nu_n(dt) \leq M.$$

Let $(\overline{\mu}_n)$ be the sequence of measures defined in Sect. 5.1. By Lemma 5.1, this sequence converges to $\mu$ with respect to the weak*−topology of $(\mathscr{M}[0, 1])^d$. Since $\overline{\mu}_n(dt) = \overline{u}_n(t)dt$ and the sequence $(\overline{u}_n)$ is bounded with respect to the usual $H^2$−norm, there exists $u \in (H^2(0, 1])^d$ such that $\mu(dt) = u(t)dt$ and

$$\liminf_{n\to\infty} \int \|D_n^2 u_n(t)\|^2 \nu_n(dt) = \liminf_{n\to\infty} \int_0^1 \|\overline{u}_n''(t)\|^2 dt \geq \int_0^1 \|u''(t)\|^2 dt.$$

Since the space $H^2(0, 1)$ is compactly embedded on $C^1[0, 1]$, the sequence $(\overline{u}_n')$ converges to $(u')$ with respect to the uniform norm over $[0, 1]$. Hence, using Lemma 5.1, We obtain

$$\|u'(t)\| = 1 \qquad \text{for every } \; t \in [0, 1]$$

then $u \in \mathscr{M}$. The proof is complete.

## 6 Conclusions

We proved a $\Gamma$-convergence result for a Hencky-type discretization of an inextensible Euler beam in large deformation regime. Future investigations should generalize the result (in a suitable form) for extensible beam models; moreover, it will be interesting to extend the convergence argument to Generalized Beam Models [42–45] and also to the dynamics of the discrete system, which should of course take into account the possibility of various kinds of dynamic instabilities [46–48]. Finally, it has to be remarked that Hencky-type discretization for *Elastica* has proven to be very effective, and is in fact used by several computational software packages (as for instance by MATLAB®). The present result gives a sound mathematical argument which this kind of numerical evidence can be based on.

## References

1. Del Vescovo, D., Giorgio, I.: Dynamic problems for metamaterials: review of existing models and ideas for further research. Int. J. Eng. Sci. **80**, 153–172 (2014)
2. dell'Isola, F., Steigmann, D., Della Corte, A.: Synthesis of fibrous complex structures: designing microstructure to deliver targeted macroscale response. Appl. Mech. Rev. **67**(6), 060804 (2016)
3. dell'Isola, F., Bucci, S., Battista, A.: Against the fragmentation of knowledge: the power of multidisciplinary research for the design of metamaterials. Advanced Methods of Continuum Mechanics for Materials and Structures, pp. 523–545. Springer, Singapore (2016)
4. Altenbach, J., Altenbach, H., Eremeyev, V.A.: On generalized cosserat-type theories of plates and shells: a short review and bibliography. Arch. Appl. Mech. **80**(1), 73–92 (2010)

5. Eremeyev, V.A., Lebedev, L.P., Altenbach, H.: Foundations of Micropolar Mechanics. Springer, Berlin (2012)
6. Pietraszkiewicz, W., Eremeyev, V.A.: On natural strain measures of the non-linear micropolar continuum. Int. J. Solids Struct. **46**(3), 774–787 (2009)
7. Yeremeyev, V.A., Zubov, L.M.: The theory of elastic and viscoelastic micropolar liquids. J. Appl. Math. Mech. **63**(5), 755–767 (1999)
8. Altenbach, H., Eremeyev, V.A.: On the linear theory of micropolar plates. ZAMM-J. Appl. Math. Mech./Zeitschrift für Angewandte Mathematik und Mechanik **89**(4), 242–256 (2009)
9. Scerrato, D., Giorgio, I., Madeo, A., Limam, A., Darve, F.: A simple non-linear model for internal friction in modified concrete. Int. J. Eng. Sci. **80**, 136–152 (2014)
10. Giorgio, I., Scerrato, D.: Multi-scale concrete model with rate-dependent internal friction. Eur. J. Environ. Civ. Eng. 1–19 (2016)
11. Yang, Y., Misra, A.: Micromechanics based second gradient continuum theory for shear band modeling in cohesive granular materials following damage elasticity. Int. J. Solids Struct. **49**(18), 2500–2514 (2012)
12. Misra, A., Yang, Y.: Micromechanical model for cohesive materials based upon pseudo-granular structure. Int. J. Solids Struct. **47**(21), 2970–2981 (2010)
13. Chang, C.S., Misra, A.: Packing structure and mechanical properties of granulates. J. Eng. Mech. **116**(5), 1077–1093 (1990)
14. Seddik, H., Greve, R., Placidi, L., Hamann, I., Gagliardini, O.: Application of a continuum-mechanical model for the flow of anisotropic polar ice to the edml core, antarctica. J. Glaciol. **54**(187), 631–642 (2008)
15. Placidi, L., Faria, S.H., Hutter, K.: On the role of grain growth, recrystallization and polygonization in a continuum theory for anisotropic ice sheets. Ann. Glaciol. **39**(1), 49–52 (2004)
16. Placidi, L., Greve, R., Seddik, H., Faria, S.H.: Continuum-mechanical, anisotropic flow model for polar ice masses, based on an anisotropic flow enhancement factor. Contin. Mech. Thermodyn. **22**(3), 221–237 (2010)
17. F dell'Isola, G Sciarra, and S Vidoli. Generalized hooke's law for isotropic second gradient materials. Proceedings of the Royal Society of London A: Mathematical, Physical and Engineering Sciences. doi:10.1098/rspa.2008.0530, (2009)
18. Andreaus, U., dell'Isola, F., Giorgio, I., Placidi, L., Lekszycki, T., Rizzi, N.L.: Numerical simulations of classical problems in two-dimensional (non) linear second gradient elasticity. Int. J. Eng. Sci. **108**, 34–50 (2016)
19. dell'Isola, F., Seppecher, P.: The relationship between edge contact forces, double forces and interstitial working allowed by the principle of virtual power. *Comptes rendus de l'Académie des sciences. Série IIb, Mécanique, physique, astronomie*, p. 7 (1995)
20. Sciarra, G., Ianiro, N., Madeo, A., et al.: A variational deduction of second gradient poroelasticity i: general theory. J. Mech. Mater. Struct. **3**(3), 507–526 (2008)
21. dell'Isola, F., Steigmann, D.: A two-dimensional gradient-elasticity theory for woven fabrics. J. Elast. **118**(1), 113–125 (2015)
22. Placidi, L.: A variational approach for a nonlinear one-dimensional damage-elasto-plastic second-gradient continuum model. Contin. Mech. Thermodyn. **28**(1–2), 119–137 (2016)
23. Alibert, J.J., Della Corte, A.: Second-gradient continua as homogenized limit of pantographic microstructured plates: a rigorous proof. Zeitschrift für angewandte Mathematik und Physik **66**(5), 2855–2870 (2015)
24. Carcaterra, A., dell'Isola, F., Esposito, R., Pulvirenti, M.: Macroscopic description of microscopically strongly inhomogenous systems: a mathematical basis for the synthesis of higher gradients metamaterials. Arch. Ration. Mech. Anal. **218**(3), 1239–1262 (2015)
25. Placidi, L., Andreaus, U., Giorgio, I.: Identification of two-dimensional pantographic structure via a linear D4 orthotropic second gradient elastic model. J. Eng. Math. **6** (2016). doi:10.1007/s10665-016-9856-8
26. Placidi, L., Greco, L., Bucci, S., Turco, E., Rizzi, N.L.: A second gradient formulation for a 2D fabric sheet with inextensible fibres. Z. für angew. Math. und Phys. **67**(5), 114 (2016)

27. Rahali, Y., Giorgio, I., Ganghoffer, J.F., dell'Isola, F.: Homogenization à la piola produces second gradient continuum models for linear pantographic lattices. Int. J. Eng. Sci. **97**, 148–172 (2015)
28. Turco, E., dell'Isola, F., Rizzi, N.L., Grygoruk, R., Müller, W.H., Liebold, C.: Fiber rupture in sheared planar pantographic sheets: numerical and experimental evidence. Mech. Res. Commun. **76**, 86–90 (2016)
29. Giorgio, I., Della Corte, A., dell'Isola, F., Steigmann, D.J.: Buckling modes in pantographic lattices. Comptes Rendus Mec. **344**(7), 487–501 (2016)
30. Cuomo, M., dell'Isola, F., Greco, L.: Simplified analysis of a generalized bias test for fabrics with two families of inextensible fibres. Z. für angew. Math. und Phys. **67**(3), 1–23 (2016)
31. dell'Isola, F., Giorgio, I., Andreaus, U.: Elastic pantographic 2d lattices: a numerical analysis on the static response and wave propagation. Proc. Est. Acad. Sci. **64**(3), 219 (2015)
32. Turco, E., Golaszewski, M., Cazzani, A., Rizzi, N.L.: Large deformations induced in planar pantographic sheets by loads applied on fibers: experimental validation of a discrete lagrangian model. Mech. Res. Commun. **76**, 51–56 (2016)
33. Hencky, H.: Über die angenäherte Lösung von Stabilitätsproblemen im Raum mittels der elastischen Gelenkkette. Ph.D. thesis, Engelmann (1921)
34. Fertis, D.J.: Nonlinear Structural Engineering. Springer, Berlin (2006)
35. Forest, S., Sievert, R.: Nonlinear microstrain theories. Int. J. Solids Struct. **43**(24), 7224–7245 (2006)
36. Abali, B.E., Müller, W.H., Eremeyev, V.A.: Strain gradient elasticity with geometric nonlinearities and its computational evaluation. Mech. Adv. Mater. Mod. Process. **1**(1), 1 (2015)
37. Milton, G.W.: Adaptable nonlinear bimode metamaterials using rigid bars, pivots, and actuators. J. Mech. Phys. Solids **61**(7), 1561–1568 (2013)
38. Braides, A.: Gamma-Convergence for Beginners, vol. 22. Clarendon Press, Oxford (2002)
39. dell'Isola, F., Giorgio, I., Pawlikowski, M., Rizzi, N.L.: Large deformations of planar extensible beams and pantographic lattices: heuristic homogenization, experimental and numerical examples of equilibrium. Proc. R. Soc. A **472**, 20150790 (2016)
40. Turco, E., dell'Isola, F., Cazzani, A., Rizzi, N.L.: Hencky-type discrete model for pantographic structures: numerical comparison with second gradient continuum models. Z. für angew. Math. und Phys. **67**(4), 1–28 (2016)
41. Alibert, J.-J., Seppecher, P., dell'Isola, F.: Truss modular beams with deformation energy depending on higher displacement gradients. Math. Mech. Solids **8**(1), 51–73 (2003)
42. Piccardo, G., Ranzi, G., Luongo, A.: A direct approach for the evaluation of the conventional modes within the gbt formulation. Thin-Walled Struct. **74**, 133–145 (2014)
43. Piccardo, G., Ranzi, G., Luongo, A.: A complete dynamic approach to the generalized beam theory cross-section analysis including extension and shear modes. Math. Mech. Solids **19**(8), 900–924 (2014)
44. Piccardo, G., Ferrarotti, A., Luongo, A.: Nonlinear generalized beam theory for open thin-walled members. Math. Mech. Solids 1081286516649990 (2016)
45. Luongo, A., Zulli, D.: A non-linear one-dimensional model of cross-deformable tubular beam. Int. J. Non-Linear Mech. **66**, 33–42 (2014)
46. Taig, G., Ranzi, G., Luongo, A.: Gbt pre-buckling and buckling analyses of thin-walled members under axial and transverse loads. Contin. Mech. Thermodyn. **28**(1–2), 41–66 (2016)
47. Luongo, A., D'Annibale, F.: A paradigmatic minimal system to explain the ziegler paradox. Contin. Mech. Thermodyn. **27**(1–2), 211–222 (2015)
48. Luongo, A., D'Annibale, F.: Double zero bifurcation of non-linear viscoelastic beams under conservative and non-conservative loads. Int. J. Non-Linear Mech. **55**, 128–139 (2013)

# Analysis of the Deformation of Cosserat Elastic Shells Using the Dislocation Density Tensor

**Mircea Bîrsan and Patrizio Neff**

**Abstract** We consider the Cosserat shell approach under finite rotations. The Cosserat shell features an additional, in principle independent orthogonal frame. In this setting we establish a novel curvature tensor which we call the shell dislocation density tensor. For this variant, we derive the equations and in a hyperelastic context we show existence of minimizers under generic convexity assumptions on the elastic energies in terms of nonlinear strain measures. The correspondence between our formulation and proposals in the literature is established.

## 1 Introduction

Thin shell-structures still represent one of the most challenging facets of problems in nonlinear elasticity. Due to their flexibility, large rotations are a commonplace observation. The modelling definitely calls for a sound finite rotation treatment. While models based on the Kirchhoff–Love normality assumption allow for a reasonable mathematical treatment in the linear, infinitesimal strain setting, this is not the case in the finite strain setting. Here, also from an engineering point of view, models with independent director fields in the spirit of the Reissner–Mindlin kinematics are more widely used for the ease with which these shell models can be coupled to beams in structural approaches.

The mathematics of these models with one independent director is, however, not yet settled. One of the reasons of this shortcoming is the lacking control of rotations. In the last decade, this fundamental problem has been understood [11] and as an answer, finite rotation Cosserat-shell models have been proposed and analysed.

M. Bîrsan (✉) · P. Neff
Lehrstuhl für Nichtlineare Analysis und Modellierung, Fakultät für Mathematik,
Universität Duisburg-Essen, Thea-Leymann Street 9, 45127 Essen, Germany
e-mail: mircea.birsan@uni-due.de

M. Bîrsan
Alexandru Ioan Cuza University of Iași, Department of Mathematics,
Blvd. Carol I, No. 11, 700506 Iași, Romania
e-mail: patrizio.neff@uni-due.de

© Springer Nature Singapore Pte Ltd. 2017
F. dell'Isola et al. (eds.), *Mathematical Modelling in Solid Mechanics*,
Advanced Structured Materials 69, DOI 10.1007/978-981-10-3764-1_2

The Cosserat kinematics endows the shell with an additional orthogonal frame, in principle independent of the deformation of the shell. For this frame, new balance equations are established. For such models, under suitable convexity assumptions on strain measures which *do not* preclude buckling, existence results can be shown when formulating the model in a hyperelastic setting. Uniqueness cannot be shown and is not to be expected. With such approaches, a quite successful modelling is possible [19]. As it turns out, the basic modelling ingredients have already been known to the Cosserat brothers [6]. However, they have never discussed any constitutive assumptions. We mention that the kinematics of Cosserat shells is equivalent to the kinematics of the so-called 6-parameter shell model, see e.g. [2, 4, 8, 16].

In this contribution we provide such a discussion with a view towards a new tensor – the shell dislocation density tensor, which seems quite appropriate to express the curvature term for the orthogonal frame-field living in SO(3). More precisely, we present first the strain and curvature measures which are commonly used in Cosserat shell models and introduce the new dislocation density tensor, as an alternative strain measure for orientation (curvature) change. We establish the extended Nye's formula, which expresses the relationship between the shell bending-curvature tensor and the shell dislocation density tensor. Then, we write the principle of virtual work and the constitutive relations using the dislocation density tensor. We formulate the minimization problem for the deformation of Cosserat-type shells and prove an existence theorem using the direct methods of the calculus of variations. As an application of these results, we investigate the special case of isotropic elastic shells. We finish with establishing the correspondence between our new formulation and more well-known representations.

## 2 Strain and Curvature Measures in the Cosserat (6-Parameter) Shell Model

We present shortly the kinematical model of Cosserat-type shells, which coincides with the kinematical model of 6-parameter shells, see e.g. [2, 4, 8].

### *2.1 Kinematics*

In this model, every material point has 6 kinematical degrees of freedom: 3 for translations and 3 for rotations. To describe the rotational motion of material points we attach a triad of orthonormal vectors (called *directors*) to every point.

Let $\mathscr{S}_\xi$ be the reference configuration of a shell and consider the midsurface $\omega_\xi \subset \mathbf{R}^3$. We denote by $(\xi_1, \xi_2, \xi_3)$ a generic point of the deformable surface $\omega_\xi$ . The Cosserat-type shell is characterized by two fields: the vectorial map $\mathbf{y}_\xi : \omega_\xi \to \omega_c$ for the deformation and the microrotation tensor $\mathbf{R}_\xi : \omega_\xi \to \mathrm{SO}(3)$.

Here, $\omega_c$ represents the deformed (current) configuration of the midsurface and SO(3) is the group of proper orthogonal $3 \times 3$ tensors.

The reference midsurface $\omega_\xi$ admits a parametric representation

$$\mathbf{y}_0 : \omega \to \omega_\xi\,, \qquad \mathbf{y}_0(x_1, x_2) = (\xi_1, \xi_2, \xi_3),$$

where $\omega \subset \mathbf{R}^2$ is a flat domain with Lipschitz boundary $\partial\omega$. We refer the domain $\omega$ to an orthogonal Cartesian frame $Ox_1x_2x_3$ such that $\omega \subset Ox_1x_2$ and let $\mathbf{e}_i$ be the unit vectors along the coordinate axes $Ox_i$.

The *deformation function* $\mathbf{y}$ and the *elastic microrotation* $\mathbf{Q}_e$ are defined by the compositions

$$\begin{aligned} \mathbf{y} = \mathbf{y}_\xi \circ \mathbf{y}_0 : \omega \to \omega_c\,, \qquad & \mathbf{y}(x_1, x_2) := \mathbf{y}_\xi\big(\mathbf{y}_0(x_1, x_2)\big), \\ \mathbf{Q}_e = \mathbf{R}_\xi \circ \mathbf{y}_0 : \omega \to \text{SO}(3), \quad & \mathbf{Q}_e(x_1, x_2) := \mathbf{R}_\xi\big(\mathbf{y}_0(x_1, x_2)\big). \end{aligned}$$

Then, the *total microrotation* $\overline{\mathbf{R}}$ is defined by

$$\overline{\mathbf{R}} : \omega \to \text{SO}(3), \qquad \overline{\mathbf{R}}(x_1, x_2) = \mathbf{Q}_e(x_1, x_2)\, \mathbf{Q}_0(x_1, x_2),$$

where $\mathbf{Q}_0 : \omega \to \text{SO}(3)$ is the *initial microrotation*, which describes the orientation of points in the reference configuration. With the help of the directors, one can express the microrotation tensors in the following way

$$\mathbf{Q}_e = \mathbf{d}_i \otimes \mathbf{d}_i^0\,, \qquad \overline{\mathbf{R}} = \mathbf{Q}_e\,\mathbf{Q}_0 = \mathbf{d}_i \otimes \mathbf{e}_i\,, \qquad \mathbf{Q}_0 = \mathbf{d}_i^0 \otimes \mathbf{e}_i\,, \tag{1}$$

where $\mathbf{d}_i^0$ stand for the initial directors (attached to points in $\omega_\xi$) and $\mathbf{d}_i$ are the directors in the deformed configuration $\omega_c$ ($i = 1, 2, 3$). In relation (1) and throughout the paper we employ the usual conventions for indices: the Latin indices $i, j, k, \ldots$ range over the set $\{1, 2, 3\}$, while the Greek indices $\alpha, \beta, \gamma, \ldots$ are confined to the range $\{1, 2\}$ ; the comma preceding an index $i$ denotes partial derivatives with respect to $x_i$ ; the Einstein summation convention over repeated indices is also used.

## 2.2 *Differential Geometry*

For the differential geometry of the reference surface $\omega_\xi$ we introduce some notations. Let $\mathbf{a}_\alpha$ be the covariant base vectors and $\mathbf{a}^\beta$ be the contravariant base vectors in the tangent plane: $\mathbf{a}_\alpha := \mathbf{y}_{0,\alpha}$ and $\mathbf{a}_\alpha \cdot \mathbf{a}^\beta = \delta_\alpha^\beta$ (the Kronecker delta). We denote

$$a_{\alpha\beta} := \mathbf{a}_\alpha \cdot \mathbf{a}_\beta\,, \qquad a^{\alpha\beta} := \mathbf{a}^\alpha \cdot \mathbf{a}^\beta, \qquad a = \sqrt{\det\big(a_{\alpha\beta}\big)_{2\times 2}} = |\mathbf{a}_1 \times \mathbf{a}_2| > 0.$$

The unit normal $\mathbf{n}_0$ to the surface is given by

$$\mathbf{n}_0 = \frac{\mathbf{a}_1 \times \mathbf{a}_2}{a}$$

and we also use $\mathbf{a}_3 := \mathbf{a}^3 := \mathbf{n}_0$ . The surface gradient $\text{Grad}_s$ and surface divergence $\text{Div}_s$ operators are defined for a vector field $\mathbf{v}$ by

$$\text{Grad}_s \mathbf{v} = \frac{\partial \mathbf{v}}{\partial x_\alpha} \otimes \mathbf{a}^\alpha = \mathbf{v}_{,\alpha} \otimes \mathbf{a}^\alpha, \qquad \text{Div}_s \mathbf{v} = \text{tr}\big[\text{Grad}_s \mathbf{v}\big] = \mathbf{v}_{,\alpha} \cdot \mathbf{a}^\alpha. \tag{2}$$

In [3] we have introduced the surface Curl operator $\text{curl}_s$ for vector fields $\mathbf{v}$ and, respectively, $\text{Curl}_s$ for tensor fields $\mathbf{T}$ by

$$\begin{aligned} \big(\text{curl}_s \mathbf{v}\big) \cdot \mathbf{k} &:= \text{Div}_s\big(\mathbf{v} \times \mathbf{k}\big) \quad \text{for all constant vectors } \mathbf{k}, \\ \big(\text{Curl}_s \mathbf{T}\big)^T \mathbf{k} &:= \text{curl}_s\big(\mathbf{T}^T \mathbf{k}\big) \quad \text{for all constant vectors } \mathbf{k}. \end{aligned} \tag{3}$$

In view of these definitions, we have the relations

$$\text{curl}_s \mathbf{v} = -\mathbf{v}_{,\alpha} \times \mathbf{a}^\alpha, \qquad \text{Curl}_s \mathbf{T} = -\mathbf{T}_{,\alpha} \times \mathbf{a}^\alpha. \tag{4}$$

Using these notations we can express the *first fundamental tensor* $\mathbf{a}$ and the *second fundamental tensor* $\mathbf{b}$ of the surface $\omega_\xi$ in the forms

$$\begin{aligned} \mathbf{a} &= a_{\alpha\beta}\mathbf{a}^\alpha \otimes \mathbf{a}^\beta = a^{\alpha\beta}\mathbf{a}_\alpha \otimes \mathbf{a}_\beta = \mathbf{a}_\alpha \otimes \mathbf{a}^\alpha, \\ \mathbf{b} = -\text{Grad}_s \mathbf{n}_0 &= -\mathbf{n}_{0,\alpha} \otimes \mathbf{a}^\alpha = b_{\alpha\beta}\, \mathbf{a}^\alpha \otimes \mathbf{a}^\beta = b^\alpha_\beta\, \mathbf{a}_\alpha \otimes \mathbf{a}^\beta, \\ \text{with} \quad b_{\alpha\beta} &= -\mathbf{n}_{0,\beta} \cdot \mathbf{a}_\alpha = b_{\beta\alpha}\,, \qquad b^\alpha_\beta = -\mathbf{n}_{0,\beta} \cdot \mathbf{a}^\alpha\,. \end{aligned}$$

*Remark 2.1* The initial directors $\mathbf{d}_i^0$ are usually chosen such that

$$\mathbf{d}_3^0 = \mathbf{n}_0\,, \qquad \mathbf{d}_\alpha^0 \cdot \mathbf{n}_0 = 0\,, \tag{5}$$

i.e. $\mathbf{d}_3^0$ is orthogonal to $\omega_\xi$ and $\mathbf{d}_\alpha^0$ belong to the tangent plane. This assumption is not necessary in general, but it will be adopted here since it simplifies many of the subsequent expressions. In the deformed configuration, the director $\mathbf{d}_3$ is no longer orthogonal to the surface $\omega_c$ (the Kirchhof–Love condition is not imposed). One convenient choice of the initial microrotation tensor $\mathbf{Q}_0 = \mathbf{d}_i \otimes \mathbf{e}_i$ such that the conditions (5) be satisfied is (see Remark 10 of [1])

$$\mathbf{Q}_0 = \text{polar}\big(\mathbf{a}_i \otimes \mathbf{e}_i\big),$$

where $\text{polar}(\mathbf{T})$ denotes the orthogonal tensor given by the polar decomposition of any tensor $\mathbf{T}$.

## 2.3 The Dislocation Density Tensor

In view of (2), the surface gradient of the deformation is

$$\mathbf{F} := \text{Grad}_s\, \mathbf{y} \;=\; \mathbf{y}_{,\alpha} \otimes \mathbf{a}^{\alpha}.$$

The *elastic shell strain tensor* $\mathbf{E}_e$ is defined by [4, 8]

$$\mathbf{E}_e \;=\; \mathbf{Q}_e^T\, \text{Grad}_s \mathbf{y} - \mathbf{a} = \big(\mathbf{Q}_e^T\, \mathbf{y}_{,\alpha} - \mathbf{y}_{0,\alpha}\big) \otimes \mathbf{a}^{\alpha} \tag{6}$$

and the *elastic shell bending-curvature tensor* $\mathbf{K}_e$ is [2, 4, 8]

$$\mathbf{K}_e = \text{axl}\big(\mathbf{Q}_e^T \mathbf{Q}_{e,\alpha}\big) \otimes \mathbf{a}^{\alpha} \;=\; \mathbf{Q}_0\big[\text{axl}\big(\overline{\mathbf{R}}^T \overline{\mathbf{R}}_{,\alpha}\big) - \text{axl}\big(\mathbf{Q}_0^T \mathbf{Q}_{0,\alpha}\big)\big], \tag{7}$$

which is a measure of orientation (curvature) change for Cosserat shells.

In [3] we have defined the *dislocation density tensor* $\mathbf{D}_e$ by

$$\mathbf{D}_e = \mathbf{Q}_e^T\; \text{Curl}_s\, \mathbf{Q}_e\ . \tag{8}$$

Using the formula $(4)_2$ , we can write this definition in the form

$$\mathbf{D}_e = \mathbf{Q}_e^T\big(-\mathbf{Q}_{e,\alpha} \times \mathbf{a}^{\alpha}\big) = -\big(\mathbf{Q}_e^T \mathbf{Q}_{e,\alpha}\big) \times \mathbf{a}^{\alpha}. \tag{9}$$

We have shown that the tensor $\mathbf{D}_e$ represents an alternative strain measure for orientation (curvature) change. Indeed, in [3] we have established the following relation

$$\mathbf{D}_e = -\mathbf{K}_e^T + \big(\text{tr}\,\mathbf{K}_e\big)\mathbf{1}_3 \quad \text{or equivalently,} \quad \mathbf{K}_e = -\mathbf{D}_e^T + \frac{1}{2}\big(\text{tr}\,\mathbf{D}_e\big)\mathbf{1}_3\,, \tag{10}$$

where $\mathbf{1}_3$ is the identity tensor in the Euclidean 3-space. We call (10) the *extended Nye's formula*, since it is a generalization of a well-known formula for infinitesimal strains in three-dimensional elasticity [14]. The formula (10) expresses the relationship between the shell bending-curvature tensor $\mathbf{K}_e$ and the dislocation density tensor $\mathbf{D}_e$ .

We present next a new proof of the extended Nye's formula (10). This proof is simpler as the one shown in [3] and is based on the relation

$$\mathbf{A} \times \mathbf{v} = \mathbf{v} \otimes \text{axl}\,\mathbf{A} - \big(\mathbf{v} \cdot \text{axl}\,\mathbf{A}\big)\mathbf{1}_3\,, \tag{11}$$

which holds true for every vector $\mathbf{v}$ and any skew-symmetric second order tensor $\mathbf{A}$.

*Proof of relation* (11): It is well-known that $\mathbf{A} = \mathbf{1}_3 \times \text{axl}\,\mathbf{A}$, where $\text{axl}\,\mathbf{A}$ is the axial vector of the skew-symmetric tensor $\mathbf{A}$. Then, we can write

$$\begin{aligned}\mathbf{A}\times\mathbf{v} &= \big(\mathbf{1}_3\times\operatorname{axl}\mathbf{A}\big)\times\mathbf{v} = \big[(\mathbf{e}_i\otimes\mathbf{e}_i)\times\operatorname{axl}\mathbf{A}\big]\times\mathbf{v}\\ &= \mathbf{e}_i\otimes\big[(\mathbf{e}_i\times\operatorname{axl}\mathbf{A})\times\mathbf{v}\big] = \mathbf{e}_i\otimes\big[(\mathbf{e}_i\cdot\mathbf{v})\operatorname{axl}\mathbf{A}-(\mathbf{v}\cdot\operatorname{axl}\mathbf{A})\,\mathbf{e}_i\big]\\ &= \big[(\mathbf{v}\cdot\mathbf{e}_i)\,\mathbf{e}_i\big]\otimes\operatorname{axl}\mathbf{A}-(\mathbf{v}\cdot\operatorname{axl}\mathbf{A})(\mathbf{e}_i\otimes\mathbf{e}_i) = \mathbf{v}\otimes\operatorname{axl}\mathbf{A}-\big(\mathbf{v}\cdot\operatorname{axl}\mathbf{A}\big)\mathbf{1}_3\,.\end{aligned}$$

*Proof of the extended Nye's formula* (10): We apply the relation (11) for the skew-symmetric tensor $\mathbf{A}=-\mathbf{Q}_e^T\mathbf{Q}_{e,\alpha}$ and the vector $\mathbf{v}=\mathbf{a}^\alpha$.
Upon summation over $\alpha=1,2$ we obtain from (7), (9) and (11) that

$$\begin{aligned}\mathbf{D}_e &= \big(-\mathbf{Q}_e^T\mathbf{Q}_{e,\alpha}\big)\times\mathbf{a}^\alpha = -\mathbf{a}^\alpha\otimes\operatorname{axl}\big(\mathbf{Q}_e^T\mathbf{Q}_{e,\alpha}\big)+\big(\mathbf{a}^\alpha\cdot\operatorname{axl}(\mathbf{Q}_e^T\mathbf{Q}_{e,\alpha})\big)\mathbf{1}_3\\ &= -\mathbf{K}_e^T+\big(\operatorname{tr}\mathbf{K}_e\big)\mathbf{1}_3\,,\end{aligned}$$

which concludes the proof.

*Remark 2.2* We can write the extended Nye's formula (10) in an equivalent form, if we make use of the following orthogonal decomposition: any second order tensor $\mathbf{X}\in\mathbb{R}^{3\times3}$ can be decomposed as direct sum in the form (the Cartan–Lie–algebra decomposition)

$$\mathbf{X} = \operatorname{dev}_3\operatorname{sym}\mathbf{X}+\operatorname{skew}\mathbf{X}+\frac{1}{3}\big(\operatorname{tr}\mathbf{X}\big)\,\mathbf{1}_3\,, \tag{12}$$

where $\operatorname{sym}\mathbf{S}$ is the symmetric part, $\operatorname{skew}\mathbf{S}$ the skew-symmetric part and $\operatorname{dev}_3\mathbf{S}:=\mathbf{S}-\frac{1}{3}(\operatorname{tr}\mathbf{S})\,\mathbf{1}_3$ is the deviatoric part of any second order tensor $\mathbf{S}$.

If we apply the operators tr, skew and $\operatorname{dev}_3\operatorname{sym}$ to the relation (10) we obtain, respectively

$$\operatorname{tr}\mathbf{D}_e = 2\operatorname{tr}\mathbf{K}_e\,,\quad \operatorname{skew}\mathbf{D}_e=\operatorname{skew}\mathbf{K}_e\,,\quad \operatorname{dev}_3\operatorname{sym}\mathbf{D}_e=-\operatorname{dev}_3\operatorname{sym}\mathbf{K}_e\,, \tag{13}$$

which express anew the relationship between $\mathbf{K}_e$ and $\mathbf{D}_e$ and are equivalent to the extended Nye's formula (10). As a direct consequence of (13) we deduce

$$\|\mathbf{D}_e\|^2=\|\mathbf{K}_e\|^2+\big(\operatorname{tr}\mathbf{K}_e\big)^2\quad\text{and}\quad\|\mathbf{K}_e\|\le\|\mathbf{D}_e\|\le2\,\|\mathbf{K}_e\|.\qquad\square \tag{14}$$

If we analyze the structure of the dislocation density tensor $\mathbf{D}_e$ we find that

$$\mathbf{D}_e = \mathbf{a}\,\mathbf{D}_e+\operatorname{tr}\big(\mathbf{a}\,\mathbf{D}_e\big)\,\mathbf{n}_0\otimes\mathbf{n}_0\,. \tag{15}$$

Thus, we see that the essential part of the tensor $\mathbf{D}_e$ is the tensor $\mathbf{a}\,\mathbf{D}_e=D_{\alpha\beta}\,\mathbf{a}^\alpha\otimes\mathbf{a}^\beta+D_{\alpha3}\,\mathbf{a}^\alpha\otimes\mathbf{n}_0$ (which has only 6 non-vanishing components $D_{\alpha i}$). Indeed, in view of (15) the two components of $\mathbf{D}_e$ in the directions $\mathbf{n}_0\otimes\mathbf{a}^\alpha$ are zero, while the component in the direction $\mathbf{n}_0\otimes\mathbf{n}_0$ is equal to $\operatorname{tr}\big(\mathbf{a}\,\mathbf{D}_e\big)$. In other words, all the information carried by the dislocation density tensor $\mathbf{D}_e$ is already contained in its part $\mathbf{a}\,\mathbf{D}_e$. For this reason, we define the new *shell dislocation density tensor* $\mathbf{D}_s$ by

$$\mathbf{D}_s := \mathbf{a}\,\mathbf{D}_e=\mathbf{a}\,\mathbf{Q}_e^T\operatorname{Curl}_s\mathbf{Q}_e=-\mathbf{a}\big(\mathbf{Q}_e^T\mathbf{Q}_{e,\alpha}\big)\times\mathbf{a}^\alpha, \tag{16}$$

which is more appropriate for the shell theory, since it is a tensor in the space $T_x \otimes E$ (where $T_x$ is the tangent plane and $E$ is the Euclidean 3-space). The definition (16) can be formulated as follows: by multiplication from the left with $\mathbf{a}$ we take the projection of the left leg of $\mathbf{D}_e$ on the tangent plane and obtain the shell dislocation density tensor $\mathbf{D}_s$ .

Next, we would like to present the relationship between the shell dislocation density tensor $\mathbf{D}_s$ and the shell bending-curvature tensor $\mathbf{K}_e$: if we multiply (10) with $\mathbf{a}$ and use the relations $\mathbf{a}\,\mathbf{D}_e = \mathbf{D}_s$ and $\mathbf{K}_e\,\mathbf{a} = \mathbf{K}_e$ we obtain

$$\mathbf{D}_s = -\mathbf{K}_e^T + (\operatorname{tr}\mathbf{K}_e)\mathbf{a} \quad \text{or equivalently,} \quad \mathbf{K}_e = -\mathbf{D}_s^T + (\operatorname{tr}\mathbf{D}_s)\mathbf{a}. \tag{17}$$

The relations (17) express the *extended Nye's formula* for the tensors $\mathbf{D}_s$ and $\mathbf{K}_e$. We observe that, in contrast to (10), the two relations (17) have a reciprocal structure: the two tensors $\mathbf{D}_s$ and $\mathbf{K}_e$ play symmetrical roles in (17).

Furthermore, we can establish an equivalent representation of the extended Nye's formula (17) using a corresponding orthogonal decomposition for these tensors (as a counterpart of Eqs. (12) and (13)):

**Lemma 2.1** *Any second order tensor* $\mathbf{X}$ *which satisfies the condition* $\langle \mathbf{X}, \mathbf{n}_0 \otimes \mathbf{n}_0 \rangle = 0$ *(i.e. the component of* $\mathbf{X}$ *in the direction* $\mathbf{n}_0 \otimes \mathbf{n}_0$ *is zero) can be decomposed as a direct sum in the form*

$$\mathbf{X} = \operatorname{dev}_s\operatorname{sym}\mathbf{X} + \operatorname{skew}\mathbf{X} + \frac{1}{2}(\operatorname{tr}\mathbf{X})\,\mathbf{a}\,, \tag{18}$$

*where we denote by* $\operatorname{dev}_s\mathbf{S} := \mathbf{S} - \frac{1}{2}(\operatorname{tr}\mathbf{S})\,\mathbf{a}$ *the* surface deviatoric part *of any second order tensor* $\mathbf{S}$.

*Proof* In view of the definition of the operator $\operatorname{dev}_s$, we have

$$\begin{aligned}\operatorname{dev}_s\operatorname{sym}\mathbf{X} + \operatorname{skew}\mathbf{X} + \tfrac{1}{2}(\operatorname{tr}\mathbf{X})\mathbf{a} &= \operatorname{sym}\mathbf{X} - \tfrac{1}{2}\operatorname{tr}(\operatorname{sym}\mathbf{X})\mathbf{a} \\ &\quad + \operatorname{skew}\mathbf{X} + \tfrac{1}{2}(\operatorname{tr}\mathbf{X})\,\mathbf{a} = \operatorname{sym}\mathbf{X} + \operatorname{skew}\mathbf{X} = \mathbf{X}\,.\end{aligned}$$

We see immediately that

$$\langle\operatorname{dev}_s\operatorname{sym}\mathbf{X}\,,\,\operatorname{skew}\mathbf{X}\rangle = 0 \quad \text{and} \quad \langle\operatorname{skew}\mathbf{X}\,,\,\frac{1}{2}(\operatorname{tr}\mathbf{X})\,\mathbf{a}\rangle = 0,$$

since $\operatorname{dev}_s\operatorname{sym}\mathbf{X}$ and $\frac{1}{2}(\operatorname{tr}\mathbf{X})\,\mathbf{a}$ are symmetric and $\operatorname{skew}\mathbf{X}$ is skew-symmetric. Finally, we employ the relations $\mathbf{a} = \mathbf{1}_3 - \mathbf{n}_0 \otimes \mathbf{n}_0$ and $\langle \mathbf{X}, \mathbf{n}_0 \otimes \mathbf{n}_0 \rangle = 0$ to get

$$\begin{aligned}&\langle\operatorname{dev}_s\operatorname{sym}\mathbf{X}\,,\,\tfrac{1}{2}(\operatorname{tr}\mathbf{X})\,\mathbf{a}\rangle = \langle\operatorname{sym}\mathbf{X} - \tfrac{1}{2}\operatorname{tr}(\operatorname{sym}\mathbf{X})\mathbf{a}\,,\,\tfrac{1}{2}(\operatorname{tr}\mathbf{X})\,\mathbf{a}\rangle \\ &= \tfrac{1}{2}(\operatorname{tr}\mathbf{X})\,\langle\mathbf{X} - \tfrac{1}{2}(\operatorname{tr}\mathbf{X})\mathbf{a}\,,\,\mathbf{a}\rangle = \tfrac{1}{2}(\operatorname{tr}\mathbf{X})\,\langle\mathbf{X}\,,\,\mathbf{a}\rangle - \tfrac{1}{4}(\operatorname{tr}\mathbf{X})^2\,\langle\mathbf{a}\,,\,\mathbf{a}\rangle \\ &= \tfrac{1}{2}(\operatorname{tr}\mathbf{X})\,\langle\mathbf{X}\,,\,\mathbf{1}_3 - \mathbf{n}_0 \otimes \mathbf{n}_0\rangle - \tfrac{1}{2}(\operatorname{tr}\mathbf{X})^2 \\ &= \tfrac{1}{2}(\operatorname{tr}\mathbf{X})^2 - \tfrac{1}{2}(\operatorname{tr}\mathbf{X})\,\langle\mathbf{X}\,,\,\mathbf{n}_0 \otimes \mathbf{n}_0\rangle - \tfrac{1}{2}(\operatorname{tr}\mathbf{X})^2 = 0.\end{aligned}$$

□

*Remark 2.3* In view of the definitions (7) and (16) we see that the tensors $\mathbf{K}_e$ and $\mathbf{D}_s$ satisfy the conditions $\langle \mathbf{K}_e, \mathbf{n}_0 \otimes \mathbf{n}_0 \rangle = 0$ and, respectively, $\langle \mathbf{D}_s, \mathbf{n}_0 \otimes \mathbf{n}_0 \rangle = 0$. Applying the above Lemma we obtain the orthogonal decompositions

$$\begin{aligned} \mathbf{K}_e &= \mathrm{dev}_S \mathrm{sym}\, \mathbf{K}_e + \mathrm{skew}\, \mathbf{K}_e + \tfrac{1}{2}\left(\mathrm{tr}\, \mathbf{K}_e\right) \mathbf{a}\,, \\ \mathbf{D}_s &= \mathrm{dev}_S \mathrm{sym}\, \mathbf{D}_s + \mathrm{skew}\, \mathbf{D}_s + \tfrac{1}{2}\left(\mathrm{tr}\, \mathbf{D}_s\right) \mathbf{a}\,. \end{aligned} \tag{19}$$

If we insert the relations (19) into (17) and compare the corresponding terms, then we obtain (in view of the uniqueness of the representation given by Lemma 2.1) the following equivalent form of the extended Nye's formula (17):

$$\mathrm{tr}\, \mathbf{D}_s = \mathrm{tr}\, \mathbf{K}_e\,, \quad \mathrm{skew}\, \mathbf{D}_s = \mathrm{skew}\, \mathbf{K}_e\,, \quad \mathrm{dev}_S \mathrm{sym}\, \mathbf{D}_s = -\mathrm{dev}_S \mathrm{sym}\, \mathbf{K}_e\,. \tag{20}$$

As a direct consequence of these equations we find

$$\|\mathbf{D}_s\| = \|\mathbf{K}_e\|\,. \tag{21}$$

Indeed, by virtue of the orthogonal decompositions (19) and the relations (20) we get

$$\begin{aligned} \|\mathbf{D}_s\|^2 &= \|\mathrm{dev}_S \mathrm{sym}\, \mathbf{D}_s\|^2 + \|\mathrm{skew}\, \mathbf{D}_s\|^2 + \tfrac{1}{2}\left(\mathrm{tr}\, \mathbf{D}_s\right)^2 \\ &= \|\mathrm{dev}_S \mathrm{sym}\, \mathbf{K}_e\|^2 + \|\mathrm{skew}\, \mathbf{K}_e\|^2 + \tfrac{1}{2}\left(\mathrm{tr}\, \mathbf{K}_e\right)^2 = \|\mathbf{K}_e\|^2. \end{aligned}$$

## 3 Governing Equations for the Equilibrium of Cosserat Shells

In what follows, we present the field equations which govern the deformation of Cosserat shells. We express the principle of virtual work using the shell dislocation density tensor $\mathbf{D}_s$. Then, we introduce the stress measure which is work-conjugate to $\mathbf{D}_s$ and deduce the corresponding constitutive equations.

Let $\mathbf{N}$ be the internal surface stress tensor and $\mathbf{M}$ be the internal surface couple tensor (of the first Piola-Kirchhoff type) for the shell. Then, the local equilibrium equations can be expressed in the form (see e.g., [1, 8])

$$\mathrm{Div}_s\, \mathbf{N} + \mathbf{f} = \mathbf{0}, \qquad \mathrm{Div}_s\, \mathbf{M} + \mathrm{axl}(\mathbf{N}\mathbf{F}^T - \mathbf{F}\mathbf{N}^T) + \mathbf{c} = \mathbf{0}, \tag{22}$$

where $\mathbf{f}$ and $\mathbf{c}$ are the external surface force and couple vectors.

Let $\boldsymbol{\nu}$ be the external unit normal vector to the boundary curve $\partial\omega_\xi$ lying in the tangent plane. We assume boundary conditions of the type [7, 16]

$$\begin{aligned} \mathbf{N}\boldsymbol{\nu} &= \mathbf{n}^*, \qquad & \mathbf{M}\boldsymbol{\nu} &= \mathbf{m}^* \quad \text{along } \partial\omega_f\,, \\ \mathbf{y} &= \mathbf{y}^*, \qquad & \overline{\mathbf{R}} &= \mathbf{R}^* \quad \text{along } \partial\omega_d\,, \end{aligned} \tag{23}$$

where $\partial\omega_f$ and $\partial\omega_d$ build a disjoint partition of $\partial\omega_\xi$ with length$(\partial\omega_d) > 0$. Here, $\mathbf{n}^*$ and $\mathbf{m}^*$ are the external boundary resultant force and couple vectors respectively, applied along the deformed boundary $\partial\omega_c$, but measured per unit length of $\partial\omega_\xi$ .

To obtain the principle of virtual work, we consider two arbitrary smooth vector fields $\mathbf{v}$ and $\mathbf{w}$ given on $\omega_\xi$ . By multiplying the Eqs. (22) and (23) with $\mathbf{v}$ and $\mathbf{w}$, we can set the integral identity

$$\int_{\omega_\xi} \big\{\big(\mathrm{Div}_s\,\mathbf{N}+\mathbf{f}\big)\cdot\mathbf{v}+\big[\mathrm{Div}_s\,\mathbf{M}+\mathrm{axl}(\mathbf{N}\mathbf{F}^T-\mathbf{F}\mathbf{N}^T)+\mathbf{c}\big]\cdot\mathbf{w}\big\}\,\mathrm{d}a$$
$$=\int_{\partial\omega_f}\big[\big(\mathbf{N}\boldsymbol{\nu}-\mathbf{n}^*\big)\cdot\mathbf{v}+\big(\mathbf{M}\boldsymbol{\nu}-\mathbf{m}^*\big)\cdot\mathbf{w}\big]\mathrm{d}\ell,$$

where $\mathrm{d}a$ is the area element on the surface $\omega_\xi$ and $\mathrm{d}\ell$ is the length element along $\partial\omega_f$. After some transformations we can rewrite this identity in the following form (see e.g., [8])

$$\int_{\omega_\xi}\Big(\big\langle\mathbf{N}\,,\,\mathrm{Grad}_s\,\mathbf{v}-\mathbf{W}\,\mathbf{F}\big\rangle+\big\langle\mathbf{M}\,,\,\mathrm{Grad}_s\,\mathbf{w}\big\rangle\Big)\mathrm{d}a=\int_{\omega_\xi}\big(\mathbf{f}\cdot\mathbf{v}+\mathbf{c}\cdot\mathbf{w}\big)\mathrm{d}a$$
$$+\int_{\partial\omega_f}\big(\mathbf{n}^*\cdot\mathbf{v}+\mathbf{m}^*\cdot\mathbf{w}\big)\mathrm{d}\ell+\int_{\partial\omega_d}\big[\big(\mathbf{N}\boldsymbol{\nu}\big)\cdot\mathbf{v}+\big(\mathbf{M}\boldsymbol{\nu}\big)\cdot\mathbf{w}\big]\mathrm{d}\ell,\quad(24)$$

where $\mathbf{W}=\mathbf{w}\times\mathbf{1}_3$ is the skew-symmetric tensor corresponding to the axial vector $\mathbf{w}$. Now, if we interpret $\mathbf{v}$ as the kinematically admissible virtual translation and $\mathbf{w}$ as the kinematically admissible virtual rotation of the shell, i.e.

$$\mathbf{v}=\delta\,\mathbf{y}\qquad\text{and}\qquad\mathbf{w}=\mathrm{axl}\big((\delta\mathbf{Q}_e)\mathbf{Q}_e^T\big),\tag{25}$$

then from the boundary conditions $(23)_{3,4}$ we obtain

$$\mathbf{v}=\mathbf{0},\qquad\mathbf{w}=\mathbf{0}\quad\text{along }\ \partial\omega_d\,,$$

which shows that the last integral in (24) vanishes. The remaining two integrals in the right-hand side of (24) describe the external virtual work and the integral in the left-hand side represents the internal virtual work. Thus, the internal virtual work power (density) $\mathscr{P}$ is given by

$$\mathscr{P}=\big\langle\mathbf{N}\,,\,\mathrm{Grad}_s\,\mathbf{v}-\mathbf{W}\,\mathbf{F}\big\rangle+\big\langle\mathbf{M}\,,\,\mathrm{Grad}_s\,\mathbf{w}\big\rangle.\tag{26}$$

This expression can be written in terms of the shell strain measures $\mathbf{E}_e$ and $\mathbf{K}_e$. Using transformations similar to those presented in [17, Sect. 4], we can put (26) in the form

$$\mathscr{P}=\big\langle\mathbf{Q}_e^T\mathbf{N}\,,\,\delta\,\mathbf{E}_e\big\rangle+\big\langle\mathbf{Q}_e^T\mathbf{M}\,,\,\delta\,\mathbf{K}_e\big\rangle.\tag{27}$$

Here, $\mathbf{Q}_e^T\mathbf{N}$ and $\mathbf{Q}_e^T\mathbf{M}$ are the shell stress measures in the material representation. They are work-conjugate to the shell strain measures $\mathbf{E}_e$ and $\mathbf{K}_e$, respectively. By virtue of (27), the principle of virtual work (24) becomes

$$\begin{aligned}&\int_{\omega_\xi}\Big(\big\langle\mathbf{Q}_e^T\mathbf{N}\,,\,\delta\,\mathbf{E}_e\big\rangle+\big\langle\mathbf{Q}_e^T\mathbf{M}\,,\,\delta\,\mathbf{K}_e\big\rangle\Big)\,\mathrm{d}a\\&\quad=\int_{\omega_\xi}(\mathbf{f}\cdot\mathbf{v}+\mathbf{c}\cdot\mathbf{w})\mathrm{d}a+\int_{\partial\omega_f}(\mathbf{n}^*\cdot\mathbf{v}+\mathbf{m}^*\cdot\mathbf{w})\mathrm{d}\ell,\end{aligned}\tag{28}$$

where $\mathbf{v}$ and $\mathbf{w}$ are given by (25).

The corresponding constitutive equations (under hyperelasticity assumptions) are

$$\mathbf{Q}_e^T\mathbf{N}=\frac{\partial\,W}{\partial\,\mathbf{E}_e}\qquad\text{and}\qquad\mathbf{Q}_e^T\mathbf{M}=\frac{\partial\,W}{\partial\,\mathbf{K}_e}\,,\tag{29}$$

where $W=W(\mathbf{E}_e,\mathbf{K}_e)$ is the elastically stored energy density for Cosserat shells.

We show next that the internal virtual work power (26) can also be expressed using the shell dislocation density tensor $\mathbf{D}_s$. On the basis of the extended Nye's formula $(17)_2$ and the relation $\big\langle\mathbf{Q}_e^T\mathbf{M}\,,\,\mathbf{a}\big\rangle=\mathrm{tr}\big(\mathbf{Q}_e^T\mathbf{M}\big)$, we get

$$\begin{aligned}\big\langle\mathbf{Q}_e^T\mathbf{M}\,,\,\delta\,\mathbf{K}_e\big\rangle&=\big\langle\mathbf{Q}_e^T\mathbf{M}\,,\,-\delta\,\mathbf{D}_s^T+\big(\mathrm{tr}\,\delta\,\mathbf{D}_s\big)\,\mathbf{a}\,\big\rangle\\&=-\big\langle\mathbf{Q}_e^T\mathbf{M}\,,\,\delta\,\mathbf{D}_s^T\,\big\rangle+\mathrm{tr}\big(\mathbf{Q}_e^T\mathbf{M}\big)\,\mathrm{tr}\big(\delta\,\mathbf{D}_s\big)\\&=\big\langle\,-\,\big(\mathbf{Q}_e^T\mathbf{M}\big)^T+\mathrm{tr}\big(\mathbf{Q}_e^T\mathbf{M}\big)\,\mathbf{a}\,\,,\,\delta\,\mathbf{D}_s\,\big\rangle.\end{aligned}$$

Inserting this expression into (27) we find

$$\mathscr{P}\,=\,\big\langle\mathbf{Q}_e^T\mathbf{N}\,,\,\delta\,\mathbf{E}_e\big\rangle+\big\langle\,-\,\big(\mathbf{Q}_e^T\mathbf{M}\big)^T+\mathrm{tr}\big(\mathbf{Q}_e^T\mathbf{M}\big)\,\mathbf{a}\,\,,\,\delta\,\mathbf{D}_s\,\big\rangle$$

and hence, the work-conjugate shell stress measures to $\mathbf{D}_s$ is $-\big(\mathbf{Q}_e^T\mathbf{M}\big)^T+\mathrm{tr}\big(\mathbf{Q}_e^T\mathbf{M}\big)\,\mathbf{a}$.

If we write the elastically stored energy density (strain energy density) $W$ as a function $W=\widehat{W}(\mathbf{E}_e,\mathbf{D}_s)$, then the corresponding constitutive equations are

$$\mathbf{Q}_e^T\mathbf{N}=\frac{\partial\,\widehat{W}}{\partial\,\mathbf{E}_e}\qquad\text{and}\qquad-\,\big(\mathbf{Q}_e^T\mathbf{M}\big)^T+\mathrm{tr}\big(\mathbf{Q}_e^T\mathbf{M}\big)\,\mathbf{a}\,=\,\frac{\partial\,\widehat{W}}{\partial\,\mathbf{D}_s}\,.\tag{30}$$

The last relation can be inverted (in the same way as the extended Nye's formula (17)) and (30) is equivalent to

$$\mathbf{N}\,=\,\mathbf{Q}_e\,\frac{\partial\,\widehat{W}}{\partial\,\mathbf{E}_e}\qquad\text{and}\qquad\mathbf{M}\,=\,\mathbf{Q}_e\Big[-\Big(\frac{\partial\,\widehat{W}}{\partial\,\mathbf{D}_s}\Big)^T+\Big(\mathrm{tr}\,\frac{\partial\,\widehat{W}}{\partial\,\mathbf{D}_s}\Big)\,\mathbf{a}\Big].\tag{31}$$

## 4 Variational Formulation and Existence of Minimizers

We consider the usual Lebesgue spaces $L^p(\omega, \mathbb{R}^3)$, $L^p(\omega, \mathbb{R}^{3\times 3})$ with $p \geq 1$ and the Sobolev spaces $H^1(\omega, \mathbb{R}^3)$, $H^1(\omega, \mathbb{R}^{3\times 3})$. With abuse of notation, we introduce the subset

$$L^p\big(\omega, \mathrm{SO}(3)\big) := \big\{\mathbf{Q} \in L^p(\omega, \mathbb{R}^{3\times 3}) \mid \mathbf{Q}(x_1, x_2) \in \mathrm{SO}(3) \ \text{ for a.e. } (x_1, x_2) \in \omega\big\}$$

with the induced strong topology of $L^p(\omega, \mathbb{R}^{3\times 3})$ and the subset

$$H^1\big(\omega, \mathrm{SO}(3)\big) := \big\{\mathbf{Q} \in H^1(\omega, \mathbb{R}^{3\times 3}) \mid \mathbf{Q}(x_1, x_2) \in \mathrm{SO}(3) \ \text{ for a.e. } (x_1, x_2) \in \omega\big\}$$

with the induced strong and weak topologies of $H^1(\omega, \mathbb{R}^{3\times 3})$.

Let us define the *admissible set* $\mathscr{A}$ by

$$\mathscr{A} = \big\{(\mathbf{y}, \overline{\mathbf{R}}) \in H^1(\omega, \mathbb{R}^3) \times H^1\big(\omega, \mathrm{SO}(3)\big) \ \big| \ \mathbf{y}\big|_{\partial\omega_d} = \mathbf{y}^*\,,\ \overline{\mathbf{R}}\big|_{\partial\omega_d} = \mathbf{R}^* \ \big\}, \tag{32}$$

where the boundary data satisfy $\mathbf{y}^* \in H^1(\omega, \mathbb{R}^3)$ and $\mathbf{R}^* \in H^1\big(\omega, \mathrm{SO}(3)\big)$.

We consider the *total energy functional*

$$\begin{aligned} \mathscr{E}(\mathbf{y}, \overline{\mathbf{R}}) &= \int_{\omega_\xi} \widehat{W}(\mathbf{E}_e, \mathbf{D}_s)\ \mathrm{d}a - \Lambda(\mathbf{y}, \overline{\mathbf{R}}) \\ &= \int_{\omega} \widehat{W}(\mathbf{E}_e, \mathbf{D}_s)\ a(x_1, x_2)\ \mathrm{d}x_1\mathrm{d}x_2 - \Lambda(\mathbf{y}, \overline{\mathbf{R}})\ , \end{aligned} \tag{33}$$

for every $(\mathbf{y}, \overline{\mathbf{R}}) \in \mathscr{A}$. Here, $\Lambda(\mathbf{y}, \overline{\mathbf{R}})$ denotes the *external loading potential*

$$\Lambda(\mathbf{y}, \overline{\mathbf{R}}) = \int_{\omega_\xi} \mathbf{f}\cdot\mathbf{u}\,\mathrm{d}a + \Pi_{\omega_\xi}(\overline{\mathbf{R}}) + \int_{\partial\omega_f} \mathbf{n}^*\cdot\mathbf{u}\,\mathrm{d}\ell + \Pi_{\partial\omega_f}(\overline{\mathbf{R}}), \tag{34}$$

where $\mathbf{u} := \mathbf{y} - \mathbf{y}_0$ is the displacement vector and we assume that $\mathbf{f} \in L^2(\omega, \mathbb{R}^3)$ and $\mathbf{n}^* \in L^2(\partial\omega_f, \mathbb{R}^3)$. The potential $\Pi_{\omega_\xi} : L^2\big(\omega, \mathrm{SO}(3)\big) \to \mathbb{R}$ of the external surface couples $\mathbf{c}$ and the potential $\Pi_{\partial\omega_f} : L^2\big(\partial\omega_f, \mathrm{SO}(3)\big) \to \mathbb{R}$ of the external boundary couples $\mathbf{m}^*$ are assumed to be continuous and bounded operators.

We formulate the following two-field minimization problem: *find the pair* $(\hat{\mathbf{y}}, \hat{\mathbf{R}}) \in \mathscr{A}$ *which realizes the minimum of the total energy functional* $\mathscr{E}(\mathbf{y}, \overline{\mathbf{R}})$ *given by* (33).

We can prove the following existence result.

**Theorem 2.1** *Assume that the reference configuration of the Cosserat shell satisfies the regularity conditions*

$$\begin{aligned} \mathbf{y}_0 \in H^1(\omega, \mathbb{R}^3), &\qquad \mathbf{Q}_0 \in H^1(\omega, \mathrm{SO}(3)), \\ \mathbf{a}_\alpha = \mathbf{y}_{0,\alpha} \in L^\infty(\omega, \mathbb{R}^3), &\qquad a(x_1, x_2) \geq a_0 > 0, \end{aligned} \tag{35}$$

*where $a_0$ is a constant. Moreover, the strain energy density $\widehat{W}(\mathbf{E}_e, \mathbf{D}_s)$ is assumed to be a quadratic convex function of $(\mathbf{E}_e, \mathbf{D}_s)$, which is also coercive, i.e. there exists a constant $C_0 > 0$ such that*

$$\widehat{W}(\mathbf{E}_e, \mathbf{D}_s) \geq C_0 \big( \|\mathbf{E}_e\|^2 + \|\mathbf{D}_s\|^2 \big). \tag{36}$$

*Then, the minimization problem (32)–(34) admits at least one minimizing solution pair $(\hat{\mathbf{y}}, \hat{\mathbf{R}}) \in \mathscr{A}$.*

*Proof* One can prove this statement using the direct methods of the calculus of variations. The procedure is very similar to the proof of Theorem 6 in [1], where we have formulated the same existence result, but expressed in terms of the shell bending-curvature tensor $\mathbf{K}_e$. By virtue of the extended Nye's formula (17) and the relation (21), we can adapt this proof to our case, i.e. when $W$ is a function of $(\mathbf{E}_e, \mathbf{D}_s)$. For the sake of brevity, we shall present only the main steps of the proof and omit further detailed explanations.

We show first that the external loading potential satisfies the estimate

$$|\,\Lambda(\mathbf{y}, \overline{\mathbf{R}})\,| \leq \; C \big( \|\mathbf{y}\|_{H^1(\omega)} + 1 \big), \qquad \forall\, (\mathbf{y}, \overline{\mathbf{R}}) \in \mathscr{A}$$

where $C > 0$ is a constant. Using this relation and the coercivity relation (36) we obtain

$$\mathscr{E}(\mathbf{y}, \overline{\mathbf{R}}) \geq C_0 \,\|\, \nabla \mathbf{y} \,\|^2_{L^2(\omega)} - C_1 \|\, \mathbf{y} \,\|_{H^1(\omega)} - C_2$$

where $C_1$, $C_2$ are some constants. To estimate the first term on the right-hand side of this inequality we apply the Poincaré–inequality and find (with $C_p > 0$ constant)

$$\mathscr{E}(\mathbf{y}, \overline{\mathbf{R}}) \geq C_p \,\|\mathbf{y} - \mathbf{y}^*\|^2_{H^1(\omega)} - C_3 \|\mathbf{y} - \mathbf{y}^*\|_{H^1(\omega)} + C_4 , \qquad \forall\, (\mathbf{y}, \overline{\mathbf{R}}) \in \mathscr{A}$$

so that the functional $\mathscr{E}(\mathbf{y}, \overline{\mathbf{R}})$ is bounded from below over $\mathscr{A}$.

Then, there exists an infimizing sequence $(\mathbf{y}_n, \mathbf{R}_n)_{n\in\mathbf{N}}$ such that

$$\lim_{n\to\infty} \mathscr{E}(\mathbf{y}_n, \mathbf{R}_n) = \inf \big\{ \mathscr{E}(\mathbf{y}, \overline{\mathbf{R}}) \,\big|\, (\mathbf{y}, \overline{\mathbf{R}}) \in \mathscr{A} \big\}. \tag{37}$$

We show that the sequences $\big(\mathbf{y}_n\big)$ and $\big(\mathbf{R}_n\big)$ are bounded in $H^1(\omega, \mathbb{R}^3)$ and $H^1(\omega, \mathbb{R}^{3\times 3})$, respectively. Then, we can extract subsequences (not relabeled) such that

$$\begin{aligned} &\mathbf{y}_n \rightharpoonup \hat{\mathbf{y}} \;\text{ in }\; H^1(\omega, \mathbb{R}^3) \qquad \text{and} \quad \mathbf{y}_n \to \hat{\mathbf{y}} \;\text{ in }\; L^2(\omega, \mathbb{R}^3), \\ &\mathbf{R}_n \rightharpoonup \hat{\mathbf{R}} \;\text{ in }\; H^1(\omega, \mathbb{R}^{3\times 3}) \quad \text{and} \quad \mathbf{R}_n \to \hat{\mathbf{R}} \;\text{ in }\; L^2(\omega, \mathbb{R}^{3\times 3}). \end{aligned}$$

The limit elements satisfy $(\hat{\mathbf{y}}, \hat{\mathbf{R}}) \in \mathscr{A}$ and we can construct the corresponding shell strain measures $\hat{\mathbf{E}}_e$, $\hat{\mathbf{D}}_s$, as well as $(\mathbf{E}_e)_n$, $(\mathbf{D}_s)_n$, using the definitions (6) and (16). Then, we show the weak convergence (on subsequences)

$$(\mathbf{E}_e)_n \rightharpoonup \hat{\mathbf{E}}_e \ \text{ in } L^2(\omega, \mathbb{R}^{3\times 3}) \qquad \text{and} \qquad (\mathbf{D}_s)_n \rightharpoonup \hat{\mathbf{D}}_s \ \text{ in } L^2(\omega, \mathbb{R}^{3\times 3}).$$

Finally, we use the convexity of the strain energy function $\widehat{W}(\mathbf{E}_e, \mathbf{D}_s)$ and deduce

$$\int_\omega \widehat{W}(\hat{\mathbf{E}}_e, \hat{\mathbf{D}}_s)\, a(x_1, x_2)\, \mathrm{d}x_1 \mathrm{d}x_2 \ \le\ \liminf_{n\to\infty} \int_\omega \widehat{W}\big((\mathbf{E}_e)_n, (\mathbf{D}_s)_n\big)\, a(x_1, x_2)\, \mathrm{d}x_1 \mathrm{d}x_2\,.$$

Fron this inequality and (33) it follows that:

$$\mathscr{E}(\hat{\mathbf{y}}, \hat{\mathbf{R}}) \ \le\ \liminf_{n\to\infty}\ \mathscr{E}(\mathbf{y}_n, \mathbf{R}_n). \tag{38}$$

In view of (37) and (38) we see that $(\hat{\mathbf{y}}, \hat{\mathbf{R}})$ is a minimizing solution pair of our minimization problem (32)–(34). □

## 5 Application: Isotropic Cosserat Shells

We employ the above results to investigate the case of isotropic elastic shells. To write the specific form of the strain energy density for isotropic shells, we use the decomposition into the *planar part* and *out-of-plane part* of the shell strain tensors $\mathbf{E}_e$ and $\mathbf{K}_e$:

$$\begin{aligned} &\mathbf{E}_e = (\mathbf{a} + \mathbf{n}_0 \otimes \mathbf{n}_0)\mathbf{E}_e = \mathbf{a}\mathbf{E}_e + \mathbf{n}_0 \otimes (\mathbf{n}_0\mathbf{E}_e), \quad \|\mathbf{E}_e\|^2 = \|\mathbf{a}\mathbf{E}_e\|^2 + \|\mathbf{n}_0\mathbf{E}_e\|^2, \\ &\mathbf{K}_e = \mathbf{a}\mathbf{K}_e + \mathbf{n}_0 \otimes (\mathbf{n}_0\mathbf{K}_e), \quad \|\mathbf{K}_e\|^2 = \|\mathbf{a}\mathbf{K}_e\|^2 + \|\mathbf{n}_0\mathbf{K}_e\|^2. \end{aligned} \tag{39}$$

In [8] the following general form of the strain energy density $W(\mathbf{E}_e, \mathbf{K}_e)$ for 6-parameter isotropic elastic shells was proposed

$$\begin{aligned} 2\,W(\mathbf{E}_e, \mathbf{K}_e) = {}& \alpha_1\big[\mathrm{tr}(\mathbf{a}\mathbf{E}_e)\big]^2 + \alpha_2\,\mathrm{tr}(\mathbf{a}\mathbf{E}_e)^2 + \alpha_3\,\|\mathbf{a}\mathbf{E}_e\|^2 + \alpha_4\,\|\mathbf{n}_0\mathbf{E}_e\|^2 \\ & + \beta_1\big[\mathrm{tr}(\mathbf{a}\mathbf{K}_e)\big]^2 + \beta_2\,\mathrm{tr}(\mathbf{a}\mathbf{K}_e)^2 + \beta_3\,\|\mathbf{a}\mathbf{K}_e\|^2 + \beta_4\,\|\mathbf{n}_0\mathbf{K}_e\|^2, \end{aligned} \tag{40}$$

where $\alpha_k$ and $\beta_k$ are constant constitutive coefficients ($k = 1, 2, 3, 4$).

We want to express the strain energy density as a function $\widehat{W}(\mathbf{E}_e, \mathbf{D}_s)$. To this aim, we decompose the shell dislocation density tensor $\mathbf{D}_s$ as

$$\mathbf{D}_s = \mathbf{D}_s(\mathbf{a} + \mathbf{n}_0 \otimes \mathbf{n}_0) = \mathbf{D}_s\,\mathbf{a} + (\mathbf{D}_s\,\mathbf{n}_0) \otimes \mathbf{n}_0, \qquad \|\mathbf{D}_s\|^2 = \|\mathbf{D}_s\,\mathbf{a}\|^2 + \|\mathbf{D}_s\,\mathbf{n}_0\|^2$$

and employ the extended Nye's formula $(17)_2$ to write

$$\mathbf{a}\mathbf{K}_e = -(\mathbf{D}_s\,\mathbf{a})^T + \mathrm{tr}\big(\mathbf{D}_s\,\mathbf{a}\big)\,\mathbf{a}, \qquad \mathbf{n}_0\mathbf{K}_e = -\mathbf{n}_0\mathbf{D}_s^T = -\mathbf{D}_s\,\mathbf{n}_0\,. \tag{41}$$

From (41) it follows

$$\text{tr}\big(\mathbf{a}\mathbf{K}_e\big) = -\text{tr}\big(\mathbf{D}_s\,\mathbf{a}\big)^T + 2\,\text{tr}\big(\mathbf{D}_s\,\mathbf{a}\big) = \text{tr}\big(\mathbf{D}_s\,\mathbf{a}\big), \qquad \|\mathbf{n}_0\mathbf{K}_e\| = \|\mathbf{D}_s\,\mathbf{n}_0\|\,. \tag{42}$$

In view of (21), $(39)_4$ and $(42)_2$ we get

$$\|\mathbf{a}\mathbf{K}_e\|^2 = \|\mathbf{K}_e\|^2 - \|\mathbf{n}_0\mathbf{K}_e\|^2 = \|\mathbf{D}_s\|^2 - \|\mathbf{D}_s\mathbf{n}_0\|^2 = \|\mathbf{D}_s\mathbf{a}\|^2. \tag{43}$$

Furthermore, from $(41)_1$ we deduce

$$(\mathbf{a}\mathbf{K}_e)^2 = (\mathbf{D}_s\,\mathbf{a})^T(\mathbf{D}_s\,\mathbf{a})^T - 2\big[\text{tr}\big(\mathbf{D}_s\,\mathbf{a}\big)\big](\mathbf{D}_s\,\mathbf{a})^T + \big[\text{tr}\big(\mathbf{D}_s\,\mathbf{a}\big)\big]^2\,\mathbf{a}$$

and applying the trace operator we find

$$\text{tr}(\mathbf{a}\mathbf{K}_e)^2 = \text{tr}(\mathbf{D}_s\,\mathbf{a})^2 - 2\big[\text{tr}\big(\mathbf{D}_s\,\mathbf{a}\big)\big]^2 + 2\big[\text{tr}\big(\mathbf{D}_s\,\mathbf{a}\big)\big]^2 = \text{tr}(\mathbf{D}_s\,\mathbf{a})^2. \tag{44}$$

If we insert the relations (42)–(44) into (40) we obtain the following expression of the strain energy density in terms of the shell strain tensor $\mathbf{E}_e$ and the shell dislocation density tensor $\mathbf{D}_s$

$$\begin{aligned}2\,\widehat{W}(\mathbf{E}_e, \mathbf{D}_s) = {} & \alpha_1\big[\text{tr}(\mathbf{a}\mathbf{E}_e)\big]^2 + \alpha_2\,\text{tr}(\mathbf{a}\mathbf{E}_e)^2 + \alpha_3\,\|\mathbf{a}\mathbf{E}_e\|^2 + \alpha_4\,\|\mathbf{n}_0\mathbf{E}_e\|^2 \\ & +\beta_1\big[\text{tr}(\mathbf{D}_s\,\mathbf{a})\big]^2 + \beta_2\,\text{tr}(\mathbf{D}_s\,\mathbf{a})^2 + \beta_3\,\|\mathbf{D}_s\,\mathbf{a}\|^2 + \beta_4\,\|\mathbf{D}_s\,\mathbf{n}_0\|^2.\end{aligned} \tag{45}$$

We can put this expression in a more convenient form. Using the orthogonal decomposition of the type (18) we derive

$$\|\,\mathbf{D}_s\,\mathbf{a}\|^2 = \|\text{dev}_s\text{sym}\,(\mathbf{D}_s\,\mathbf{a})\|^2 + \|\text{skew}\,(\mathbf{D}_s\,\mathbf{a})\|^2 + \frac{1}{2}\,\big[\text{tr}\,(\mathbf{D}_s\,\mathbf{a})\big]^2 \tag{46}$$

and

$$\text{tr}(\mathbf{D}_s\,\mathbf{a})^2 = \|\text{dev}_s\text{sym}\,(\mathbf{D}_s\,\mathbf{a})\|^2 - \|\text{skew}\,(\mathbf{D}_s\,\mathbf{a})\|^2 + \frac{1}{2}\,\big[\text{tr}\,(\mathbf{D}_s\,\mathbf{a})\big]^2. \tag{47}$$

Similar expressions can be obtained for $\|\mathbf{a}\mathbf{E}_e\|^2$ and $\text{tr}(\mathbf{a}\mathbf{E}_e)^2$. In view of relations (46) and (47) we can finally write the strain energy density (45) in the form

$$\begin{aligned}2\,\widehat{W}(\mathbf{E}_e, \mathbf{D}_s) = {} & (\alpha_2+\alpha_3)\|\text{dev}_s\text{sym}\,(\mathbf{a}\mathbf{E}_e)\|^2 + (\alpha_3-\alpha_2)\|\text{skew}\,(\mathbf{a}\mathbf{E}_e)\|^2 \\ & +\big(\alpha_1 + \frac{\alpha_2+\alpha_3}{2}\big)\big[\text{tr}\,(\mathbf{a}\mathbf{E}_e)\big]^2 + \alpha_4\|\mathbf{n}_0\mathbf{E}_e\|^2 \\ & +(\beta_2+\beta_3)\|\text{dev}_s\text{sym}\,(\mathbf{D}_s\,\mathbf{a})\|^2 + (\beta_3-\beta_2)\|\text{skew}\,(\mathbf{D}_s\,\mathbf{a})\|^2 \\ & +\big(\beta_1 + \frac{\beta_2+\beta_3}{2}\big)\big[\text{tr}\,(\mathbf{D}_s\,\mathbf{a})\big]^2 + \beta_4\|\mathbf{D}_s\,\mathbf{n}_0\|^2.\end{aligned} \tag{48}$$

By virtue of Lemma 2.1 we see that the quadratic function (48) is coercive if and only if the constitutive coefficients satisfy the inequalities

$$\begin{aligned} &\alpha_1 + \frac{\alpha_2+\alpha_3}{2} > 0, \qquad \alpha_2+\alpha_3 > 0, \qquad \alpha_3-\alpha_2 > 0, \qquad \alpha_4 > 0,\\ &\beta_1 + \frac{\beta_2+\beta_3}{2} > 0, \qquad \beta_2+\beta_3 > 0, \qquad \beta_3-\beta_2 > 0, \qquad \beta_4 > 0. \end{aligned} \tag{49}$$

Under these conditions, all the hypotheses on the strain energy density $W$ required by Theorem 2.1 are fulfilled. Thus, we can apply the Theorem 2.1 to prove the existence of minimizers for isotropic Cosserat shells.

*Remark 2.4* To apply the model in practical situations it is useful to express the constitutive coefficients $\alpha_k$ and $\beta_k$ in terms of the material parameters of the elastic continuum and the thickness of the shell. In the case of an isotropic Cauchy elastic material (characterized by the Young modulus $E$ and the Poisson ratio $\nu$), a particular (simplified) expression of the strain energy density $W(\mathbf{E}_e, \mathbf{K}_e)$ has been proposed in [4, 5]

$$\begin{aligned} 2\,W(\mathbf{E}_e, \mathbf{K}_e) &= C\nu\big[\text{tr}(\mathbf{aE}_e)\big]^2 + C(1-\nu)\,\|\,\mathbf{aE}_e\|^2 + \kappa_s\,C(1-\nu)\,\|\,\mathbf{n}_0\mathbf{E}_e\|^2\\ &+ D\nu\big[\text{tr}(\mathbf{aK}_e)\big]^2 + D(1-\nu)\,\|\,\mathbf{aK}_e\|^2 + \kappa_t\,D(1-\nu)\,\|\,\mathbf{n}_0\mathbf{K}_e\|^2, \end{aligned} \tag{50}$$

where $C = \frac{E\,h}{1-\nu^2}$ is the stretching (membrane) stiffness of the shell, $D = \frac{E\,h^3}{12(1-\nu^2)}$ is the bending stiffness, $h$ is the thickness of the shell, and $\kappa_s$ , $\kappa_t$ are two shear correction factors. In [5], the values of the shear correction factors have been set to $\alpha_s = \frac{5}{6}$ , $\alpha_t = \frac{7}{10}$ using the numerical treatment of some non-linear shell problems. The same values have been also proposed previously in the literature, see e.g. [9, 15, 18].

In view of (42), (43), we can write the function (50) in terms of the shell dislocation density tensor $\mathbf{D}_s$

$$\begin{aligned} 2\,\widehat{W}(\mathbf{E}_e, \mathbf{D}_s) &= C\nu\big[\text{tr}(\mathbf{aE}_e)\big]^2 + C(1-\nu)\,\|\,\mathbf{aE}_e\|^2 + \kappa_s\,C(1-\nu)\,\|\,\mathbf{n}_0\mathbf{E}_e\|^2\\ &+ D\nu\big[\text{tr}(\mathbf{D}_s\,\mathbf{a})\big]^2 + D(1-\nu)\,\|\,\mathbf{D}_s\,\mathbf{a}\|^2 + \kappa_t\,D(1-\nu)\,\|\,\mathbf{D}_s\,\mathbf{n}_0\|^2. \end{aligned} \tag{51}$$

This corresponds to the following choice of the constitutive coefficients $\alpha_k$ and $\beta_k$

$$\begin{aligned} &\alpha_1 = C\,\nu, \qquad \alpha_2 = 0, \qquad \alpha_3 = C\,(1-\nu), \qquad \alpha_4 = \kappa_s\,C\,(1-\nu),\\ &\beta_1 = D\,\nu, \qquad \beta_2 = 0, \qquad \beta_3 = D\,(1-\nu), \qquad \beta_4 = \kappa_t\,D\,(1-\nu). \end{aligned} \tag{52}$$

To verify the conditions (49) we compute using (52)

$$
\begin{aligned}
&\alpha_1 + \frac{\alpha_2+\alpha_3}{2} = C\cdot\frac{1+\nu}{2} = h\cdot\frac{E}{2(1-\nu)} = h\cdot\frac{\mu(3\lambda+2\mu)}{\lambda+2\mu}\,,\\
&\alpha_2+\alpha_3 = \alpha_3-\alpha_2 = h\cdot\frac{E}{1+\nu} = 2h\,\mu, \qquad \alpha_4 = 2h\,\kappa_s\,\mu,\\
&\beta_1 + \frac{\beta_2+\beta_3}{2} = D\cdot\frac{1+\nu}{2} = \frac{h^3}{24}\cdot\frac{E}{1-\nu} = \frac{h^3}{12}\cdot\frac{\mu(3\lambda+2\mu)}{\lambda+2\mu}\,,\\
&\beta_2+\beta_3 = \beta_3-\beta_2 = \frac{h^3}{12}\cdot\frac{E}{1+\nu} = \frac{h^3}{6}\cdot\mu, \qquad \beta_4 = \frac{h^3}{6}\cdot\kappa_t\,\mu,
\end{aligned}
\tag{53}
$$

where $\lambda$ and $\mu$ are the Lamé constants of the isotropic and homogeneous elastic material. If we insert the relations (53) into the general expression (48), then we obtain the appropriate form of the strain energy density in this model

$$
\begin{aligned}
\widehat{W}(\mathbf{E}_e,\mathbf{D}_s) = \mu\,h\Big[&\|\text{dev}_s\text{sym}\,(\mathbf{a}\mathbf{E}_e)\|^2 + \|\text{skew}\,(\mathbf{a}\mathbf{E}_e)\|^2 + \frac{3\lambda+2\mu}{2(\lambda+2\mu)}\big[\text{tr}\,(\mathbf{a}\mathbf{E}_e)\big]^2\\
&+\kappa_s\,\|\mathbf{n}_0\mathbf{E}_e\|^2\Big] + \mu\,\frac{h^3}{12}\Big[\|\text{dev}_s\text{sym}\,(\mathbf{D}_s\,\mathbf{a})\|^2 + \|\text{skew}\,(\mathbf{D}_s\,\mathbf{a})\|^2\\
&+\frac{3\lambda+2\mu}{2(\lambda+2\mu)}\big[\text{tr}\,(\mathbf{D}_s\,\mathbf{a})\big]^2 + \kappa_t\,\|\mathbf{D}_s\,\mathbf{n}_0\|^2\Big].
\end{aligned}
\tag{54}
$$

We see that the inequalities (49) are satisfied in this case, provided

$$
E > 0, \qquad -1 < \nu < \frac{1}{2}\,,
$$

or equivalently, in terms of the Lamé constants,

$$
\mu > 0, \qquad 3\lambda+2\mu > 0.
$$

These inequalities are satisfied, in view of the positive definiteness of the three-dimensional quadratic elastic strain energy for isotropic materials.

Since the conditions (49) hold, we are able to apply the Theorem 2.1 and we obtain the existence of minimizers also for this special constitutive model.

## 6 Open Problems

The presented existence result based on strict convexity in the employed strain and curvature measures does not exhaust all possibilities. Indeed, from a modelling point of view it is pertinent to consider a generalized stress-strain relation for which only an estimate of the elastic energy in terms of symmetrized elastic strains is available, i.e. only an estimate of the type

$$\widehat{W}(\mathbf{E}_e, \mathbf{D}_s) \geq C\left( \|\mathrm{sym}\, \mathbf{E}_e\|^2 + \|\mathbf{D}_s\|^2 \right) \tag{55}$$

is available (instead of (36)). Whether this situation remains well-posed is presently not known.

In our example for isotropic Cosserat shells, this case corresponds to $\alpha_2 = \alpha_3$, since the contribution of $\|\mathrm{skew}\, \mathbf{E}_e\|^2$ in (48) would then vanish. If we consider isotropic elastic shells made of an Cosserat material [10], this situation corresponds to the case $\mu_c = 0$, where $\mu_c$ denotes the Cosserat couple modulus of the three-dimensional continuum [12]. Similar problems have been dealt with in [13], where the case of vanishing Cosserat couple modulus has been investigated.

## References


1. Bîrsan, M., Neff, P.: Existence of minimizers in the geometrically non-linear 6-parameter resultant shell theory with drilling rotations. Math. Mech. Solids **19**(4), 376–397 (2014)
2. Bîrsan, M., Neff, P.: Shells without drilling rotations: A representation theorem in the framework of the geometrically nonlinear 6-parameter resultant shell theory. Int. J. Eng. Sci. **80**, 32–42 (2014)
3. Bîrsan, M., Neff, P.: On the dislocation density tensor in the Cosserat theory of elastic shells. In: Naumenko, K., Assmus, M. (eds.) Advanced Methods of Continuum Mechanics for Materials and Structures, Advanced Structured Materials 60, pp. 391–413. Springer Science+Business Media, Singapore (2016)
4. Chróścielewski, J., Makowski, J., Pietraszkiewicz, W.: Statics and Dynamics of Multifold Shells: Nonlinear Theory and Finite Element Method (in Polish). Wydawnictwo IPPT PAN, Warsaw (2004)
5. Chróścielewski, J., Pietraszkiewicz, W., Witkowski, W.: On shear correction factors in the non-linear theory of elastic shells. Int. J. Solids Struct. **47**, 3537–3545 (2010)
6. Cosserat, E., Cosserat, F.: Théorie des corps déformables. Hermann et Fils (reprint 2009), Paris (1909)
7. Eremeyev, V., Pietraszkiewicz, W.: The nonlinear theory of elastic shells with phase transitions. J. Elast. **74**, 67–86 (2004)
8. Eremeyev, V., Pietraszkiewicz, W.: Local symmetry group in the general theory of elastic shells. J. Elast. **85**, 125–152 (2006)
9. Naghdi, P.: The theory of shells and plates. In: Flügge, S. (ed.) Handbuch der Physik, Mechanics of Solids, pp. 425–640. Springer, Berlin (1972)
10. Neff, P.: A geometrically exact Cosserat-shell model including size effects, avoiding degeneracy in the thin shell limit. Cont. Mech. Thermodyn. **16**, 577–628 (2004)
11. Neff, P.: Geometrically exact Cosserat theory for bulk behaviour and thin structures. Modelling and mathematical analysis. Signatur HS 7/0973. Habilitationsschrift, Universitäts- und Landesbibliothek, Technische Universität Darmstadt, Darmstadt (2004)
12. Neff, P.: The Cosserat couple modulus for continuous solids is zero viz the linearized Cauchy-stress tensor is symmetric. Z. Angew. Math. Mech. **86**, 892–912 (2006)
13. Neff, P.: A geometrically exact planar Cosserat shell-model with microstructure: existence of minimizers for zero Cosserat couple modulus. Math. Mod. Meth. Appl. Sci. **17**, 363–392 (2007)
14. Nye, J.: Some geometrical relations in dislocated crystals. Acta Metall. **1**, 153–162 (1953)
15. Pietraszkiewicz, W.: Finite Rotations and Langrangian Description in the Non-linear Theory of Shells. Polish Scientific Publishers, Warsaw-Poznań (1979)

16. Pietraszkiewicz, W.: Refined resultant thermomechanics of shells. Int. J. Eng. Sci. **49**, 1112–1124 (2011)
17. Pietraszkiewicz, W., Eremeyev, V.: On natural strain measures of the non-linear micropolar continuum. Int. J. Solids Struct. **46**, 774–787 (2009)
18. Reissner, E.: The effect of transverse shear deformation on the bending of elastic plates. J. Appl. Mech. Trans. ASME **12**, A69–A77 (1945)
19. Sander, O., Neff, P., Bîrsan, M.: Numerical treatment of a geometrically nonlinear planar Cosserat shell model. Comput. Mech. **57**, 817–841 (2016)

# Flow Relations and Yield Functions for Dissipative Strain-Gradient Plasticity

**Carsten Carstensen, François Ebobisse, Andrew T. McBride, B. Daya Reddy and Paul Steinmann**

**Abstract** In this work we carry out a theoretical investigation of a dissipative model of rate-independent strain-gradient plasticity. The work builds on the investigation in [1], which in turn was inspired by the investigations in [4] of responses to non-proportional loading in the form of surface passivation. We recall the global nature of the flow relation when expressed in terms of the Cauchy stress and dissipation function. We highlight the difficulties encountered in attempts to obtain dual forms of the flow relation and associated yield functions, for the continuous and discrete problems, and derive upper bounds on the global yield function.

## 1 Introduction

This work is concerned with some aspects of the small-strain, rate-independent theory of strain-gradient plasticity. The model studied is based on that first proposed for rate-dependent materials in [9], and subsequently developed and analyzed for the rate-independent case in [11, 12]. The works [6, 7] present and analyze closely related rate-independent and -dependent theories. Further background to developments in

C. Carstensen
Humboldt-Universität Zu Berlin, Berlin, Germany
e-mail: cc@math.hu-berlin.de

F. Ebobisse · B.D. Reddy (✉)
University of Cape Town, Cape Town, South Africa
e-mail: daya.reddy@uct.ac.za

F. Ebobisse
e-mail: francois.ebobissebille@uct.ac.za

A.T. McBride
The University of Glasgow, Glasgow, UK
e-mail: andrew.mcbride@glasgow.ac.uk

P. Steinmann
University of Erlangen-Nuremberg, Erlangen, Germany
e-mail: paul.steinmann@ltm.uni-erlangen.de



F. dell'Isola et al. (eds.), *Mathematical Modelling in Solid Mechanics*, Advanced Structured Materials 69, DOI 10.1007/978-981-10-3764-1_3

strain-gradient plasticity may be found, for example, in the references cited above and in [3, 8, 9].

In the aforementioned models, gradient effects are accounted for either through their inclusion in the free energy, or in an extension of the flow law. These are referred to respectively as energetic and dissipative models, and differ substantially in their structure and predicted behaviour. For example, as indicated in [4], for energetic theories it is possible to construct the flow relation in a manner that mimics that for the classical problem, in that one is able to express stress rates in terms of plastic strain rates. For this reason the energetic models are referred to in [4] as incremental theories. On the other hand, local forms of the flow relation for fully dissipative theories lead to the expression of microscopic stresses – not their increments or rates – in terms of the rates of plastic strain and its gradient. The latter are referred to in [4] as non-incremental theories. These differences are explored theoretically and computationally in [4] in analyses of two problems that involve non-proportional loading. The main distinguishing feature in the two examples is, in the case of the dissipative theory, an elastic gap: that is, elastic behaviour associated with non-proportional loading following loading into the plastic range. This phenomenon has been further investigated in [5].

In the recent work [1], some features of the dissipative theory have been investigated further, spurred on by the investigations in [4, 5]. In particular, the global nature of the flow relation, when formulated in terms of the dissipation function and expressed in terms of the Cauchy stress, is reviewed. This important result addresses the apparent dilemma that presents itself when considering the flow law in its local form: in the local case, the extension of the classical normality law is perforce expressed in terms of the microstresses, which are not known in the elastic region and so cannot be used to determine whether yield has occurred. The dilemma is removed through a global approach to the flow law.

It is shown in [1] though that the task of determining the corresponding global yield function and normality relation is a complex one. For this reason a fully discrete version of the problem is investigated there. For this discrete approximation the process of obtaining a dualized form of the flow law, in terms of a yield function and normality relation, also presents significant mathematical difficulties, and at best an upper bound for the yield function is obtained. The expression for the yield function as a supremum of a quantity involving the dissipation function makes it possible to show the feasibility of an elastic gap in the case of problems in which non-proportional loading is effected through a change in micro-boundary conditions from Neumann (or micro-free) to Dirichlet (or micro-hard). All of these issues are revisited in this work with a view to further exploration of their implications.

The plan of the rest of this work is as follows. We summarize the governing relations for the flow law in Sect. 2, and formulate these in weak form. In Sect. 3 the formulation of the variational problem is completed by adding the weak form of the equilibrium equation. We then state an existence and uniqueness result that is valid when hardening is present. Section 4 is concerned with the fully discrete problem.

## 2 Governing Equations and Inequalities

We adopt the model of strain-gradient plasticity proposed in [9], and specialized to rate-independent plasticity in [11]. Small strains are assumed. The displacement is denoted by $\mathbf{u}$, the total strain by $\boldsymbol{\varepsilon}$, and the stress by $\boldsymbol{\sigma}$.

The strain is decomposed into elastic and plastic components $\boldsymbol{\varepsilon}^e$ and $\boldsymbol{\varepsilon}^p$ according to

$$\boldsymbol{\varepsilon} = \boldsymbol{\varepsilon}^e + \boldsymbol{\varepsilon}^p , \tag{1}$$

with the total strain given in terms of the displacement by

$$\boldsymbol{\varepsilon}(\mathbf{u}) = \frac{1}{2}(\nabla\mathbf{u} + [\nabla\mathbf{u}]^T) . \tag{2}$$

The strain-gradient theory makes provision for a 2nd-order microscopic stress tensor $\boldsymbol{\pi}$ and a 3rd-order microscopic stress $\boldsymbol{\Pi}$. The quantity $\boldsymbol{\pi}$ is symmetric and deviatoric, while $\boldsymbol{\Pi}$ is symmetric and deviatoric in its first two indices, in the sense that $\Pi_{ijk} = \Pi_{jik}$, $\Pi_{ppk} = 0$. Here and elsewhere the summation convention on repeated indices is invoked.

We define the generalized stress $\mathsf{S}$ and plastic strain $\Gamma$ to be the ordered pairs

$$\mathsf{S} = (\boldsymbol{\pi}, \ell^{-1}\boldsymbol{\Pi}), \qquad \Gamma = (\boldsymbol{\varepsilon}^p, \ell\nabla\boldsymbol{\varepsilon}^p) . \tag{3}$$

Here $\ell$ is a length parameter, assumed constant, and the inner product of the two generalized quantities is denoted by

$$\mathsf{S} \diamond \Gamma := \boldsymbol{\pi} : \boldsymbol{\varepsilon}^p + \boldsymbol{\Pi} \circ \nabla\boldsymbol{\varepsilon}^p = \pi_{ij}\varepsilon^p_{ij} + \Pi_{ijk}\varepsilon^p_{ij,k} .$$

Assuming quasistatic behaviour, the equation of equilibrium is given by

$$-\operatorname{div}\boldsymbol{\sigma} = \mathbf{b} , \tag{4}$$

where $\mathbf{b}$ is the body force. In addition, the stress and microscopic stresses are related to each other through the microforce balance equation

$$\operatorname{dev}\boldsymbol{\sigma} = \boldsymbol{\pi} - \operatorname{div}\boldsymbol{\Pi} \quad \text{or, in index form,} \quad (\operatorname{dev}\boldsymbol{\sigma})_{ij} = \pi_{ij} - \Pi_{ijk,k} . \tag{5}$$

The macroscopic boundary conditions are assumed for simplicity to be homogeneous Dirichlet; that is,

$$\mathbf{u} = \mathbf{0} \text{ on } \partial\Omega . \tag{6}$$

In addition we assume homogeneous micro-hard and micro-free boundary conditions on complementary parts $\partial\Omega_H$ and $\partial\Omega_F$ of the boundary; that is,

$$\boldsymbol{\varepsilon}^p = \mathbf{0} \text{ on } \partial\Omega_H , \tag{7a}$$

$$\boldsymbol{\Pi n} = \mathbf{0} \text{ on } \partial\Omega_F . \tag{7b}$$

Given the free energy $\psi$, the free-energy imbalance takes the form

$$\dot{\psi} - \boldsymbol{\sigma} : \dot{\boldsymbol{\varepsilon}}^e - \boldsymbol{\pi} : \dot{\boldsymbol{\varepsilon}}^p - \boldsymbol{\Pi} \circ \nabla\dot{\boldsymbol{\varepsilon}}^p \leq 0 . \tag{8}$$

We restrict attention to free energy functions of the form

$$\psi = \psi^e(\boldsymbol{\varepsilon}^e) = \frac{1}{2}\boldsymbol{\varepsilon}^e : \mathbb{C}\boldsymbol{\varepsilon}^e \tag{9}$$

in which the elasticity tensor $\mathbb{C}$ is given, for isotropic materials, by

$$\mathbb{C}\boldsymbol{\varepsilon} = \lambda(\operatorname{tr}\boldsymbol{\varepsilon})\boldsymbol{I} + 2\mu\,\boldsymbol{\varepsilon} \quad \text{for any symmetric tensor } \boldsymbol{\varepsilon} . \tag{10}$$

Here $\lambda$ and $\mu$ are the Lamé parameters. Substitution of (9) in (8) and use of the usual Coleman–Noll procedure leads to the elastic relation

$$\boldsymbol{\sigma} = \mathbb{C}(\boldsymbol{\varepsilon} - \boldsymbol{\varepsilon}^p) \tag{11}$$

and the reduced dissipation inequality

$$\boldsymbol{\pi} : \dot{\boldsymbol{\varepsilon}}^p + \boldsymbol{\Pi} \circ \nabla\dot{\boldsymbol{\varepsilon}}^p \geq 0 \quad \text{or} \quad \mathsf{S} \diamond \dot{\mathsf{\Gamma}} \geq 0 . \tag{12}$$

**Flow relation**. Based on the reduced dissipation inequality (12) we postulate the existence of a smooth yield function $f$, which is a function of the generalized stress $\mathsf{S}$, and a flow relation that takes the form of a normality law; that is,

$$f(\mathsf{S}) \leq 0 , \tag{13a}$$

$$\dot{\mathsf{\Gamma}} = \lambda\frac{\partial f}{\partial \mathsf{S}} , \tag{13b}$$

$$\lambda \geq 0, \quad f \leq 0, \quad \lambda f = 0 . \tag{13c}$$

Equivalently,

$$\dot{\mathsf{\Gamma}} \in N_{\mathscr{E}}(\mathsf{S}) \quad \text{or} \quad \dot{\mathsf{\Gamma}} \diamond (\mathsf{T} - \mathsf{S}) \leq 0 \qquad \text{for all } \mathsf{T} \in \mathscr{E} := \{\mathsf{T} \mid f(\mathsf{T}) \leq 0\} \tag{14}$$

where $\mathscr{E}$ is the convex elastic region and $N_{\mathscr{E}}(\mathsf{S})$ denotes the normal cone to $\mathscr{E}$ at $\mathsf{S}$. The dual of this relation is given by

$$\mathsf{S} \in \partial D(\dot{\mathsf{\Gamma}}) \quad \Longleftrightarrow \quad D(\dot{\mathsf{Q}}) - D(\dot{\mathsf{\Gamma}}) - \mathsf{S} \diamond (\mathsf{Q} - \dot{\mathsf{\Gamma}}) \geq 0 \tag{15}$$

where $\partial D$ denotes the subdifferential of $D$, the support function of $\mathscr{E}$, or in the context of plasticity, the dissipation function. The function $D$ is convex and positively homogeneous, with the property $D(\mathsf{0}) = 0$. For the special but important case in which

$$f(\mathsf{S}) = |\mathsf{S}| - Y = \sqrt{|\boldsymbol{\pi}|^2 + \ell^{-2}|\boldsymbol{\Pi}|^2} - Y \leq 0 \tag{16}$$

and where $Y$ is the initial yield stress, we have

$$D(\dot{\mathsf{\Gamma}}) = Y|\dot{\mathsf{\Gamma}}| = Y\sqrt{|\dot{\boldsymbol{\varepsilon}}^p|^2 + \ell^2|\nabla\dot{\boldsymbol{\varepsilon}}^p|^2}\,. \tag{17}$$

**The global problem**. In order to formulate the microforce balance and flow relations in global form we introduce or recall some definitions. We introduce the space of pairs

$$\mathscr{Q} = \{(\mathbf{q}, \ell\nabla\mathbf{q}) \mid q_{ij} = q_{ji},\ q_{ii} = 0,\ q_{ij} \in L^2(\Omega),\ q_{ij,k} \in L^2(\Omega),\ q_{ij} = 0 \text{ on } \partial\Omega_H\}\,, \tag{18}$$

and denote by $\mathscr{Q}^\star$ the topological dual space of $\mathscr{Q}$, that is, the space of continuous linear functionals on $\mathscr{Q}$. Duality pairing is written $\langle\mathsf{S}, \mathsf{\Gamma}\rangle$ for $\mathsf{\Gamma} \in \mathscr{Q}$ and $\mathsf{S} = (\boldsymbol{\pi}, \ell^{-1}\boldsymbol{\Pi}) \in \mathscr{Q}^\star$.

The global dissipation function is the functional $j : \mathscr{Q} \to \mathbb{R}$ defined by

$$j(\mathsf{Q}) = \int_\Omega D(\mathsf{Q})\, dx\,. \tag{19}$$

Taking the inner product of (5) with arbitrary $\mathbf{q} \in \mathscr{Q}$, integrating by parts, and imposing the microscopic boundary conditions (7a) and (7b) we obtain the weak formulation

$$\int_\Omega \operatorname{dev}\boldsymbol{\sigma} : \mathbf{q}\, dx = \int_\Omega \mathsf{S} \diamond \mathsf{Q}\, dx \tag{20}$$

of the microforce balance equation.

We obtain a weak or global form for the flow relation by integrating $(15)_2$ and adding to this the weak form of the microforce balance equation (20), to get

$$\int_\Omega \left[D(\mathsf{Q}) - D(\dot{\mathsf{\Gamma}}) - \operatorname{dev}\boldsymbol{\sigma} : (\mathbf{q} - \dot{\boldsymbol{\varepsilon}}^p)\right] \geq 0 \tag{21}$$

or

$$\int_\Omega \left[D(\mathsf{Q}) - D(\dot{\mathsf{\Gamma}}) - \Sigma \diamond (\mathsf{Q} - \dot{\mathsf{\Gamma}})\right] \geq 0 \tag{22}$$

where $\Sigma := (\operatorname{dev}\boldsymbol{\sigma}, \mathbf{0}) \in \mathscr{Q}^\star$.

The subdifferential $\partial j(\mathsf{\Gamma})$ of $j$ at $\mathsf{\Gamma}$ is the subset of $\mathscr{Q}^\star$ defined by[1]

---

[1] See, for example, [2, 13] for this and other definitions in the context of convex analysis on topological vector spaces.

$$\partial j(\Gamma) = \{\Sigma \in \mathcal{Q}^\star \mid j(\mathsf{Q}) \geq j(\Gamma) + \langle \Sigma, Q - \Gamma \rangle \geq 0\}. \tag{23}$$

The conjugate $j^\star : \mathcal{Q}^\star \to \mathbb{R}$ of $j$ is given by

$$\begin{aligned} j^\star(\Sigma) &= \sup_{\mathsf{Q}\in\mathcal{Q}} \{\langle \Sigma, Q \rangle \mid \mathsf{Q} \in \mathcal{Q}\} \\ &= \mathrm{I}_{\mathscr{E}_{\text{glob}}}(\Sigma), \end{aligned} \tag{24}$$

in which the indicator function appearing in the last line is defined, for a subset $\mathscr{E}_{\text{glob}}$ of $\mathcal{Q}^\star$, by

$$\mathrm{I}_{\mathscr{E}_{\text{glob}}}(\Sigma) = \begin{cases} 0 & \Sigma \in \mathscr{E}_{\text{glob}}, \\ +\infty & \text{otherwise}. \end{cases} \tag{25}$$

We recognize (22) as the inclusion

$$\Sigma \in \partial j(\dot{\Gamma}) \tag{26}$$

and note the important duality relation

$$\Sigma \in \partial j(\dot{\Gamma}) \quad \Longleftrightarrow \quad \dot{\Gamma} \in \partial I_{\mathscr{E}_{\text{glob}}}. \tag{27}$$

The latter relation in turn is equivalent to

$$\dot{\Gamma} \in N_{\mathscr{E}_{\text{glob}}}(\Sigma) \quad \text{or} \quad \langle \dot{\Gamma}, \mathsf{T} - \Sigma \rangle \leq 0, \quad \mathsf{T} \in \mathcal{Q}^\star, \tag{28}$$

in which $N_{\mathscr{E}_{\text{glob}}}(\Sigma)$ is the normal cone to $\mathscr{E}_{\text{glob}}$ at $\Sigma$.

Note that the microstress $\mathsf{S}$ has been eliminated from the global flow relation. This quantity is *indeterminate* in the elastic region and could not be used to determine when yield occurs. On the other hand, yield is now a global issue, and (24) would require that the global yield function whose normal cone appears in this inclusion be found.

We have an expression for $D$ and $j$. The matter of finding an explicit expression for $\mathscr{E}_{\text{glob}}$ is not a simple one. The complication arises from the structure of the space $\mathcal{Q}$ and of the local dissipation function $j$, which depends on a function and its gradient. The matter may be approached by formulating the problem in mixed form, by introducing an independent variable $\mathbf{P} = \ell \nabla \boldsymbol{\varepsilon}^p$. This has been considered in [1], where it is shown that the result is a mixed variational problem of non-standard type. In view of these difficulties we explore the nature of the global flow relation further in Sect. 4 in the form of a fully discrete approximation.

## 3 Weak Form of Equilibrium and the Elastic–Plastic Initial-Boundary Value Problem

The formulation of the problem is completed with a statement of global equilibrium. For this purpose the space of displacements $\mathscr{V}$ is defined by

$$\mathscr{V} = \{\mathbf{v} \mid v_i \in L^2(\Omega),\ v_{i,j} \in L^2(\Omega),\ \mathbf{v} = \mathbf{0} \text{ on } \partial\Omega_D\}\,. \tag{29}$$

Then the weak form of the equilibrium equation (4) is as follows: for $\mathbf{f} \in L^2(\Omega)^d$, find $\mathbf{u} \in \mathscr{V}$ that satisfies

$$\int_\Omega \boldsymbol{\sigma}(\mathbf{u}, \boldsymbol{\varepsilon}^p) : \boldsymbol{\varepsilon}(\mathbf{v})\, dx = \int_\Omega \mathbf{f} \cdot \mathbf{v}\, dx \quad \text{for all } \mathbf{v} \in \mathscr{V}\,, \tag{30}$$

the expression for the stress $\boldsymbol{\sigma}$ being given by (11). The weak form of the flow relation is given by (22) or (27). Existence of a unique solution for the purely dissipative problem requires some hardening behaviour. This may be incorporated in the model by introducing the accumulated plastic strain $\eta \in L^2(\Omega)$ as a hardening variable; that is,

$$\eta(t) = \int_0^t |\dot{\boldsymbol{\varepsilon}}^p(s)|\, ds\,. \tag{31}$$

The space of hardening variables is denoted by $\mathscr{M} = L^2(\Omega)$. Without repeating all of the details (see for example [10, 11]), with the use of isotropic hardening two modifications need to be made to the governing relations: first, the dissipation function $D$ in (17) now becomes, with the additional independent variable,

$$D_{\text{hard}}(\dot{\mathsf{\Gamma}}, \dot{\eta}) = \begin{cases} Y|\dot{\mathsf{\Gamma}}| & |\dot{\mathsf{\Gamma}}| \le \dot{\eta}, \\ +\infty & \text{otherwise}\,. \end{cases} \tag{32}$$

The second modification is an addition to the variational inequality (22) of a term that takes account of hardening, assumed here to be linear: the resulting inequality is now

$$\int_\Omega \left[D_{\text{hard}}(\mathsf{Q}) - D_{\text{hard}}(\dot{\mathsf{\Gamma}}) - \mathsf{\Sigma} \diamond (\mathsf{Q} - \dot{\mathsf{\Gamma}}) - g\eta(\mu - \dot{\eta})\right] \ge 0 \tag{33}$$

for arbitrary $\mu \in \mathscr{M}$, and $g$ is the hardening parameter with $g \in L^\infty(\Omega)$ and $g(\mathbf{x}) > 0$ a.e. We then have the following result [11, Theorem 2].

**Theorem** *The problem given by (30) and (33) has a unique solution* $\mathbf{u}(t), \epsilon^p(t), \eta(t) \in \mathscr{V} \times \mathscr{Q} \times \mathscr{M}$.

## 4 The Fully Discrete Problem

For convenience we return to the problem without hardening. We discretize in time by partitioning the time interval $[0, T]$ as $0 = t_1 < t_2 < \cdots < t_n < \cdots t_N = T$, set $w_n := w(t_n)$ and $\Delta w = w_{n+1} - w_n$ for any function $w$, and replace the time derivative $\dot{w}$ by its backward Euler approximation $\Delta w/\Delta t$.

Assume that the domain is polygonal (in two dimensions) or polyhedral (in three) and covered by a mesh of NE triangular or tetrahedral elements denoted by $\Omega_e$ $(e = 1, \ldots, \text{NE})$. We introduce finite-dimensional spaces of functions in which a discrete version of the variational problem will be sought. Thus, for definiteness define

$$\mathscr{V}^h \subset \mathscr{V}, \quad \text{and} \quad \mathscr{Q}^h \subset \mathscr{Q} \tag{34}$$

to be spaces of continuous functions whose restrictions to $\Omega_e$ are polynomials of degree 1.

We adopt Voigt notation in this section, so that the plastic strain $\boldsymbol{\varepsilon}^p$ is written in two and three dimensions respectively as

$$\begin{aligned} \boldsymbol{\varepsilon}^p &= \left[\varepsilon_{11}^p \ \varepsilon_{22}^p \ \varepsilon_{12}^p\right] && \text{in two dimensions,} \\ \boldsymbol{\varepsilon}^p &= \left[\varepsilon_{11}^p \ \varepsilon_{22}^p \ \varepsilon_{33}^p \ \varepsilon_{12}^p \ \varepsilon_{13}^p \ \varepsilon_{23}^p\right] && \text{in three dimensions.} \end{aligned}$$

Note from the definition of $\mathscr{Q}$ that there are only two and five independent components of plastic strain, by virtue of the condition of plastic incompressibility. The components of total strain $\boldsymbol{\varepsilon}$ are organized using Voigt notation, in the same way.

Denote by $\mathbf{u}_h$ and $\boldsymbol{\varepsilon}_h^p$ members of $\mathscr{V}^h$ and $\mathscr{Q}^h$ respectively, and their global degrees of freedom by $\mathsf{d}$ and $\mathsf{p}$ respectively. Then

$$\boldsymbol{\varepsilon}_h^p = \mathsf{Np}, \quad \nabla\boldsymbol{\varepsilon}_h^p = \mathsf{Bp}, \quad \mathbf{u}_h = \overline{\mathsf{N}}\mathsf{d}, \quad \boldsymbol{\varepsilon}(\mathbf{u}_h) = \overline{\mathsf{B}}\mathsf{d} \tag{35}$$

with matrices of shape functions $\mathsf{N}$ and $\overline{\mathsf{N}}$ and of shape function derivatives $\mathsf{B}$ and $\overline{\mathsf{B}}$.

Here and elsewhere we drop the subscript $n$ that denotes quantities at time $t_n$. We proceed using the example (17) of a dissipation function. Since

$$|\boldsymbol{\varepsilon}_h^p| = \sqrt{\mathsf{p}^T\mathsf{N}^\mathsf{T}\mathsf{Np}}, \qquad |\nabla\boldsymbol{\varepsilon}_h^p| = \sqrt{\mathsf{p}^T\mathsf{B}^\mathsf{T}\mathsf{Bp}}$$

we have, with $\Gamma_h = (\boldsymbol{\varepsilon}_h^p, \ell\nabla\boldsymbol{\varepsilon}_h^p)$,

$$\begin{aligned} D(\Gamma_h) = D_h(\mathsf{p}) &= Y\left[\sqrt{\mathsf{p}^T\mathsf{N}^\mathsf{T}\mathsf{Np} + \ell^2\mathsf{p}^T\mathsf{B}^\mathsf{T}\mathsf{Bp}}\right] \\ &= Y\sqrt{\mathsf{p}^T\mathsf{Kp}} \end{aligned} \tag{36}$$

where

$$\mathsf{K} = \mathsf{N}^\mathsf{T}\mathsf{N} + \ell^2\mathsf{B}^\mathsf{T}\mathsf{B}.$$

Set

$$\mathscr{J}(\mathsf{q}) := \int_{\Omega} D_h(\mathsf{q})\, dx\,; \tag{37}$$

then (22) or (26) becomes, for the incremental problem,

$$\mathsf{s} \in \partial \mathscr{J}(\Delta\mathsf{p}) \quad \text{or} \quad \mathscr{J}(\mathsf{q}) - \mathscr{J}(\Delta\mathsf{p}) - (\mathsf{q} - \Delta\mathsf{p})^T \mathsf{s} \geq 0 \quad \text{for all } \mathsf{q}\,, \tag{38}$$

where

$$\mathsf{s} := \int_{\Omega} \mathsf{N}^{\mathsf{T}} \operatorname{dev} \boldsymbol{\sigma}_h\, dx \tag{39}$$

and $\boldsymbol{\sigma}_h$ is an appropriate discrete approximation of the stress $\boldsymbol{\sigma}$.

The dual of this discrete inclusion is, from (24),

$$\Delta\mathsf{p} \in N_{\mathscr{E}^h_{\text{glob}}}(\mathsf{s}) \quad \text{or} \quad (\Delta\mathsf{p})^T(\mathsf{t} - \mathsf{s}) \leq 0 \quad \text{for all } \mathsf{t} \in \mathscr{E}^h_{\text{glob}}\,, \tag{40}$$

in which the discrete global elastic region $\mathscr{E}^h_{\text{glob}}$ is a subset of the space of discrete stresses.

The task is now one of constructing the unknown $\mathscr{E}^h_{\text{glob}}$ and corresponding yield function from the known $\mathscr{J}$. From a result in convex analysis (see for example [10, p. 109]), given a dissipation function $\mathscr{J}$, one may construct a function $\phi(\mathsf{s})$, known as the canonical yield function, with the properties

$$\phi \text{ is convex}\,, \tag{41a}$$

$$\mathscr{E}^h_{\text{glob}} = \{\mathsf{s} \mid \phi(\mathsf{s}) \leq 1\}\,, \tag{41b}$$

$$\phi(\mathsf{s}) = \sup_{\mathsf{q} \neq 0} \frac{\mathsf{q}^T \mathsf{s}}{\mathscr{J}(\mathsf{q})}\,. \tag{41c}$$

It is seen immediately, given the positive homogeneity of the dissipation function in the plastic strain, that the canonical yield function is positively homogeneous in the discrete stresses. The yield and dissipation functions are referred to as polar conjugates in this context. It follows that $\phi$ can be constructed if we are able to evaluate the supremum in (41c).

This is not a simple task. From the expression

$$\mathscr{J}(\mathsf{q}) = Y \int_{\Omega} |\mathsf{K}^{1/2}\mathsf{q}|\, dx$$

and taking $|\Omega| = 1$ for convenience it follows that

$$\begin{aligned} \mathsf{s}^T \mathsf{q} &= \int_{\Omega} ([\mathsf{K}(\mathbf{x})]^{-1/2}\mathsf{s})^T ([\mathsf{K}(\mathbf{x})]^{1/2}\mathsf{q})\, dx \\ &\leq Y^{-1}\big(\max\nolimits_{\mathbf{x}\in\Omega} |[\mathsf{K}(\mathbf{x})]^{-1/2}\mathsf{s}|\big)\, \mathscr{J}(\mathsf{q})\,. \end{aligned} \tag{42}$$

Hence we have the upper bound

$$\phi(\mathsf{s}) \leq Y^{-1}\max_{\mathbf{x}\in\Omega}|[\mathsf{K}(\mathbf{x})]^{-1/2}\mathsf{s}|\,. \tag{43}$$

The upper bound is, however, not achieved in general, as shown in [1].

**Some comments on the elastic gap**. In [4], the consequences have been investigated of changes in direction in the loading path as a result of a change in the boundary conditions for the plastic strain. Specifically, two example problems are considered in that work: one of a strip in tension and the other a beam in bending. In both cases the micro-free boundary condition of the form (7b) is maintained until loading extends well into the plastic range. Then, the boundary condition is changed to a micro-hard one (7a); physically, this process is referred to as passivation. The result is that the behaviour following passivation is elastic for a marked range of the loading, after which plastic behaviour resumes. This phenomenon, referred to in [4] as the elastic gap, has motivated the objective of finding explicit expressions for the yield function earlier in this section and, for the continuous problem, in Sect. 3. With the availability of such a yield function it might then be possible to determine constructively the change in conditions, mathematically, that account for the elastic gap.

While it has not been possible to find expressions for the yield function, some insight may be gained by a qualitative exploration using the definition (41c) of the yield function $\phi$ corresponding to the discrete problem. For a problem such as those considered in [4], or more generally, any problem characterized by a change in boundary conditions in the plastic range from micro-free to micro-hard, denote by $N_{\text{pre}}$ the total number of nodal degrees of freedom of plastic strain. Given the micro-free boundary condition, this will be equal to the total number of nodes. Likewise, denote by $N_{\text{pass}}$ the number of degrees of freedom in the passivation phase. Clearly $N_{\text{pass}} < N_{\text{pre}}$ as the plastic strain is set equal to zero on a finite number of the boundary nodes, in line with the Dirichlet boundary condition. Assuming that the difference between the vectors $\mathsf{s}$ of nodal stresses just before and after initiation of passivation is negligible, it follows from (41c) that the yield function $\phi_{\text{pass}}$ in the passivation phase is found from

$$\begin{aligned}\phi_{\text{pass}}(\mathsf{s}) &= \sup \frac{\mathsf{s}^{\mathsf{T}}\mathsf{q}_{\text{pass}}}{\mathscr{J}(\mathsf{q}_{\text{pass}})}\\ &\leq \sup \frac{\mathsf{s}^{\mathsf{T}}\mathsf{q}_{\text{pre}}}{\mathscr{J}(\mathsf{q}_{\text{pre}})}\\ &= \phi_{\text{pre}}(\mathsf{s}) = 1\,.\end{aligned} \tag{44}$$

Here we have denoted by $\mathsf{q}_{\text{pre}}$ and $\mathsf{q}_{\text{pass}}$ arbitrary vectors in the uniform and passivated phases, respectively. Also, in the last line we use the assumption that the material is in the plastic range in the uniform phase just before passivation. The inequality in the second line follows from the fact that $\dim \mathsf{q}_{\text{pass}} = N_{\text{pass}} < N_{\text{pre}} = \dim \mathsf{q}_{\text{pre}}$, so that the supremum is being taken over a larger set. This bound permits the possibility that $\phi_{\text{pass}} < 1$ and, hence, of elastic behaviour in the initial passivation phase.

## 5 Concluding Remarks

This work has been concerned with properties of the flow rule for the case of dissipative strain-gradient plasticity. The work builds on the investigation in [1], which in turn was inspired by the investigations in [4] of responses to non-proportional loading in the form of surface passivation.

The global nature of the flow rule, when expressed in terms of the Cauchy stress, is reiterated, and the conditions for well-posedness of the resulting initial-boundary value problem summarized. In particular, it is noted that well-posedness requires the presence of hardening for the dissipative problem.

This study returns to a difficulty encountered in [1], viz. an inability to derive an explicit expression for the dual form of the global flow relation as a normality law, from the statement of the flow rule in terms of the dissipation function. The difficulty is encountered also for the fully discrete approximation of the problem, for which only an upper bound to the yield function is achievable. That the elastic gap constitutes a feasible response following passivation is shown through a simple bound on the expression for the yield function in terms of the dissipation function.

Issues that await further investigation include experimental investigation of the existence or otherwise of the elastic gap; the development of alternative approaches that would allow the global yield functions, for both the continuous and discrete problems, to be determined and explored; and an analysis of convergence of finite element approximations. These issues are all currently receiving attention.

**Acknowledgements** The work reported in this paper was carried out with support through the South African Research Chair in Computational Mechanics to BDR and ATMcB. This support is gratefully acknowledged. PS acknowledges support through the Collaborative Research Center 814.

## References

1. Carstensen, C., Ebobisse, F., McBride, A.T., Reddy, B.D., Steinmann, P.: Some properties of the dissipative model of strain-gradient plasticity. Phil. Mag., in press (2017)
2. Ekeland, I., Temam, R.: Convex Analysis and Variational Problems. North-Holland, Amsterdam (1976)
3. Fleck, N.A., Hutchinson, J.W.: A reformulation of strain gradient plasticity. J. Mech. Phys. Solids **49**, 2245–2271 (2001)
4. Fleck, N.A., Hutchinson, J.W., Willis, J.R.: Strain-gradient plasticity under non-proportional loading. Proc. R. Soc. A **470**, 20140267 (2015)
5. Fleck, N.A., Hutchinson, J.W., Willis, J.R.: Guidelines for constructing strain gradient plasticity theories. J. Appl. Mech. **82**, 071002-1–10 (2015)
6. Fleck, N.A., Willis, J.R.: A mathematical basis for strain-gradient plasticity - Part I: scalar plastic multiplier. J. Mech. Phys. Solids **57**, 151–177 (2009)
7. Fleck, N.A., Willis, J.R.: A mathematical basis for strain-gradient plasticity - Part II: tensorial plastic multiplier. J. Mech. Phys. Solids **57**, 1045–1057 (2009)
8. Gudmundson, P.: A unified treatment of strain gradient plasticity. J. Mech. Phys. Solids **52**, 1379–1406 (2004)

9. Gurtin, M.E., Anand, L.: A theory of strain-gradient plasticity for isotropic, plastically irrotational materials. Part I: small deformations. J. Mech. Phys. Solids **53**, 1624–1649 (2005)
10. Han, W., Reddy, B.D.: Plasticity: Mathematical Theory and Numerical Analysis. Second Edition, Springer, New York (2013)
11. Reddy, B.D.: The role of dissipation and defect energy in variational formulations of problems in strain-gradient plasticity. Part 1: polycrystalline plasticity. Contin. Mech. Thermodyn. **23**, 527–549 (2011)
12. Reddy, B.D., Ebobisse, F., McBride, A.T.: Well-posedness of a model of strain gradient plasticity for plastically irrotational materials. Int. J. Plast. **24**, 55–73 (2008)
13. Rockafellar, R.T.: Convex Analysis. Princeton University Press, Princeton (1970)

# Finite Elasto-Plastic Models for Lattice Defects in Crystalline Materials

**Sanda Cleja-Ţigoiu**

**Abstract** Elasto-plastic models for continuous distributed defects are provided for materials endowed with Cartan-Riemannian differential geometric structures. The geometrical measure of defects, dislocations and disclinations, are related to the incompatibilities of the so-called plastic distortion and plastic connection, respectively. The coupling between defects is described through the non-local evolution equations which are compatible with the free energy imbalance principle.

## 1 Introduction

The continuum elasto-plastic constitutive models proposed in this paper describe the behaviour of crystalline material with microstructural defects in terms of three configurations:

Let $k$ be a fixed reference configuration of the body $\mathscr{B}$, $k(\mathscr{B}) \subset E$, and $\mathscr{B}$ will be identified with $k(\mathscr{B})$;
$\chi(\cdot, t)$ the deformed configuration at time $t$, for any motion of the body $\mathscr{B}$, $\chi : \mathscr{B} \times R \longrightarrow E$, and $\mathbf{F}(\mathbf{X}, t) = \nabla\chi((\mathbf{X}, t)$ denotes the deformation gradient;
there exists $\mathscr{K}$, a time dependent anholonomic configuration (so- called configuration with torsion), defined by the pair $(\mathbf{F}^p, \overset{(p)}{\boldsymbol{\Gamma}})$, $\mathbf{F}^p$-plastic distortion and $\overset{(p)}{\boldsymbol{\Gamma}}$-plastic connection.

The reference and deformed (actual) configurations, which are global configurations of the elasto-plastic body, characterize the material within Riemannian geometry, while the local configurations, attached to the material points of the body, are

S. Cleja-Ţigoiu (✉)
Faculty of Mathematics and Computer Science, University of Bucharest,
Str. Academiei 14, 010014 Bucharest, Romania
e-mail: tigoiu@fmi.unibuc.ro

S. Cleja-Ţigoiu
Institute of Solid Mechanics, Romanian Academy, Str. Constantin Mille 15,
010141 Bucharest, Romania



F. dell'Isola et al. (eds.), *Mathematical Modelling in Solid Mechanics*,
Advanced Structured Materials 69, DOI 10.1007/978-981-10-3764-1_4

*anholonomic configuration*, and the Riemann-Cartan geometry describes geometrically the measures of defects, see Yavari and Goriely [19]. The geometrical measure of defects, dislocations and disclinations, are related to the incompatibilities of the so-called plastic distortion and plastic connection, respectively. The incompatibility of the plastic distortion, i.e. $curl\,\mathbf{F}^p \neq 0$, which means the presence of dislocations, and the incompatibility of the so-called plastic connection, i.e. $\overset{(p)}{\boldsymbol{\Gamma}} \neq (\mathbf{F}^p)^{-1}\nabla\mathbf{F}^p$, which means the presence of the disclinations, see for instance de Wit [10], Cleja-Ţigoiu et al. [9], Fressengeas et al. [13]. The interplay between the defects such as dislocations and disclinations is described through the Cartan torsion attached to the plastic connection, denoted by $\mathbf{S}^p$. The dislocation density tensor (called also the geometrically necessary dislocation) and disclination density tensor characterize non-zero Burgers and Frank vector, defined at the end of this section. The physical motivations for the defects such as disclinations can be found in [18], see also [4, 13] In the present paper the plastic connection which is $\mathbf{C}^p$-metric compatible has been considered as in the previous papers [5, 6], apart from the paper by Clayton et al. [4], devoted to finite miscropolar elastoplasticity, where a connection defined by Minagawa [15] has been introduced. Let us remark here that when $\mathbf{Qu}$ for all vector $\mathbf{u}$ is a skew-symmetric second order tensor, the coefficients of this connection and our plastic connection coincide.

The energetic arguments are necessary to complete the description of the elasto-plastic models with defects, such as dislocations and disclinations. The balance equations for the micro forces have been revised in this paper, starting from the basic hypothesis concerning the expression of the internal dissipation power during the elasto-plastic material with microstructural defects, in conjunction with the virtual power assumptions. As the principle of the virtual power expending during the plastic and disclination mechanisms have been formulated in terms of the incompatible second order virtual rates, contrary to the virtual power principle considered by Fosdick [12] where the virtual second order velocity field $\tilde{\mathbf{L}}$ is compatible, i.e. $\tilde{\mathbf{L}} = \nabla\tilde{\mathbf{v}}$, only one balance equation for appropriate micro forces has been provided. For the macro balance equations similar to those proposed by Fleck et al. [11], see also [8], have been adopted.

Clayton et al. [3, 4] introduced the balance equations for micro forces similar to Fleck et al. [11].

Fosdick [12] considered the following example of the *principle of virtual power*

$$\int_{\mathscr{P}} (\tilde{\mathbf{T}} \cdot \nabla\mathbf{u} + \mathscr{J} \cdot \nabla(\nabla\mathbf{u})dV = \int_{\partial\mathscr{P}} (\mathbf{t} \cdot \mathbf{u} + \mathbf{J} \cdot \nabla\mathbf{u})dA \quad \forall\,\mathscr{P} \subset \mathscr{B} \quad \forall\mathbf{u},$$

where the left hand side represents the internal virtual power and the right hand side is the external virtual power on $\mathscr{P}$, in which the body force term was neglected. $\mathbf{t}$ and $\mathbf{J}$ depend on the normal to the surface, $\mathbf{n}$, and have independent physical significance, being restricted to different, possible conditions. The principle is valid for any arbitrary part $\mathscr{P}$ of a body and for any arbitrary virtual velocity field $\mathbf{u}$. Fosdick analyzed all the consequences that can be drawn and remarked that the principle of virtual power has been applied for models by considering various length scales, because of "several major consequences," that follows from the assumption that the principle of the virtual work holds for "arbitrary parts of the body," "have not been clearly expressed" in certain papers.

**The postulate** of the free energy imbalance expresses the restriction on the elasto-plastic material to be satisfied in the configuration with torsion, $\mathcal{K}$, as an *imbalanced free energy condition*, see Cleja-Ţigoiu [5, 6], as well as Gurtin et al. [14], for the initial original ideas related to the free energy imbalance.

**List of the Notations**:

$\mathcal{V}$-the vector space of translations of the three dimensional Euclidean space $\mathcal{E}$;
Lin-the set of the linear mappings from $\mathcal{V}$ to $\mathcal{V}$, i.e. the set of second order tensors; $\mathbf{u}\cdot\mathbf{v}$, $\mathbf{u}\times\mathbf{v}$, $\mathbf{u}\otimes\mathbf{v}$ denote scalar, cross and tensorial products of vectors;
$(\mathbf{u},\mathbf{v},\mathbf{z}) := (\mathbf{u}\times\mathbf{v})\cdot\mathbf{z}$ is the mixt product of the vectors from $\mathcal{V}$;
$\mathbf{a}\otimes\mathbf{b}$ and $\mathbf{a}\otimes\mathbf{b}\otimes\mathbf{c}$ are defined to be a second order tensor and a third order tensor by $(\mathbf{a}\otimes\mathbf{b})\mathbf{u} = \mathbf{a}(\mathbf{b}\cdot\mathbf{u})$, $(\mathbf{a}\otimes\mathbf{b}\otimes\mathbf{c})\mathbf{u} = (\mathbf{a}\otimes\mathbf{b})(\mathbf{c}\cdot\mathbf{u})$, for all vectors $\mathbf{u}$;
For any second order tensor $\mathbf{A}\in Lin$ we use the notations $\{\mathbf{A}\}^S$, $\{\mathbf{A}\}^a$ for its symmetric and skew-symmetric part;
$\mathbf{I}$ the identity tensor in Lin, $\mathbf{A}^T$ denotes the transpose of $\mathbf{A}\in Lin$;
$\partial_{\mathbf{A}}\phi(x)$ denotes the partial differential of the function $\phi$ with respect to the field $\mathbf{A}$;
$\mathbf{A}\cdot\mathbf{B} := \mathrm{tr}(\mathbf{A}\mathbf{B}^T) = A_{ij}B_{ij}$ is the scalar product of $\mathbf{A},\mathbf{B}\in Lin$;

$curl\mathbf{A}$, a second-order tensor field, is defined by

$$(curl\mathbf{A})(\mathbf{u}\times\mathbf{v}) := (\nabla\mathbf{A}(\mathbf{u}))\mathbf{v} - (\nabla\mathbf{A}(\mathbf{v}))\mathbf{u} \quad \forall\mathbf{u},\mathbf{v}\in\mathcal{V} \quad \text{and}$$
$$(curl\mathbf{A})_{pi} = \in_{ijk}\frac{\partial A_{pk}}{\partial X^j}, \quad \text{are the component of } curl\mathbf{A} \text{ given in a Cartesian basis;}$$

$\in$ and $\in_{ijk}$ denote Ricci permutation tensor and its components, respectively;
$\nabla\mathbf{A}$ the derivative (or the gradient) of the field $\mathbf{A}$ in a coordinate system $\{\mathbf{x}^a\}$ (with respect to the reference configuration), $\nabla\mathbf{A} = \frac{\partial A_{ij}}{\partial X^k}\mathbf{e}^i\otimes\mathbf{e}^j\otimes\mathbf{e}^k$;
$\nabla_\chi\mathbf{L} \equiv \frac{\partial}{\partial x^k}(\frac{\partial v_i}{\partial x^j})\mathbf{e}^i\otimes\mathbf{e}^j\otimes\mathbf{e}^k$, where the dual basis $\mathbf{e}^a$, is defined by the inner product $\mathbf{e}^b\cdot\mathbf{e}_a = \delta^b{}_a$.

We also denote:

$Lin(\mathcal{V}, Lin) = \{\mathbf{N} : \mathcal{V} \longrightarrow Lin, \quad \text{linear}\}$ - the space of all third-order tensors; an element of this pspace is given by $\mathbf{N} = N_{ijk}\mathbf{i}^i\otimes\mathbf{i}^j\otimes\mathbf{i}^k$;

$\mathbf{N}\cdot\mathbf{M} = N_{ijk}M_{ijk}$ - the scalar product of third-order tensors expressed in a Cartesian basis.

The differential of any smooth tensor field $\bar{\mathbf{F}}$, defined on $k(\mathcal{B})$, with respect to the configuration with torsion $\mathcal{K}$ is given by

$$(\nabla_{\mathcal{K}}\bar{\mathbf{F}})\tilde{\mathbf{u}} = (\nabla\bar{\mathbf{F}})(\mathbf{F}^p)^{-1}\tilde{\mathbf{u}}, \quad \forall, \tilde{\mathbf{u}}\in\mathbf{F}^{\mathbf{p}}\mathcal{V}.$$

Moreover, $\boldsymbol{\Lambda}_1 \times \boldsymbol{\Lambda}_2$ denotes the third-order tensor generated by the second order (covariant) tensors $\boldsymbol{\Lambda}_j$ j $= 1, 2$ defined by

$$((\boldsymbol{\Lambda}_1 \times \boldsymbol{\Lambda}_2)\mathbf{u})\mathbf{v} = \boldsymbol{\Lambda}_1\mathbf{u} \times \boldsymbol{\Lambda}_2\mathbf{v}, \quad \forall \mathbf{u}, \mathbf{v}$$

and $Skw\mathcal{N}$ is the third-order tensor associated with $\mathcal{N}$, defined by

$$((Skw\mathcal{N})\mathbf{u})\mathbf{v} = (\mathcal{N}\mathbf{u})\mathbf{v} - (\mathcal{N}\mathbf{v})\mathbf{u}, \ \forall\, \mathbf{u}, \mathbf{v}, \ \text{i.e.}((Skw\mathcal{N})\mathbf{u})\mathbf{v} = -((Skw\mathcal{N})\mathbf{v})\mathbf{u}.$$

Finally, the Frank vector is defined in terms of the disclination tensor (following the definitions introduced by Cleja-Ţigoiu [7]) by

$$\boldsymbol{\omega}_{\mathcal{K}} = \int_{C_0} \widetilde{\boldsymbol{\Lambda}}\mathbf{F}^p \, d\mathbf{X} = \int_{\mathcal{A}_0} \text{curl}(\widetilde{\boldsymbol{\Lambda}}\mathbf{F}^p)\mathbf{N} dA.$$

In the model we consider here, *the disclination tensor* $\widetilde{\boldsymbol{\Lambda}}$ is introduced as independent measure of certain defects, and we consider the expression of the disclination density in terms of the disclination tensor in the anholonomic configuration, i.e.

$$\boldsymbol{\alpha}^{\Lambda}_{\mathcal{K}} = \frac{1}{\det\mathbf{F}^p}\text{curl}(\widetilde{\boldsymbol{\Lambda}}\mathbf{F}^p)(\mathbf{F}^p)^T.$$

The Burgers vector associated with the circuit $C_0$ is defined (following Cleja-Ţigoiu [5]) by

$$\mathbf{b}_{\mathcal{K}} = \int_{C_0} \mathbf{F}^p \, d\mathbf{X} = \int_{\mathcal{A}_0} (\text{curl}\mathbf{F}^p)\mathbf{N} dA,$$

see also Acharya [1], Bilby [2] and Fressengeas et al. [13].

The *dislocation density tensor* $\boldsymbol{\alpha}_{\mathcal{K}}$ is expressed by

$$\boldsymbol{\alpha}_{\mathcal{K}} := \frac{1}{\det\mathbf{F}^p}(\text{curl } \mathbf{F}^p)(\mathbf{F}^p)^T,$$

in a configuration with torsion, is called *Noll's dislocation density*, and was introduced by Noll [17].

In what follows, the anholonomic basis vectors are related to *the crystal* and is defined by $\mathbf{e}_j = \mathbf{F}^p\mathbf{G}_j$, where $\{\mathbf{G}_j\}_{j=1,2,3}$ is a basis in the reference configuration and the Christoffel symbols are represented by $\overset{(p)}{\boldsymbol{\Gamma}}(\mathbf{G}_j, \mathbf{G}_k) = \overset{(p)}{\Gamma}{}^i_{jk}\,\mathbf{G}_i$.

## 2 Geometry and Kinematics of Elasto-Plastic Body

The behaviour of the elasto-plastic material with microstructural defects will be described based on three configurations: the initial, $k$, and the deformed configurations, $\chi$ which are global configurations, and so-called configuration with torsion associated with any material particle, $X$, of the body, formally denoted by $\mathscr{K}$.

We assume the multiplicative decomposition

$$\mathbf{F} = \mathbf{F}^e \mathbf{F}^p \tag{1}$$

As a consequence of the multiplicative decomposition of the deformation gradient (1), we obtain

$$\begin{aligned} &\mathbf{L} = \mathbf{L}^e + \mathbf{F}^e \mathbf{L}^p (\mathbf{F}^e)^{-1}, \quad \text{where} \\ &\mathbf{L} = \dot{\mathbf{F}}(\mathbf{F})^{-1}, \quad \mathbf{L}^p = \dot{\mathbf{F}}^p (\mathbf{F}^p)^{-1}, \quad \mathbf{L}^e = \dot{\mathbf{F}}^e (\mathbf{F}^e)^{-1}, \end{aligned} \tag{2}$$

$\mathbf{L}$ is the velocity gradient in the actual configuration.

The geometrical structure of the configuration $\mathscr{K}$ is characterized by the pair of plastic distortion, $\mathbf{F}^p$, and the so-called plastic connection $\overset{(p)}{\boldsymbol{\Gamma}}$, which is $\mathbf{C}^p$-metric connection. Following [5] the plastic connection with metric property is represented under the form

$$\begin{aligned} &\overset{(p)}{\boldsymbol{\Gamma}} = \overset{(p)}{\mathscr{A}} + (\mathbf{C}^p)^{-1} (\boldsymbol{\Lambda} \times \mathbf{I}), \quad \text{with} \\ &\mathbf{C}^p = (\mathbf{F}^p)^T \mathbf{F}^p, \quad \overset{(p)}{\mathscr{A}} = (\mathbf{F}^p)^{-1} \nabla \mathbf{F}^p, \end{aligned} \tag{3}$$

where the second order tensor $\boldsymbol{\Lambda}$ is called the *disclination* tensor and $\overset{(p)}{\mathscr{A}}$ is a Bilby type plastic connection, see [2].

As a direct consequence of the multiplicative decomposition of the deformation gradient (1), the material connection $\boldsymbol{\Gamma}$ is represented in terms of the Bilby type elastic and plastic connection by

$$\begin{aligned} &\boldsymbol{\Gamma} = (\mathbf{F}^p)^{-1} \overset{(e)}{\mathscr{A}}_{\mathscr{K}} [\mathbf{F}^p, \mathbf{F}^p] + \overset{(p)}{\mathscr{A}}, \quad \text{with} \\ &\overset{(e)}{\mathscr{A}}_{\mathscr{K}} = (\mathbf{F}^e)^{-1} \nabla_{\mathscr{K}} \mathbf{F}^e, \end{aligned} \tag{4}$$

where $\overset{(e)}{\mathscr{A}}_{\mathscr{K}}$ represent the Bilby type elastic connection with respect to the configuration with torsion.

We introduce the notation $\mathbf{S}^p$ for the third order Cartan torsion, associated to the plastic connection $\overset{(p)}{\boldsymbol{\Gamma}}$, which can be expressed as a consequence of (3) by

$$\mathbf{S}^p = Skw\overset{(p)}{\mathscr{A}} + Skw((\mathbf{C}^p)^{-1}(\boldsymbol{\Lambda} \times \mathbf{I})), \quad \text{where} \quad \overset{(p)}{\mathscr{A}} = (\mathbf{F}^p)^{-1} \nabla \mathbf{F}^p. \tag{5}$$

Let us introduce the notation

$$\mathscr{S}^p_{\mathscr{K}} = -\mathbf{F}^p\mathbf{S}^p[(\mathbf{F}^p)^{-1}, (\mathbf{F}^p)^{-1}], \tag{6}$$

and write the formula

$$\mathscr{S}^p_{\mathscr{K}} = Skw\overset{(p)}{\mathscr{A}}_{\mathscr{K}} + Skw(\widetilde{\boldsymbol{\Lambda}} \times \mathbf{I}), \quad \text{where} \quad \overset{(p)}{\mathscr{A}}_{\mathscr{K}} = \mathbf{F}^p\, \nabla_{\mathscr{K}} (\mathbf{F}^p)^{-1}, \tag{7}$$

which holds for $\boldsymbol{\Lambda}$ and $\widetilde{\boldsymbol{\Lambda}}$, related by

$$\widetilde{\boldsymbol{\Lambda}} = \frac{1}{\det\mathbf{F}^p}\, \mathbf{F}^p \boldsymbol{\Lambda}(\mathbf{F}^p)^{-1}, \quad \widetilde{\rho}\det\mathbf{F}^p = \rho_0. \tag{8}$$

*Remark* $Skw\overset{(p)}{\mathscr{A}}_{\mathscr{K}}$ can be viewed as a measure of dislocation, motivated by the relationships between the fields referred to the configurations $\mathscr{K}$ and k, which is expressed under the form

$$((Skw\, \overset{(p)}{\mathscr{A}}_{\mathscr{K}})\tilde{\mathbf{u}})\tilde{\mathbf{v}} = (\overset{(p)}{\mathscr{A}}_{\mathscr{K}}\, \tilde{\mathbf{u}})\tilde{\mathbf{v}} - (\overset{(p)}{\mathscr{A}}_{\mathscr{K}}\, \tilde{\mathbf{v}})\tilde{\mathbf{u}} = (curl\, \mathbf{F}^p)(\mathbf{u} \times \mathbf{v}), \tag{9}$$

$\forall\, \tilde{\mathbf{u}} = \mathbf{F}^p\mathbf{u}, \quad \tilde{\mathbf{v}} = \mathbf{F}^p\mathbf{v}.$

We put into evidence the rates of the above geometrical fields

$$\begin{aligned}
\frac{d}{dt}(\mathbf{S}^p_{\mathscr{K}}) \quad &= Skw\left\{\frac{d}{dt}(\overset{(p)}{\mathscr{A}}_{\mathscr{K}})\right\} + Skw\left\{\left(\frac{d}{dt}(\widetilde{\boldsymbol{\Lambda}}) \times \mathbf{I}\right)\right\},\\
\frac{d}{dt}(\, \overset{(e)}{\mathscr{A}}_{\mathscr{K}}\, ) &= (\mathbf{F}^p)^{-1}\nabla_{\mathscr{K}}\mathbf{L}^p[\mathbf{F}^p, \mathbf{F}^p] - \nabla_{\mathscr{K}}\mathbf{L}^p + \\
&\quad + \mathbf{L}^p\, \overset{(e)}{\mathscr{A}} - \overset{(e)}{\mathscr{A}}_{\mathscr{K}}\, \mathbf{L}^p - \overset{(e)}{\mathscr{A}}_{\mathscr{K}}\, [\mathbf{I}, \mathbf{L}^p],\\
\frac{d}{dt}(\, \overset{(p)}{\mathscr{A}}_{\mathscr{K}}\, ) &= -\nabla_{\mathscr{K}}\mathbf{L}^p + \mathbf{L}^p\, \overset{(p)}{\mathscr{A}} - \overset{(p)}{\mathscr{A}}_{\mathscr{K}}\, \mathbf{L}^p - \overset{(p)}{\mathscr{A}}_{\mathscr{K}}\, [\mathbf{I}, \mathbf{L}^p].
\end{aligned} \tag{10}$$

The time-derivative of $\nabla_{\mathscr{K}}\widetilde{\boldsymbol{\Lambda}}$ is expressed by

$$\frac{d}{dt}(\nabla_{\mathscr{K}}\widetilde{\boldsymbol{\Lambda}}) = \nabla_{\mathscr{K}}\left(\frac{d}{dt}\widetilde{\boldsymbol{\Lambda}}\right) - (\nabla_{\mathscr{K}}\widetilde{\boldsymbol{\Lambda}})\mathbf{L}^p. \tag{11}$$

# 3 Energetic Assumptions

## 3.1 Micro Balance Equations

Concerning the micro forces we assume that they generate internal power and satisfy their own balance equations. In order to put into evidence their appropriate balance equations, we start from the expression of the internal power, expended during the elasto-plastic process and expressed with respect to the configuration $\mathscr{K}$, at which level the presence of the defects can be emphasized.

**Ax.internal power** The internal power in the configuration with torsion is postulated to be given by the expression

$$(\mathscr{P}_{int})_{\mathscr{K}} = \frac{1}{\rho}\mathbf{T}\cdot\mathbf{L}^e + \frac{1}{\widetilde{\rho}}\boldsymbol{\mu}_{\mathscr{K}}\cdot((\mathbf{F}^e)^{-1}(\nabla_\chi\mathbf{L})[\mathbf{F}^e,\mathbf{F}^e] - \nabla_{\mathscr{K}}\mathbf{L}^p) + \frac{1}{\widetilde{\rho}}\boldsymbol{\Upsilon}^p\cdot\mathbf{L}^p + \\ + \frac{1}{\widetilde{\rho}}\boldsymbol{\mu}^p\cdot\nabla_{\mathscr{K}}\mathbf{L}^p + \frac{1}{\widetilde{\rho}}\boldsymbol{\Upsilon}^\lambda\cdot\frac{d}{dt}\widetilde{\boldsymbol{\Lambda}} + \frac{1}{\widetilde{\rho}}\boldsymbol{\mu}^\lambda\cdot\nabla_{\mathscr{K}}\frac{d}{dt}\widetilde{\boldsymbol{\Lambda}}, \tag{12}$$

- $\boldsymbol{\Upsilon}^p$ and $\boldsymbol{\mu}^p$ denote the plastic micro stress and micro stress momentum, with respect to the configuration with torsion $\mathscr{K}$, which are power conjugated with $\mathbf{L}^p$ and its gradient $\nabla_{\mathscr{K}}\mathbf{L}^p$, respectively,
- $\boldsymbol{\Upsilon}^\lambda$ and $\boldsymbol{\mu}^\lambda$ represent the micro stress and micro stress momentum, which are related with the disclination mechanism and power conjugated with the rate $\frac{d}{dt}\widetilde{\boldsymbol{\Lambda}}$ and the gradient of the appropriate rate $\nabla_{\mathscr{K}}\frac{d}{dt}\widetilde{\boldsymbol{\Lambda}}$, respectively.

Micro balance equations will be derived from the formulated principle of virtual power relative to the disclination and plastic mechanisms, as independent. Here a basic role is played by the supposition, (12), concerning the virtual internal power postulated within the constitutive framework.

**Ax.disclination virtual power** We assume that the virtual internal power related to the disclination mechanism is equal to the external virtual power, produced by the virtual variation associated with the disclination

$$\int_{\mathscr{K}(\mathscr{B}^p,t)}\{\boldsymbol{\Upsilon}^\lambda\cdot\delta\widetilde{\boldsymbol{\Lambda}} + \boldsymbol{\mu}^\lambda\cdot\nabla_{\mathscr{K}}\delta\widetilde{\boldsymbol{\Lambda}}\}dV_{\mathscr{K}} = \\ = \int_{\partial\mathscr{K}(\mathscr{B}^p,t)}\mathbf{M}^\lambda(\mathbf{n}_{\mathscr{K}})\cdot\delta\widetilde{\boldsymbol{\Lambda}}\,dA_{\mathscr{K}} + \int_{\mathscr{K}(\mathscr{B}^p,t)}\tilde{\rho}\mathbf{B}^\lambda\cdot\delta\widetilde{\boldsymbol{\Lambda}}\,dV_{\mathscr{K}}, \tag{13}$$

holds for any virtual rate associated with the disclination, $\delta\widetilde{\boldsymbol{\Lambda}}$, which is an *incompatible second order field*. Here $\mathbf{n}_{\mathscr{K}}$ the unit vector of the normal. The integral is taken over arbitrary part of the plastically deformed domain $\mathscr{K}(\mathscr{B}^p,t)$.

We can apply "the common tetrahedron argument," and the linearity of the mapping $\mathbf{n} \in \mathscr{V} \longrightarrow \mathbf{M}^{\lambda}(\mathbf{n}) \in Lin$ follows, namely there exists $\widetilde{\boldsymbol{\mu}}$ such that $\widetilde{\boldsymbol{\mu}}\mathbf{n}_{\mathscr{K}} = \mathbf{M}^{\lambda}(\mathbf{n}_{\mathscr{K}})$. Using Green's formula first the equality $\widetilde{\boldsymbol{\mu}} = \boldsymbol{\mu}^{\lambda}$ yields and the following result can be proved:

**Proposition 4.1** *The micro balance equation for micro forces associated with the disclination is written in the configuration with torsion $\mathscr{K}$, under the form*

$$\boldsymbol{\Upsilon}^{\lambda} = div_{\mathscr{K}}\ \boldsymbol{\mu}^{\lambda} + \tilde{\rho}\mathbf{B}^{\lambda}, \quad or$$
$$\frac{1}{\tilde{\tilde{\rho}}}\boldsymbol{\Upsilon}^{\lambda} = div\left(\frac{1}{\tilde{\tilde{\rho}}}\boldsymbol{\mu}^{\lambda}\ (\mathbf{F}^{p})^{-1}\right) + \mathbf{B}^{\lambda}. \tag{14}$$

Here $\tilde{\rho}\mathbf{B}^{\lambda}$ is mass density of the couple body force, $\boldsymbol{\Upsilon}^{\lambda}$ is micro stress and $\boldsymbol{\mu}^{\lambda}$ is micro momentum associated with the disclinations.

**Ax.(plastic virtual power)** We assume that the virtual internal power related to the plastic mechanism is equal to the external virtual power, produced by the virtual variation of the rate of plastic distortion

$$\int_{\mathscr{K}(\mathscr{B}^{p},t)} \{(\boldsymbol{\Upsilon}^{p} - \Sigma_{\mathscr{K}})\cdot\widetilde{\mathbf{L}}^{p} + (\boldsymbol{\mu}^{p} - \boldsymbol{\mu}_{\mathscr{K}})\cdot\nabla_{\mathscr{K}}\widetilde{\mathbf{L}}^{p}\}dV_{\mathscr{K}} =$$
$$= \int_{\partial\mathscr{K}(\mathscr{B}^{p},t)} \mathbf{M}^{p}(\mathbf{n}_{\mathscr{K}})\cdot\widetilde{\mathbf{L}}^{p}\,dA_{\mathscr{K}} + \int_{\mathscr{K}(\mathscr{B}^{p},t)} \tilde{\rho}\mathbf{B}^{p}\cdot\widetilde{\mathbf{L}}^{p}\,dV_{\mathscr{K}}, \tag{15}$$
$$\text{where } \Sigma_{\mathscr{K}} = (\det\mathbf{F}^{e})(\mathbf{F}^{e})^{T}\mathbf{T}(\mathbf{F}^{e})^{-T},$$

holds for any virtual rate of plastic distortion, $\widetilde{\mathbf{L}}^{p}$, which is an *incompatible second order field.* The integral is taken over arbitrary part of the plastically deformed domain, generically denoted here by $\mathscr{K}(\mathscr{B}^{p}, t)$.

*Remark* In the formula (15) the Mandel type stress tensor associated with the Cauchy stress tensor has been introduced, as a direct consequence of the power expression.

By a similar argument the local form of the balance equation for micro forces associated with plastic mechanism can be proved.

**Proposition 4.2** *The micro balance equation for micro forces associated with the plastic mechanism is written in the configuration with torsion $\mathscr{K}$, under the form*

$$\boldsymbol{\Upsilon}^{p} - \boldsymbol{\Sigma}_{\mathscr{K}} = div_{\mathscr{K}}\ (\boldsymbol{\mu}^{p} - \boldsymbol{\mu}_{\mathscr{K}}) + \tilde{\rho}\mathbf{B}^{p}, \quad or$$
$$\frac{1}{\tilde{\tilde{\rho}}}(\boldsymbol{\Upsilon}^{p} - \boldsymbol{\Sigma}_{\mathscr{K}}) = div\left(\frac{1}{\tilde{\tilde{\rho}}}(\boldsymbol{\mu}^{p} - \boldsymbol{\mu}_{\mathscr{K}})(\mathbf{F}^{p})^{-1}\right) + \mathbf{B}^{p}. \tag{16}$$

Here $\tilde{\rho}\mathbf{B}^p$ is mass density of the couple body force, $\boldsymbol{\Upsilon}^p$ is micro stress and $\boldsymbol{\mu}^p$ is micro momentum associated with the plastic mechanism, while $\boldsymbol{\mu}_{\mathcal{K}}$ is the macro stress momentum. The appropriate boundary condition for micro stress momentum has to be associated.

*Remark* The virtual principles as they were formulated above, namely (13) and (15), have been applied for any virtual rates $\tilde{\mathbf{L}}^p$ and $\delta\boldsymbol{\Lambda}$, respectively, which are described by *incompatible* second order tensors. Consequently only one balance equation characterizes the peculiar *microstructural mechanism*, namely one balance equation for micro forces associated with disclination mechanism and another one associated with the plastic mechanism.

We now pass to the reference configuration and we derive the transformed balance equation in terms of the micro forces in reference configuration.

*Remark* We proved here the appropriate micro balance equations for disclination mechanism. For micro balance equations related with plastic behaviour we make references to Cleja-Ţigoiu [5], or to Cleja-Ţigoiu and Ţigoiu [8] in a paper concerning a strain gradient finite elasto-plastic model. The micro balance equation (13) contains only the micro forces, a similar point of view appears in Clayton et al. [4], apart from the micro balance equation (15) which contain the difference between macro and micro forces (like in the models developed by Gurtin [14], Cleja-Ţigoiu and Ţigoiu [8]).

An alternative formulation of the internal power in $\mathcal{K}$ has been postulated, see [7, 9], as the free energy has been postulated through an appropriate expression in the reference configuration, k. In the aforementioned papers the variation in time of the disclination with respect to the configuration with torsion and its gradient, respectively, were introduced in terms of the rate of disclination tensor with respect to the reference configuration, and its gradient, respectively, pushed away to the configuration with torsion.

## 3.2 Free Energy Density Function

We introduce now the expression of the free energy density postulated with respect to the configuration with torsion. We assume that

**Ax.1:** The *free energy density* is postulated to be dependent on the second order elastic deformation, in terms of $(\mathbf{C}^e, \overset{(e)}{\mathcal{A}}_{\mathcal{K}})$, and on the defects through $(\mathbf{S}^p_{\mathcal{K}}, \widetilde{\boldsymbol{\Lambda}}, \nabla_{\mathcal{K}}\, \widetilde{\boldsymbol{\Lambda}})$, as

$$\psi_{\mathcal{K}} = \psi(\mathbf{C}^e, \overset{(e)}{\mathcal{A}}_{\mathcal{K}}, \mathbf{S}^p_{\mathcal{K}}, \widetilde{\boldsymbol{\Lambda}}, \nabla_{\mathcal{K}}\, \widetilde{\boldsymbol{\Lambda}}), \quad \mathbf{C}^e = (\mathbf{F}^e)^T \mathbf{F}^e. \tag{17}$$

The elements which enter the free energy density function in $\mathcal{K}$ have been represented in terms of the appropriate expressions.

*Remark* In Cleja-Ţigoiu [7] the free energy density was postulated to be dependent on the second order elastic deformation, in terms of $(\mathbf{C}^e, \overset{(e)}{\mathscr{A}}_{\mathscr{K}})$, and being dependent on the second order plastic deformation through $(\mathbf{S}^e_{\mathscr{K}}, \widetilde{\boldsymbol{\Lambda}})$,

$$\psi = \psi_{\mathscr{K}}(\mathbf{C}^e, \overset{(e)}{\mathscr{A}}_{\mathscr{K}}, \mathbf{S}^e_{\mathscr{K}}, \widetilde{\boldsymbol{\Lambda}}), \quad \mathbf{C}^e = (\mathbf{F}^e)^T\mathbf{F}^e, \tag{18}$$

while in [6] the free energy density has been postulated as

$$\psi = \psi_{\mathscr{K}}(\mathbf{C}^e, \overset{(e)}{\mathscr{A}}_{\mathscr{K}}, (\mathbf{F}^p)^{-1}, \overset{(p)}{\mathscr{A}}_{\mathscr{K}}, \widetilde{\boldsymbol{\Lambda}}, \nabla_{\mathscr{K}}\widetilde{\boldsymbol{\Lambda}}), \quad \mathbf{C}^e = (\mathbf{F}^e)^T\mathbf{F}^e, \tag{19}$$

where

$$\overset{(p)}{\mathscr{A}}_{\mathscr{K}} \equiv -\mathbf{F}^p \overset{(p)}{\mathscr{A}} [(\mathbf{F}^p)^{-1}, (\mathbf{F}^p)^{-1}]. \tag{20}$$

In [7] thermomechanic restrictions imposed on the elastic type constitutive functions show that the macro stress momentum is not vanishing, since $\frac{1}{\tilde{\rho}}\boldsymbol{\mu}_{\mathscr{K}} = \partial_{\mathscr{A}^e}\psi + \partial_{\mathbf{S}^e}\psi$. We conclude that the macro stress momentum is involved in the constitutive models if, for instance, $\mathscr{A}^e$ is involved in the free energy density function. The evolution equation for plastic distortion, i.e. $\mathbf{L}^p$, and for the disclination tensor, i.e. $\dot{\widetilde{\boldsymbol{\Lambda}}}$ were defined to be compatible with the reduced dissipation inequality. As the micro momentum related to plastic mechanism is vanishing, $\frac{1}{\tilde{\rho}}\boldsymbol{\mu}^p = 0$, then the microforce $\boldsymbol{\Upsilon}^p$ can be identified with the Mandel stress measure, which is power conjugate with the rate of plastic strain $\mathbf{L}^p$, namely $\frac{1}{\tilde{\rho}}\boldsymbol{\Upsilon}^p = \frac{1}{\tilde{\rho}}\boldsymbol{\Sigma}^p = \frac{1}{\rho}(\mathbf{F}^e)^T\{\mathbf{T}\}^S(\mathbf{F}^e)^{-T}$. The micro stress associated with the disclination mechanism remains undefined in the model performed in Cleja-Ţigoiu [7], $\widetilde{\boldsymbol{\Lambda}}$ can be viewed as internal variable. Consequently, it is possible to take $\boldsymbol{\Upsilon}^\lambda = \tilde{\rho}\partial_{\tilde{\Lambda}}\psi$, see for instance the discussion concerning this issue in [16]. To avoid the above mentioned disadvantage the free energy density function has been reconsidered to be given by (17).

## 4 The Postulate of the Free Energy Imbalance

Within the constitutive framework developed herein, see [5, 6], the second law of thermodynamics is formulated as a postulate of the free energy imbalance on isothermal processes.

The postulate of the free energy imbalance expresses the restriction on the elasto-plastic material to be satisfied in the configuration with torsion, $\mathscr{K}$, under the form: the internal power has to be grater or equal to the rate of the free density energy

$$-\dot{\psi}_{\mathscr{K}} + (\mathscr{P}_{int})_{\mathscr{K}} \geq 0, \tag{21}$$

for an appropriate definition for the internal power $(\mathscr{P}_{int})_{\mathscr{K}}$ and for any virtual (isothermal) processes, when free energy density, $\psi_{\mathscr{K}}$, is given.

In order to investigate the consequences of the free energy imbalance, we proceed as follows

i. We emphasize a set of independent kinematic variables and their gradients, namely $\mathbf{L}, \mathbf{L}^p, \dot{\Lambda}$ and $\nabla_\chi \mathbf{L}, \nabla_{\mathscr{K}} \mathbf{L}^p, \nabla \dot{\boldsymbol{\Lambda}}$, respectively;
ii. Under the assumption that no evolution of the plastic distortion and of the disclination mechanism occurs, which means that $\mathbf{L}^p = 0$ and $\dot{\boldsymbol{\Lambda}} = 0$, the elastic type constitutive equations for macro forces are derived;
iii. The reduced dissipation inequality is derived;
iv. The possible consequences on the evolution for plastic distortion and disclination tensor and for their derivative are provided and analyzed to ensure their compatibility with the reduced dissipation inequality.

The time derivative of the free energy density function (17) is expressed through

$$\begin{aligned}\dot{\psi}_{\mathscr{K}} = \partial_{\mathbf{C}^e}\psi \cdot \frac{d}{dt}(\mathbf{C}^e) + \partial_{\mathscr{A}^e_{\mathscr{K}}}\psi \cdot \frac{d}{dt}(\overset{(e)}{\mathscr{A}}_{\mathscr{K}}) + \partial_{\mathbf{S}^p_{\mathscr{K}}}\psi \cdot \Bigg( Skw \frac{d}{dt} \overset{(p)}{\mathscr{A}}_{\mathscr{K}} + \\ + Skw\left(\left(\frac{d}{dt}\widetilde{\boldsymbol{\Lambda}}\right) \times \mathbf{I}\right)\Bigg) + \partial_{\widetilde{\boldsymbol{\Lambda}}}\psi \cdot \frac{d}{dt}(\widetilde{\boldsymbol{\Lambda}}) + \partial_{\nabla\widetilde{\boldsymbol{\Lambda}}}\psi \cdot \frac{d}{dt}(\nabla_{\mathscr{K}}\widetilde{\boldsymbol{\Lambda}}).\end{aligned} \tag{22}$$

In (22) the rate of the appropriate fields have to be replaced by the formulae (10), and the formula

$$\frac{d}{dt}(\mathbf{C}^e) = 2(\mathbf{F}^e)^T \mathbf{D}^e \mathbf{F}^e, \quad \text{where} \quad \mathbf{D}^e = \{\mathbf{L}^e\}^S. \tag{23}$$

**Proposition 4.3** *The free energy imbalance is satisfied for any virtual process, if the following inequality holds*

$$\begin{aligned}&\left\{\frac{1}{\rho}\{\mathbf{T}\}^S - 2\mathbf{F}\partial_{\bar{\mathbf{C}}^e}\psi\mathbf{F}^T\right\} \cdot \mathbf{D}^e + \frac{1}{\tilde{\rho}}\boldsymbol{\Upsilon}^p \cdot \mathbf{L}^p + \left(\frac{1}{\tilde{\rho}}\boldsymbol{\mu}^p + \partial_{\mathbf{S}^p_{\mathscr{K}}}\psi\right) \cdot \nabla_{\mathscr{K}}\mathbf{L}^p + \\ &+ \left(\frac{1}{\tilde{\rho}}\boldsymbol{\mu}_{\mathscr{K}} - \partial_{\mathscr{A}^e_{\mathscr{K}}}\psi\right) \cdot ((\mathbf{F}^e)^{-1}(\nabla_\chi \mathbf{L})[\mathbf{F}^e, \mathbf{F}^e] - \nabla_{\mathscr{K}}\mathbf{L}^p) + \\ &+ \left(\frac{1}{\tilde{\rho}}\boldsymbol{\Upsilon}^\lambda - \partial_{\widetilde{\boldsymbol{\Lambda}}}\psi\right) \cdot \dot{\widetilde{\boldsymbol{\Lambda}}} - \partial_{\mathbf{S}^p_{\mathscr{K}}}\psi \cdot Skw\{\dot{\widetilde{\boldsymbol{\Lambda}}} \times \mathbf{I}\} + \left(\frac{1}{\tilde{\rho}}\boldsymbol{\mu}^\lambda - \partial_{\nabla_{\mathscr{K}}\widetilde{\boldsymbol{\Lambda}}}\psi\right) \cdot \nabla_{\mathscr{K}}\dot{\widetilde{\boldsymbol{\Lambda}}} + \\ &+ \partial_{\nabla_{\mathscr{K}}\widetilde{\boldsymbol{\Lambda}}}\psi \cdot (\nabla_{\mathscr{K}}\widetilde{\boldsymbol{\Lambda}})\mathbf{L}^p - \partial_{\mathscr{A}^e_{\mathscr{K}}}\psi \cdot \left(\mathbf{L}^p \overset{(e)}{\mathscr{A}}_{\mathscr{K}} - \overset{(e)}{\mathscr{A}}_{\mathscr{K}} \mathbf{L}^p - \overset{(e)}{\mathscr{A}}_{\mathscr{K}} [\mathbf{I}, \mathbf{L}^p]\right) - \\ &- \partial_{\mathbf{S}^p_{\mathscr{K}}}\psi \cdot \left(\mathbf{L}^p \overset{(p)}{\mathscr{A}}_{\mathscr{K}} - \overset{(p)}{\mathscr{A}}_{\mathscr{K}} \mathbf{L}^p - \overset{(p)}{\mathscr{A}}_{\mathscr{K}} [\mathbf{I}, \mathbf{L}^p]\right) \geq 0.\end{aligned} \tag{24}$$

We introduce three types of second order tensors that can be associated with any pair of third order tensors, $\mathscr{A}$, $\mathscr{B}$, following the rules written below

$$\begin{aligned}
(\mathscr{A} \odot \mathscr{B}) \cdot \mathbf{L} &= \mathscr{A}[\mathbf{I}, \mathbf{L}] \cdot \mathscr{B} = \mathscr{A}_{isk} L_{sn} \mathscr{B}_{ink} \\
(\mathscr{A}\ {}_r\!\odot \mathscr{B}) \cdot \mathbf{L} &= \mathscr{A} \cdot (\mathbf{L}\mathscr{B}) = \mathscr{A}_{ijk} L_{in} \mathscr{B}_{njk} \\
(\mathscr{A} \odot_l \mathscr{B}) \cdot \mathbf{L} &= \mathscr{A} \cdot (\mathscr{B}\mathbf{L}) = \mathscr{A}_{ijk} \mathscr{B}_{ijn} L_{kn}.
\end{aligned} \tag{25}$$

for all $\mathbf{L} \in Lin$, in order to put into evidence the linear dependence on $\mathbf{L}^p$ in the dissipation inequality (24).

We use the above representation and we get

$$\partial_{\mathbf{S}^p_{\mathscr{K}}} \psi \cdot Skw(\dot{\tilde{\boldsymbol{\Lambda}}} \times \mathbf{I}) = -2(\in \odot_l \partial_{\mathbf{S}^p_{\mathscr{K}}} \psi) \cdot \dot{\tilde{\boldsymbol{\Lambda}}}. \tag{26}$$

If we suppose that no evolution of the plastic distortion and of the disclination mechanism occurs, which means that $\mathbf{L}^p = 0$ and $\dot{\tilde{\boldsymbol{\Lambda}}} = 0$, then $\mathbf{L}^e = \mathbf{L}$, and we get from the free energy imbalance, (24), that the following inequality holds

$$\left\{ \frac{1}{\rho} \{\mathbf{T}\}^S - 2\mathbf{F} \partial_{\mathbf{C}^e} \psi \mathbf{F}^T \right\} \cdot \mathbf{L} + \left( \frac{1}{\tilde{\rho}} \boldsymbol{\mu}_{\mathscr{K}} - \partial_{\mathscr{A}^e_{\mathscr{K}}} \psi \right) \cdot (\mathbf{F}^e)^{-1} (\nabla_\chi \mathbf{L})[\mathbf{F}^e, \mathbf{F}^e] \geq 0 \tag{27}$$

for any virtual process, i.e. $\forall\ \mathbf{L}, \nabla_\chi \mathbf{L}$.

**Theorem 4.1** *1. The thermomechanic restrictions imposed on the elastic type constitutive functions are*

$$\begin{aligned}
\frac{1}{\rho} \{\mathbf{T}\}^S &= 2\mathbf{F} \partial_{\bar{\mathbf{C}}} \psi \mathbf{F}^T \\
\frac{1}{\tilde{\rho}} \boldsymbol{\mu}_{\mathscr{K}} &= \partial_{\mathscr{A}^e_{\mathscr{K}}} \psi.
\end{aligned} \tag{28}$$

*2. The dissipative inequality (24) is reduced to the following inequality*

$$\begin{aligned}
&\frac{1}{\tilde{\rho}} \boldsymbol{\Upsilon}^p \cdot \mathbf{L}^p + \left( \frac{1}{\tilde{\rho}} \boldsymbol{\mu}^p + \partial_{\mathbf{S}^p_{\mathscr{K}}} \psi \right) \cdot \nabla_{\mathscr{K}} \mathbf{L}^p + \\
&+ \left( \frac{1}{\tilde{\rho}} \boldsymbol{\Upsilon}^\lambda - \partial_{\tilde{\boldsymbol{\Lambda}}} \psi + 2(\in \odot_l \partial_{\mathbf{S}^p_{\mathscr{K}}} \psi) \right) \cdot \dot{\tilde{\boldsymbol{\Lambda}}} + \left( \frac{1}{\tilde{\rho}} \boldsymbol{\mu}^\lambda - \partial_{\nabla_{\mathscr{K}} \tilde{\boldsymbol{\Lambda}}} \psi \right) \cdot \nabla_{\mathscr{K}} \dot{\tilde{\boldsymbol{\Lambda}}} + \\
&+ \left( \nabla_{\mathscr{K}} \tilde{\boldsymbol{\Lambda}} \odot_l \partial_{\nabla_{\mathscr{K}} \tilde{\boldsymbol{\Lambda}}} \psi \right) \cdot \mathbf{L}^p - \frac{1}{\tilde{\rho}} \boldsymbol{\mu}_{\mathscr{K}} \cdot \left( \mathbf{L}^p \overset{(e)}{\mathscr{A}}_{\mathscr{K}} - \overset{(e)}{\mathscr{A}}_{\mathscr{K}} \mathbf{L}^p - \overset{(e)}{\mathscr{A}}_{\mathscr{K}} [\mathbf{I}, \mathbf{L}^p] \right) - \\
&- \partial_{\mathbf{S}^p_{\mathscr{K}}} \psi \cdot \left( \mathbf{L}^p \overset{(p)}{\mathscr{A}}_{\mathscr{K}} - \overset{(p)}{\mathscr{A}}_{\mathscr{K}} \mathbf{L}^p - \overset{(p)}{\mathscr{A}}_{\mathscr{K}} [\mathbf{I}, \mathbf{L}^p] \right) \geq 0.
\end{aligned} \tag{29}$$

## 5 Viscoplastic Type Evolution Equations

Based on the dissipation inequality written in (29), we formulate the *constitutive hypotheses* in plastically deformed configuration:

**Ax.5** The plastic micro stress momentum and micro stress momentum related with the disclination mechanism are represented through certain energetic relationships

$$\frac{1}{\tilde{\rho}}\boldsymbol{\mu}^p = -\partial_{\mathbf{S}^p_{\mathcal{K}}}\psi$$
$$\frac{1}{\tilde{\rho}}\boldsymbol{\mu}^\lambda = \partial_{\nabla_{\mathcal{K}}\tilde{\boldsymbol{\Lambda}}}\psi. \tag{30}$$

Using the definition of the operators introduced by (25) the linear terms in $\mathbf{L}^p$ can be grouped together as follows

$$\frac{1}{\tilde{\rho}}\boldsymbol{\Upsilon}^p\cdot\mathbf{L}^p+\left(-\frac{1}{\tilde{\rho}}\boldsymbol{\mu}_{\mathcal{K}}\ {}_r\odot\overset{(e)}{\mathscr{A}}_{\mathcal{K}}+\overset{(e)}{\mathscr{A}}_{\mathcal{K}}\odot_l\frac{1}{\tilde{\rho}}\boldsymbol{\mu}_{\mathcal{K}}+\frac{1}{\tilde{\rho}}\boldsymbol{\mu}_{\mathcal{K}}\odot\overset{(e)}{\mathscr{A}}_{\mathcal{K}}\right)\cdot\mathbf{L}^p+$$
$$+\left(\frac{1}{\tilde{\rho}}\boldsymbol{\mu}^p\ {}_r\odot\overset{(p)}{\mathscr{A}}_{\mathcal{K}}-\overset{(p)}{\mathscr{A}}_{\mathcal{K}}\odot_l\frac{1}{\tilde{\rho}}\boldsymbol{\mu}^p-\frac{1}{\tilde{\rho}}\boldsymbol{\mu}^p\odot\overset{(p)}{\mathscr{A}}_{\mathcal{K}}\right)\cdot\mathbf{L}^p+ \tag{31}$$
$$+\left(\frac{1}{\tilde{\rho}}\boldsymbol{\Upsilon}^\lambda-\partial_{\tilde{\boldsymbol{\Lambda}}}\psi+2(\in\odot_l\partial_{\mathbf{S}^p_{\mathcal{K}}}\psi)\right)\cdot\dot{\tilde{\boldsymbol{\Lambda}}}+\left(\nabla_{\mathcal{K}}\tilde{\boldsymbol{\Lambda}}\odot_l\partial_{\nabla_{\mathcal{K}}\tilde{\boldsymbol{\Lambda}}}\psi\right)\cdot\mathbf{L}^p\geq 0.$$

We introduce now the evolution equation for the plastic distortion, $\mathbf{F}^p$, and for the disclination tensor, $\tilde{\boldsymbol{\Lambda}}$, in the configuration with torsion.

**Ax.6** The rate of plastic distorsion in the configuration with torsion is characterized in terms of micro and macro forces by

$$\xi_1\mathbf{L}^p=\frac{1}{\tilde{\rho}}\boldsymbol{\Upsilon}^p-\frac{1}{\tilde{\rho}}\boldsymbol{\mu}_{\mathcal{K}}\ {}_r\odot\overset{(e)}{\mathscr{A}}_{\mathcal{K}}+\overset{(e)}{\mathscr{A}}_{\mathcal{K}}\odot_l\frac{1}{\tilde{\rho}}\boldsymbol{\mu}_{\mathcal{K}}+\frac{1}{\tilde{\rho}}\boldsymbol{\mu}_{\mathcal{K}}\odot\overset{(e)}{\mathscr{A}}_{\mathcal{K}}+$$
$$+\frac{1}{\tilde{\rho}}\boldsymbol{\mu}^p\ {}_r\odot\overset{(p)}{\mathscr{A}}_{\mathcal{K}}-\overset{(p)}{\mathscr{A}}_{\mathcal{K}}\odot_l\frac{1}{\tilde{\rho}}\boldsymbol{\mu}^p-\frac{1}{\tilde{\rho}}\boldsymbol{\mu}^p\odot\overset{(p)}{\mathscr{A}}_{\mathcal{K}}+\nabla_{\mathcal{K}}\tilde{\boldsymbol{\Lambda}}\odot_l\frac{1}{\tilde{\rho}}\boldsymbol{\mu}^\lambda \tag{32}$$

**Ax.7** The variation in time of the disclination tensor, $\tilde{\boldsymbol{\Lambda}}$, is characterized by the micro forces as

$$\xi_2\dot{\tilde{\boldsymbol{\Lambda}}}=\frac{1}{\tilde{\rho}}\boldsymbol{\Upsilon}^\lambda-\partial_{\tilde{\boldsymbol{\Lambda}}}\psi-2\left(\in\odot_l\frac{1}{\tilde{\rho}}\boldsymbol{\mu}^p\right) \tag{33}$$

in terms of the micro forces.

*Remark* If the viscous parameters $\xi_1$ and $\xi_2$ are scalar positive functions, then the reduced dissipation inequality (31) is satisfied

$$\xi_2 \dot{\widetilde{\boldsymbol{\Lambda}}} \cdot \dot{\widetilde{\boldsymbol{\Lambda}}} + \xi_1 \mathbf{L}^p \cdot \mathbf{L}^p \geq 0. \tag{34}$$

# 6 Concluding Remarks

We end our work with the following concluding remarks.

1. The rate of plastic distortion described by (32) is strongly dependent on the macro stress momentum $\boldsymbol{\mu}_{\mathscr{K}}$ and the micro force related to the plastic mechanism, $(\frac{1}{\tilde{\rho}}\boldsymbol{\Upsilon}^p, \frac{1}{\tilde{\rho}}\boldsymbol{\mu}^p)$, as well as on the micro stress momentum related to the disclination mechanism $\frac{1}{\tilde{\rho}}\boldsymbol{\mu}^\lambda$.

2. A coupling term in the evolution equation shows that the disclination mechanism is influenced by the plastic mechanism.

3. By eliminating the micro stresses from the evolution Eqs. (32) and (33), via the balance equation for micro forces (14) and (16) in the absence of the mass densities of the couple body forces, the differential type evolution equations for $\mathbf{F}^p$ and $\widetilde{\boldsymbol{\Lambda}}$ become

$$\begin{aligned} \xi_1 \mathbf{L}^p &= \boldsymbol{\Sigma}_{\mathscr{K}} + \operatorname{div}\left(\frac{1}{\tilde{\rho}}(\boldsymbol{\mu}^p - \boldsymbol{\mu}_{\mathscr{K}})(\mathbf{F}^p)^{-1}\right) + \\ &- \frac{1}{\tilde{\rho}}\boldsymbol{\mu}_{\mathscr{K}}\ {}_r\!\odot \overset{(e)}{\mathscr{A}}_{\mathscr{K}} + \overset{(e)}{\mathscr{A}}_{\mathscr{K}} \odot_l \frac{1}{\tilde{\rho}}\boldsymbol{\mu}_{\mathscr{K}} + \frac{1}{\tilde{\rho}}\boldsymbol{\mu}_{\mathscr{K}} \odot \overset{(e)}{\mathscr{A}}_{\mathscr{K}} + \\ &+ \frac{1}{\tilde{\rho}}\boldsymbol{\mu}^p\ {}_r\!\odot \overset{(p)}{\mathscr{A}}_{\mathscr{K}} - \overset{(p)}{\mathscr{A}}_{\mathscr{K}} \odot_l \frac{1}{\tilde{\rho}}\boldsymbol{\mu}^p - \frac{1}{\tilde{\rho}}\boldsymbol{\mu}^p \odot \overset{(p)}{\mathscr{A}}_{\mathscr{K}} + \nabla_{\mathscr{K}}\widetilde{\boldsymbol{\Lambda}} \odot_l \frac{1}{\tilde{\rho}}\boldsymbol{\mu}^\lambda \end{aligned} \tag{35}$$

and

$$\xi_2 \dot{\widetilde{\boldsymbol{\Lambda}}} = \operatorname{div}\left(\frac{1}{\tilde{\rho}}\boldsymbol{\mu}^\lambda\ (\mathbf{F}^p)^{-1}\right) - \partial_{\widetilde{\boldsymbol{\Lambda}}}\psi - 2\left(\in \odot_l \frac{1}{\tilde{\rho}}\boldsymbol{\mu}^p\right) \tag{36}$$

3. The evolution equation for the disclination tensor $\widetilde{\boldsymbol{\Lambda}}$ is characterized by the micro stress momentum only, i.e. no direct influence of the macro forces has been emphasized in the proposed model.

**Acknowledgements** This research was supported by the Romanian National Authority for Scientific Research (CNCS-UEFISCDI), project number PN II-ID-PCE-2011-3-0521.

## References


1. Acharya, A.: Constitutive analysis of finite deformation field dislocation mechanics. J. Mech. Phys. Solid **52**, 301–316 (2004)
2. Bilby, B.A.: Continuous distribution of dislocations. In: Sneddon, I.N., Hill, R. (eds.) Progress in Solid Mechanics, pp. 329–398. North-Holland, Amsterdam (1960)
3. Clayton, J.D., Bammann, D.J., McDowell, D.L.: Anholonomic configuration spaces and metric tensors in finite elastoplasticity. Int. J. Non-linear Mech. **39**, 1039–1049 (2004)
4. Clayton, J.D., McDowell, D.L., Bammann, D.J.: Modeling dislocations and disclinations with finite micropolar elastoplasticity. Int. J. Plast. **22**, 210–256 (2006)
5. Cleja-Ţigoiu, S.: Material forces in finite elasto-plasticity with continuously distributed dislocations. Int. J. Fract. **147**, 67–81 (2007)
6. Cleja-Ţigoiu, S.: Elasto-plastic materials with lattice defects modeled by second order deformations with non-zero curvature. Int. J. Fract. **166**, 61–75 (2010)
7. Cleja-Ţigoiu, S.: Dislocations and disclinations: continuously distributed defects in elasto-plastic crystalline materials. Arch. Appl. Mech. **84**, 1293–1306 (2014)
8. Cleja-Ţigoiu, S., Ţigoiu, V.: Strain gradient effects in finite elasto-plastic damaged materials. Int. J. damage Mech. **20**, 484–577 (2011)
9. Cleja-Ţigoiu, S., Paşcan, R., Ţigoiu, V.: Interplay between continuous dislocations and disclinations in elasto-plasticity. Int. J. Plast. **79**, 68–110 (2016)
10. de Wit, R.: A view of the relation between the continuum theory of lattice defects and non-Euclidean geometry in the linear approximation. Int. J. Eng. Sci. **19**, 1475–1506 (1981)
11. Fleck, N.A., Muller, G.M., Ashby, M.F., Hutchinson, J.W.: Strain gradient plasticity: theory and experiment. Acta Metall. Mater. **42**, 475–487 (1994)
12. Fosdick, R.: Observations concerning virtual power. Math. Mech. Solids **16**(6), 573–585 (2011)
13. Fressengeas, C., Taupin, V., Capolungo, L.: An elasto-plastic theory of dislocation and disclination field. Int. J. Solids Struct. **48**, 3499–3509 (2011)
14. Gurtin, M.E., Fried, E., Anand, L.: The Mechanics and Thermodynamics of Continua. University Press, Cambridge (2010)
15. Minagawa, S.: A non-Riemannian geometrical theory of imperfections in a Cosserat continuum. Arch. Mech. **31**, 783–792 (1979)
16. Maugin, G.A.: Internal variables and dissipative structures. J. Non-Equilib. Thermodyn. **15**, 173–192 (1990)
17. Noll, W.: Materially uniform simple bodies with inhomogeneities. Arch. Rat. Mech. Anal. 1967; The Foundations of Mechanics and Thermodynamics, Selected papers. Springer, Berlin (1974)
18. Romanov, A.E.: Mechanics and physics of disclinations in solids. Eur. J. Mech. A/Solids **22**, 727–741 (2003)
19. Yavari, A., Goriely, A.: Math. Mech. Solids **18**, 91–102 (2012)

# Modeling Deformable Bodies Using Discrete Systems with Centroid-Based Propagating Interaction: Fracture and Crack Evolution

**Alessandro Della Corte, Antonio Battista, Francesco dell'Isola and Ivan Giorgio**

**Abstract** We use a simple discrete system in order to model deformation and fracture within the same theoretical and numerical framework. The model displays a rich behavior, accounting for different fracture phenomena, and in particular for crack formation and growth. A comparison with standard Finite Element simulations and with the basic Griffith theory of fracture is provided. Moreover, an 'almost steady' state, i.e. a long apparent equilibrium followed by an abrupt crack growth, is obtained by suitably parameterizing the system. The model can be easily generalized to higher order interactions corresponding, in the homogenized limit, to higher gradient continuum theories.

## 1 Introduction

### 1.1 Motivation and Basic Ideas

Modeling fracture in an effective way has always been a major challenge for solid mechanics. Many sophisticated theoretical and numerical tools have been developed, and considerable progresses have been obtained in recent years in the framework of

A. Della Corte
Department of Mechanical and Aerospace Engineering, Università di
Roma La Sapienza, Via Eudossiana 18, 00184 Rome, Italy
e-mail: alessandro.dellacorte@uniroma1.it

A. Battista
Laboratory of Science for Environmental Engineering, Université de
la Rochelle, La Rochelle, France
e-mail: antonio.battista@univ-lr.fr

F. dell'Isola (✉) · I. Giorgio
MeMoCS, International Research Center for the Mathematics & Mechanics
of Complex Systems, Università dell'Aquila, L'Aquila, Italy
e-mail: francesco.dellisola@uniroma1.it

I. Giorgio
e-mail: ivan.giorgio@uniroma1.it

© Springer Nature Singapore Pte Ltd. 2017
F. dell'Isola et al. (eds.), *Mathematical Modelling in Solid Mechanics*,
Advanced Structured Materials 69, DOI 10.1007/978-981-10-3764-1_5

both classical and generalized continuum models (see [1–5]). However, the richness of the phenomenology is leading to theoretical formulations that are becoming increasingly complex and may be computationally expensive and/or involve a stochastic approach in order to capture the peculiar characteristics displayed by real fractures. The idea behind the present work is to explore a very simple discrete system characterized by a centroid-based propagating interaction, and evolving through actual configurations connected by virtual (i.e., not visible) ones; the system, in its basic version and without including fracture, was introduced in [6], while some preliminary results on fracture have been shown in [7]. The model presents some similarities with Molecular Dynamics (MD) but, as it will be shown, its specific features cannot be recovered within standard MD models.

The potential advantages of the proposed approach are mainly linked to the simplicity of the model. First of all this results in a significantly low computational cost; moreover, deformation and fracture are covered here within the same simple model; finally as we will see, more general continuum theories can be easily numerically investigated by means of slight modifications of the algorithm. Since generalized continua are one of the most promising and rapidly evolving areas in modern mechanics, this last feature of the proposed model may be particularly appealing (for a theoretical coverage on higher gradient theories, see e.g.: [8–11], and specifically for an approach combining higher gradient theories with lattice models, see [12]). A closely related topic is that of micromorphic/microstructured continua (see [13–16] for classical references and [17–22] for interesting applications), which can be viewed as a generalization of higher gradient theories and may benefit as well from the development of new discrete approaches. What is making these subjects critical, in the opinion of the authors, is the advancements in manufacturing possibilities in the last years, as 3D printing and other computer-aided manufacturing techniques, which are resulting in new metamaterials requiring a suitable theoretical description (and related numerical techniques) for objects whose richness at the micro-scale cannot be captured by classical continuum models (see e.g.: [23] for a review of recent results).

Discrete systems are, of course, very frequently studied in order to address the aforementioned problems. In particular, Molecular Dynamics (MD) is by now a very large research area with specific methods and very sound results (see e.g.: [24–27]). The model presented herein, while sharing certain basic features with MD ones, is characterized by some relevant differences in the approach. MD is indeed based on the numerical study of systems constituted by a very large number $N$ of elements. The numerical computation of the trajectories of particles in the ordinary $6N$-dimensional phase space of positions and momenta employs the classical mechanics laws of motion (for a sample of the numerical method the reader can see: [28–30]). The system investigated in the present paper, on the other hand, does not consider any explicit equation of motion. Instead, in order to simplify the model to the maximum, the elements of our discrete system move according to an interaction law which:

- is purely geometrically based;
- propagates along square frames centered in special elements (the 'leaders' of the system);
- drives the elements through virtual configurations which are needed for the computation of the evolution, but are not visible in the final output.

An important class of discrete systems for continuous deformable bodies, and in fact a particular case of MD models, is given by masses-springs or beads-springs models, which can be characterized in different ways by changing the geometry of the springs and their type (extensional, torsional). The mass-spring systems are common tools in Computer Graphics for the simulation of soft bodies [31, 32].

There are several relevant differences between these systems and the proposed model. The two probably most important ones are the following:

i. In case of mass-spring system, a uniform motion of the leaders results in a global motion $r(t) = r^*(t) + r^{**}(t)$, with $r^{**}(t)$ being a periodic function and $r^*(t)$ a transient term such that $||r^*(t)|| \to 0$ as $t \to \infty$ [33]. As we will see, no such decomposition makes sense in case of the proposed system. For instance, it can be proven that a uniform motion imposed to a leader results, asymptotically, in a rigid motion of the system.
ii. It can also be proven that uniformly accelerated motion of a set of leaders results in our case in a disaggregation of the system (i.e. there is no $R \in \mathbb{R}$ such that for every $t \in [0, \infty[$ the system is contained in $\mathscr{B}_R(x, y)$ for some $x, y \in \mathbb{R}^2$).

The characteristics of our model have, as we will see, several advantages, but at the same time make harder a standard variational formulation. As we will show, energy-based investigations are possible in the discrete context here considered, but the identification of an explicit Lagrangian whose minimization leads to the exact dynamics displayed by the discrete system is far from trivial, and will be one of the main objectives of future investigations. One of the main checks we will perform will concern the systematyc comparison with Finite Element simulations. Nowadays FE analysis is indeed probably the most reliable tool for the numerical simulation of the behavior of deformable bodies, also thanks to the possibility of adapting it to the features of the problem studied. Isogeometric analysis (see for instance [34–37]), in particular, can be especially convenient for shape optimization problems that easily arise in the study of multi-agent systems moving in unbounded domains and starting form arbitrary configurations, as is the case in our context.

## 1.2 A Summary of the Algorithm and of the Formalism Employed

In this section, we will briefly describe the model studied in the present chapter. The algorithm and the formalism will be summarized in their most relevant features; the reader can find a fully detailed description in [6].

Let us consider a discrete system $\mathscr{S}$ constituted by a finite number of points which are characterized by their position in the real plane; we will call them the 'elements' of $\mathscr{S}$. The elements are set, in the initial configuration $C^0$, in the nodes of a squared grid sized $L \times L$. We consider a set of discrete time steps $\mathscr{T}_m = \{0, t_1, ..., t_m, ...\}$ and an orthonormal reference system with axes parallel to the sides of the system, the unit length being equal to the cell side in $C^0$. Each Lagrangian element $(i, j)$ of $\mathscr{S}$, placed in $(x^1_{ij}, x^2_{ij})(t_m)$, has a set of neighbors

$$N_n(\bar{i}, \bar{j}) := \left\{(i, j) \in C^0 : \rho[(\bar{i}, \bar{j}), (i, j)] = n\right\}$$

where $\rho$ is the $\mathbb{R}^2$-Chebyshev distance, i.e. the distance in $\mathbb{R}^2$ given by

$$\rho((x^1, x^2), (y^1, y^2)) = \max\left\{|x^1 - y^1|, |x^2 - y^2|\right\}.$$

With this definition, selecting $n = 1$ and $n = 2$, we have respectively the first and second neighbors as shown in Fig. 1.

Let us select a leader element $\mathscr{L}$ whose actual position is defined by a prescribed motion $\mathscr{M} : t_m \in \mathscr{T}_m \longrightarrow (x^1_{\bar{i}\bar{j}}, x^2_{\bar{i}\bar{j}})(t_m) \in \mathbb{R}^2$. We are now ready to describe the interactions between the elements. We will consider virtual steps in between two actual states, i.e. states that are invisible in the real displacements of the system but are necessary for computing its evolution. Let us consider a configuration $V_0^{t_1}$ such that the leader $\mathscr{L}$ is positioned in $\mathscr{M}(t_1)$: this is defined as the first virtual configuration. Then, recalling that by centroid of a set of points $P_1(x^1_1, ..., x^1_n), ..., P_m$

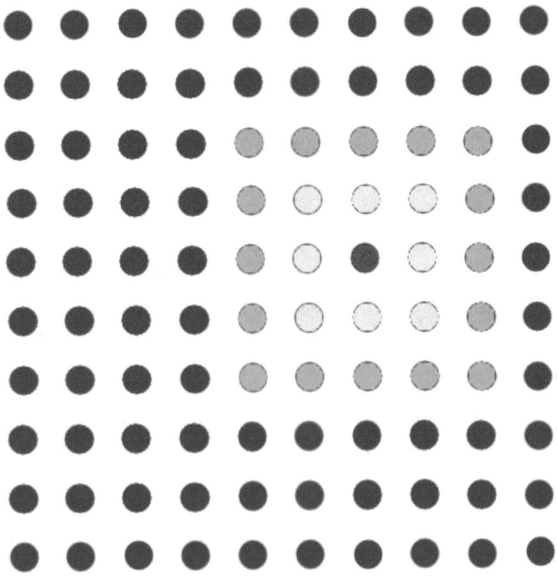

**Fig. 1** Graphical representation of neighbors: the first and second neighbors of the *red* element are the ones respectively in *yellow* and *green*

$(x_1^m, \dots, x_n^m) \in \mathbb{R}^n$ one means the point $P(\sum_{i=1}^{m} \frac{x_1^i}{m}, \sum_{i=1}^{n} \frac{x_2^i}{m}, \dots \sum_{i=1}^{n} \frac{x_n^i}{m})$, the next virtual configuration $V_1^{t_1}$ will be defined as the one in which:

i. the leader $\mathscr{L}$ remains in $V_0^{t_1}$;
ii. every one of its first neighbors $N_1(\bar{i}, \bar{j})$ moves to the centroid of its own first neighbors in $V_0^{t_1}$;
iii. all others elements remain in the position they had in $V_0^{t_1}$.

Iterating and generalizing to the $n$-th virtual step, $V_n^{t_1}$ will be the virtual configuration in which:

i. the leader $\mathscr{L}$ together with its first $(n-1)$-th neighbors are in the same position they had in $V_{(n-1)}^{t_1}$;
ii. every one of the $n$-th neighbors of $\mathscr{L}$ has moved to the centroid of its own first neighbors in $V_{(n-1)}^{t_1}$;
iii. all others elements are in the same position they had in $V_{(n-1)}^{t_1}$.

We will get the actual configuration $C^{t_1}$ when $n$ equals the maximum Lagrangian–Chebyshev distance of the elements from the leader (see Fig. 2).

It is easy to see that in the model so described edge effects will arise, since spontaneous shrinking will concern boundary elements because of the non-symmetric placement of their neighbors. In order to overcome this problem several standard possibilities can be considered. Choosing probably the most simple one, we introduce a 'fictitious' boundary constituted by an external frame of elements which simply follow, at a fixed distance and always in the same direction, their closest 'true' element. The fictitious elements move only in a specific virtual time step which follows the other ones. In this way, every true element will have a complete

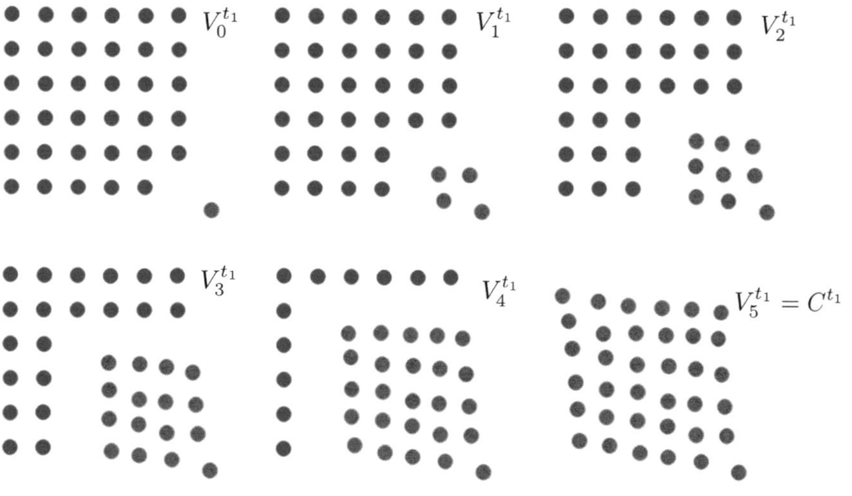


**Fig. 2** Graphical representation of the virtual configurations. In *red* the elements which are already in the position they will have in $C^{t_1}$

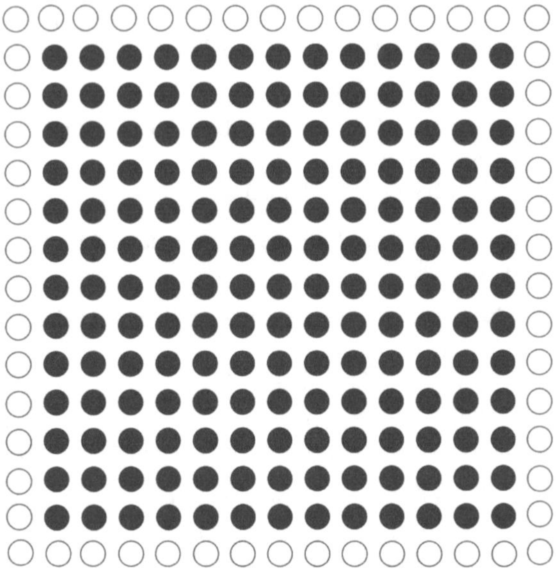

**Fig. 3** Graphical representation of fictitious elements. Every fictitious element (empty *red dots*) moves in rigid translation with the closest true one (*blue dots*); every vertex true element carries in rigid translation the vertex fictitious one and also its two fictitious neighbors

set of 8 neighbors (see Fig. 3 for a graphical representation. An explicit description of the motion of the fictitious elements is in [6]). The distance at which the fictitious elements are positioned represents a key characteristic length, determining the lattice step of the system in the equilibrium configuration. In the following numerical simulations, unless differently specified, this length is equal to 1 length unit for all the fictitious elements.

The previous algorithm can be easily generalized to the case of second and, in general, $n$-th neighbors interaction, by simply computing the centroid of sets of points having maximum Lagrangian–Chebyshev distance of 2 (in general, $n$). In this case, additional sets of fictitious elements have also to be introduced in order to have, for every element of the system, a full set of neighbors. In [6] the system is also generalized to the case of multiple leaders, which will be used throughout the present paper.

The geometric centroid of a given set of points $P_1$, ..., $P_n$ minimizes the sum of the squared distances, i.e. the function:

$$f(P) = \sum_{i=1}^{n} ||P_i - P||^2. \tag{1}$$

It is therefore possible to see that for a first-(second-)neighbor interaction, a natural deformation energy density can be written as a function of first-(second-)order finite differences of the placement $\chi$ for a given point, and therefore, as the step length goes to zero, a second gradient homogenized energy $\mathscr{E}[\nabla\chi(P), \nabla\nabla\chi(P)]$ can be conjectured for second neighbors interaction systems (for more details see again [6]).

The algorithm described before has much in common with theoretical models conceived for describing collective behaviors (e.g. the evolution of swarms, flocks etc.). Such models have already been used for mechanical investigations, in cases in which the potential energy of a mechanical system is minimal in correspondence with stable equilibrium positions of a swarm system [38, 39]. However, in the present work we prefer not to use the name 'swarm' system, since one usually associates that name, and in particular the expression Particle Swarm Optimization (PSO), to cases in which each particle of the swarm represents a potential solution of a given (a priori formulated) optimization problem [40–42]. Our aim, in fact, is not the study of numerical tools for the solution of well defined optimization problems, but rather the development of a new model directly accounting for the description of the phenomenology; other features distinguishing the proposed discrete model from PSO ones is the presence of leaders (whereas PSO models are usually anarchic) and the absence of any randomness.

It should also be pointed out that some basic features of our model (in particular the Lagrangian character of the neighborhood) are quite different from standard swarm robotic modeling. In modeling swarms, flocks and schools dynamics, a topological concept of neighborhood (rather than a metric one) is emerging as one of the most promising in order to account for the observed phenomenology (see [43, 44]). The model proposed in the present paper mixes these two ideas, since it is based on a concept of neighborhood which depends on the topological distance between the elements, while the way in which the interaction works depends on the metric (Euclidean) distance in the actual configuration. Finally, another feature of the proposed model which is often met in swarm modeling is that the elements, due to the presence of the fictitious boundary, do not behave all in the same way; this has been proposed as one of the possible discriminating factors between crowds and swarms/flocks/schools models (see e.g. [45, 46]), but also in this case it can be pointed out that the property of being (or not) a boundary element, in the model considered herein, is Lagrangian, i.e. it is preserved during the time evolution of the system.

## *1.3 Short Summary of Preliminary Numerical Results*

We summarize in this section the numerical results obtained in [6]. With the aim of comparing the discrete models characterized by (respectively) first and second neighbor interaction (FNI and SNI) with first and second gradient continuum theories, some simple cases were investigated.

Two squared systems FNI and SNI were considered, in which the leaders $\mathscr{L}$, situated in a vertex of the square, were pulled outside (or pushed inside) in the diagonal direction at 45 degrees with respect to the sides of the system in $C^*$ with a uniform motion. The numerical results obtained were compared with Finite Element simulations of a 2D continuous squared body, at the vertex of which a prescribed displacement was imposed. Two cases were considered: a standard energy (for the

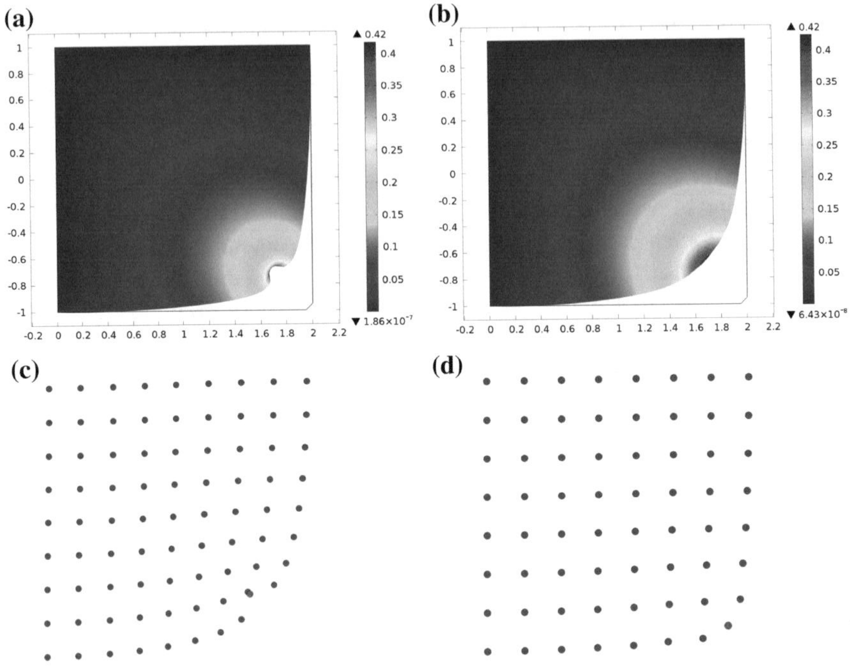


**Fig. 4** A comparison between FE simulation of first (**a**) or second (**b**) gradient continua and first (**c**) or second (**d**) neighbors interaction with a similar external action

first gradient case) and the Mindlin general form [15] for the second gradient case. In Fig. 4 we can appreciate one of the similarities between the continuum models and the discrete system here described (for other features see [6]). As we can see, in the FNI/first gradient continuum cases a loss of boundary convexity around the vertex is observed, whereas this behavior is not present in SNI/second gradient continuum cases; this is highlighted in Fig. 5.

Releasing the vertex at a certain time step, i.e. letting the system evolve while the leader is stopped, one observes the return to the original configuration, in agreement with the behavior expected in elastically deformed bodies. As can be seen in Fig. 6, the return to the initial configuration is not instantaneous, which implies the presence of inertia effects even if no explicit variable accounting for the elements mass was introduced. It is possible to prove that actually we have an asymptotic convergence, which means that some viscous effect has to be considered if one want to identify the Lagrangian system exactly corresponding to the evolution of the proposed system [6].

These results (among others) suggest that the evolution of the presented discrete system resembles that of elastic deformable bodies, and that varying the order of the interaction, specific characteristics of higher gradient theories are also recovered. In the following we will see a more direct comparison with 2D continuum simulations performed with COMSOL.

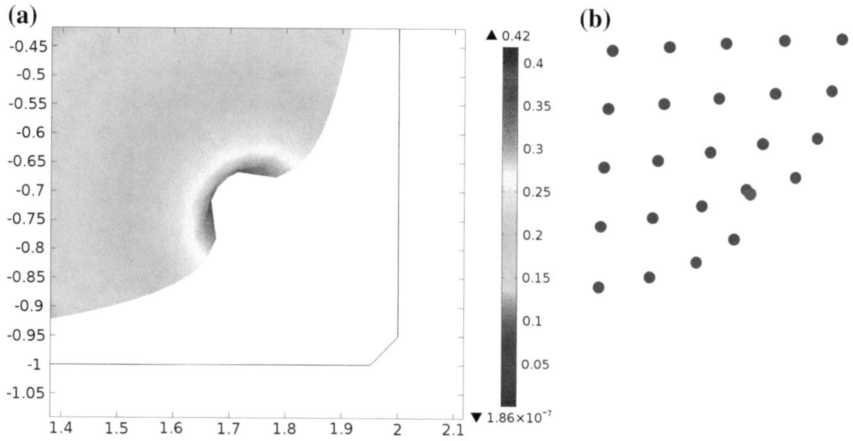


**Fig. 5** Zoom on the loss of convexity in first gradient (**a**) and first neighbors interaction (**b**) cases

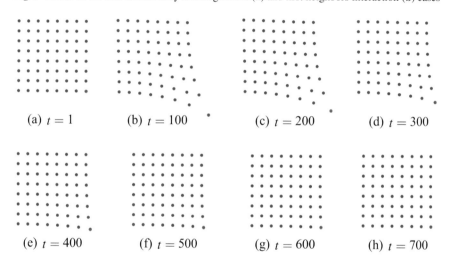

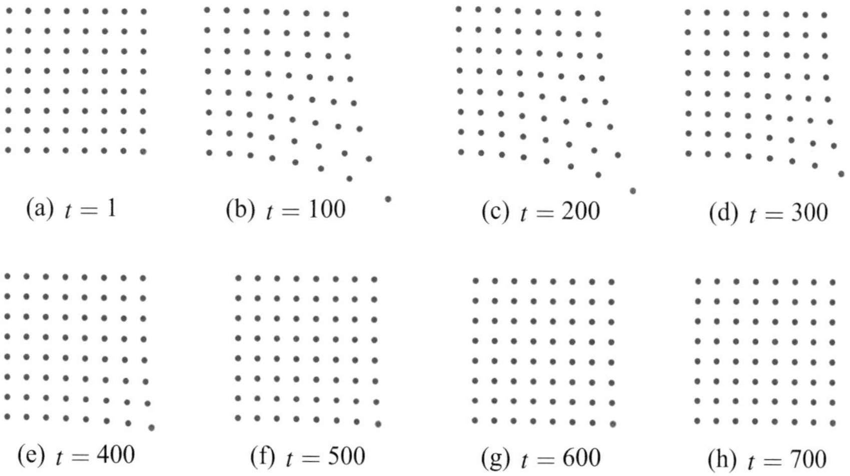


**Fig. 6** Imposing a stop to the leader (at $t = 150$) the system tends to return to the original configuration

Finally, we want to devote a few words to the nonlinear character of the evolution of the system. In the following sections, a fracture algorithm (obviously entailing nonlinearities) will be introduced, but even the basic form of the model as described up to this point exhibits a nonlinear behavior. In Fig. 7, we compare two identical systems in which different actions are imposed to one single leader. In the left panel, a speed of 1 length units per unit time is imposed for a total of 10 time steps; in the right panel, we imposed a speed of 0.001 length units per unit time for a total of 10000 time steps. As one can see, the resulting configurations are clearly different, thus implying the nonlinearity with respect to the imposed external action.

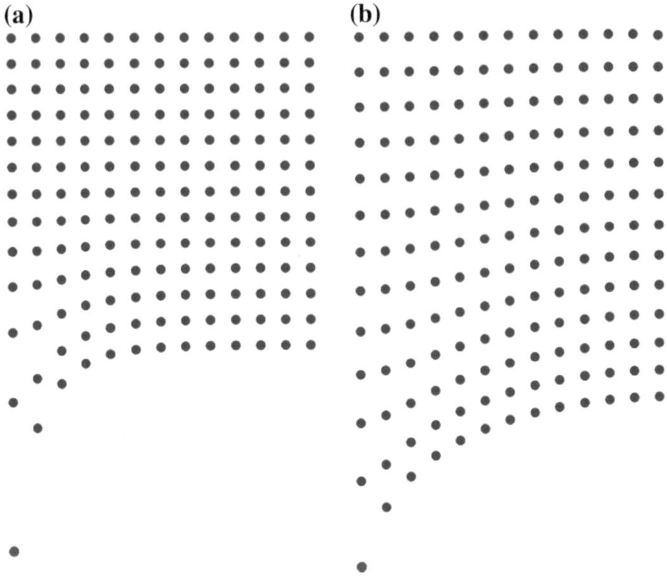


**Fig. 7** Nonlinearity of the system: in **a**, a speed $s = 1$ length units per time step is imposed to the leader for an interval $I$ of 10 time steps; in **b**, the $s = 0.001$ and $I = 10000$. The two resulting configurations are different

## *1.4 Further Comparison with Finite Element Simulations*

In this section we want to directly compare the results of our discrete model with the ones obtained by standard Finite Element simulations.

In Fig. 8 we show the superposition of:

- a simulation with an imposed external action consisting in pulling in opposite directions two opposed vertexes of our system;
- a standard FE simulation (performed with COMSOL) in which a similar action (imposed displacement) is applied on a squared 2D continuum.

For the continuous simulation, we considered both a classical Cauchy continuum (left) and a second gradient one (right). In color map the modulus of the displacement is shown. The corresponding discrete simulation involve first- (left) and second-neighbors (right) interaction. As it can be seen, in both cases the shape of the sample is accurately approximated by the discrete system, in the specific case employing only a very limited number of elements ($12 \times 12 = 144$ elements in total). This means that our system produces reliable results with a very limited computational cost, at least with this kind of external action.

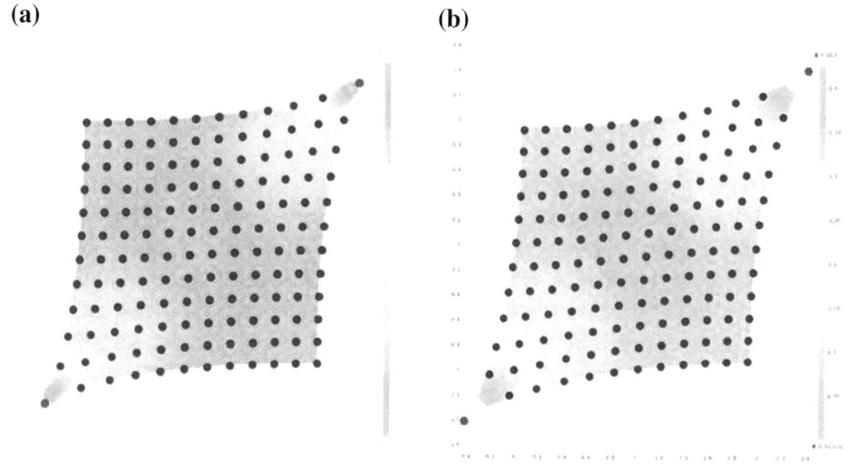


**Fig. 8** Superposition of the discrete system with a standard FE continuum simulation involving first gradient/first-neighbors (*left*) and second gradient/second neighbors models (*right*)

## 2 Spontaneous Evolution and Discrete Energy Equivalent

The previously described algorithm can be adapted to the case of a spontaneously evolving system, i.e. systems in which we do not impose a particular motion to any of the elements. To do so, we start from an arbitrary initial configuration $C^0$, in which in general the elements do not lie in the centroid of their neighbors. We apply the algorithm starting from a selected element $\mathscr{P}$ (which we will call the 'pseudo-leader') and then proceeding through concentric square frames. The only difference with the case of a 'true' leader is that the pseudo-leader $\mathscr{P}$ does not obey to an imposed motion, but simply goes, in the first virtual step, in the centroid of its first neighbors as all other elements do.

The observations made at the end of Sect. 1.2 suggest the introduction of a quantity $\mathscr{E}^D(t)$, which we will call Discrete Energy Equivalent (DEE), defined on a geometrical basis. The DEE will represent a measure of the deformation energy stored in the actual configuration of the system in case of first neighbors interaction. Due to the minimum property of the centroid above recalled (see Eq. (1)), and considering that an (Euclidean) distance equal to 1 or to $\sqrt{2}$ corresponds to the pairs of first neighbors elements in the equilibrium configuration $C^*$, it is natural to define $\mathscr{E}^D$ as follows:

$$\mathscr{E}^D = \sum_{S_1}(d^t - 1)^2 + \sum_{S_2}\frac{(d^t - \sqrt{2})^2}{2}. \tag{2}$$

Here the sums are extended over all the pairs of first neighbors 'true' elements. The sets $S_1$ and $S_2$ contain the pairs that in $C^*$ lie respectively orthogonally or diagonally with respect to the sides. A unit length is subtracted from $d^t$ ($\sqrt{2}$ in case of diagonal pairs) because the step length is unitary; more precisely, because the fictitious elements are at a fixed unitary distance from the true boundary. Since the sums have the dimension of an area, a coefficient $\alpha$ having the dimension of a force unit over a length unit has to be introduced to define the energy of the system: $\mathscr{E} = \alpha\mathscr{E}^D$.

The first set of numerical simulations will concern the study of the DEE decay in systems evolving without leaders, and since in our simulations $\mathscr{E}^D(C^0) \neq 0$, $C^0$ can be seen as a pre-stressed configuration. A basic case, in which the center element is selected as the pseudo-leader, is represented in Fig. 9. A sharper case in which the initial configuration is more stressed is shown in Fig. 10. In both cases, the energy decay is well interpolated by an exponential behavior; in Fig. 11 the interpolation of the form $e^{ax^b}$, with $a \approx -0.59$ and $b \approx 0.52$ is shown. This behavior is well known for several elastic systems in which viscous dissipation occurs in both linear and nonlinear cases (see e.g. [47–50]).

One may wonder whether the system always asymptotically converges to the same limit configuration (i.e., the one with $\mathscr{E}^D = 0$) independently on the choice of the pseudo-leader. This seems a rather natural request if the system is intended to model the behavior of a deformable body. In the Appendix (see 4) we prove that in the 1D case the answer is yes. In 2D, the conjecture that the asymptotic configuration is independent on the pseudo-leader choice remains more than reasonable, but the proof is more difficult, since in this case it is not true anymore that the total discrete energy decreases in every virtual step. Numerical evidence of this somewhat surprising statement is shown in Fig. 12, where one can observe that, after a first approximately exponential phase, the energy increases slightly before its eventual decay.

We want now to perform some quantitative analysis in order to evaluate some magnitudes concerning the evolution of our system on the basis of the DEE above defined. Let us first consider a very simple case in which uniaxial motion is imposed to a set of leaders constituted by the elements belonging to two opposite sides of the system. In particular, the fictitious elements relative to the left side are treated as leaders, as they are motionless by definition (from a mechanical point of view, the side has an imposed displacement equal to zero). Moreover, we impose a motion to the elements belonging to the right side; the motion is uniform and directed along the $x$ axis up to a certain time $t_s$, after which they stop. The system is then left evolving until equilibrium is achieved. The result, with relative energy versus time plotting, is shown in Fig. 13 (here and in all the following simulations, the leaders are represented by red dots). Since our external action is an imposed motion, it is not immediate to derive from it a discrete version of the applied uniaxial force $\sigma^D$. We can however reasonably define it in various equivalent ways; for instance, we can use the identity

$$\frac{(\sigma^D)^2 A}{2E} = \mathscr{E} = \alpha\mathscr{E}^D$$

where $A$ is the area of the system at the equilibrium, $E$ is Young's modulus and $\alpha$ the previously introduced constant. We will not lose generality by selecting the unit for

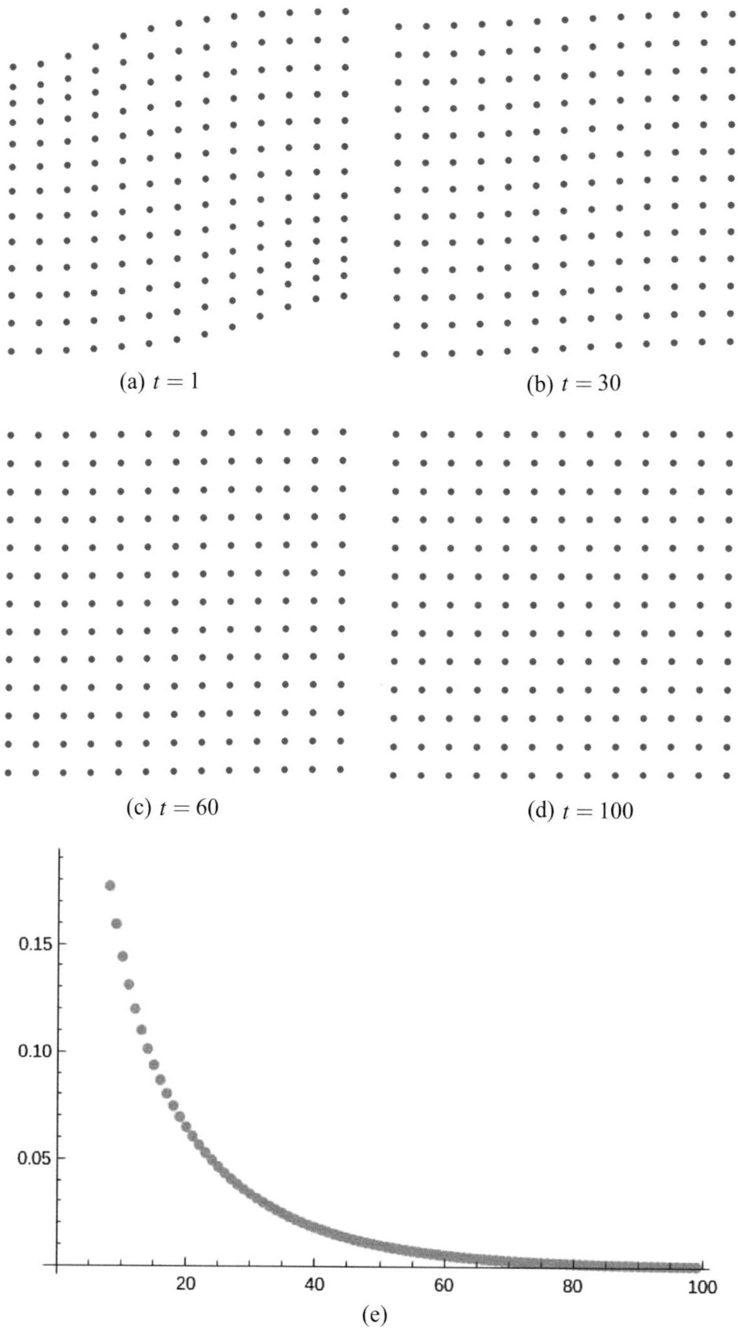


**Fig. 9** **a–d** Spontaneuos evolution of the system from a prestressed configuration; the pseudoleader is the central element. **e** Time history of the Discrete Energy Equivalent

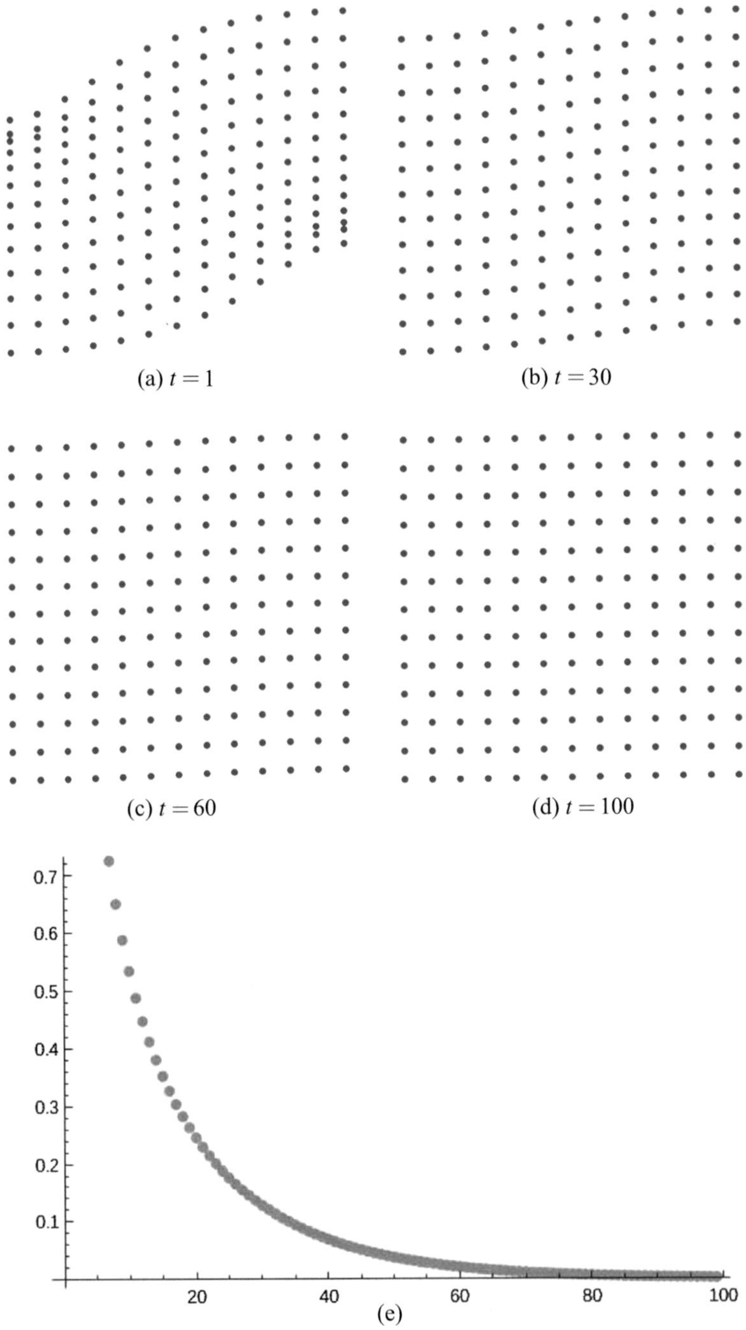


**Fig. 10** **a–d** The same simulation of Fig. 9 with a more stressed initial configuration. **e** time history of the Discrete Energy Equivalent

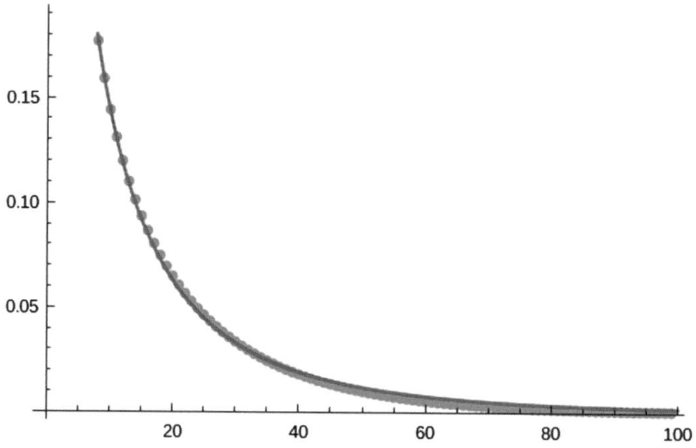


**Fig. 11** Exponential energy decay: in *red* it is represented the curve $e^{-0{,}5912x^{0{,}5155}}$

$\sigma^D$ so as to have $2E = 1$, and therefore we get $(\sigma^D)^2 = \frac{\alpha\mathscr{E}^D}{A}$. The numerical data for $A$ and $\mathscr{E}^D$ give:

$$(\sigma^D)^2 = \frac{\alpha\mathscr{E}^D}{A} \approx \frac{12.5\alpha}{382} \approx 0.03\alpha. \tag{3}$$

Using the reasonable assumption that a certain imposed motion will approximately correspond to the same $\sigma^D$ if only small and local changes are considered in $C^*$, we will employ the value now obtained in the simulations on the crack formation and growth considered below.

# 3 Fracture and Crack Formation and Evolution

## *3.1 Introduction of the Fracture*

In the proposed model, the fracture is intended as a loss of interaction between neighboring elements. Specifically, when the Euclidean distance (evaluated in actual configurations) between two interacting elements overcomes a certain threshold $s_f$, the two elements do not interact anymore. Obviously, when this happens, the computation of the centroid relative to the considered elements is ill defined, in the same sense as intended for what concerns boundary elements, as seen in Sect. 1.2. We solve the problem in a similar way as done before. Indeed, we introduce for every bond break a new pair of fictitious elements evolving in the same way as the fictitious boundary elements introduced above, i.e. each of them following one of the two elements whose bond has been broken. In this way, the two elements among

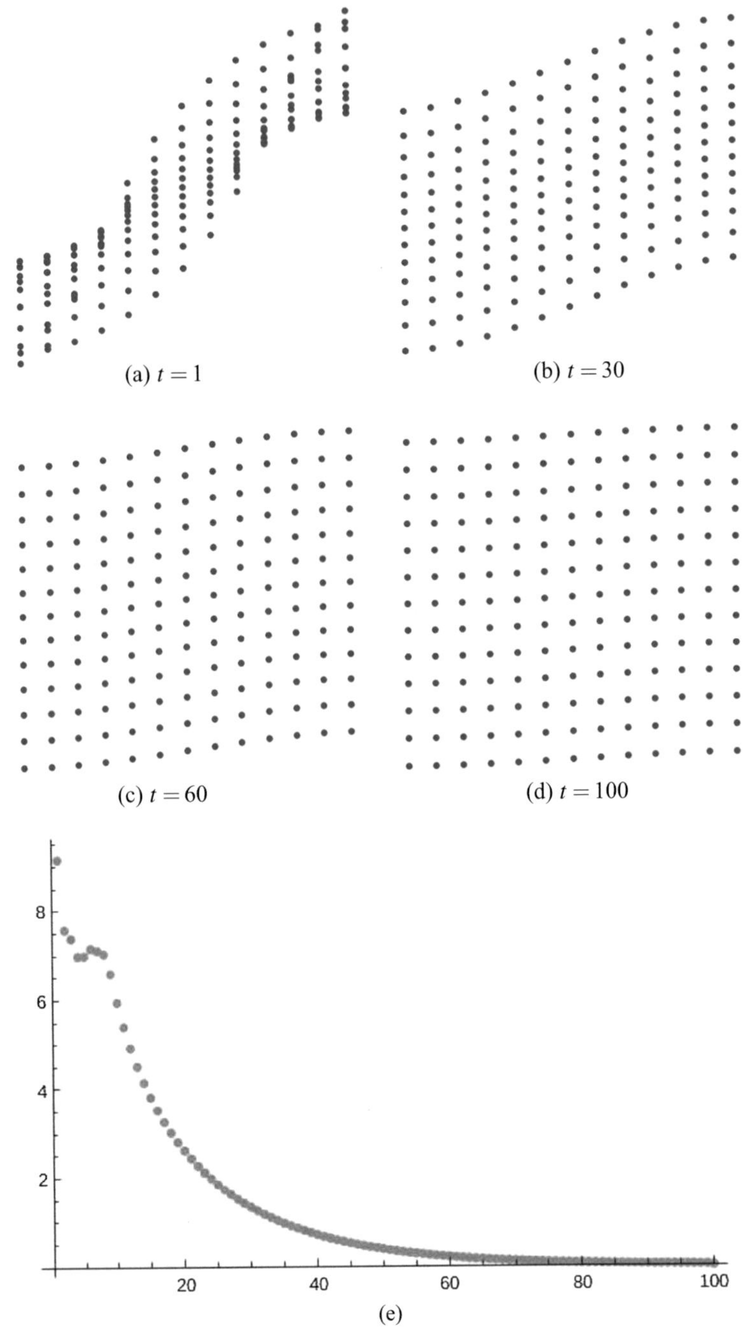


**Fig. 12** Spontaneous evolution starting from a severely prestressed configuration: the energy is not a monotonically decreasing function

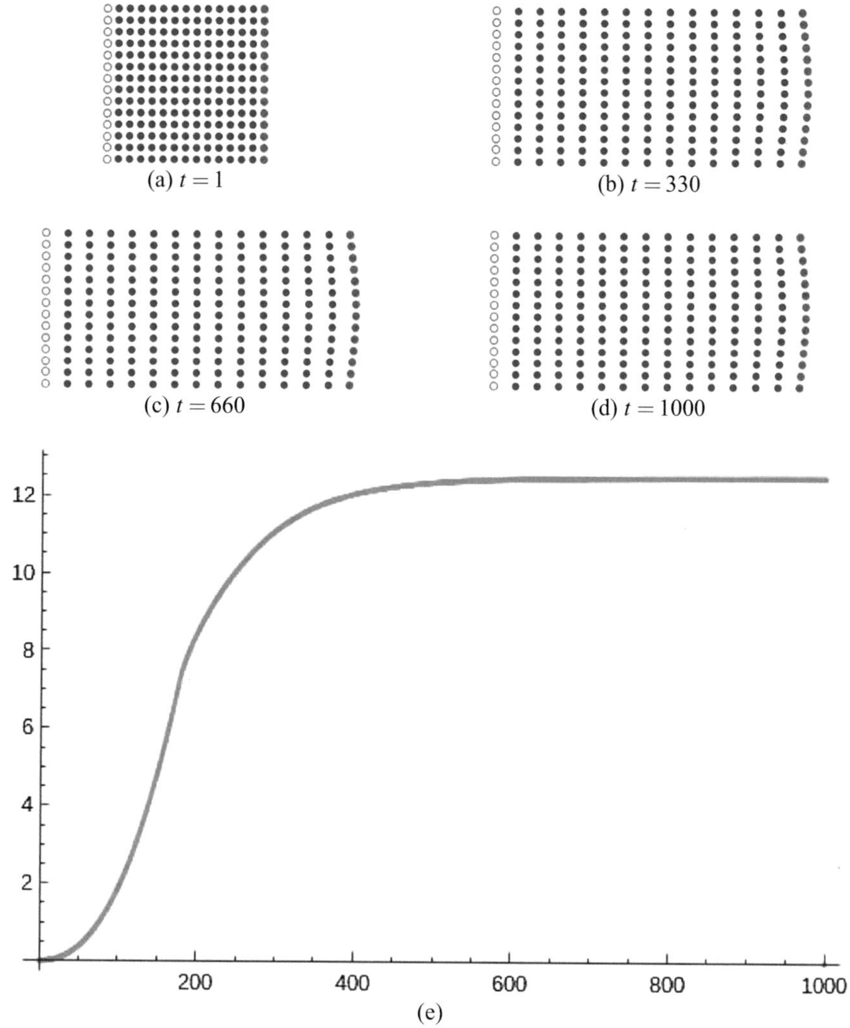


**Fig. 13** Uniaxial imposed motion to the leaders. The leaders (*solid red dots*) have a uniform motion parallel to the $x$ axis and then are stopped at $t = 180$. An equilibrium configuration is reached around $t = 450$. We use empty *red dots* to indicate that the fictitious elements relative to the *left side* have an imposed null velocity, i.e. the *left side* has an imposed displacement equal to zero

which the fracture occurred become boundary elements in the same sense as the preexisting ones.

The fracture algorithm works as follows: we introduce a prescribed order in the $n$-th neighbors of a given non-fictitious element starting, for example, from the neighbors in the left upper corner of the frame square centered in the considered element and proceeding clockwise. The labeling of neighbor elements allows the introduction of an adjacency matrix $J$, that in this case is a tensor of boolean variables $J_{ijk}$, the first two indexes determining the particle $(i, j)$ of the system, and the third one $k$, ranging from 1 to 8 in FNI, identifying one of its neighbors. When the distance between two elements is larger than $s_f$, the value of the corresponding element of $J$ is set to $FALSE = 0$ (otherwise it is $TRUE = 1$). When the centroid is computed, each neighbor coordinate is multiplied by the relative element of $J$. Moreover, another term is considered in the centroid computation, consisting in the coordinate of fictitious elements, each multiplied by $\bar{J}$ (where the bar indicates logical complement). It is clear that in this way, when the centroid is evaluated, a true or a fictitious element enters the calculation according to the fact that the interaction is present or broken. The irreversibility of the fracture, which is a desirable feature when modeling solid bodies, is verified since once an element of $J$ is set to 0, it can not become 1 anymore.

Another degree of freedom of the model, concerning the fictitious elements appearing after the fracture, naturally emerges form the described algorithm. Indeed, one can choose to place the new fictitious elements at a distance $\rho_f$ that does not correspond to the lattice step in $C^*$ (more precisely, we should say that it would not correspond to the fixed distance at which the fictitious boundary is from the true elements. It is indeed this last distance that determines the lattice step at which the system is in equilibrium). In this way, as we will see in the following, relevant features of fracture phenomena can be modeled.

## 3.2 *Basic Fracture and Crack Evolution*

A well known variational approach to the fracture was developed by Griffith in the '20. From Griffith's model (in 2D case), we have at the equilibrium:

$$\mathscr{E}_{TOT} = 2\gamma a + \frac{\sigma^2}{2E}A - \frac{\sigma^2}{2E}\beta a^2 \tag{4}$$

where $\gamma$ represents the energy per unit line required to break atomic bonds, $E$ the Young Modulus, $A$ the area of the sample, $a$ is the crack length, $\sigma$ the stress and $\beta$ an non-dimensional parameter accounting for the measure of the part of the surface relaxing as a consequence of the crack opening.

The most relevant qualitative aspect of Griffith's theory is that the dependence of the energy on the crack length forecasts that below a critical value of $a$ the system is stable, and a crack growth is possible only providing additional energy to the system. If this critical value is reached, the system becomes unstable and the crack evolves spontaneously. In the next groups of simulations we selected two leaders belonging to the bottom side (red elements) and imposed to them a uniform velocity, with components (0.03, 0.03) and (−003, 0.03) respectively. The leaders were stopped at $t = 80$. In Fig. 14 we set the fracture threshold $s_f = 1.17$ length units; we can see that the crack keeps growing even after the leaders have stopped, that means that the crack length is reaching the equilibrium value for the given imposed action. The crack length equilibrium value is reached around $t = 160$, after which the cracks remain stable while the system keeps relaxing and releasing deformation energy.

In Fig. 15 the same simulation was performed with the fracture threshold set at $s_f = 1.12$. It can be seen that the crack growth does not stop and indeed goes up to the complete split of the system, which is reached around $t = 250$. This means that stopping the motion at $t = 80$ is sufficient for furnishing to the system the energy needed to reach the length threshold after which the crack grows spontaneously.

## 3.3 Uniaxial External Action

The previous results indicate that qualitative features of Griffith's theory are recovered through our discrete model. However, in order to be able to compare quantitatively our numerical results with Griffith's theory, we need to apply a simpler external action, i.e. a uniaxial one, as done in the simulation shown in Fig. 13. Let us first formulate Eq. (4) by means of the discrete variables considered here. In our model, the energy can be written as:

$$\mathscr{E}_{TOT} = 3\alpha(s_{eff})^2 l + (\sigma^D)^2 A - (\sigma^D)^2 \beta l^2 \tag{5}$$

where $s_{eff} = s_f - 1$ is the DEE lost when a single bond is broken, $A$ represents the area of the system (in squared unit lengths) and $l$ is the number of broken pairs of elements in the crack.

In the simulation shown in Fig. 16 we imposed to the leaders the same external action used in the simulation of Fig. 13, i.e. we imposed an uniaxial displacement of the leaders in the $x$ direction up to 180 time steps followed by a stop, and a zero displacement to the fictitious elements relative to the left vertical side. In order to see a crack formation and evolution we created a 'defect' by removing 6 bonds close to the middle points of the bottom side. We can observe that a crack indeed opens in correspondence with the defect, and that it reaches a stable length already at $t = 1000$.

The DEE $\mathscr{E}^D$ that we obtained as an output (≈5) times the dimensional constant $\alpha$ is the discrete estimate of the deformation energy of the system, and therefore represents the quantity given by the last two terms in the right hand side of Eq. (5). This allows us to provide an estimate for $\beta$. Indeed, substituting the value for $\sigma^D$ taken from the simulation shown in Fig. 13, we have:

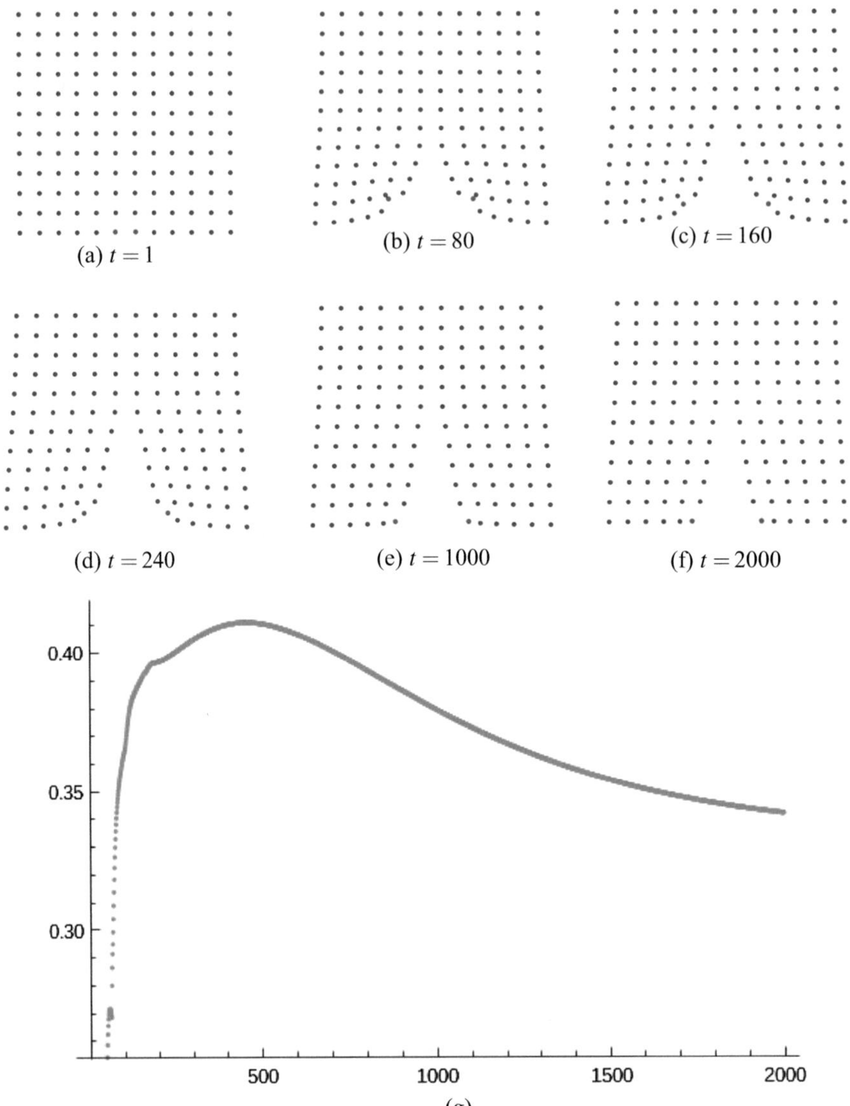


**Fig. 14** Crack growth as a consequence of two leaders pushing from inside at 45 degrees with uniform velocity; with $s_f = 1.17$ the crack length stabilizes around $t = 160$

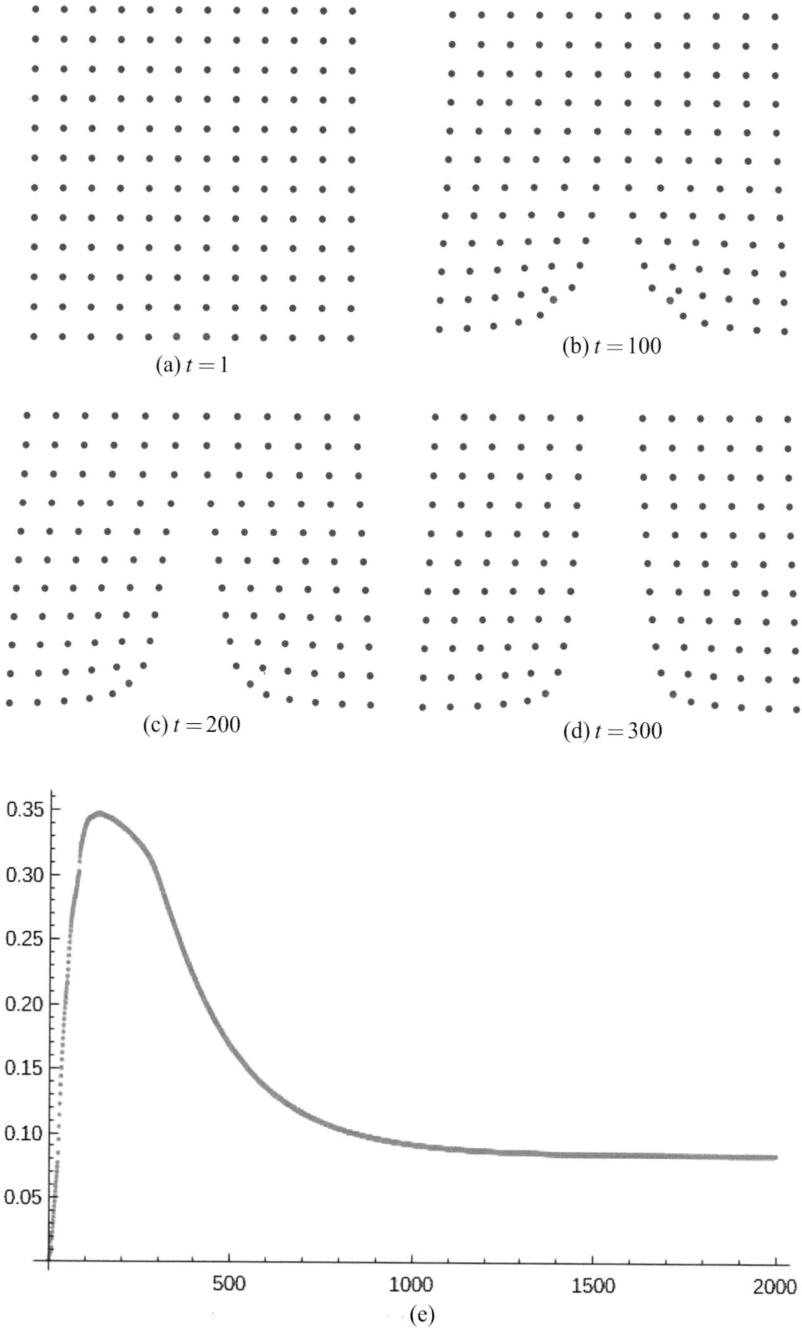


**Fig. 15** The same simulation of Fig. 15 is performed with $s_f = 1.12$; in this case, the critical length is overcome and the crack growth keeps up to the complete split of the system

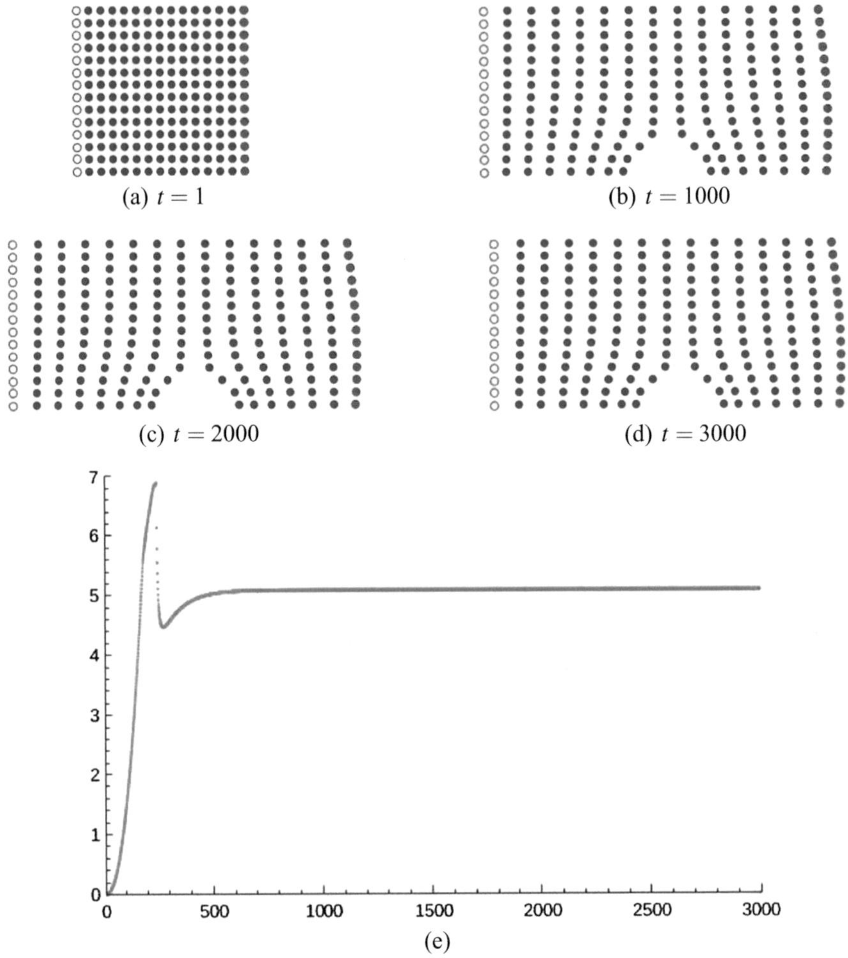


**Fig. 16** Uniaxial external action imposed to a set of boundary leaders with a defect consisting in the removal of six bounds around the middle elements of the bottom side ($s_f = 2.62$). The leadres stop at $t = 180$; a crack opens in correspondence with the defect and reaches a stable length (as before, we use empty *red dots* to indicate that the fictitious elements relative to the *left side* have an imposed null velocity, i.e. the *left side* has an imposed displacement equal to zero)

$$\beta = \frac{(\sigma^D)^2 \mu n^2 - \alpha \mathscr{E}^D}{(\sigma^D)^2 l^2} \approx 25.7. \tag{6}$$

According to the standard interpretation in Griffith's fracture theory, this means that an area measuring approximately $\beta \times l^2 = 25.7 \times 9$ is relaxed as a consequence of the crack formation; this area corresponds to a fraction of approximately 0.58 of the total area.

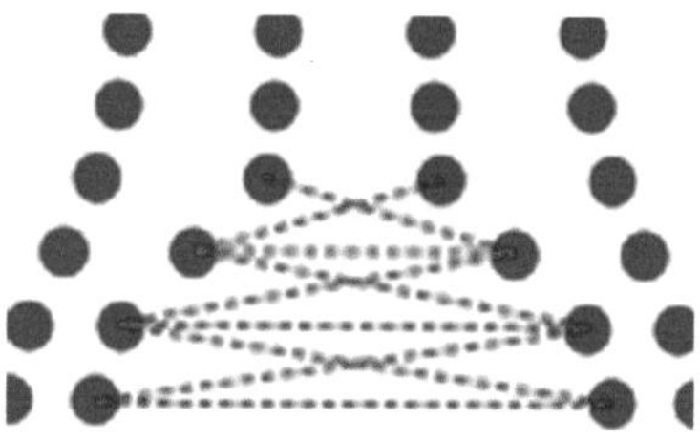

**Fig. 17** Graphical representation of the bonds broken in the simulation shown in Fig. 16

Let us now consider the first term in the right hand side of Eq. (5). The numerical output tells us that a total of 9 bonds, in Fig. 16, are stretched beyond the fracture threshold (see Fig. 17 for a graphical representation). As we recall, we introduced a 'defect' consisting in removing six interactions close to the edge; this means that a total of three elementary bonds were broken by applying the external action. Since the threshold was set at $s_f = 2.62$, the corresponding DEE, i.e. the energy $\mathscr{E}_b$ required to break the bonds can be estimated by:

$$\mathscr{E}_b = \alpha(s_f - 1)^2 \times 6 = 7.68\alpha. \tag{7}$$

The sum $\mathscr{E}_b + \alpha\mathscr{E}^D \approx 12.68\alpha$ is an estimate of the total energy in the system. As one can see, it is very close to the total deformation energy ($\approx$12.5$\alpha$, see Eq. (3)) with the same external action in case no fracture threshold is considered. This is consistent since, in the proposed model, no additional dissipated energy is associated to the break of the interactions, and therefore the total energy should coincide with the deformation energy. The small difference has to be related with the fact that we introduced a local defect in the simulation shown in Fig. 16; since we imposed an external action consisting in imposed displacements, this means that, assuming the same value for $\sigma^D$ as the one measured in the simulation of Fig. 13, we are slightly overestimating it. Actually, the ratio $\frac{12.68-12.5}{12.68} \approx 0.014$ is very close to the ratio between the number of removed bonds over the total bonds present in the system ($\frac{6}{520} \approx 0.012$).

## 3.4 *Almost-Steady State*

In the next simulation we want to underline a peculiar behavior shown by the model which can be seen as an 'almost steady' state followed by a catastrophic evolution of the crack. We considered the same system of the previous section, making only a small change in the fracture threshold $s_f$. Tuning very finely the threshold (i.e. setting $s_f = 2.5741$), one can obtain that a significantly long almost steady phase is followed by a nearly spontaneous crack opening around 1000 time steps (Fig. 18). We recall that the leaders were stopped at $t = 180$, which means that the abrupt opening of the crack around $t = 900$ is a consequence of an internal evolution of the

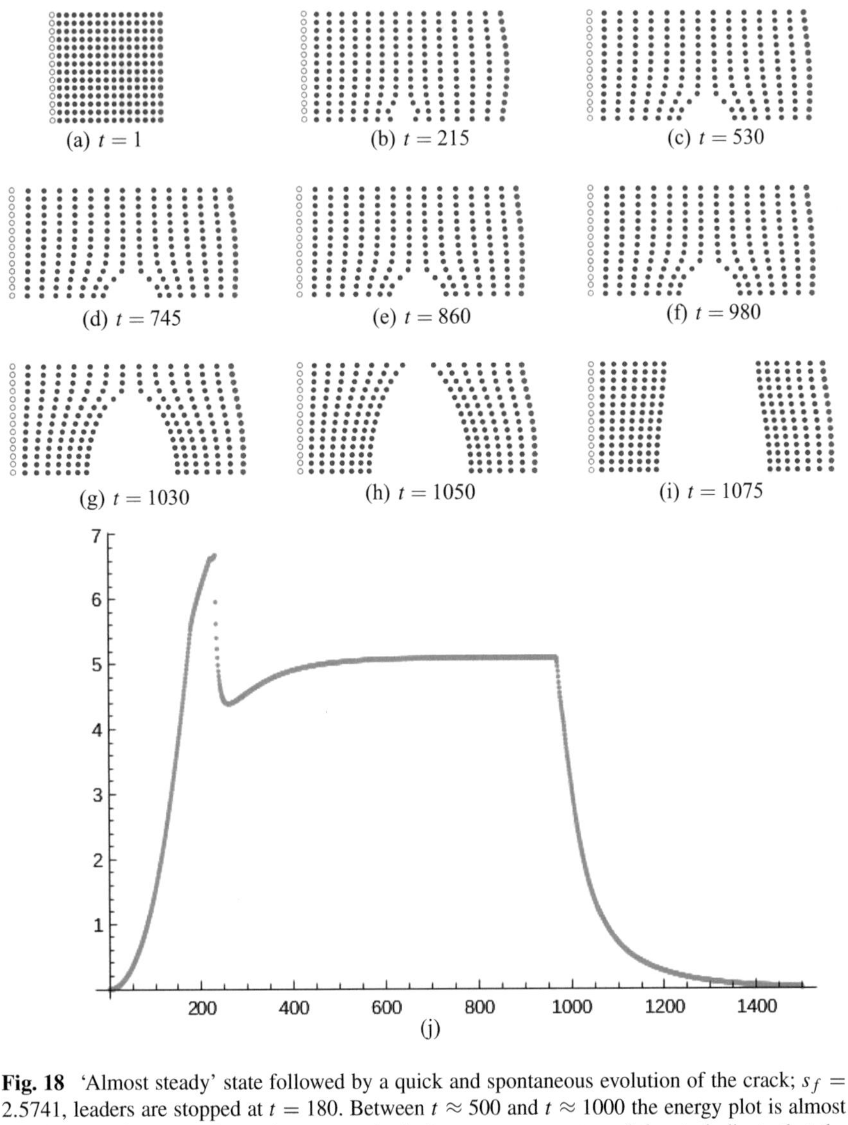


**Fig. 18** 'Almost steady' state followed by a quick and spontaneous evolution of the crack; $s_f = 2.5741$, leaders are stopped at $t = 180$. Between $t \approx 500$ and $t \approx 1000$ the energy plot is almost flat, then an abrupt crack opening occurs (as before, we use empty *red dots* to indicate that the fictitious elements relative to the *left side* have an imposed null velocity, i.e. the *left side* has an imposed displacement equal to zero).

system, though it is hardly visible since only very tiny changes in the positions of the elements (invisible at the 'macro' level) occur for a long time. In the opinion of the authors, this behavior, which is experimentally well known [51], is quite interesting, especially considering the simplicity of the discrete model employed.

The sensitivity of the model with respect to very fine changes in the parameters suggests that non trivial instability issues may arise from the study of its homogenized form. This can be a very interesting aspect to be investigated, useful theoretical tools being available in recent literature (the interested reader can see, for instance, [52–56]).

## 4 Conclusion and Further Objectives

Herein a new simple discrete system (introduced in [6]) with a centroid-based propagating interaction has been used in order to account for fracture phenomena in deformable bodies. Some results on pure deformation including a direct comparison with standard Finite Element simulations are presented. A discrete equivalent of the deformation energy is also defined and used to prove the independence of the asymptotic configuration on the choice of the element from which the algorithm starts in case of spontaneous evolution. A fracture algorithm is then introduced, with the definition of a suitable adjacency matrix specifying the pairs of elements among which no interaction exists anymore because the two elements overcame a prescribed threshold in the actual distance. The discrete equivalent of an external stress is also defined in the uniaxial case, and used to compare the numerical results with Griffith's theory of fracture. Different examples of crack formation and growth are considered.

The results indicate that the proposed model is promising for developing new and computationally advantageous tools for the study of fracture, and that the model is rich enough to produce interesting behaviors such as a spontaneous crack evolution after a long almost steady state.

Future investigations will concern how, by suitably weighting the contribution of the neighbors in the computation of the centroid, one can obtain different material models, and in particular different kinds of anisotropies. Moreover, one may try to generalize the model by suitably coupling the coordinates in order to account for Poisson's ratios different from 0 (while using directly the centroid-based algorithm we described this is the only possible value). Also interesting will be the study of more cases in which higher order interactions, reproducing behaviors typical of higher order continua, are considered. Finally theoretical results, among which the investigation of the homogenized limit of the considered system and the relative variational formulation, are needed in order to make further and sound progresses in the subject.

### Proof of the Independence of the Asymptotic Behavior on the Choice of the Pseudo-leader for 1D Systems

In Sect. 2 we considered the problem of the uniqueness of the asymptotic configuration of the system evolving spontaneously (i.e., without leaders), independently on the selected pseudo-leader. As this represents an important objectivity requirement for the model, in the present appendix we deserve a more formal treatment to the 1D case.

Let us consider a 1-dimensional system $\mathscr{S}$ constituted by $N$ elements mathematically represented by points in the real line, and an infinite set of discrete time steps $\mathscr{T} = \{t_0, t_1, ..., t_m, ...\}$. The system will evolve exactly as explained in the 2D case, with the set of first neighbors being constituted, for every element, by the two closest ones. We call *configuration of* $\mathscr{S}$ a strictly increasing function $C$ mapping $i$ into $x_i \in \mathbb{R}$, with the integer $i : 0 < i \leq N$ indicating the Lagrangian label of the elements of the system. We indicate by $C^{t_j}$ the configuration at time $t_j \in \mathscr{T}$, and by $\mathscr{S}_{\bar{i}}$ the system evolving with the chosen pseudo-leader $(\bar{i})$. We say that $\mathscr{S}_{\bar{i}}$ asymptotically converges to the configuration $C^{t_\infty}$, and write $\mathscr{S}_{\bar{i}} \to C^{t_\infty}$, if for every $\varepsilon > 0$ there exists a positive integer $M$ such that $|x_i^{t_m} - x_i^{t_\infty}| < \varepsilon$ for every integer $i : 0 < i \leq N$ and every $t_m \in \mathscr{T} : m > M$ (we can notice that the given definition, which is natural when considering finite systems, would correspond in the homogenized limit to a uniform functional convergence). Let $\mathscr{C}^*$ be the set of configurations in which $|x_{i+1} - x_i| = 1$ for every integer $i$ such that $0 < i < N$; obviously one has $\mathscr{E}^D(C^*) = 0$ for every $C^* \in \mathscr{C}^*$ (see Eq. (2)), and the representative $C^*$ is unique up to a translation. We have the following result.

**Proposition** *A system $\mathscr{S}_i$ evolving spontaneously tends to an asymptotic configuration $C^{t_\infty}$,and it is $C^{t_\infty} \in \mathscr{C}^*$ $\forall i \in \mathbb{N} : 1 \leq i \leq N$.*

**Proof** Let us first prove that the limit configuration exists. Then we will prove the Lemma by showing that, if $C^{t_\infty}$ is the asymptotic configuration, one has $\mathscr{E}^D(C^{t_\infty}) = 0$ (the thesis follows then immediately).

We start by noticing that in every virtual time step one element or two elements which are not neighbors move to the centroid of their first neighbors, while all other elements do not move. Since the centroid locally minimizes the discrete energy, this means that $\mathscr{E}^D(V_h, t_j) \geq \mathscr{E}^D(V_{h+1}, t_j)$ and $\mathscr{E}^D(V_h, t_j) \geq \mathscr{E}^D(V_k, t_{j+1})$ for all integers $j$, and for all $h$ and $k$ for which virtual configurations are defined. Since actual configurations of the system are a subset (preserving the order relation) of the virtual configurations, one has that $\mathscr{E}^D(C^{t_j})$ $(t_j \in \mathscr{T})$ is a monotonically decreasing function of $j$. Since $\mathscr{E}^D$ is non negative, this means that there exists $\lim_{m\to\infty} \mathscr{E}^D(C^{t_m}) = \mathscr{E}^D(C^{t_\infty})$. Let us now enumerate all virtual configurations (including the ones identified by definition with actual ones): $V_1, V_2, ... V_n, ...$. Since $\mathscr{E}^D$ converges, one has:

$$\Delta\mathscr{E}^D(V_h) := \mathscr{E}^D(V_h) - \mathscr{E}^D(V_h + 1) \to 0 \text{ for } m \to \infty \tag{8}$$

Let $\Delta s_h$ be the sum of the *moduli* of the displacements of the elements that move in the virtual time step $V_h$. Since $\Delta\mathscr{E}^D(V_h)$ is a continuous function of $\Delta s_h$ vanishing if and only if $\Delta s_h = 0$, Eq. (8) implies that $\lim_{h\to\infty} \Delta s_h = 0$, i.e. all virtual displacements of the elements (and thus the actual ones) tend to zero as time goes to infinity. Therefore, there exists a limit configuration $C^{t_\infty}$.

Let us now suppose by contradiction that $\mathscr{E}^D(C^{t_\infty}) = \tilde{\mathscr{E}} > 0$ for some element $\bar{i}$ chosen as the pseudo-leader.

Let now $d_{max}$ and $d_{min}$ be respectively the maximum and minimum distance between (non fictitious) elements which are first neighbors in $C^{t_\infty}$. Since $\tilde{\mathscr{E}} > 0$, one has that at least one between the two inequalities $d_{max} > 1$ and $d_{min} < 1$ holds; without losing generality, we select the first possibility. Let $\tilde{i}$ and $\tilde{i}+1$ be the two Lagrangian elements such that

$$\lim_{m\to\infty} (|x_{\tilde{i}+1}^{t_m} - x_{\tilde{i}}^{t_m}|) = d_{max} \tag{9}$$

Recalling that the centroid algorithm moves the element $\tilde{i}$ to the centroid of its neighbors, it easily follows from Eq. (9) that:

$$\lim_{m\to\infty} (|x_{\tilde{i}}^{t_m} - x_{\tilde{i}-1}^{t_m}|) = d_{max} \tag{10}$$

Let $x_F^{t_m}$ be the coordinate of the left fictitious element of the system at time $t_m$. Iterating the previous reasoning, one gets eventually

$$\lim_{m\to\infty} |x_0^{t_m} - x_F^{t_m}| = d_{max} > 1 \tag{11}$$

and since by definition $|x_0 - x_F| = 1$ at every time, the contradiction completes the proof.

**Remark** The result is easily generalized to system with second or higher order interaction. As mentioned in the paper, however, the generalization to 2D systems is not straightforward as in that case it is not true that the energy is decreasing.

## References


1. Francfort, G.A., Marigo, J.-J.: Revisiting brittle fracture as an energy minimization problem. J. Mech. Phys. Solids **46**(8), 1319–1342 (1998)
2. Bourdin, B., Francfort, G.A., Marigo, J.J.: The variational approach to fracture. J. Elast. **91**(1–3), 5–148 (2008)
3. Ching, W.Y., Rulis, P., Misra, A.: Ab initio elastic properties and tensile strength of crystalline hydroxyapatite. Acta Biomaterialia **5**(8), 3067–3075 (2009)
4. Kulkarni, M.G., Pal, S., Kubair, D.V.: Mode-3 spontaneous crack propagation in unsymmetric functionally graded materials. Int. J. Solids Struct. **44**(1), 229–241 (2007)
5. Rinaldi, A., Placidi, L.: A microscale second gradient approximation of the damage parameter of quasi-brittle heterogeneous lattices. ZAMM-Journal of Applied Mathematics and Mechanics/Zeitschrift für Angewandte Mathematik und Mechanik **94**(10), 862–877 (2014)
6. Della Corte, A., Battista, A.: Referential description of the evolution of a 2D swarm of robots interacting with the closer neighbors: perspectives of continuum modeling via higher gradient continua. Int. J. Non-Linear Mech. **80**, 209–220 (2016)
7. Battista, A., Rosa, L., dell'Erba, R., Greco, L.: Numerical investigation of a particle system compared with first and second gradient continua: Deformation and fracture phenomena. Mathematics and Mechanics of Solids (2016). doi:10.1177/1081286516657889

8. Mindlin, R.D.: Second gradient of strain and surface-tension in linear elasticity. Int. J. Solids Struct. **1**(4), 417–438 (1965)
9. dell'Isola, F., Seppecher, P.: Edge contact forces and quasi-balanced power. Meccanica **32**(1), 33–52 (1997)
10. dell'Isola, F., Sciarra, G., Vidoli, S.: Generalized Hooke's law for isotropic second gradient materials. In: Proceedings of the Royal Society of London A: Mathematical, Physical and Engineering Sciences (pp. rspa-2008). The Royal Society. (2009)
11. Placidi, L.: A variational approach for a nonlinear 1-dimensional second gradient continuum damage model. Contin. Mech. Thermodyn. **27**(4–5), 623–638 (2015)
12. Sunyk, R., Steinmann, P.: On higher gradients in continuum-atomistic modelling. Int. J. Solids Struct. **40**(24), 6877–6896 (2003)
13. Eringen, A.C.: Microcontinuum Field Theories: I. Foundations and Solids. Springer Science & Business Media (2012)
14. Germain, P.: The method of virtual power in continuum mechanics. Part 2: Microstructure. SIAM J. Appl. Math. **25**(3), 556–575 (1973)
15. Mindlin, R.D.: Micro-structure in linear elasticity. Arch. Ration. Mech. Anal. **16**(1), 51–78 (1964)
16. Steinmann, P.: A micropolar theory of finite deformation and finite rotation multiplicative elastoplasticity. Int. J. Solids Struct. **31**(8), 1063–1084 (1994)
17. Altenbach, J., Altenbach, H., Eremeyev, V.A.: On generalized Cosserat-type theories of plates and shells: a short review and bibliography. Arch. Appl. Mech. **80**(1), 73–92 (2010)
18. Placidi, L., Faria, S.H., Hutter, K.: On the role of grain growth, recrystallization and polygonization in a continuum theory for anisotropic ice sheets. Ann. Glaciol. **39**(1), 49–52 (2004)
19. Eremeyev, V.A., Ivanova, E.A., Morozov, N.F., Strochkov, S.E.: The spectrum of natural oscillations of an array of micro-or nanospheres on an elastic substrate. In: Doklady Physics, vol. 52, pp. 699–702. MAIK Nauka/Interperiodica (2007)
20. Madeo, A., Placidi, L., Rosi, G.: Towards the design of metamaterials with enhanced damage sensitivity: second gradient porous materials. Research in Nondestructive Evaluation **25**(2), 99–124 (2014)
21. Eremeyev, V.A.: Acceleration waves in micropolar elastic media. In: Doklady Physics, vol. 50, pp. 204–206. MAIK Nauka/Interperiodica (2005)
22. Chang, C.S., Misra, A.: Application of uniform strain theory to heterogeneous granular solids. J. Eng. Mech. **116**(10), 2310–2328 (1990)
23. Dell'Isola, F., Steigmann, D., Della Corte, A.: Synthesis of Fibrous Complex Structures: Designing Microstructure to Deliver Targeted Macroscale Response. Appl. Mech. Rev. **67**(6), 21 (2016)
24. Thiagarajan, G., Misra, A.: Fracture simulation for anisotropic materials using a virtual internal bond model. Int. J. Solids Struct. **41**(11), 2919–2938 (2004)
25. Koh, S.J.A., Lee, H.P., Lu, C., Cheng, Q.H.: Molecular dynamics simulation of a solid platinum nanowire under uniaxial tensile strain: temperature and strain-rate effects. Phys. Rev. B **72**(8), 085414 (2005)
26. Misra, A., Roberts, L.A., Levorson, S.M.: Reliability analysis of drilled shaft behavior using finite difference method and Monte Carlo simulation. Geotech. Geol. Eng. **25**(1), 65–77 (2007)
27. Kim, S.P., Van Duin, A.C., Shenoy, V.B.: Effect of electrolytes on the structure and evolution of the solid electrolyte interphase (SEI) in Li-ion batteries: a molecular dynamics study. J. Power Sources **196**(20), 8590–8597 (2011)
28. Tuckerman, M.E.: Ab initio molecular dynamics: basic concepts, current trends and novel applications. J. Phys.: Conden. Matter **14**(50), R1297 (2002)
29. Payne, M.C., Teter, M.P., Allan, D.C., Arias, T.A., Joannopoulos, J.D.: Iterative minimization techniques for ab initio total-energy calculations: molecular dynamics and conjugate gradients. Rev. Mod. Phys. **64**(4), 1045 (1992)
30. Allen, M.P.: Introduction to molecular dynamics simulation. Comput. Soft Matter Synth. Polym. Proteins **23**, 1–28 (2004)

31. Levitan, E.S.: Forced Oscillation of a Spring-Mass System having Combined Coulomb and Viscous Damping. J. Acoust. Soc. Am. **32**(10), 1265–1269 (1960)
32. Kot, M., Nagahashi, H., Szymczak, P.: Elastic moduli of simple mass spring models. Vis. Comput. **31**(10), 1339–1350 (2015)
33. Bishop, R.E.D., Johnson, D.C.: The Mechanics of Vibration. Cambridge University Press (2011)
34. Hughes, T.J., Cottrell, J.A., Bazilevs, Y.: Isogeometric analysis: CAD, finite elements, NURBS, exact geometry and mesh refinement. Comput. Methods Appl. Mech. Eng. **194**(39), 4135–4195 (2005)
35. Wall, W.A., Frenzel, M.A., Cyron, C.: Isogeometric structural shape optimization. Comput. Methods Appl. Mech. Eng. **197**(33), 2976–2988 (2008)
36. Greco, L., Cuomo, M.: An implicit G1 multi patch B-spline interpolation for Kirchhoff-Love space rod. Comput. Methods Appl. Mech. Eng. **269**, 173–197 (2014)
37. Cazzani, A., Malagù, M., Turco, E.: Isogeometric analysis: a powerful numerical tool for the elastic analysis of historical masonry arches. Contin. Mech. Thermodyn. **28**(1–2), 139–156 (2016)
38. Toklu, Y.C.: Nonlinear analysis of trusses through energy minimization. Comput. Struct. **82**(20), 1581–1589 (2004)
39. Temür, R., Türkan, Y.S., Toklu, Y.C.: Geometrically nonlinear analysis of trusses using particle swarm optimization. In: Recent Advances in Swarm Intelligence and Evolutionary Computation, pp. 283–300. Springer International Publishing (2015)
40. Kaveh, A., Talatahari, S.: Hybrid algorithm of harmony search, particle swarm and ant colony for structural design optimization. In: Harmony Search Algorithms for Structural Design Optimization, pp. 159–198. Springer Berlin Heidelberg (2009)
41. Clerc, M.: Particle Swarm Optimization, vol. 93. John Wiley & Sons (2010)
42. Vaz Jr., M., Cardoso, E.L., Stahlschmidt, J.: Particle swarm optimization and identification of inelastic material parameters. Eng. Comput. **30**(7), 936–960 (2013)
43. Ballerini, M., Cabibbo, N., Candelier, R., Cavagna, A., Cisbani, E., Giardina, I., Lecomte, V., Orlandi, A., Parisi, G., Procaccini, A., et al.: Interaction ruling animal collective behavior depends on topological rather than metric distance: Evidence from a field study. Proc. Natl. Acad. Sci. **105**(4), 1232–1237 (2008)
44. Bellomo, N., Soler, J.: On the mathematical theory of the dynamics of swarms viewed as complex systems. Math. Models Methods Appl. Sci. **22**(supp01), 1140006 (2012)
45. Bellomo, N., Dogbe, C.: On the modelling crowd dynamics from scaling to hyperbolic macroscopic models. Math. Models Methods Appl. Sci. **18**(supp01), 1317–1345 (2008)
46. Bellomo, N., Knopoff, D., Soler, J.: On the difficult interplay between life, "complexity", and mathematical sciences. Math. Models Methods Appl. Sci. **23**(10), 1861–1913 (2013)
47. Berrimi, S., Messaoudi, S.A.: Exponential decay of solutions to a viscoelastic equation with nonlinear localized damping. Electron. J. Differ. Eq. **88**(2004), 1–10 (2004)
48. Berrimi, S., Messaoudi, S.A.: Existence and decay of solutions of a viscoelastic equation with a nonlinear source. Nonlinear analysis: theory. Methods Appl. **64**(10), 2314–2331 (2006)
49. Messaoudi, S.A., Tatar, N.E.: Exponential and polynomial decay for a quasilinear viscoelastic equation. Nonlinear analysis: theory. Methods Appl. **68**(4), 785–793 (2008)
50. Liang, F., Gao, H.: Exponential energy decay and blow-up of solutions for a system of nonlinear viscoelastic wave equations with strong damping. Bound. Value Prob. **2011**(1), 1 (2011)
51. Sih, G.C.: Mechanics of Fracture Initiation and Propagation: Surface and Volume Energy Density Applied as Failure Criterion, vol. 11. Springer Science & Business Media (2012)
52. Luongo, A.: A unified perturbation approach to static/dynamic coupled instabilities of nonlinear structures. Thin-Walled Struct. **48**(10), 744–751 (2010)
53. Luongo, A.: On the use of the multiple scale method in solving 'difficult'bifurcation problems. Mathematics and Mechanics of Solids (2015). doi:10.1177/1081286515616053
54. Luongo, A.: Mode localization by structural imperfections in one-dimensional continuous systems. J. Sound Vib. **155**(2), 249–271 (1992)

55. Piccardo, G., Pagnini, L.C., Tubino, F.: Some research perspectives in galloping phenomena: critical conditions and post-critical behavior. Contin. Mech. Thermodyn. **27**(1–2), 261–285 (2015)
56. Rizzi, N.L., Varano, V., Gabriele, S.: Initial postbuckling behavior of thin-walled frames under mode interaction. Thin-Walled Struct. **68**, 124–134 (2013)

# A Review on Wave Propagation Modeling in Band-Gap Metamaterials via Enriched Continuum Models

Angela Madeo, Patrizio Neff, Gabriele Barbagallo, Marco Valerio d'Agostino and Ionel-Dumitrel Ghiba

**Abstract** In the present contribution we show that the relaxed micromorphic model is the only non-local continuum model which is able to account for the description of band-gaps in metamaterials for which the kinetic energy accounts separately for micro and macro-motions without considering a micro-macro coupling. Moreover, we show that when adding a gradient inertia term which indeed allows for the description of the coupling of the vibrations of the microstructure to the macroscopic motion of the unit cell, other enriched continuum models of the micromorphic type

A. Madeo (✉) · M.V. d'Agostino
LGCIE, INSA-Lyon, Université de Lyon, 20 Avenue Albert Einstein, 69621
Villeurbanne Cedex, France
e-mail: angela.madeo@insa-lyon.fr

M.V. d'Agostino
e-mail: marco-valerio.dagostino@insa-lyon.fr

A. Madeo
IUF, Institut Universitaire de France, 1 rue Descartes, 75231 Paris Cedex 05, France

P. Neff
Head of Chair for Nonlinear Analysis and Modelling, Fakultät für Mathematik,
Universität Duisburg-Essen, Mathematik-Carrée, Thea-Leymann-Straße 9,
45127 Essen, Germany
e-mail: patrizio.neff@uni-due.de

G. Barbagallo
LaMCoS-CNRS \& LGCIE, INSA-Lyon, Universitité de Lyon, 20 Avenue
Albert Einstein, 69621 Villeurbanne Cedex, France
e-mail: gabriele.barbagallo@insa-lyon.fr

I.-D. Ghiba
Lehrstuhl Für Nichtlineare Analysis und Modellierung, Fakultät für Mathematik,
Universität Duisburg-Essen, Thea-Leymann Str. 9, 45127 Essen, Germany
e-mail: dumitrel.ghiba@uni-due.de, dumitrel.ghiba@uaic.ro

I.-D. Ghiba
Department of Mathematics, Alexandru Ioan Cuza University of Iaşi, Blvd. Carol I,
no. 11, 700506 Iaşi, Romania

I.-D. Ghiba
Octav Mayer Institute of Mathematics of the Romanian Academy, Iaşi Branch,
700505 Iaşi, Romania

© Springer Nature Singapore Pte Ltd. 2017
F. dell'Isola et al. (eds.), *Mathematical Modelling in Solid Mechanics*,
Advanced Structured Materials 69, DOI 10.1007/978-981-10-3764-1_6

may allow the description of the onset of band-gaps. Nevertheless, the relaxed micromorphic model proves to be yet the most effective enriched continuum model which is able to describe multiple band-gaps in non-local metamaterials.

## 1 Introduction

In the last years, a lot of interest has been raised from a class of microscopically heterogeneous materials which show exotic behaviors such as that of "stopping" the propagation of elastic waves. In some cases, the waves lose some of the energy due to micro-diffusion phenomena (Bragg scattering) or even local resonance of the microstructure (Mie resonance). These effects can be exploited to design innovative materials whose dynamical behavior differs completely from the classical materials usually employed in engineering sciences.

As a matter of fact, classical Cauchy continuum theories, are not always well adapted to cover the wealth of experimental evidences on the dynamical behavior of real materials. As a first point, in fact, real materials commonly show dispersive behaviors, which means that the speed of propagation of the traveling wave changes when considering smaller wavelengths. Such phenomenon is not astonishing if one thinks that the structure of matter changes when observing it at smaller scales. It suffices to go down to the scale of the crystals or molecules to be aware of the heterogeneity of matter. It is for this reason that waves with wavelengths which are small enough to "sense" the presence of the microstruture will propagate at a different speed than other waves with higher wavelengths. Cauchy continuum theories are not able in any case to account for dispersive phenomena and are a good approximation of reality only for those materials which do not exhibit their heterogeneity at the scale of interest. As far as one wants to model dispersive behaviors, Cauchy continuum theories are no longer adapted and more refined models need to be introduced. One possibility is to introduce second or higher order theories so allowing the description of dispersion for the acoustic modes (see e.g. [4, 21]). Nevertheless, if second gradient theories may, on the one hand, be of use for the description of some dispersive behaviors, on the other hand they are often insufficient to describe more complex behaviors of metamaterials in which the microstructure can have its own vibrational modes independently of the motion of the unit cell. In order to describe the complex dynamical behavior of such metamaterials in a continuum framework, the introduction of continuum models with enriched kinematics (micromorphic models) is a mandatory requirement [5, 7, 8, 11, 13]. Continuum models of the micromorphic type, in fact, allow for the description of microstructure-related vibrational modes thanks to the introduction of extra degrees of freedom with respect to the displacement field alone.

The insufficiency of Cauchy continuum theories becomes even more evident when considering more complex metamaterials which are able to inhibit wave propagation, i.e. so called band-gap metamaterials. To catch the complex dynamical behavior of such materials, even some of the available micromorphic models are not sufficiently adapted. Indeed, it has been shown in previous contributions that the relaxed micromorphic model is the only continuum model of the micromorphic type which is able to account for the description of band-gaps when considering a kinetic energy in which the macroscopic and microscopic motions are completely uncoupled [7, 8, 11, 12]. In this contribution we will show, following what done in [10], that the addition of kinetic energy terms which couple the motions of the microstructure to the macro-motions of the unit cells may have a deep impact on the ability of describing band-gaps behaviors.

## 1.1 Notations

In this contribution, we denote by $\mathbb{R}^{3\times 3}$ the set of real $3 \times 3$ second order tensors, written with capital letters. We denote respectively by $\cdot$ , : and $\langle \cdot, \cdot \rangle$ a simple and double contraction and the scalar product between two tensors of any suitable order.[1] Everywhere we adopt the Einstein convention of sum over repeated indices if not differently specified. The standard Euclidean scalar product on $\mathbb{R}^{3\times 3}$ is given by $\langle X, Y\rangle_{\mathbb{R}^{3\times 3}} = \operatorname{tr}(X \cdot Y^T)$, and thus the Frobenius tensor norm is $\|X\|^2 = \langle X, X\rangle_{\mathbb{R}^{3\times 3}}$. In the following we omit the index $\mathbb{R}^3, \mathbb{R}^{3\times 3}$. The identity tensor on $\mathbb{R}^{3\times 3}$ will be denoted by $\mathbb{1}$, so that $\operatorname{tr}(X) = \langle X, \mathbb{1}\rangle$.

We consider a body which occupies a bounded open set $B_L$ of the three-dimensional Euclidian space $\mathbb{R}^3$ and assume that its boundary $\partial B_L$ is a smooth surface of class $C^2$. An elastic material fills the domain $B_L \subset \mathbb{R}^3$ and we refer the motion of the body to rectangular axes $Ox_i$.

For vector fields $v$ with components in $\mathrm{H}^1(B_L)$, i.e. $v = (v_1, v_2, v_3)^T$ , $v_i \in \mathrm{H}^1(B_L)$, we define $\nabla v = \left((\nabla v_1)^T, (\nabla v_2)^T, (\nabla v_3)^T\right)^T$, while for tensor fields $P$ with rows in $\mathrm{H}(\operatorname{curl}; B_L)$, resp. $\mathrm{H}(\operatorname{div}; B_L)$, i.e. $P = \left(P_1^T, P_2^T, P_3^T\right)$, $P_i \in \mathrm{H}(\operatorname{curl}; B_L)$ resp. $P_i \in \mathrm{H}(\operatorname{div}; B_L)$ we define $\operatorname{Curl} P = \left((\operatorname{curl} P_1)^T, (\operatorname{curl} P_2)^T, (\operatorname{curl} P_3)^T\right)^T$, $\operatorname{Div} P = (\operatorname{div} P_1, \operatorname{div} P_2, \operatorname{div} P_3)^T$ .

A subscript $_{,t}$ will indicate derivation with respect to time and, analogously a subscript $_{,tt}$ stands for the second derivative of the considered quantity with respect to time.

As for the kinematics of the considered micromorphic continua, we introduce the functions

$$\chi(X, t) : B_L \to \mathbb{R}^3, \qquad P(X, t) : B_L \to \mathbb{R}^{3\times 3},$$

which are known as *placement* vector field and *micro-distortion* tensor, respectively. The physical meaning of the placement field is that of locating, at any instant $t$, the current position of the material particle $X \in B_L$, while the micro-distortion field describes deformations of the microstructure embedded in the material particle $X$.

[1]For example, $(A \cdot v)_i = A_{ij} v_j$, $(A \cdot B)_{ik} = A_{ij} B_{jk}$, $A : B = A_{ij} B_{ji}$, $(C \cdot B)_{ijk} = C_{ijp} B_{pk}$, $(C : B)_i = C_{ijp} B_{pj}$, $\langle v, w\rangle = v \cdot w = v_i w_i$, $\langle A, B\rangle = A_{ij} B_{ij}$ etc.

**Table 1** Values of the elastic parameters used in the numerical simulations (*left*), characteristic lengths and inertiae (*center*) and corresponding values of the Lamé parameters and of the Young modulus and Poisson ratio (*right*), for the formulas needed to calculate the homogenized macroscopic parameters starting from the microscopic ones, see [1]

| Parameter | Value | Unit |
|---|---|---|
| $\mu_e$ | 200 | $MPa$ |
| $\lambda_e = 2\mu_e$ | 400 | $MPa$ |
| $\mu_c = 5\mu_e$ | 1000 | $MPa$ |
| $\mu_{\text{micro}}$ | 100 | $MPa$ |
| $\lambda_{\text{micro}}$ | 100 | $MPa$ |

| Parameter | Value | Unit |
|---|---|---|
| $L_c$ | 1 | $mm$ |
| $\rho$ | 2000 | $kg/m^3$ |
| $\eta$ | $10^{-2}$ | $kg/m$ |
| $\overline{\eta}_i$ | $10^{-1}$ | $kg/m$ |

| Parameter | Value | Unit |
|---|---|---|
| $\lambda_{\text{macro}}$ | 82.5 | $MPa$ |
| $\mu_{\text{macro}}$ | 66.7 | $MPa$ |
| $E_{\text{macro}}$ | 170 | $MPa$ |
| $\nu_{\text{macro}}$ | 0.28 | – |

As it is usual in continuum mechanics, the displacement field can also be introduced as the function $u(X,t) : B_L \to \mathbb{R}^3$ defined as

$$u(X,t) = \chi(X,t) - X.$$

In the remainder of the paper, the following acronyms will be used to refer to the branches of the dispersion curves:

- TRO: transverse rotational optic,
- TSO: transverse shear optic,
- TCVO: transverse constant-volume optic,
- LA: longitudinal acoustic,
- $LO_1$-$LO_2$: $1^{st}$ and $2^{nd}$ longitudinal optic,
- TA: transverse acoustic,
- $TO_1$-$TO_2$: $1^{st}$ and $2^{nd}$ transverse optic.

If not differently specified, the results presented in this paper are obtained for values of the elastic coefficients chosen as in Table 1 (see Eqs. (1), (2), (5), (8), (11) and (14) for their definition).

## 1.2 *The Fundamental Role of Micro-Inertia in Enriched Continuum Mechanics*

As far as enriched continuum models are concerned, a central issue which is also an open scientific question is that of identifying the role of so-called micro-inertia terms on the dispersive behavior of such media. As a matter of fact, enriched continuum models usually provide a richer kinematics, with respect to the classical macroscopic displacement field alone, which is related to the possibility of describing the motions of the microstructure inside the unit cell. The adoption of such enriched kinematics (given by the displacement field $u$ and the micro-distortion tensor $P$, see e.g. [5–9, 11, 12, 14, 19]) as we will see, allows for the introduction of constitutive laws for the strain energy density that are able to describe the mechanical behavior of some metamaterials in the static regime. When the dynamical regime is considered, things become even more delicate since the choice of micro-inertia terms to be introduced in the kinetic energy density must be carefully based on

- a compatibility with the chosen kinematics and constitutive laws used for the description of the static regime,
- the specific inertial characteristics of the metamaterial that one wants to describe (e.g. eventual coupling of the motion of the microstructure with the macro motions of the unit cell, specific resistance of the microstructure to independent motion, etc.).

In the present paper, we will suppose that the kinetic energy takes the following general form (see [10]):

$$J = \underbrace{\frac{1}{2}\rho \left\|u_{,t}\right\|^2}_{\text{Cauchy inertia}} + \underbrace{\frac{1}{2}\eta \left\|P_{,t}\right\|^2}_{\text{free micro-inertia}} + \underbrace{\frac{1}{2}\overline{\eta}_1 \left\| \text{dev sym } \nabla u_{,t}\right\|^2 + \frac{1}{2}\overline{\eta}_2 \left\| \text{skew } \nabla u_{,t}\right\|^2 + \frac{1}{6}\overline{\eta}_3 \,\text{tr}\left(\nabla u_{,t}\right)^2}_{\text{new gradient micro-inertia}}, \tag{1}$$

where $\rho$ is the value of the average macroscopic mass density of the considered metamaterial, $\eta$ is the free micro-inertia density and the $\overline{\eta}_i$, $i = \{1, 2, 3\}$ are the gradient micro-inertia densities associated to the different terms of the Cartan-Lie decomposition of $\nabla u$. We will be hence able to explicitly show which is the specific role of the gradient micro-inertia on the onset of band-gaps in continuum models of the micromorphic type. More precisely, we will highlight which is the effect of the introduction of gradient micro-inertia terms on different enriched continuum models, namely:

- the classical relaxed micromorphic model,
- the relaxed micromorphic model with curvature $\|\text{Div}P\|^2 + \|\text{Curl}P\|^2$,
- the relaxed micromorphic model with curvature $\|\text{Div}P\|^2$,
- the standard Mindlin-Eringen model,
- the internal variable model.

## 2 The Classical Relaxed Micromorphic Model

Our relaxed micromorphic model endows Mindlin-Eringen's representation with the second order **dislocation density tensor** $\alpha = -\,\text{Curl}P$ instead of the third order tensor $\nabla P$.[2] In the isotropic case the energy of the relaxed micromorphic model reads

[2]The dislocation tensor is defined as $\alpha_{ij} = -\left(\text{Curl}P\right)_{ij} = -P_{ih,k}\varepsilon_{jkh}$, where $\varepsilon$ is the Levi-Civita tensor and Einstein notation of sum over repeated indices is used.

$$W = \underbrace{\mu_e \,\|\, \mathrm{sym}\,(\nabla u - P)\|^2 + \frac{\lambda_e}{2}\,(\mathrm{tr}\,(\nabla u - P))^2}_{\text{isotropic elastic} - \text{energy}} + \underbrace{\mu_c \,\|\, \mathrm{skew}\,(\nabla u - P)\|^2}_{\text{rotational elastic coupling}} \tag{2}$$

$$+ \underbrace{\mu_{\mathrm{micro}} \,\|\, \mathrm{sym}\, P\|^2 + \frac{\lambda_{\mathrm{micro}}}{2}\,(\mathrm{tr} P)^2}_{\text{micro} - \text{self} - \text{energy}} + \underbrace{\frac{\mu_e L_c^2}{2}\,\|\, \mathrm{Curl} P\|^2}_{\text{isotropic curvature}}\,,$$

where the parameters and the elastic stress are analogous to the standard Mindlin-Eringen micromorphic model. The model is well-posed in the static and dynamical case including when $\mu_c = 0$, see [6, 18].

Here, the complexity of the classical Mindlin-Eringen micromorphic model has been decisively reduced featuring basically only symmetric gradient micro-like variables and the Curl of the micro-distortion $P$. However, the relaxed model is still general enough to include the full micro-stretch as well as the full Cosserat micro-polar model, see [19]. Furthermore, well-posedness results for the static and dynamical cases have been provided in [19] making decisive use of recently established new coercive inequalities, generalizing Korn's inequality to incompatible tensor fields [2, 3, 15–17, 20].

The relaxed micromorphic model counts 6 constitutive parameters in the isotropic case ($\mu_e, \lambda_e, \mu_{\mathrm{micro}}, \lambda_{\mathrm{micro}}, \mu_c, L_c$). The characteristic length $L_c$ is intrinsically related to non-local effects due to the fact that it weights a suitable combination of first order space derivatives of $P$ in the strain energy density (2). For a general presentation of the features of the relaxed micromorphic model in the anisotropic setting, we refer to [1].

The associated equations of motion in strong form, obtained by a classical least action principle take the form (see [7, 8, 12, 18])

$$\rho\, u_{,tt} - \mathrm{Div}[\,\mathscr{I}\,] = \mathrm{Div}\,[\,\widetilde{\sigma}\,], \qquad \eta\, P_{,tt} = \widetilde{\sigma} - s - \mathrm{Curl}\, m, \tag{3}$$

where

$$\begin{aligned}
\mathscr{I} &= \overline{\eta}_1\ \mathrm{dev\,sym}\ \nabla u_{,tt} + \overline{\eta}_2\ \mathrm{skew}\ \nabla u_{,tt} + \frac{1}{3}\overline{\eta}_3\,\mathrm{tr}\left(\nabla u_{,tt}\right),\\
\widetilde{\sigma} &= 2\,\mu_e\ \mathrm{sym}\,(\nabla u - P) + \lambda_e\ \mathrm{tr}\,(\nabla u - P)\ \mathbb{1} + 2\,\mu_c\ \mathrm{skew}\,(\nabla u - P),\\
s &= 2\,\mu_{\mathrm{micro}}\ \mathrm{sym}\, P + \lambda_{\mathrm{micro}}\ \mathrm{tr}\,(P)\ \mathbb{1},\\
m &= \mu_e L_c^2\ \mathrm{Curl} P.
\end{aligned} \tag{4}$$

The fact of adding a gradient micro-inertia in the kinetic energy (1) modifies the strong-form PDEs of the relaxed micromorphic model with the addition of the new term $\mathscr{I}$. Of course, boundary conditions would also be modified with respect to the ones presented in [9, 12]. The study of the new boundary conditions induced by

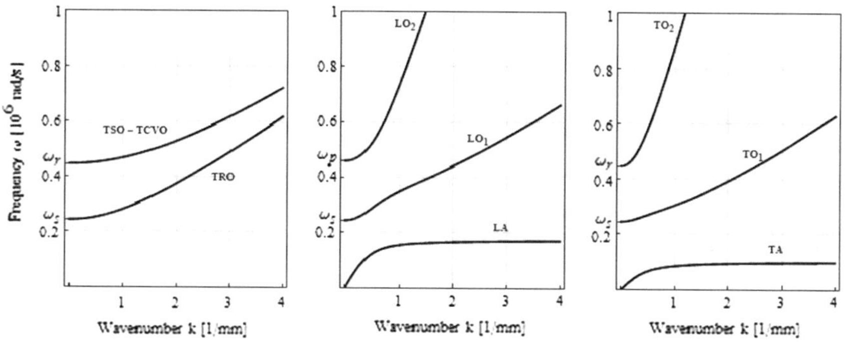


**Fig. 1** Dispersion relations $\omega = \omega(k)$ of the **relaxed micromorphic model** for the uncoupled (*left*), longitudinal (*center*) and transverse (*right*) waves with vanishing gradient micro-inertia $\overline{\eta} = 0$. One complete band-gap is possible

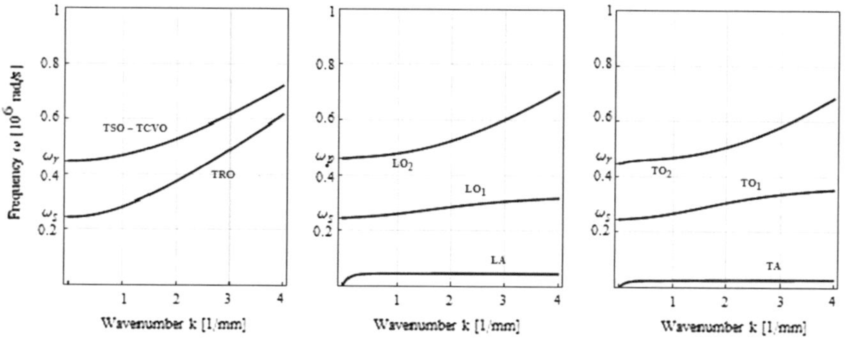


**Fig. 2** Dispersion relations $\omega = \omega(k)$ of the **relaxed micromorphic model** for the uncoupled (*left*), longitudinal (*center*) and transverse (*right*) waves with non-vanishing gradient micro-inertia $\overline{\eta} \neq 0$. Two band-gaps are possible

gradient micro-inertia will be the object of a subsequent paper where the effect of such extra terms on the conservation of energy will also be analyzed (Fig. 1).

As it has been shown in previous contributions [7, 8, 11], the relaxed micromorphic model is able to capture band-gap behaviors thanks to the fact that the acoustic branches have a horizontal asymptote. We show in Fig. 2 the dispersion relations obtained in previous work which are recovered here setting the gradient micro inertia to be vanishing ($\overline{\eta} = 0$).

Things are different when adding a gradient micro-inertia $\overline{\eta} \neq 0$. Surprisingly, the combined effect of the free micro-inertia $\eta$ with the gradient micro-inertia can lead to the onset of a second longitudinal and transverse band gap, due to the fact that the first longitudinal and transverse acoustic branches ($LO_1$ and $TO_1$) are flattened. Moreover, it is possible to notice that the addition of gradient micro-inertiae $\overline{\eta}_1$, $\overline{\eta}_2$ and $\overline{\eta}_3$ has no effect on the cut-off frequencies, which only depend on the free micro-inertia $\eta$ (and of course on the constitutive parameters).

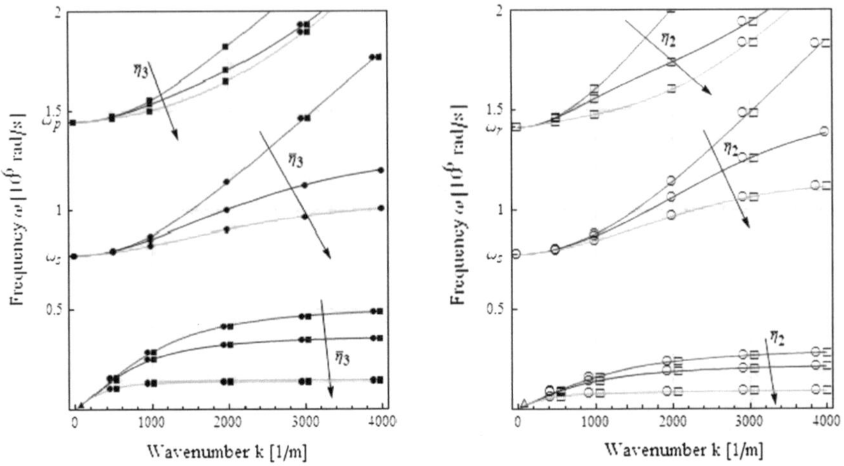


**Fig. 3** Dispersion relations $\omega = \omega(k)$ of the **relaxed micromorphic model** for longitudinal (*left*) and transverse (*right*) waves with increasing gradient micro-inertia. The markers indicate the main mode of vibration considering: *black triangle* $u_1$, *black circle* $P^S$, *black square* $P^D$, *empty triangle* $u_\xi$, *empty circle* $P_{(1\xi)}$ and *empty square* $P_{[1\xi]}$ with $\xi = 2, 3$. When two markers are present it means that there is no clear main mode

In Fig. 3 we show more explicitly the flattening effect of the gradient inertia parameters on longitudinal and transverse waves. In the same Figure we indicate the main mode of vibration associated to each branch of the dispersion curves. In contrast to Cauchy models, the modes of vibration change when changing the wavenumbers.

In particular, it is possible to notice that the main mode of the acoustic branches is the longitudinal or transverse displacement (as it is the case for Cauchy media) only for very small wavenumbers $k$ (large wavelengths). Increasing the wavenumber (decreasing the wavelength), the longitudinal and transverse vibrations are characterized by a coupling of the modes $P^S$ and $P^D$, and $P_{(1\xi)}$ and $P_{[1\xi]}$, respectively. Moreover, it can be seen that the optic branches are characterized by one main microstructure-related vibrational mode until relatively high values of the wavenumber $k$. The coupling occurs for higher values of $k$ and it is strongly influenced by the effects of the gradient micro-inertia.

## 3 The Micromorphic Model with Curvature $\|\mathrm{Div}P\|^2 + \|\mathrm{Curl}P\|^2$

We consider now an extension of the relaxed micromorphic model obtained considering the energy (see [11]):

$$W = \underbrace{\mu_e \,\|\, \text{sym}\,(\nabla u - P)\|^2 + \frac{\lambda_e}{2}\,(\text{tr}\,(\nabla u - P))^2}_{\text{isotropic elastic} - \text{energy}} + \underbrace{\mu_c \,\|\, \text{skew}\,(\nabla u - P)\|^2}_{\text{rotational elastic coupling}} \quad (5)$$
$$+ \underbrace{\mu_{\text{micro}} \,\|\, \text{sym}\,P\|^2 + \frac{\lambda_{\text{micro}}}{2}\,(\text{tr}P)^2}_{\text{micro} - \text{self} - \text{energy}} + \underbrace{\frac{\mu_e L_c^2}{2}\left(\|\text{Div}P\|^2 + \|\,\text{Curl}P\|^2\right)}_{\text{augmented isotropic curvature}}.$$

The dynamical equilibrium equations are:

$$\rho\, u_{,tt} - \text{Div}[\,\mathscr{I}\,] = \text{Div}\,[\,\widetilde{\sigma}\,], \qquad \eta\, P_{,tt} = \widetilde{\sigma} - s - M, \quad (6)$$

where

$$\begin{aligned}
\mathscr{I} &= \overline{\eta}_1 \,\text{dev sym}\, \nabla u_{,tt} + \overline{\eta}_2 \,\text{skew}\, \nabla u_{,tt} + \frac{1}{3}\overline{\eta}_3 \,\text{tr}\left(\nabla u_{,tt}\right),\\
\widetilde{\sigma} &= 2\,\mu_e \,\text{sym}\,(\nabla u - P) + \lambda_e \,\text{tr}\,(\nabla u - P)\, \mathbb{1} + 2\,\mu_c \,\text{skew}\,(\nabla u - P), \quad (7)\\
s &= 2\,\mu_{\text{micro}} \,\text{sym}\,P + \lambda_{\text{micro}} \,\text{tr}\,(P)\, \mathbb{1},\\
M &= -\mu_e L_c^2 \underbrace{(\nabla\,(\text{Div}P) - \text{Curl}\,\text{Curl}P)}_{=\text{Div}\nabla P = \Delta P}.
\end{aligned}$$

Note that the structure of the equation is equivalent to the one obtained in the standard micromorphic model with curvature $\frac{1}{2}\|\nabla P\|^2$, see Eq. (12) in Sect. 5.

We present the dispersion relations obtained with a vanishing gradient inertia (Fig. 4) and for a non-vanishing gradient micro-inertia (Fig. 5). We conclude that when considering the model with micromorphic medium with $\|\text{Div}P\|^2 + \|\,\text{Curl}P\|^2$ with vanishing gradient micro-inertia, there always exist waves which propagate inside the considered medium independently of the value of frequency even if

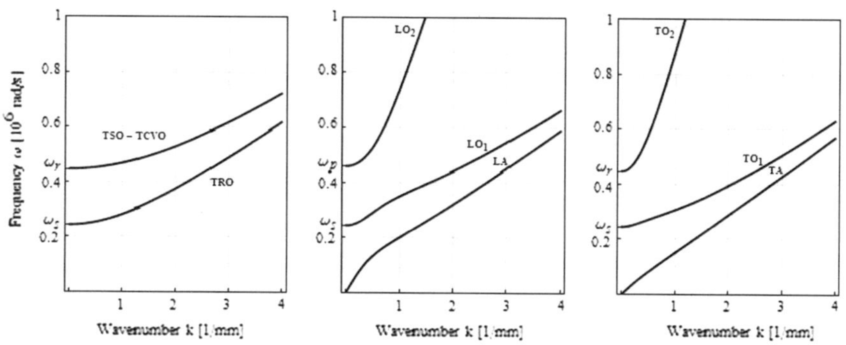


**Fig. 4** Dispersion relations $\omega = \omega(k)$ of the **relaxed micromorphic model with curvature** $\|\text{Div}P\|^2 + \|\,\text{Curl}P\|^2$ for the uncoupled (*left*), longitudinal (*center*) and transverse (*right*) waves with vanishing gradient micro-inertia. No band-gap is possible

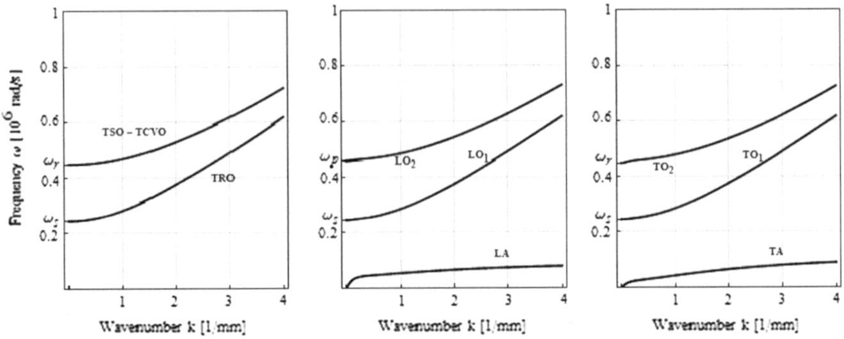


**Fig. 5** Dispersion relations $\omega = \omega(k)$ of the **relaxed micromorphic model with curvature** $\|\text{Div}P\|^2 + \|\text{Curl}P\|^2$ for the uncoupled (*left*), longitudinal (*center*) and transverse (*right*) waves with non-vanishing gradient micro-inertia. One band-gap is possible

considering separately longitudinal, transverse and uncoupled waves. On the other hand, switching on the gradient inertia it is possible to obtain a total band-gap.

## 4 The Micromorphic Model with Curvature $\|\text{Div}P\|^2$

The isotropic micromorphic model with $\|\text{Div}P\|^2$ is yet another variant of the classical relaxed micromorphic model (see [11]) with energy:

$$W = \underbrace{\mu_e \,\|\, \text{sym}\,(\nabla u - P)\|^2 + \frac{\lambda_e}{2}\,(\text{tr}\,(\nabla u - P))^2}_{\text{isotropic elastic} - \text{energy}} + \underbrace{\mu_c \,\|\, \text{skew}\,(\nabla u - P)\|^2}_{\text{rotational elastic coupling}} \tag{8}$$

$$+ \underbrace{\mu_{\text{micro}} \,\|\, \text{sym}\, P\|^2 + \frac{\lambda_{\text{micro}}}{2}\,(\text{tr}P)^2}_{\text{micro} - \text{self} - \text{energy}} + \underbrace{\frac{\mu_e L_d^2}{2}\,\|\text{Div}P\|^2}_{\text{isotropic curvature}} .$$

The dynamical equilibrium equations are:

$$\rho\, u_{,tt} - \text{Div}[\,\mathscr{I}\,] = \text{Div}\,[\,\widetilde{\sigma}\,], \qquad\qquad \eta\, P_{,tt} = \widetilde{\sigma} - s - M, \tag{9}$$

where

$$
\begin{aligned}
\mathscr{I} &= \overline{\eta}_1 \operatorname{dev}\operatorname{sym} \nabla u_{,tt} + \overline{\eta}_2 \operatorname{skew} \nabla u_{,tt} + \frac{1}{3}\overline{\eta}_3 \operatorname{tr}\left(\nabla u_{,tt}\right), \\
\widetilde{\sigma} &= 2\,\mu_e \operatorname{sym}\left(\nabla u - P\right) + \lambda_e \operatorname{tr}\left(\nabla u - P\right)\mathbb{1} + 2\,\mu_c \operatorname{skew}\left(\nabla u - P\right), \qquad (10) \\
s &= 2\,\mu_{\text{micro}} \operatorname{sym} P + \lambda_{\text{micro}} \operatorname{tr}(P)\,\mathbb{1}, \\
M &= -\mu_e L_c^2\, \nabla\,(\operatorname{Div} P)\,.
\end{aligned}
$$

We present the **dispersion relations** obtained with a non-vanishing gradient inertia (Fig. 7) and for a vanishing gradient inertia (Fig. 6). In the Figures we consider uncoupled waves (a), longitudinal waves (b) and transverse waves (c). Even in this case, when considering the micromorphic model with only $\|\operatorname{Div} P\|^2$ with vanishing gradient inertia, there always exist waves which propagate inside the considered medium independently of the value of the frequency and the uncoupled waves assume a peculiar behavior in which the frequency is independent of the wavenumber k. On the other hand, when switching on the gradient inertia, a behavior analogous to the relaxed micromorphic model appears: it is possible to model the onset of two complete band-gaps. The uncoupled waves remain unaffected by the introduction of the gradient micro-inertia and they keep their characteristic of being independent of the wavenumber in strong contrast to what happen for the relaxed micromorphic model in which the uncoupled waves are dispersive.

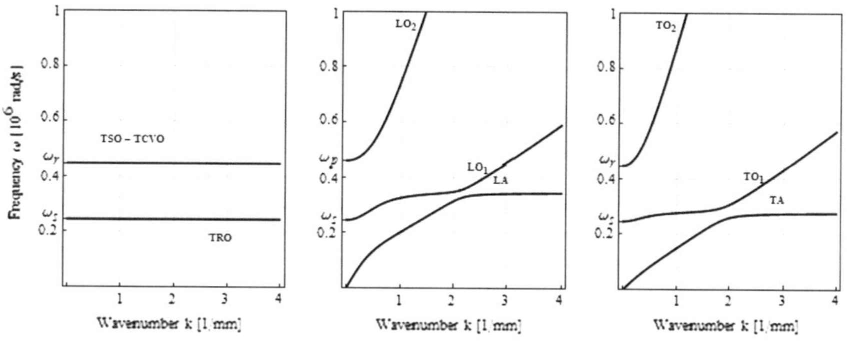


**Fig. 6** Dispersion relations $\omega = \omega(k)$ of the **relaxed micromorphic model with curvature** $\|\operatorname{Div} P\|^2$ for the uncoupled (*left*), longitudinal (*center*) and transverse (*right*) waves with vanishing gradient micro-inertia. No band-gap is possible

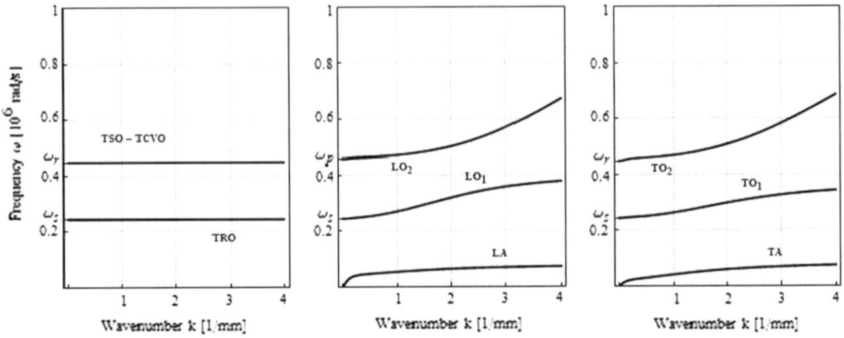


**Fig. 7** Dispersion relations $\omega = \omega(k)$ of the **relaxed micromorphic model with curvature** $\|\mathrm{Div}P\|^2$ for the uncoupled (*left*), longitudinal (*center*) and transverse (*right*) waves with non-vanishing gradient micro-inertia. Two band-gaps are possible

## 5 The Standard Mindlin-Eringen Model

In this section. we discuss the effect on the Mindlin-Eringen of the addition of the gradient micro-inertia $\overline{\eta}\| \nabla u_{,t}\|^2$ to the classical terms $\rho\|u_{,t}\|^2 + \eta\|P_{,t}\|^2$. We recall that the strain energy density for this model in the isotropic case takes the form:

$$W = \underbrace{\mu_e \,\| \operatorname{sym} \left( \nabla u - P\right)\|^2 + \frac{\lambda_e}{2} \left(\operatorname{tr} \left( \nabla u - P\right)\right)^2}_{\text{isotropic elastic} - \text{energy}} + \underbrace{\mu_c \,\| \operatorname{skew} \left( \nabla u - P\right)\|^2}_{\text{rotational elastic coupling}} \tag{11}$$

$$+ \underbrace{\mu_{\text{micro}} \,\| \operatorname{sym} P\|^2 + \frac{\lambda_{\text{micro}}}{2} \left(\operatorname{tr}P\right)^2}_{\text{micro} - \text{self} - \text{energy}} + \underbrace{\frac{\mu_e L_c^2}{2} \,\|\nabla P\|^2}_{\text{isotropic curvature}} .$$

The dynamical equilibrium equations are:

$$\rho\, u_{,tt} - \mathrm{Div}[\,\mathscr{I}\,] = \mathrm{Div}\,[\,\widetilde{\sigma}\,], \qquad \eta\, P_{,tt} = \widetilde{\sigma} - s - M, \tag{12}$$

where

$$\begin{aligned} \mathscr{I} &= \overline{\eta}_1 \operatorname{dev}\operatorname{sym} \nabla u_{,tt} + \overline{\eta}_2 \operatorname{skew} \nabla u_{,tt} + \frac{1}{3}\overline{\eta}_3 \operatorname{tr}\left(\nabla u_{,tt}\right), \\ \widetilde{\sigma} &= 2\,\mu_e \operatorname{sym}\left(\nabla u - P\right) + \lambda_e \operatorname{tr}\left(\nabla u - P\right) \mathbb{1} + 2\,\mu_c \operatorname{skew}\left(\nabla u - P\right), \\ s &= 2\,\mu_{\text{micro}} \operatorname{sym} P + \lambda_{\text{micro}} \operatorname{tr}(P)\, \mathbb{1}, \\ M &= -\mu_e L_c^2 \underbrace{\mathrm{Div}\nabla P}_{=\Delta P}. \end{aligned} \tag{13}$$

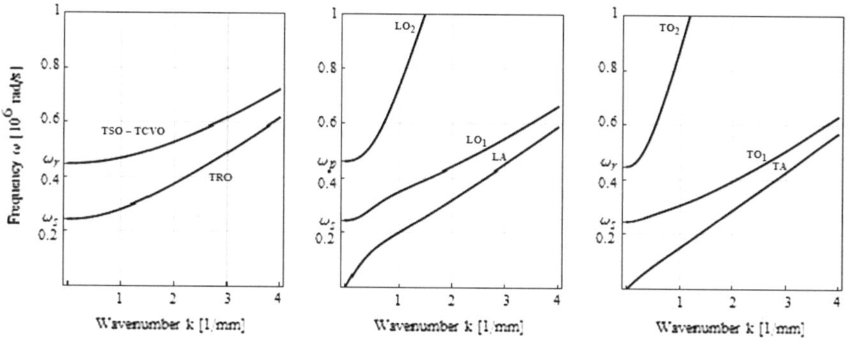


**Fig. 8** Dispersion relations $\omega = \omega(k)$ of the **standard Mindlin-Eringen micromorphic model** for the uncoupled (*left*), longitudinal (*center*) and transverse (*right*) waves with vanishing gradient micro-inertia. No band-gap is possible

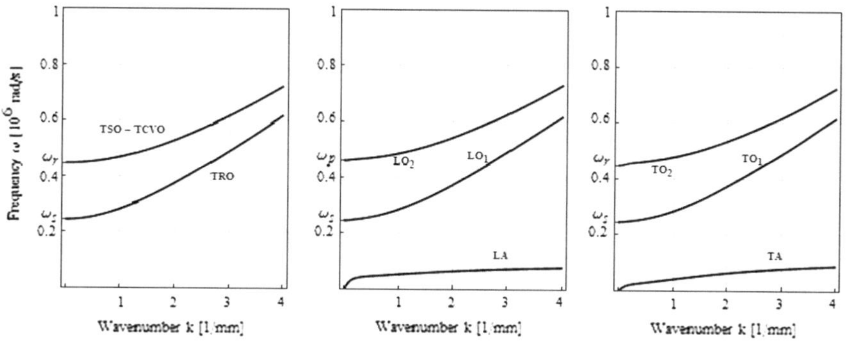


**Fig. 9** Dispersion relations $\omega = \omega(k)$ of the **standard Mindlin-Eringen micromorphic model** for the uncoupled (*left*), longitudinal (*center*) and transverse (*right*) waves with non-vanishing gradient micro-inertia. One band-gap is possible

Recalling the results of [7], we remark that when the gradient micro-inertia is vanishing ($\overline{\eta}_1 = \overline{\eta}_2 = \overline{\eta}_3 = 0$) the Mindlin-Eringen model does not allow the description of band-gaps (see Fig. 8), due to the presence of straight acoustic waves. On the other hand, when switching on the parameters $\overline{\eta}_2$ and $\overline{\eta}_3$ , the acoustic branches are flattened, so that the first band-gap can be created (see Fig. 9). The analogous case for the relaxed micromorphic model (Fig. 1) allowed instead for the description of 2 band gaps.

As already pointed out and as shown in [11], the classical Mindlin-Eringen model can be considered to be equivalent to the relaxed micromorphic model with curvature $\|\mathrm{Div}P\|^2 + \|\,\mathrm{Curl}P\|^2$.

## 6 The Internal Variable Model

Figure 10 shows the behavior of the addition of the gradient micro-inertia $\overline{\eta}\| \nabla u_{,t}\|^2$ in the internal variable model. We recall (see [19]) that the energy for the internal variable model does not include higher space derivatives of the micro-distortion tensor $P$ and, in the isotropic case, takes the form:

$$W = \underbrace{\mu_e \| \operatorname{sym} (\nabla u - P)\|^2 + \frac{\lambda_e}{2} (\operatorname{tr} (\nabla u - P))^2}_{\text{isotropic elastic} - \text{energy}} + \underbrace{\mu_c \| \operatorname{skew} (\nabla u - P)\|^2}_{\text{rotational elastic coupling}} \tag{14}$$

$$+ \underbrace{\mu_{\text{micro}} \| \operatorname{sym} P\|^2 + \frac{\lambda_{\text{micro}}}{2} (\operatorname{tr} P)^2}_{\text{micro} - \text{self} - \text{energy}} .$$

The dynamical equilibrium equations are:

$$\rho\, u_{,tt} - \operatorname{Div}[\mathscr{I}] = \operatorname{Div}[\widetilde{\sigma}], \qquad \eta\, P_{,tt} = \widetilde{\sigma} - s, \tag{15}$$

where

$$\begin{aligned} \mathscr{I} &= \overline{\eta}_1 \operatorname{dev} \operatorname{sym} \nabla u_{,tt} + \overline{\eta}_2 \operatorname{skew} \nabla u_{,tt} + \frac{1}{3}\overline{\eta}_3 \operatorname{tr} \left(\nabla u_{,tt}\right), \\ \widetilde{\sigma} &= 2\,\mu_e \operatorname{sym} (\nabla u - P) + \lambda_e \operatorname{tr} (\nabla u - P)\, \mathbb{1} + 2\,\mu_c \operatorname{skew} (\nabla u - P), \\ s &= 2\,\mu_{\text{micro}} \operatorname{sym} P + \lambda_{\text{micro}} \operatorname{tr} (P)\, \mathbb{1}. \end{aligned} \tag{16}$$

We present the dispersion relations obtained for the internal variable model with a non-vanishing gradient inertia (Fig. 11) and for a vanishing gradient inertia (Fig. 10). We start noticing in Fig. 10 that the internal variable model with vanishing gradient micro-inertia allows for the description of two complete band-gap even if it is not able to account for the presence of non-localities in metamaterials. Moreover, by direct observation of Fig. 11, we can notice that, when switching on the gradient micro-inertia, suitably choosing the relative position of $\omega_r$ and $\omega_p$, the internal variable model allows to account for 3 band gaps. We thus have an extra band-gap with respect to the case with vanishing gradient inertia (Fig. 10) and to the analogous case for the relaxed micromorphic model (see Fig. 2), but we are not able to consider non-local effects. The fact of excluding the possibility of describing non-local effects in metamaterials can be sometimes too restrictive. For example, flattening the curve which originates from $\omega_r$ and which is associated to rotational modes of the microstructure is nonphysical for the great majority of metamaterials.

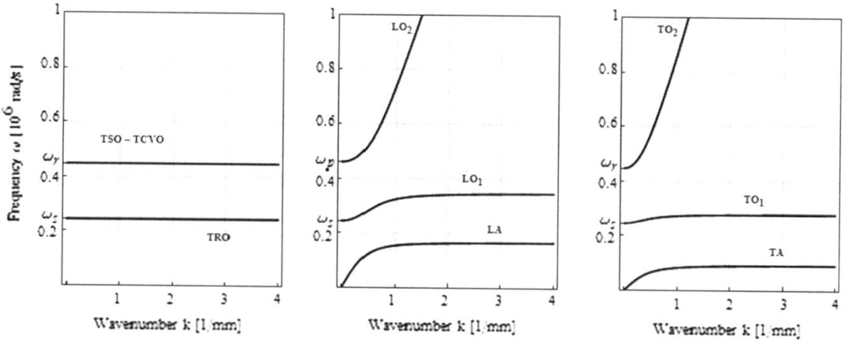


**Fig. 10** Dispersion relations $\omega = \omega(k)$ of the **internal variable model** for the uncoupled (*left*), longitudinal (*center*) and transverse (*right*) waves with vanishing gradient micro-inertia. Two band-gaps are possible but non-local effects cannot be described

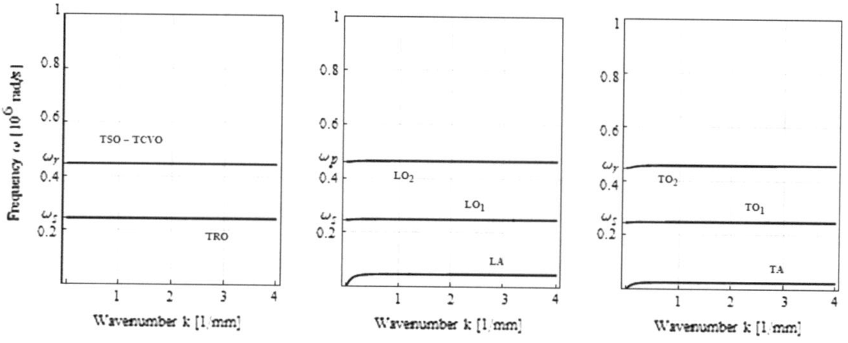


**Fig. 11** Dispersion relations $\omega = \omega(k)$ of the **internal variable model** for the uncoupled (*left*), longitudinal (*center*) and transverse (*right*) waves with non-vanishing gradient micro-inertia. Three band-gap are possible but non-local effects cannot be described. The overall trend of the dispersion curves is unrealistic for the great majority of metamaterials

## 7 Conclusions

In this paper we make a review of some of the available isotropic, linear-elastic, enriched continuum models for the description of the dynamical behavior of metamaterials. We show that the relaxed micromorphic model previously introduced by the authors is the only non-local enriched model which is able to describe band-gaps when considering a kinetic energy independently accounting for micro and macro motions. As far as an inertia term which couples the micro-motions to the macroscopic motions of the unit cell is introduced, also other non-local models exhibit the possibility of describing band-gap behaviors. Nevertheless, the relaxed micromorphic model is still the more effective one to describe (multiple) band-gaps and non-local effects in a realistic way. In fact, even with the addition of the new micro-inertia term, the relaxed model is able to account for the description of two band-gaps, in contrast with to the single band-gap allowed by the Mindlin-Eringen model. The micromorphic model with curvature $\|\text{Div}P\|^2$ also allows for the description of two band-gaps when considering an augmented kinetic energy, but the uncoupled waves are forced to be non-dispersive: this fact can be considered to be a limitation for the realistic description of a wide class of band-gap metamaterials. Finally, the internal variable model with the new kinetic energy terms allow for the description of up to three band gaps. Nevertheless, the overall trends shown by the dispersion curves turn to be quite unrealistic due to the fact that all the branches of the dispersion curves show very low or no dispersion at all.

**Acknowledgements** The work of Ionel-Dumitrel Ghiba was supported by a grant of the Romanian National Authority for Scientific Research and Innovation, CNCS-UEFISCDI, project number PN-II-RU-TE-2014-4-0320. Angela Madeo thanks INSA-Lyon for the funding of the BQR 2016 "Caractérisation mécanique inverse des métamatériaux: modélisation, identification expérimentale des paramètres et évolutions possibles", as well as the CNRS-INSIS for the funding of the PEPS project.

## References

1. Barbagallo, G., d'Agostino, M.V., Abreu, R., Ghiba, I.-D., Madeo, A., Neff, P.: Transparent anisotropy for the relaxed micromorphic model: macroscopic consistency conditions and long wave length asymptotics. Preprint arXiv:1601.03667 (2016)
2. Bauer, S., Neff, P., Pauly, D., Starke, G.: New Poincaré-type inequalities. Comptes Rendus Mathematique **352**(2), 163–166 (2014)
3. Bauer, S., Neff, P., Pauly, D., Starke, G.: Dev-Div- and DevSym-DevCurl-inequalities for incompatible square tensor fields with mixed boundary conditions. ESAIM: Control Optim. Calc. Var. **22**(1), 112–133 (2016)
4. dell'Isola, F., Madeo, A., Placidi, L.: Linear plane wave propagation and normal transmission and reflection at discontinuity surfaces in second gradient 3D continua. Zeitschrift für Angewandte Mathematik und Mechanik **92**(1), 52–71 (2012)
5. Eringen, A.C.: Microcontinuum Field Theories. Springer-Verlag, New York (1999)
6. Ghiba, I.-D., Neff, P., Madeo, A., Placidi, L., Rosi, G.: The relaxed linear micromorphic continuum: existence, uniqueness and continuous dependence in dynamics. Math. Mech. Solids **20**(10), 1171–1197 (2014)
7. Madeo, A., Neff, P., Ghiba, I.-D., Placidi, L., Rosi, G.: Band gaps in the relaxed linear micromorphic continuum. Zeitschrift für Angewandte Mathematik und Mechanik **95**(9), 880–887 (2014)
8. Madeo, A., Neff, P., Ghiba, I.-D., Placidi, L., Rosi, G.: Wave propagation in relaxed micromorphic continua: modeling metamaterials with frequency band-gaps. Contin. Mech. Thermodyn. **27**(4–5), 551–570 (2015)
9. Madeo, A., Barbagallo, G., d'Agostino, M.V., Placidi, L., Neff, P.: First evidence of non-locality in real band-gap metamaterials: determining parameters in the relaxed micromorphic model. Proc. R. Soc. A Math. Phys. Eng. Sci. **472**(2190), 20160169 (2016)
10. Madeo, A., Neff, P., Aifantis, E.C., Barbagallo, G., d'Agostino, M.V.: On the role of micro-inertia in enriched continuum mechanics. Preprint arXiv:1607.07385 (2016)
11. Madeo, A., Neff, P., d'Agostino, M.V., Barbagallo, G.: Complete band gaps including non-local effects occur only in the relaxed micromorphic model. Preprint arXiv:1602.04315 (2016)
12. Madeo, A., Neff, P., Ghiba, I.-D., Rosi, G.: Reflection and transmission of elastic waves in non-local band-gap metamaterials: a comprehensive study via the relaxed micromorphic model. J. Mech. Phys. Solids (2016)

13. Mindlin, R.D.: Microstructure in Linear Elasticity. Technical report, Office of Naval Research (1963)
14. Mindlin, R.D.: Micro-structure in linear elasticity. Arch. Ration. Mech. Anal. **16**(1), 51–78 (1964)
15. Neff, P.: On Korn's first inequality with non-constant coefficients. Proc. R. Soc. Edinb. Sect. A Math. **132**(01), 221 (2002)
16. Neff, P., Pauly, D., Witsch, K.-J.: A canonical extension of Korn's first inequality to H(Curl) motivated by gradient plasticity with plastic spin. Comptes Rendus Mathematique **349**(23), 1251–1254 (2011)
17. Neff, P., Pauly, D., Witsch, K.-J.: Maxwell meets Korn: a new coercive inequality for tensor fields in $R^{n\times n}$ with square-integrable exterior derivative. Math. Methods Appl. Sci. **35**(1), 65–71 (2012)
18. Neff, P., Ghiba, I.-D., Lazar, M., Madeo, A.: The relaxed linear micromorphic continuum: well-posedness of the static problem and relations to the gauge theory of dislocations. Q. J. Mech. Appl. Math. **68**(1), 53–84 (2014)
19. Neff, P., Ghiba, I.-D., Madeo, A., Placidi, L., Rosi, G.: A unifying perspective: the relaxed linear micromorphic continuum. Contin. Mech. Thermodyn. **26**(5), 639–681 (2014)
20. Neff, P., Pauly, D., Witsch, K.-J.: Poincaré meets Korn via Maxwell: extending Korn's first inequality to incompatible tensor fields. J. Differ. Equ. **258**(4), 1267–1302 (2015)
21. Placidi, L., Rosi, G., Giorgio, I., Madeo, A.: Reflection and transmission of plane waves at surfaces carrying material properties and embedded in second-gradient materials. Math. Mech. Solids **19**(5), 555–578 (2014)

# A Continuous Model for the Wave Scattering by a Bounded Defective Domain

**Loïc Le Marrec and Lalaonirina R. Rakotomanana**

**Abstract** Elastic wave propagation and scattering in a media containing a continuous density of defect is modeled with a geometrical approach. The material is supposed to be a Riemann–Cartan manifold with a connection enriched by a nonzero torsion. The study is followed until to reveal analytical solutions. The scattering of a defective domain shows explicitly some non-classical phenomena.

## 1 Introduction

Analysis of wave propagating through a media containing continuous distribution density of defect is performed at a mesoscopic scale in the framework of generalized continua. The main idea is to introduce the material manifold concept in the context of differential geometry e.g. [16, 19]. In a Riemann–Cartan (RC) manifold, the transformation of a continuum is described by both the metric (measuring the shape change) and a connection $\nabla$ (the basic tool for differential operator on a manifold) [1]. For nonvanishing torsion the affine connection differs from the Levi-Civita connection classically used in theory of elasticity $\overline{\nabla}$. Torsion field is related to the incompatible part of a deformation [3] and may be interpreted as the density of Burgers vectors. Modification of classical differential tools engendered by torsion may be seen as the mesoscopic illustration of the microstructural defects [17].

In the framework of elastoplasticity, torsion and metric are the unknown fields that we have to look for in order to entirely describe the transformation of matter [4, 6]. Then, metric and torsion may change in the course of time. However, we focus in this work on a particular case since we only consider a superimposed evolution of a continuum containing defects. The superimposed perturbation is assumed to not modify the torsion field. It is worth to remind that a continuous distribution of defect is present even in linear elasticity and that it plays a significant role on the

L. Le Marrec (✉) · L.R. Rakotomanana
IRMAR UMR 6625 CNRS, Université de Rennes 1, Rennes, France
e-mail: loic.lemarrec@univ-rennes1.fr

L.R. Rakotomanana
e-mail: lalaonirina.rakotomanana-ravelonarivo@univ-rennes1.fr



F. dell'Isola et al. (eds.), *Mathematical Modelling in Solid Mechanics*,
Advanced Structured Materials 69, DOI 10.1007/978-981-10-3764-1_7

value of mesoscopic elastic modulus [18]. Then it seems relevant to investigate the effect of such defect density by means of superimposed linear elastic perturbation. One advantage is to avoid nonlinear difficulties arising for large deformation, and anyway allows us to investigate most of the dynamical behavior in more details.

The continuum is considered to be self-equilibrated and the dynamical perturbation is assumed to be small and elastic. The reference state is modeled by a RC manifold in order to take into account the presence of defects whereas the superimposed elastic perturbation remains a diffeomorphism. Conforming to the generalized continuum method, the conservation equations are formulated with the material connection attached to the reference state. They are expected to hopefully capture the influence of the defects during a superimposed wave. We obtain a three-dimensional wave equation which extends the Navier equation on an arbitrary RC manifold [2]. In the case of a uniform density of screw-type dislocation the dispersion equations are related to one quasi-longitudinal and two quasi-transversal waves.

The second objective of this contribution is an attempt to give a practical application of such theoretical model. Indeed such dynamical investigation may be of practical interest in order to investigate density of structural effects by means of elastic waves [7, 9]. The example analyses the three-dimensional scattering phenomena induced by vertical shear wave propagating normally to a straight cylinder containing uniform density of screw-type defects. This problem has been studied by other authors with different model. Comparison of such approach with others highlights advantages and drawbacks of the model. From this investigation some future work can be considered in order to enlarge the physical versatility of RC model.

## 2 Differential Tools on Riemann–Cartan Geometry

We consider manifolds $\mathbb{E}$ and $\mathbb{M}$ as two states of the same body. The $\mathbb{E}$-state is considered as a homogeneous continuum [15] whereas the $\mathbb{M}$-state is a defective one. Local coordinates $(X^i)_{i=1\ldots 3}$ and metric $g_{ij}$ are associated to $\mathbb{E}$ and the reference material manifold $\mathbb{M}$ is described by local coordinates $(x^\mu)_{\mu=1\ldots 3}$ and metric $g_{\mu\nu}$ (Latin letters refer to $\mathbb{E}$ whereas Greek refer $\mathbb{M}$). Vector basis of the tangent space of $\mathbb{E}$ (resp. $\mathbb{M}$) are $\mathbf{e_i}$ (resp. $\mathbf{e}_\mu$). The transition between both states is performed by the triads $e_i^\mu$ which are the local (not necessarily plastic) transformation of the $\mathbb{E}$-state into the $\mathbb{M}$-state: $\mathrm{d}x^\mu = e_i^\mu\, \mathrm{d}X^i$, for smooth transformation $e_i^\mu = \frac{\partial x^\mu}{\partial X^i} = \partial_i x^\mu$. The material metric $g_{\mu\nu}$ and the reciprocal triad $e_\mu^i$ are given by $g_{\mu\nu} = e_\mu^i e_\nu^j g_{ij}$ and $e_i^\mu e_\mu^j = \delta_i^j$ respectively, where $\delta_i^j$ is the Kronecker symbol. In $\mathbb{M}$, the covariant derivative $\nabla_\alpha \mathbf{e}_\beta$ of vector basis $\mathbf{e}_\beta$ along the direction $\mathbf{e}_\alpha$ is defined by the affine connection $\nabla$: $\nabla_\alpha \mathbf{e}_\beta = \Gamma_{\alpha\beta}^\gamma \mathbf{e}_\gamma$ , where the connection coefficient $\Gamma_{\alpha\beta}^\gamma$ can be written in terms of triads: $\Gamma_{\alpha\beta}^\gamma = e_i^\gamma \partial_\alpha e_\beta^i = -e_\beta^i \partial_\alpha e_i^\gamma$. The material connection is assumed to be metric compatible: $\nabla g = 0$ [14]. In this context the torsion is:

$$S_{\alpha\beta}^\gamma := \Gamma_{\alpha\beta}^\gamma - \Gamma_{\beta\alpha}^\gamma = e_i^\gamma \left(\partial_\alpha e_\beta^i - \partial_\beta e_\alpha^i\right), \quad \text{then} \quad S_{\alpha\beta}^\gamma = -S_{\beta\alpha}^\gamma\,.$$

The relation $\partial_\alpha e^i_\beta - \partial_\beta e^i_\alpha = 0$ being a Schwartz integrability condition for the triads, nonholonomic transformations are thus defined by a nonzero torsion. For zero torsion $\nabla$ becomes the Levi-Civita connection $\overline{\nabla}$ for which coefficients are the Christoffel symbols $\overline{\Gamma}^{\gamma}_{\alpha\beta}$ uniquely defined by the metric. The connection coefficient casts into two parts: $\Gamma^{\gamma}_{\alpha\beta} = \overline{\Gamma}^{\gamma}_{\alpha\beta} + K^{\gamma}_{\alpha\beta}$, with the contortion tensor $K$:

$$K^{\gamma}_{\alpha\beta} = (1/2)\left(S^{\gamma}_{\alpha\beta} + g^{\gamma\lambda} S^{\kappa}_{\lambda\beta} g_{\kappa\alpha} + g^{\gamma\lambda} S^{\kappa}_{\lambda\alpha} g_{\kappa\beta}\right). \tag{1}$$

For a given set of metric and connection, Bianchi's identities are satisfied. The local coordinates attached to matter give tensors as if we live in the $\mathbb{M}$-space. However, for an external observer the tensor fields are generally still expressed in the coordinates of the ambient space which is Euclidean. We can always express tensors in the Euclidian coordinates by using the triads and reciprocal triads.

# 3 Elastic Wave in a Riemann–Cartan Manifold

## 3.1 Constitutive Model

Let $\mathbf{u}$, $\varepsilon$ and $\sigma$ be respectively the displacement field, the small strain tensor and the symmetric stress tensor associated to the elastic superimposed perturbation. Then the conservation laws include the mass conservation $\rho = \rho_0 \det(e^{\alpha}_i)$ and the momentum conservation [16]. The latter is (in absence of external force):

$$\rho_0\, \partial^2_t \mathbf{u} = \nabla \cdot \sigma \tag{2}$$

We consider linear constitutive law of an homogeneous isotropic elastic medium,

$$\sigma = \lambda \mathrm{Tr}(\varepsilon)\mathbf{G} + 2\mu\varepsilon\ , \tag{3}$$

with Lamé coefficients $\lambda$ and $\mu$ and $\mathbf{G}$ the identity tensor (metric). The small-strain tensor is defined by:

$$\varepsilon := (1/2)\left(\nabla\mathbf{u} + (\nabla\mathbf{u})^t\right)\ . \tag{4}$$

Again, the small strain is defined thanks to the covariant derivative $\nabla$ according to the RC geometry of the reference state (see [2] for more detail). In that sense the strain describes how the material is modified.

### 3.2 Navier Equation

By injecting Eq. 4 into Eq. 3 we obtain

$$\sigma = \lambda\,(\nabla\cdot\mathbf{u})\,\mathbf{G} + \mu\left(\nabla\mathbf{u} + (\nabla\mathbf{u})^t\right), \tag{5}$$

since $\mathrm{Tr}(\varepsilon) = \nabla\cdot\mathbf{u}$. The stress can be computed with the following convention: $\nabla\mathbf{u} = \overline{\nabla}\mathbf{u} + K\mathbf{u}$, with $K\mathbf{u} = K^i_{jk}u^k\mathbf{e_i}\otimes\mathbf{e^j}$ and $\nabla\cdot\mathbf{u} = \overline{\nabla}\cdot\mathbf{u} + K\cdot\mathbf{u}$ with $K\cdot\mathbf{u} = K^i_{ik}u^k$. For $\nabla\cdot\sigma$ in Eq. 2, the general form of the divergence of a full contravariant tensor $T$ is synthesized as: $\nabla\cdot T = \overline{\nabla}T + K\cdot T$ with $K\cdot T = (K^i_{ik}T^{ji} + K^j_{ik}T^{ki})\mathbf{e_j}$. Hence, divergence of Eq. 5 writes in terms of $\overline{\nabla}$ and $K$ according to:

$$\begin{aligned}
&\nabla\cdot[(\nabla\cdot\mathbf{u})\mathbf{G}] = \overline{\nabla}\big(\overline{\nabla}\cdot\mathbf{u}\big) + (\overline{\nabla}\cdot\mathbf{u})\big(K\cdot\mathbf{G}\big) + \overline{\nabla}\big(K\cdot\mathbf{u}\big) + (K\cdot\mathbf{u})\big(K\cdot\mathbf{G}\big)\,,\\
&\nabla\cdot[\nabla\mathbf{u}] = \overline{\nabla}\cdot\big(\overline{\nabla}\mathbf{u}\big) + K\cdot(\overline{\nabla}\mathbf{u}) + \overline{\nabla}\cdot\big(K\mathbf{u}\big) + K\cdot(K\mathbf{u})\,,\\
&\nabla\cdot\left[(\nabla\mathbf{u})^t\right] = \overline{\nabla}\cdot(\overline{\nabla}\mathbf{u})^t + K\cdot(\overline{\nabla}\mathbf{u})^t + \overline{\nabla}\cdot(K\mathbf{u})^t + K\cdot(K\mathbf{u})^t\,.
\end{aligned}$$

The wave equation becomes:

$$\rho_0\partial_t^2\mathbf{u} = \overline{\nabla}\cdot\overline{\sigma} + \mathbf{F}. \tag{6}$$

Here $\overline{\sigma}$ is the stress in a torsion-free medium and $\mathbf{F}$ is a configurational force density [8]:

$$\begin{aligned}
\mathbf{F} = \lambda\Big((\overline{\nabla}\cdot\mathbf{u})\big(K\cdot\mathbf{G}\big) + \overline{\nabla}\big(K\cdot\mathbf{u}\big) + (K\cdot\mathbf{u})\big(K\cdot\mathbf{G}\big)\Big) + \ldots\\
\ldots\mu\Big(K\cdot\big((\overline{\nabla}\mathbf{u}) + (\overline{\nabla}\mathbf{u})^t\big) + \overline{\nabla}\cdot\big(K\mathbf{u} + (K\mathbf{u})^t\big) + K\cdot\big(K\mathbf{u} + (K\mathbf{u})^t\big)\Big).
\end{aligned} \tag{7}$$

## 4 Screw-Type Torsion in Cylindrical Coordinate System

We consider uniform torsion, with screw-type behavior in $(\mathbf{e_1}, \mathbf{e_2})$-plane [2]. The only non-zero components are $S^3_{12} = -S^3_{21} = rS$, where the coefficient $r$ is due to the covariant base for the plane component and $S$ is the (uniform) magnitude of the torsion. According to Eq. 1, the contortion has non-zero components: $K^3_{12} = K^1_{32} = K^1_{23} = \frac{rS}{2}$, $K^3_{21} = -\frac{rS}{2}$ and $K^2_{13} = K^2_{31} = -\frac{S}{2r}$.

Equations 6 and 7 are valid for any coordinate system, however contravariant and covariant bases $\{\mathbf{e_i}\}$ and $\{\mathbf{e^i}\}$ are not orthonormal bases. In the following we privilege physical cylindrical bases $(\mathbf{e_r}, \mathbf{e_\theta}, \mathbf{e_z})$ for which final results are more easily interpretable. We refer to classical books on differential geometry for the definition of associated metric, Christoffel coefficients and relations between contravariant, covariant and physical components of a vector [14, 19].

### 4.1 Stress Tensor

The stress tensor is defined by Eq. 5: $\sigma = \overline{\sigma} + \lambda\,(K \cdot \mathbf{u})\,\mathbf{G} + \mu\,\left(K\mathbf{u} + (K\mathbf{u})^t\right)$. In this domain, the nonzero component of contortion has no repeated indices, then $K \cdot \mathbf{u} = K^i_{ik} u^k = 0$. After straightforward calculation of $K\mathbf{u} = g^{jl} K^i_{lk} u^k \mathbf{e_i} \otimes \mathbf{e_j}$ the additional stress imposed by torsion is an anti-plane stress proportional to the shear modulus $\mu$ and torsion density $S$. It can be written in cylindrical physical base:

$$\mu\left(K\mathbf{u} + (K\mathbf{u})^t\right) = \mu S \begin{pmatrix} 0 & 0 & u_\theta \\ 0 & 0 & -u_r \\ u_\theta & -u_r & 0 \end{pmatrix}. \tag{8}$$

### 4.2 Configurational Force

In the Navier Eq. 6, we compute only $\mathbf{F}$ (Eq. 7) as the other terms can be found in all classical books on elasticity. In our case $\mathbf{F}$ reduces to:

$$\mathbf{F} = \mu\,\left(K \cdot \left(\overline{\nabla}\mathbf{u} + (\overline{\nabla}\mathbf{u})^t\right) + \nabla \cdot \left(K\mathbf{u} + (K\mathbf{u})^t\right)\right). \tag{9}$$

First $K \cdot T = K^j_{ik} T^{ki} \mathbf{e_j}$ with symmetric $T = \overline{\nabla}\mathbf{u} + (\overline{\nabla}\mathbf{u})^t$ is:

$$K \cdot \left(\overline{\nabla}\mathbf{u} + (\overline{\nabla}\mathbf{u})^t\right) = rS(\partial_3 u^2 + \frac{1}{r^2}\partial_2 u^3)\mathbf{e_1} - \frac{S}{r}(\partial_3 u^1 + \partial_1 u^3)\mathbf{e_2}. \tag{10}$$

For $\nabla \cdot \left(K\mathbf{u} + (K\mathbf{u})^t\right)$ in Eq. 9, we compute each terms thanks Eq. 8. This calculation is straightforward and gives after algebraic manipulation:

$$\begin{aligned}\nabla \cdot \left(K\mathbf{u} + (K\mathbf{u})^t\right) = \left(S\partial_3(ru_2) - S^2 u_1\right)\mathbf{e_1} - \left((S/r)\partial_3 u_1 + S^2 u_2\right)\mathbf{e_2} + \dots \\ S\left(\partial_1(ru_2) + u^2 - (1/r)\partial_2 u_1\right)\mathbf{e_3}.\end{aligned} \tag{11}$$

Collecting Eqs. 10 and 11, the configuration force is written on the physical base:

$$\mathbf{F} = \mu S\Big[(2\partial_z u_\theta + \frac{\partial_\theta u_z}{r} - Su_r)\mathbf{e_r} - (2\partial_z u_r + \partial_r u_z + Su_\theta)\mathbf{e_\theta} + (\partial_r u_\theta + \frac{u_\theta}{r} - \frac{\partial_\theta u_r}{r})\mathbf{e_z}\Big].$$

## 5 General Time-Harmonic Solutions

We consider infinite domain invariant along $\mathbf{e_z}$ for which we are looking for time-harmonic solution: here $\partial_t^2\mathbf{u} = -\omega^2\mathbf{u}$ and $\partial_z\mathbf{u} = 0$. Let write $\mathbf{u} = \mathbf{u_z} + \mathbf{w}$, with vertical component $\mathbf{u_z} = u_z\mathbf{e_z}$ and in-plane field $\mathbf{w} = u_r\mathbf{e_r} + u_\theta\mathbf{e_\theta}$. As $\mathbf{u}$ is independent of

$z$ we have: $\mathbf{F} = \mu S \overline{\nabla} \times \mathbf{u} - \mu S^2 \mathbf{w}$. Owing calculus identities with $\overline{\nabla}$, Eq. 6 becomes:

$$-\omega^2 \mathbf{u} = c_l^2 \overline{\nabla}(\overline{\nabla} \cdot \mathbf{u}) - c_t^2 \overline{\nabla} \times (\overline{\nabla} \times \mathbf{u}) + c_t^2 S \overline{\nabla} \times \mathbf{u} - (c_t S)^2 \mathbf{w} \tag{12}$$

where $c_t = \sqrt{\mu/\rho_0}$ and $c_l = \sqrt{(\lambda + 2\mu)/\rho_0}$.

## 5.1 Solution in a Perfect Medium

The 2D problem in a torsion-free medium is the superposition of a anti-plane problem and in-plane problem (respectively SH and PSV in geophysical terminology). By projection of Eq. 12 (with $S = 0$) along $\mathbf{e_z}$ and along $(\mathbf{e_r}, \mathbf{e_\theta})$-plane we obtain the following uncoupled systems:

$$\text{SH-problem: } \overline{\Delta} u_z = -\frac{\omega^2}{c_t^2} u_z, \quad \text{PSV-problem: } -\omega^2 \mathbf{w} = c_l^2 \overline{\nabla}(\overline{\nabla} \cdot \mathbf{w}) - c_t^2 \overline{\nabla} \times (\overline{\nabla} \times \mathbf{w}).$$

Applying Helmholtz decomposition: $\mathbf{w} = \overline{\nabla}\varphi + \overline{\nabla} \times (\psi \mathbf{e_z})$, longitudinal potential $\varphi$ and shear potential $\psi$ are respectively solution of $\overline{\Delta}\varphi = -(\omega/c_l)^2 \varphi$ and $\overline{\Delta}\psi = -(\omega/c_t)^2 \psi$, respectively. A general solution $f(r, \theta)$ of Helmholtz equation $\overline{\Delta} f = -k^2 f$ is

$$f(r, \theta) = \sum_n \left(a_n J_n(kr) + b_n H_n(kr)\right) e^{in\theta} \tag{13}$$

where $J_n$ is the $n^{th}$ Bessel function and $H_n$ is the $n^{th}$ Hankel function of first type. For each mode $n$, amplitudes $a_n$ and $b_n$ are defined by source or boundary conditions. In conclusion, vertical displacement and potentials have the form (13), with wavenumber $k_t = \omega/c_t$ for $u_z$ and $\psi$ and $k_l = \omega/c_l$ for $\varphi$.

## 5.2 Solution in a Defective Medium

Applying the spatial divergence $\overline{\nabla}\cdot$ on each side of the Eq. 12 gives

$$(\omega^2 - (c_t S)^2) \overline{\nabla} \cdot \mathbf{w} + c_l^2 \overline{\Delta}(\overline{\nabla} \cdot \mathbf{w}) = 0$$

because $\overline{\nabla} \cdot \mathbf{u} = \overline{\nabla} \cdot \mathbf{w}$ for $z$-invariant displacement. This suggests to invoke the Helmholtz decomposition $\mathbf{w} = \overline{\nabla}\varphi + \overline{\nabla} \times (\psi \mathbf{e_z})$ where $\varphi$ and $\psi$ differ from the perfect medium case. We directly observe that $\varphi$ solves $\overline{\Delta}\varphi = -(\omega^2 - (c_t S)^2 / c_l^2)\varphi$ and the modal decomposition (13) can be used with the wavenumber:

$$(k^o)^2 = k_l^2 \left(1 - (S/k_t)^2\right). \tag{14}$$

Consider the divergence-free displacement, $\mathbf{u} = u_z\mathbf{e_z} + \mathbf{w}$ with $\mathbf{w} = \overline{\nabla} \times (\psi\mathbf{e_z})$. In one hand the problem in the $(\mathbf{e_r}, \mathbf{e_\theta})$-plane can be written in the form:

$$-\omega^2\mathbf{w} = c_l^2\overline{\nabla}(\overline{\nabla}\cdot\mathbf{w}) - c_t^2\overline{\nabla}\times(\overline{\nabla}\times\mathbf{w}) + c_t^2 S\overline{\nabla}\times\mathbf{u_z} - (c_t S)^2\mathbf{w}.$$

Because $\overline{\nabla}\times(\overline{\nabla}\times\mathbf{w}) = \overline{\nabla}(\overline{\nabla}\cdot\mathbf{w}) - \overline{\Delta}\mathbf{w} = -\overline{\Delta}(\overline{\nabla}\times(\psi\mathbf{e_z}))$ we have:

$$(\omega^2 - (c_t S)^2)\overline{\nabla}\times(\psi\mathbf{e_z}) + c_t^2\overline{\Delta}(\overline{\nabla}\times(\psi\mathbf{e_z})) = -c_t^2 S\,\overline{\nabla}\times(u_z\mathbf{e_z})$$

that can be written in a scalar form: $(k_t^2 - S^2)\psi + \overline{\Delta}\psi = -Su_z$. In other hand the projection of the wave equation onto $\mathbf{e_z}$ gives:

$$k_t^2 u_z + \overline{\Delta}u_z = -S\left(\overline{\nabla}\times\mathbf{w}\right)\cdot\mathbf{e_z}.$$

Using $(\overline{\nabla}\times\mathbf{w})\cdot\mathbf{e_z} = -\overline{\Delta}\psi$, the two previous equations define a system where the divergence-free fields are coupled.

$$\begin{aligned}(k_t^2 - S^2)\psi + \overline{\Delta}\psi &= -Su_z,\\ k_t^2 u_z + \overline{\Delta}u_z &= S\overline{\Delta}\psi.\end{aligned}$$

We focus on harmonics solutions $\overline{\Delta}f = -k^2 f$ for which this system reduces to an eigenvalue problem, where $k$ is unknown:

$$\begin{pmatrix} k_t^2 - S^2 - k^2 & S \\ Sk^2 & k_t^2 - k^2 \end{pmatrix}\begin{pmatrix}\psi \\ u_z\end{pmatrix} = \begin{pmatrix}0\\0\end{pmatrix}. \tag{15}$$

Solutions exist if and only if the determinant of the matrix is null, implying:

$$k^2 = k_t^2\,(1 \pm (S/k_t)) := (k^{\pm})^2 \tag{16}$$

For instance, we can write a modal decomposition for $\psi$:

$$\psi = \psi^+ + \psi^- := \sum_n \left(a_n^+ J_n(k^+ r) + a_n^- J_n(k^- r)\right) e^{in\theta}. \tag{17}$$

Of course Hankel functions can be used too. It is not written explicitly in order to simplify notations. The modal form for $u_z$ is obtained with any line of Eq. 15:

$$u_z = S\psi + k_t\left(\psi^+ - \psi^-\right). \tag{18}$$

Figure 1 reports the dispersion curves associated to dispersion relations (14) and (16). They are compatible with results obtained in Cartesian coordinate [2].

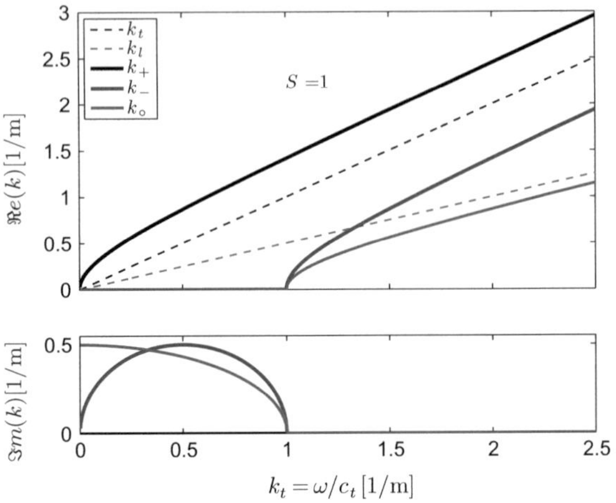


**Fig. 1** Dispersion *curves* for perfect medium (*dot*) and media with torsion (*line*) $S = 1\,\mathrm{m}^{-1}$ and steel-like media: $c_l/c_t = 2$. At high frequency or low torsion density $S/k_t \ll 1$ we have $k^\circ = k_l + \mathcal{O}(S^2)$ and $k^\pm = k_t \pm \frac{S}{2} + \mathcal{O}(S^2)$. In the opposite situation, $S/k_t \gg 1$, we have $k^\circ \sim iSc_t/c_l$ (no frequency dependence) and $k^\pm \sim 0$ in first approximation. In that case modified Bessel functions can be used. If $k$ is purely imaginary, the wave is evanescent

# 6 Scattering Problem

We consider infinite domain with homogeneous elastic properties $\rho_0$, $\lambda$ and $\mu$. The torsion field is confined in a cylinder $\Omega_1$ oriented along $\mathbf{e_z}$ and with radius $a$. The center of $\Omega_1$ is the origin of cylindrical coordinates system. In the remaining torsion-free domain, $\Omega_0$, the wave field $\mathbf{u^0}$ is the superposition of an incoming wave $\mathbf{u^i}$ and an outgoing scattered wave $\mathbf{u^s}$: $\mathbf{u^0} = \mathbf{u^i} + \mathbf{u^s}$. In this work we focus on an incoming plane shear wave propagating along $\mathbf{e_x}$ with vertical polarity: $\mathbf{u^i} = Ae^{ik_tx}\mathbf{e_z}$; according to Jacobi-Anger expansion $u_z^i = A\sum_n i^n J_n(k_tr)e^{in\theta}$. The scattered field $\mathbf{u^s} = \overline{\nabla}\varphi^s + \overline{\nabla} \times (\psi^s\mathbf{e_z}) + u_z^s\mathbf{e_z}$ is decomposed with Hankel function in order to satisfy Sommerfeld radiation condition; then:

$$\varphi^s = \sum_n a_n^l H_n(k_lr)e^{in\theta}, \quad \psi^s = \sum_n a_n^t H_n(k_tr)e^{in\theta}, \quad u_z^s = k_t\sum_n a_n^z H_n(k_tr)e^{in\theta}. \tag{19}$$

The factor $k_t$ for $u_z^s$ preserves the same dimension for all modal coefficients. In the scatterer $\Omega_1$ the transmitted field $\mathbf{u^1}$ is regular: $J_n$ function are used. We recall $\mathbf{u^1} = \overline{\nabla}\varphi^1 + \overline{\nabla} \times (\psi^1\mathbf{e_z}) + u_z^1\mathbf{e_z}$ with $u_z^1$ defined by Eq. 18 and $\psi^1 = \psi^+ + \psi^-$ with

$$\varphi^1 = \sum_n b_n^\circ J_n(k^\circ r)e^{in\theta}, \; \psi^+ = \sum_n b_n^+ J_n(k^+r)e^{in\theta}, \; \psi^- = \sum_n b_n^- J_n(k^-r)e^{in\theta}. \tag{20}$$

## 6.1 Boundary Conditions

At the interface, continuity of displacement $\mathbf{u^0} = \mathbf{u^1}$ and normal stress $\sigma^0(\mathbf{e_r}) = \sigma^1(\mathbf{e_r})$ are imposed. In $\Omega_0$ $\sigma^0 \equiv \overline{\sigma}^0$ but in $\Omega_1$ $\sigma^1$ contains the additional term Eq. 8. For this linear problem $\sigma^0$ can be decomposed as $\sigma^0 = \sigma^i + \sigma^s$, too. The six scalar boundary conditions are, for $r = a$ and any $\theta$:

$$\begin{array}{lll} u_r^1 - u_r^s = u_r^i, & u_\theta^1 - u_\theta^s = u_\theta^i, & u_z^1 - u_z^s = u_z^i, \\ \sigma_{rr}^1 - \sigma_{rr}^s = \sigma_{rr}^i, & \sigma_{r\theta}^1 - \sigma_{r\theta}^s = \sigma_{r\theta}^i, & \sigma_{rz}^1 - \sigma_{zr}^s = \sigma_{rz}^i \end{array} \tag{21}$$

The fields are written with the decomposition (13). The modal contribution are uncoupled by multiplying equations by $e^{-in\theta}$ and integration over $\theta \in [0, 2\pi]$. The previous system is then written for each mode $n$. In the following the potentials, fields are provided for mode $n$, and the suffix $_n$ is omitted for the sake of clarity.

## 6.2 Displacement and Stress

For any displacement $\mathbf{u}$ ($\equiv \mathbf{u^0}, \mathbf{u^i}, \mathbf{u^s}$) in $\Omega_0$, the displacement are written with the corresponding potential and according Helmholtz decomposition:

$$u_r = (1/r)\left(in\psi + k_l r\varphi'\right), \qquad u_\theta = (1/r)\left(in\varphi - k_t r\psi'\right), \quad r \geq a \tag{22}$$

where $\partial_r f = kf'$. In $\Omega_1$ the displacement components are, for $r \leq a$:

$$\begin{gathered} u_r^1 = \frac{1}{r}\left(in\psi^+ + in\psi^- + k^\circ r\varphi^{1\prime}\right), \quad u_\theta^1 = \frac{1}{r}\left(in\varphi^1 - k^+ r\psi^{+\prime} - k^- r\psi^{-\prime}\right), \\ u_z^1 = (S + k_t)\psi^+ + (S - k_t)\psi^-. \end{gathered} \tag{23}$$

In $\Omega_0$, the components of the stress tensors are:

$$\begin{aligned} \sigma_{rr} &= -\frac{2\mu}{r^2}\left(\left(\frac{\lambda + 2\mu}{\mu}\frac{(k_l r)^2}{2} - n^2\right)\varphi + k_l r\varphi' + in\left(\psi - k_t r\psi'\right)\right), \\ \sigma_{r\theta} &= -\frac{2\mu}{r^2}\left(in\left(\varphi - k_l r\varphi'\right) - \left(\frac{(k_t r)^2}{2} - n^2\right)\psi - k_t r\psi'\right), \\ \sigma_{rz} &= \mu k_t u_z'. \end{aligned} \tag{24}$$

In the scatterer $\Omega_1$ the relation is similar (with replacement rules $\{k_l, \varphi\} \rightarrow \{k^\circ, \varphi^1\}$ and $\{k_t, \psi\} \rightarrow \{k^\pm, \psi^\pm\}$) but an additional term appears in $\sigma_{rz}^1$ due to Eq. 8:

$$\sigma_{rz}^1 = \mu k_t\left((inS/k_t r)\varphi^1 + k^+(\psi^+)' - k^-(\psi^-)'\right). \tag{25}$$

The system defined by modal Eq. 21 can be written in a matrix form [11]:

$$\mathbb{A}\mathbf{X} = \mathbf{Y} \tag{26}$$

where $\mathbf{Y} = Ai^n(0, 0, J_n(k_t a), 0, 0, \mu k_t J_n'(k_t a))^t$ is related to the incoming wave and $\mathbf{X} = (a_n^t, a_n^z, b_n^+, b_n^-, a_n^l, b_n^\circ)^t$ is unknown. The components of $\mathbb{A}$ are obtained thanks to expressions provided in this Sect. 6.2 where $r$ is fixed to $a$. Solving the system (26) allows us to compute the fields in each domain.

**Special case: far-field radiation** Far from the scatterer, the scattered field can be computed with large argument asymptotics: $H_n(x) \sim \frac{1-i}{\sqrt{\pi}} \frac{1}{i^n} \frac{e^{ix}}{\sqrt{x}}$. Keeping only the leading term of Eq. 19, the displacement field far from the object is

$$\begin{aligned}
u_r &= f_{zl}(\theta)\frac{e^{ik_l r}}{\sqrt{k_l r}} \quad \text{with} \quad f_{zl}(\theta) = k_l \frac{1+i}{\sqrt{\pi}} \sum_n \frac{a_n^l}{i^n} e^{in\theta}, \\
u_\theta &= f_{zt}(\theta)\frac{e^{ik_t r}}{\sqrt{k_t r}} \quad \text{with} \quad f_{zt}(\theta) = -k_t \frac{1+i}{\sqrt{\pi}} \sum_n \frac{a_n^t}{i^n} e^{in\theta}, \\
u_z &= f_{zz}(\theta)\frac{e^{ik_t r}}{\sqrt{k_t r}} \quad \text{with} \quad f_{zz}(\theta) = k_t \frac{1-i}{\sqrt{\pi}} \sum_n \frac{a_n^z}{i^n} e^{in\theta}.
\end{aligned} \tag{27}$$

The angular function $f_{z\#}$ is the far-field pattern of the scattering field with polarization # due to an incoming wave with polarization $z$.

## 7 Asymptotics

The problem is governed by three independent space scales: (i) the dimension of the scatterer is controlled by $a$, (ii) the incoming wave introduces a wavelength $L = 2\pi/k_t$ for the shear wave (iii) the effective Burger vector $b$ in $\Omega_1$: $b := \int_{\Omega_1} S\, dxdy$; Because $S$ is uniform in $\Omega_1$ we have $S = b/(\pi a^2)$.

If $b$ is held fixed and $a$ tends toward zero, $S$ behaves as 2D-Dirac distribution $b\delta(x)\delta(y)$. A priori the scattering behavior of a discrete dislocation line may be recovered. However the presence of quadratic term for $S$ in Eq. 12 induces a drastic limitation because the square of a Dirac-$\delta$-function is not defined. This means, that according to this model, the defective core is always finite. Note that generalized continuum model is *per se* not adapted for discrete lattice then a discrete defect do not have sense for such model that is limited to (at least) mesoscale. As a consequence two asymptotics will be analyzed. The first concerns the hierarchy $b \ll a \ll L$ corresponding to a core containing a low density of defect. For larger defect density, we consider $b \sim a \ll L$ according to the previous remark. For these two cases, the wavelength is large compared to the typical dimension of the defective core: in some sense the wave still considers the core as a single line.

## 7.1 Low Defect's Density

If $b \ll a \ll L$, all the non-dimensional wavenumbers present in $\mathbb{A}$ are small: $|\kappa_*| \ll 1$ with $\kappa_* = k^* a$ and $* = \{t, l, +, -, \circ\}$. Asymptotic expansion of Bessel functions are used to derive analytic approximation. For example: $J_0(\kappa) = 1 - \kappa^2/4$ and $H_0(\kappa) = 2i/\pi \left[g(\kappa) + (\kappa^2/4)(1 - g(\kappa))\right]$ where $g(\kappa) = \gamma - i\pi/2 + \ln(\kappa/2)$ and $\gamma \sim 0.577$ is the Euler-Mascheroni constant. The overall methodology is the following: first $\mathbb{A}$ is reformulated according to these asymptotics and rewritten in terms of only three non-dimensional parameters: $\varepsilon = k_t a$, $o = S/k_t$ and $\alpha = c_l/c_t$. Here $\varepsilon \ll 1$ and $\alpha = \sqrt{(\lambda + 2\mu)/\mu} = \mathscr{O}(1)$ is a material constant. However $o = S/k_t \sim bL/a^2$ can be small or large in regards to unity. Hence $\kappa_\circ$ and $\kappa_-$ may be real or purely imaginary (but $|\kappa_\circ| \ll 1$ and $|\kappa_-| \ll 1$).

Symbolic computation is used to solve Eq. 26 in order to provide scattering and transmitting amplitudes in the form of $\varepsilon$ polynomial. The $\mathscr{O}(\varepsilon^2)$-terms of scattering modes are

$$a_0^l = a_0^t = a_0^z = a_1^z = a_{-1}^z = \mathscr{O}(\varepsilon^4), \qquad a_{\pm 1}^t = \frac{\pi}{8} o\varepsilon^2 \frac{A}{k_t}, \qquad a_{\pm 1}^l = \mp i \frac{\pi}{8} \frac{o\varepsilon^2}{\alpha} \frac{A}{k_t}.$$

The $\mathscr{O}(\varepsilon^2)$-terms of the transmission modes are $b_0^+ = A/(2k_t(o+1))$, $b_0^- = A/(2k_t(o-1))$ and $b_0^o = 0$ and for $n = \pm 1$:

$$b_{\pm 1}^+ = i\frac{A}{k_t}\left(\frac{o+2}{4(1+o)^{\frac{3}{2}}} + \frac{o(o-2)\left(g(\frac{\varepsilon}{\alpha}) + \alpha^2 g(\varepsilon)\right) - 4\alpha^2 g(\varepsilon)}{32\alpha^2(1+o)^{\frac{3}{2}}} o\varepsilon^2\right),$$

$$b_{\pm 1}^- = i\frac{A}{k_t}\left(\frac{o-2}{4(1-o)^{\frac{3}{2}}} + \frac{o(o+2)\left(g(\frac{\varepsilon}{\alpha}) + \alpha^2 g(\varepsilon)\right) - 4\alpha^2 g(\varepsilon)}{32\alpha^2(1+o)^{\frac{3}{2}}} o\varepsilon^2\right),$$

$$b_{\pm 1}^\circ = \mp\frac{A}{k_t}\left(\alpha\frac{o}{2(1-o^2)^{\frac{3}{2}}} + \frac{o^2\left(g(\frac{\varepsilon}{\alpha}) + \alpha^2 g(\varepsilon)\right) - 4g(\frac{\varepsilon}{\alpha})}{16\alpha(1-o^2)^{\frac{3}{2}}} o\varepsilon^2\right).$$

Displacement in $\Omega_0$ and $\Omega_1$ is computed according to Eqs. 19, 22 and 23 with potentials given by Eqs. 19 and 20. In $\Omega_1$ $k_* r \leq \kappa_*$, hence spatial field (Eq. 20) is also given thanks to low-argument asymptotics too:

$$\mathbf{u^1} = \mathbf{u^i} + (i/4\alpha^2)\left(g(\frac{\varepsilon}{\alpha}) + \alpha^2 g(\varepsilon)\right) A o\varepsilon^2 \mathbf{e_y} + \mathscr{O}(\varepsilon^3). \tag{28}$$

$\mathscr{O}(\varepsilon^2)$-terms introduce an interesting phenomenon: the perturbation $\mathbf{u^1} - \mathbf{u^i}$ due to the scatterer is mainly a uniform translation along $\mathbf{e_y}$ (no vertical contribution). The scattering, for its part, is described by the far-field (up to $\mathscr{O}(\varepsilon^2)$):

$$f_{zt} = \frac{1-i}{4}\sqrt{\pi} A o\varepsilon^2 \cos\theta\,, \quad f_{zl} = \frac{1}{\alpha^2}\frac{1-i}{4}\sqrt{\pi} A o\varepsilon^2 \sin\theta\,, \quad f_{zz} = 0? \tag{29}$$

The shear-vertical incoming wave induces a pure in-plane scattering. Hence this defective zone (hereafter the core) acts as a discrete mode-converter between SH-incoming wave and PSV scattering. Both longitudinal and transverse scattering are di-polar but the direction of maximum radiation differs by a $\pi/2$-angle. Note that in presence of free surface normal to the defective cylinder, this type of radiation may induce a quadripolar radiation of surface wave as observed theoretically by Maurel et al. with a different model [10].

From a quantitative point of view, the magnitudes of the perturbations in Eqs. 28 and 29 are proportional to $A\sigma\varepsilon^2 = 2Ab/L$. Then the displacement of the core and the scattering are not sensitive to the size of the defective area if $b \ll a \ll L$. In the linear regime the magnitude of the incoming wave must respect $\omega A \ll c_t$ or equivalently $A \ll L$. This means that the core's displacement is always lower than the effective Burger vector $b$ of the core.

## 7.2 *Higher Defect Density*

If $b \sim a \ll L$, the non-dimensional wavenumbers $\kappa_l$ and $\kappa_t$ are $\mathscr{O}(\varepsilon)$, but in the core:

$$\kappa^\circ \sim i\frac{Sa}{\alpha} \sim i\frac{b}{a} = \mathscr{O}(1), \quad \kappa^+ \sim \sqrt{\frac{k_t b}{\pi}} = \mathscr{O}(\sqrt{\varepsilon}), \quad \kappa^- \sim i\sqrt{\frac{k_t b}{\pi}} = \mathscr{O}(\sqrt{\varepsilon}).$$

In this context the previous methodology can be used too. However, the final results are no more synthetic formulas, exhibiting simple behavior. The investigation of this regime is then carried out with numerical computation.

For $b$ and $L$ fixed the maximum magnitude of the in-plane $u_y$ and vertical $u_z = u_z^1 - u_z^i$ perturbation are reported in Fig. 2 for various size, $a$, of the core. If $b \ll a$ the core is not vertically excited by the incoming wave; the displacement is purely in-plane as observed in the previous section. However, for $b \sim a$ the vertical displacement reach a maximum and is of the same order of magnitude than $u_y$. More precisely, we have $u_y$ and $u_z = \mathscr{O}(b/L)$ for large defect density (i.e. if $b \sim a$).

Displacement of the core is reported in Fig. 3 for $b = a$. In a qualitative point of view, $u_z(x, y) \propto xb/L$ whereas $u_y = A\vartheta$ with an order of magnitude $\vartheta = \mathscr{O}(b/L)$ as in Eq. 28. The in-plane scattering is of the same form as reported in Eq. 29 still with $\mathscr{O}(b/L)$ order of magnitude. The vertical radiation $f_{zz}$ is dipolar with the same directivity as $f_{zt}$: $f_{zz} = A\beta\cos\theta$. The sensitivity of the coefficient $\beta$ versus $b/L$ is investigated numerically in Fig. 4. It turns out that $\chi = \beta/\vartheta = \mathscr{O}(b/L)$, then $f_{zz} = \mathscr{O}((b/L)^2)$. In other words, the vertical radiation is low even for high density of defect. In first approximation, this screw type defect acts again as a mode converter between SH-incoming wave and PSV scattering.

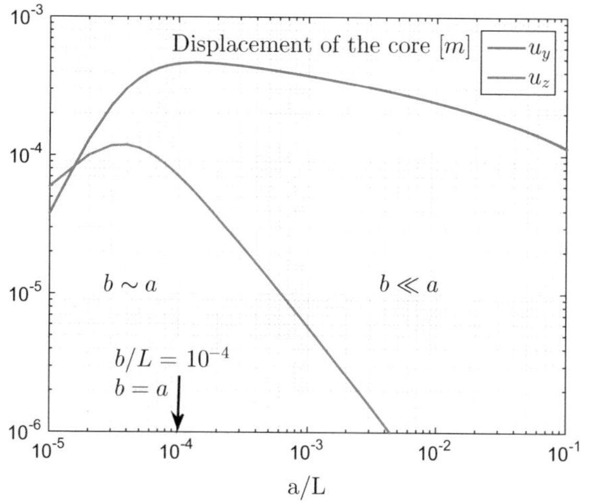


**Fig. 2** Magnitude of the in-plane displacement $u_y$ (*blue*) and *vertical* perturbation $u_z = u_z^1 - u_z^i$ (*red*) of the defective core, for various size $a$ of the defective cylinder. Computation are performed with $A = 1$, $L = 1\,\mathrm{m}$ and $b = 10^{-4}\,\mathrm{m}$, last $\alpha = 2$. The domain of variation of $a$ is chosen in such a way that both regimes $b \sim a$ and $b \ll a$ are present

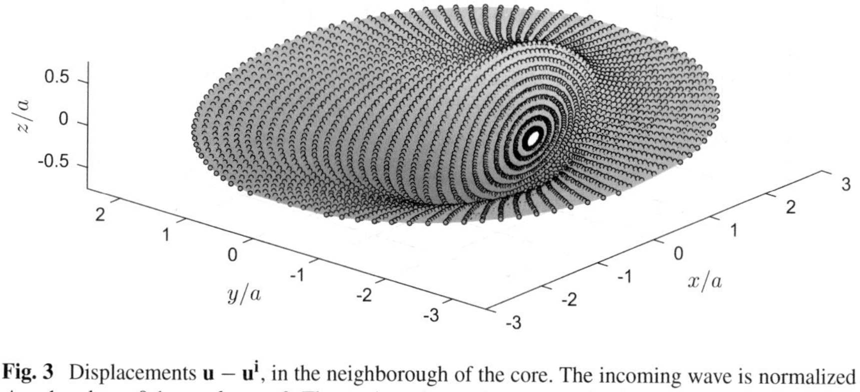


**Fig. 3** Displacements $\mathbf{u} - \mathbf{u}^{\mathbf{i}}$, in the neighborough of the core. The incoming wave is normalized $A = 1$ and $\varepsilon = 0.1$, $a = b$, $\alpha = 2$. The motion of a core's cross-section is an in-plane displacement along $\mathbf{e_y}$ combined with a rigid rotation around $\mathbf{e_x}$

## 7.3 Discussion

The present model can be compared with previous work concerning the scattering of shear wave by a screw dislocation [9]. In this reference the displacement of the core and scattering pattern [9] are in accordance with previous works [12, 13]. All these works are focused on defects in crystal lattice and mainly concerns phenomena at

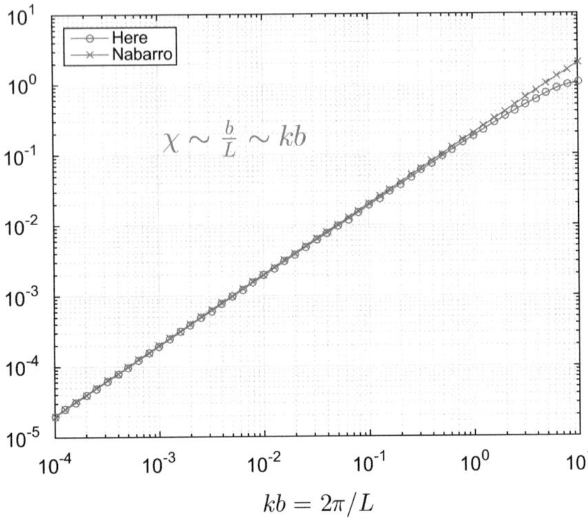


**Fig. 4** Ratio $\chi$ between $\beta$ the magnitude of the *vertical* scattering function and $\vartheta$ the magnitude of the in-plane displacement of the core. The ratio is presented versus $kb = 2\pi b/L$ with $a = b$. Calculation are independent of $A$, the material is steel-like and $\alpha = 2$

microscopic scale. Even if different formalisms are used, these references share the same hypothesis: (i) the dislocation is modeled as a single line with zero core-width and (ii) this defect is able to move in the material background. This last point is a huge difference with the model presented in this paper, as we consider that the torsion density is fixed in the material background. In an other hand, the introduction of Sect. 7 shows that the *zero core-width* hypothesis is maybe too restrictive.

The same ratio $\chi$ is computed with results Eqs. 3.1 and 3.3 of [9]. Despite the difference between this model and ours, the same relation between in plane displacement and vertical scattering is found in Fig. 4.

Fixing for example $a = A = b = 1$ nm, the magnitude of the velocity $v_y = i\omega u_y$ and displacement $u_y$ of the core are computed in a large frequency range, namely for $0.5\,\text{nm} < L < 50\,\mu\text{m}$, with the model presented in this paper and the one in [9] (Fig. 5). The magnitude of the in-plane motion of the core is very different according to the model used. If the dislocation is supposed to move in the lattice, the velocity is close to speed of sound. This has been observed experimentally at high frequency ($f = 500\,\text{MHz}$) in $LiNbO_3$ crystals [20]. The model presented here suggests a fairly low velocity: around $10^{-5}$ m/s. This order of magnitude is not so far to value obtained in quasi-static regime. For example Kruml observes experimentally dislocation velocities of $10^{-7}$ m/s under $10 - 50\,\text{MPa}$ static loading on pure Ge [5].

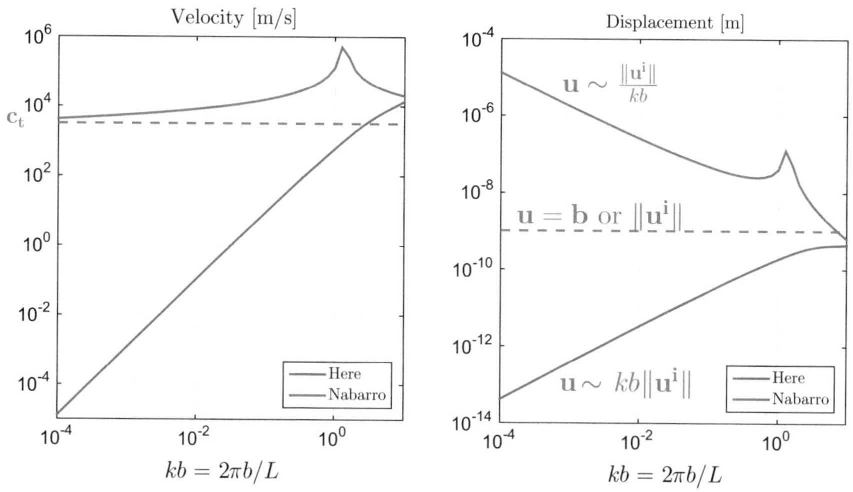


**Fig. 5** Magnitude of the in-plane velocity (*left*) and displacement (*right*) of the dislocation core versus $kb = 2\pi b/L$. Comparaison between the model presented here for defect fixed to the material background (*blue*) and obtained by Nabarro's model for moving dislocation (*red*). Calculation are performed with $a = A = b = 1$ nm, the material is steel-like $\alpha = 2$

# 8 Conclusion

The present work is an attempt to take the advantages of theoretical approach based on the differential geometry of continua in order to present reliable analytical solutions for the problem of waves in a defective elastic medium. This has been performed for a canonical configuration, but it looks now attainable to investigate the effect of other incoming polarisation or defect property with the same methodology.

According to this work, we observe that the presence of a defective zone induces a scattering effect that can not be obtained in the case of an heterogeneity filled by different material. This non-classical scattering behavior is particularly characterized by a unusual mode conversion. For the presented example the defect is invisible in direction of the incoming wave. The relation between displacement of the defective zone and the scattering behavior is similar to other works dedicated to defect in cristal lattice. However the amplitude of the displacement of the core differs by a large order of magnitude according to the model used. It is not surprising that the present work, for which the torsion has been imposed to be fixed in space and time, does not match with model based on moving dislocation in a lattice. In an other hand the correspondence with previous experimental static result is promising, but has to be investigated more deeply. Moreover the physical application of this continuous model encompasses a large scale of applications that does not reduce to phenomena of discrete defects.

In future works, the authors wish to extend this promising approach and will try to confront this modeling of defective media to other phenomena observable at a larger scale in order to avoid some difficulty inherent to the comparison between mesoscale model (here) and micro-scale model (moving dislocation). The other challenge will be to extend the approach to the case of time dependent torsion fields.

## References


1. Cartan, E.: On Manifolds with an Affine Connection and the Theory of General Relativity. Bibliopolis, Napoli (1986)
2. Futhazar, G., Le Marrec, L., Rakotomanana, L.R.: Covariant gradient continua applied to wave propagation within defective material. Arch. App. Mech. **84**(9), 1339–1356 (2014)
3. Kröner, E.: Continuum theory of defects, Physics of defects. **35**, 217–315 (1981)
4. Kröner, E.: Dislocation: a new concept in the continuum theory of plasticity. J. Math. Phys. **42**(1-4), 27–37 (1963)
5. Kruml, T., Caillard, D., Dupas, C., Martin, J.L.: A transmission electron microscopy in situ study of dislocation mobility in Ge. J. Phys. Condens. Matter **14**(48), 12897–12902 (2002)
6. Lazar, M.: An elastoplastic theory of dislocations as a physical field with torsion. J. Phys. **35**(8), 1983–2004 (2002)
7. Lund, F.: Response of a stringlike dislocation loop to an external stress. J. Mater. Res. **3**(2), 280–297 (1988)
8. Maugin, G.A.: Configurational Forces: Thermomechanics, Physics, Mathematics, and Numerics. Taylor & Francis, London (2010)
9. Maurel, A., Mercier, J.-F., Lund, F.: Scattering of an elastic wave by a single dislocation. J. Acoust. Soc. Am. **115**(6), 2773–2780 (2004)
10. Maurel, A., Pagneux, V., Barra, F., Lund, F.: Surface acoustic waves in interaction with a dislocation. Ultrasonics **50**(2), 161–166 (2010)
11. Morse, P.M., Feshbach, H.: Methods of Theoretical Physics. Cambridge University Press, Cambridge (1953)
12. Mura, T.: Continuous distribution of moving dislocations. Philos. Mag. **8**(89), 843–857 (1963)
13. Nabarro, F.R.N.: The interaction of screw dislocations and sound waves. Proc. R. Soc. Lond. A **209**(1097), 278–290 (1951)
14. Nakahara, M.: Geometry, Topology and Physics. CRC Press, IOP Publishing, England (1996)
15. Noll, W.: A mathematical theory of the mechanical behavior of continuous media. Arch. Rat. Mech. Anal. **2**, 197–226 (1958)
16. Rakotomanana, L.R.: A Geometric Approach to Thermomechanics of Dissipating Continua. Birkhauser, Boston (2003)
17. Rakotomanana, L.R.: Contribution à la modélisation géométrique et thermodynamique d'une classe de milieux faiblement continus. Arch. Rat. Mech. Anal. **141**, 199–236 (1998)
18. Salvetat, J.-P., Briggs, G., Andrew, D., Bonard, J.-M., Bacsa, R.R., Kulik, A.J., Stöckli, T., Burnham, N.A., Forró, L.: Elastic and shear moduli of single-walled carbon nanotube ropes. Phys. Rev. Lett. **82**(5), 944–947 (1999)
19. Schouten, J.A.: Ricci-Calculus. Springer, Berlin Heidelberg (1954)
20. Shilo, D., Zolotoyabko, E.: Stroboscopic X-ray imaging of vibrating dislocations excited by 0.58 GHz phonons. Phys. Rev. Lett. **91**(11) (2003)

# A History-Dependent Variational-Hemivariational Inequality in Contact Mechanics

**Stanisław Migórski and Mircea Sofonea**

**Abstract** This chapter is closely related with Chap. 16 of this book. It deals with a mathematical model which describes the frictional contact between a viscoelastic body and a rigid foundation covered by a deformable layer made of soft material and a thin crust. We describe the model, list the assumption on the data and derive a variational formulation of the problem. Then, we prove the unique weak solvability of the model. The proof is based on arguments of history-dependent variational-hemivariational inequalities that we recall in an abstract functional framework.

## 1 Introduction

History-dependent operators represent an important class of operators which arises in functional analysis, mechanics, theory of ordinary and partial differential equations. Some simple examples are the integral operator and the Volterra-type operators. In Contact Mechanics, history-dependent operators could arise both in the constitutive law of the material and in the frictional contact conditions, as shown in [9, 14–16]. On the other hand, the mathematical analysis of contact models leads to the study of variational and hemivariational inequalities. References in the field include the books [3–5, 8, 11–13, 16]. For this reason, a large number of contact problems lead to inequalities which involve history-dependent operators, the so-called history-dependent inequalities. Such inequalities could be variational (if they are governed by convex functions), hemivariational (if they are governed by nonconvex locally Lipschitz functions) and variational-hemivariational (if they are governed by both convex and nonconvex functions).

S. Migórski
Faculty of Mathematics and Computer Science, Jagiellonian University in Krakow, ul. Łojasiewicza 6, 30348 Krakow, Poland
e-mail: migorski@ii.uj.edu.pl

M. Sofonea (✉)
Laboratoire de Mathématiques et Physique, Université de Perpignan Via Domitia, 52 Avenue Paul Alduy, 66860 Perpignan, France
e-mail: sofonea@univ-perp.fr



F. dell'Isola et al. (eds.), *Mathematical Modelling in Solid Mechanics*, Advanced Structured Materials 69, DOI 10.1007/978-981-10-3764-1_8

Currently, there is a growing interest in the study of history-dependent inequalities. For instance, a class of variational inequalities with history-dependent operators was considered in [15], where abstract existence, uniqueness and regularity results were proved. These results were extended in [18] to a more general class of variational inequalities and were completed in [6] with error estimate and convergence results. Various results on hemivariational and variational-hemivariational inequalities with history-dependent operators, formulated in Sobolev-type spaces, could be found in [7, 9, 14]. An abstract study of such kind of inequalities, considered in the framework of real reflexive Banach spaces, has been performed in [17]. The results presented in all of the papers mentioned in this paragraph have been used in order to prove the unique solvability of various mathematical models which describe the contact between a deformable body and a foundation.

The purpose of this chapter is to introduce a new model of frictional contact for viscoelastic materials and to illustrate the use of history-dependent variational-hemivariational inequality in its variational analysis. Thus, in Sect. 2, we introduce the contact problem, in which the material's behavior is modeled by a nonlinear viscoelastic constitutive law with long memory, the process is quasistatic, the contact is frictional and the contact conditions are in a subdifferential form with unilateral conditions for the displacement. Then, in Sect. 3, we list the assumptions on the data and derive the variational formulation of the problem. It is in a form of a history-dependent variational-hemivariational inequality in which the unknown is the displacement field. Next, we state our main existence and uniqueness result, Theorem 8.1. The proof of the theorem is presented in Sect. 5. It is based on an abstract existence and uniqueness result recently obtained in [17], that we recall in Sect. 4.

## 2 The Contact Model

The physical setting and the notation we use are similar to those introduced in the Chap. 16 and are resumed as follows. Consider a viscoelastic body which occupies the domain $\Omega$, in the reference configuration. We assume that $\Omega$ is an open, bounded and connected set in $\mathbb{R}^d$ $(d = 2, 3)$. The boundary $\Gamma = \partial\Omega$ is Lipschitz continuous and is partitioned into three disjoint and measurable parts $\Gamma_1$, $\Gamma_2$ and $\Gamma_3$ such that $\mathrm{meas}\,(\Gamma_1) > 0$. The body is fixed on $\Gamma_1$ and, therefore, the displacement field vanishes there. Moreover, it is acted upon by time-dependent forces of density $\mathbf{f}_0$ in $\Omega$ and surface tractions of density $\mathbf{f}_2$ on $\Gamma_2$ and, in addition, is in frictional contact on $\Gamma_3$ with a foundation. We denote by $\mathbf{u}$, $\boldsymbol{\sigma}$ and $\boldsymbol{\varepsilon}(\mathbf{u})$ the displacement field, the stress field and the linearized strain tensor, respectively, and let $\boldsymbol{\nu}$ be the unit outward normal vector to $\Gamma$. Here and below, we sometimes do not indicate explicitly the dependence of various functions on the spatial variable $\mathbf{x} \in \Omega \cup \Gamma$. For a vector field $\mathbf{v}$, we use notation $v_\nu = \mathbf{v} \cdot \boldsymbol{\nu}$ and $\mathbf{v}_\tau = \mathbf{v} - v_\nu \boldsymbol{\nu}$ for the normal and tangential components of $\mathbf{v}$ on $\Gamma$. Similarly, for the stress field $\boldsymbol{\sigma}$, its normal and tangential components on the boundary are defined by equalities $\sigma_\nu = (\boldsymbol{\sigma}\boldsymbol{\nu}) \cdot \boldsymbol{\nu}$ and $\boldsymbol{\sigma}_\tau = \boldsymbol{\sigma}\boldsymbol{\nu} - \sigma_\nu \boldsymbol{\nu}$, respectively. Finally, we use $\mathbb{S}^d$ for the space of second order symmetric tensors on

$\mathbb{R}^d$ and "$\cdot$", $\|\cdot\|$ will represent the canonical inner product and the Euclidean norm on the spaces $\mathbb{R}^d$ and $\mathbb{S}^d$, respectively.

With these preliminaries, the mathematical model which describes the equilibrium of the viscoelastic body in the physical setting above is the following.

PROBLEM $\mathscr{P}$. *Find a displacement field* $\mathbf{u}\colon \Omega \times \mathbb{R}_+ \to \mathbb{R}^d$, *a stress field* $\boldsymbol{\sigma} : \Omega \times \mathbb{R}_+ \to \mathbb{S}^d$ and two interface forces $\eta_\nu : \Gamma_3 \times \mathbb{R}_+ \to \mathbb{R}$, $\xi_\nu : \Gamma_3 \times \mathbb{R}_+ \to \mathbb{R}$ *such that*

$$\boldsymbol{\sigma}(t) = \mathscr{F}\boldsymbol{\varepsilon}(\mathbf{u}(t)) + \int_0^t \mathscr{R}(t-s)\boldsymbol{\varepsilon}(\mathbf{u}(s))\,ds \qquad \text{in } \Omega, \tag{1}$$

$$\mathrm{Div}\,\boldsymbol{\sigma}(t) + \mathbf{f}_0(t) = \mathbf{0} \qquad \text{in } \Omega, \tag{2}$$

$$\mathbf{u}(t) = \mathbf{0} \qquad \text{on } \Gamma_1, \tag{3}$$

$$\boldsymbol{\sigma}(t)\boldsymbol{\nu} = \mathbf{f}_2 \qquad \text{on } \Gamma_2, \tag{4}$$

$$\left.\begin{array}{l} u_\nu(t) \le g, \quad \sigma_\nu(t) + \xi_\nu(t) + \eta_\nu(t) \le 0, \\ (u_\nu(t) - g)(\sigma_\nu(t) + \xi_\nu(t) + \eta_\nu(t)) = 0, \\ |\eta_\nu(t)| \le F_m\Big(\displaystyle\int_0^t u_\nu^+(s)\,ds\Big), \\ \eta_\nu(t) = \begin{cases} 0 & \text{if } \ u_\nu(t) < 0, \\ F_m\Big(\displaystyle\int_0^t u_\nu^+(s)\,ds\Big) & \text{if } \ u_\nu(t) \ge 0, \end{cases} \\ \xi_\nu(t) \in \partial j_\nu(u_\nu(t)) \end{array}\right\} \qquad \text{on } \Gamma_3, \tag{5}$$

$$\left.\begin{array}{l} \|\boldsymbol{\sigma}_\tau(t)\| \le F_b(u_\nu(t)), \\ -\boldsymbol{\sigma}_\tau(t) = F_b(u_\nu(t))\frac{\mathbf{u}_\tau(t)}{\|\mathbf{u}_\tau(t)\|} \quad \text{if } \ \mathbf{u}_\tau(t) \ne \mathbf{0} \end{array}\right\} \qquad \text{on } \Gamma_3. \tag{6}$$

*for all* $t \in \mathbb{R}_+$.

The equations and boundary conditions in Problem $\mathscr{P}$ are similar to those used in the study of Problem $P$ in Chap. 16. The difference arises in the fact that now some memory terms are involved into the model, and therefore, the problem is time-dependent. For the convenience of the reader, we provide in what follows a short description of (1)–(6).

First, Eq. (1) is the constitutive law for viscoelastic materials in which $\mathscr{F}$ represent the elasticity operator and $\mathscr{R}$ represents the relaxation tensor. Various comments and mechanical interpretation related to such kind of equations could be found in [8, 16].

Equation (2) is the equilibrium equation that we use here since we assume that the process is quasistatic. Conditions (3) and (4) represent the displacement and traction conditions, respectively. Condition (5) represents the contact condition in which $g > 0$, $j_\nu$ and $F_m$ are given functions and $\partial j_\nu$ represents the Clarke subdifferential of $j_\nu$. Note also that here and below we use the notation $r^+$ for the positive part of $r$, that is $r^+ = \max\{0, r\}$. Finally, relations (6) represent the static version of Coulomb's law of dry friction. Here $F_b$ denotes a positive function, the friction bound, assumed to depend on the normal displacement $u_\nu$. This dependence is reasonable from the physical point of view, as explained in [14].

We return now to the contact condition (5) which represents the trait of novelty of our model. Note that this condition models the contact with a foundation made of a rigid body covered by a layer made of soft material and a thin crust with memory effects. It is obtained by using the assumptions (a)–(e) below, in which the various equalities and inequalities are valid on the contact $\Gamma_3$, at any time $t \in \mathbb{R}_+$.

(a) The soft material is deformable and allows penetration. Nevertheless, the later is restricted by the rigid body. Therefore, the normal displacement satisfies the unilateral restriction

$$u_\nu(t) \le g, \tag{7}$$

where $g > 0$ represents the thickness of the soft layer.

(b) The normal stress on the contact surface has an additive decomposition of the form

$$\sigma_\nu(t) = \sigma_\nu^D(t) + \sigma_\nu^M(t) + \sigma_\nu^R(t), \tag{8}$$

where the functions $\sigma_\nu^D$, $\sigma_\nu^M$ and $\sigma_\nu^R$ describe the reaction of the soft layer, the crust and the rigid body, respectively.

(c) The function $\sigma_\nu^D$ satisfies a normal compliance contact condition of the form

$$-\sigma_\nu^D(t) \in \partial j_\nu(u_\nu(t)). \tag{9}$$

Such kind of condition have been already used in literature, see [8] and the references therein.

(d) The function $\sigma_\nu^M$ satisfy the conditions

$$|\sigma_\nu(t)| \le F_m\Big(\int_0^t u_\nu^+(s)\,ds\Big), \tag{10}$$

$$-\sigma_\nu^M(t) = \begin{cases} 0 & \text{if } \ u_\nu(t) < 0, \\ F_m\Big(\displaystyle\int_0^t u_\nu^+(s)\,ds\Big) & \text{if } \ u_\nu(t) > 0, \end{cases} \tag{11}$$

where $F_m$ is a given positive function and its argument represents the accumulated penetration at the moment $t$. These conditions show that $\sigma_\nu^M$ is bounded, vanishes when there is separation between the body and the foundation (i.e., when $u_\nu < 0$) and is towards the body when there is penetration (i.e. when $u_\nu > 0$). Assume that $\sigma_\nu(t) \neq 0$. Then, using (10) and (11), we have

$$|\sigma_\nu(t)| < F_m\Big(\int_0^t u_\nu^+(s)\,ds\Big) \quad \Longrightarrow \quad u_\nu(t) \le 0,$$

$$|\sigma_\nu(t)| = F_m\Big(\int_0^t u_\nu^+(s)\,ds\Big) \quad \Longrightarrow \quad u_\nu(t) \ge 0.$$

This conditions describe the contact with an obstacle which behaves like a rigid as far as the inequality $|\sigma_\nu(t)| < F_m(\cdot)$ holds, and offers no additional resistence to penetration when $|\sigma_\nu(t)| = F_m(\cdot)$. We conclude from here that conditions (10), (11) model the contact with a rigid-plastic obstacle. Moreover, the dependence of the bound $F_m$ on the accumulated penetration describes hardening and softening properties. For all these reasons, we choose condition (10), (11) in order to model the behaviour of the thin crust which, recall, is supposed to have memory effects.

(e) Finally, $\sigma_\nu^R$ satisfies the Signorini unilateral condition in a form with a gap,

$$\sigma_\nu^R(t) \le 0, \qquad \sigma_\nu^R(u_\nu(t) - g) = 0. \tag{12}$$

This condition is widely used in the literature in order to model the contact with rigid obstacles, as explained in [3, 13, 16] and the reference therein.

Denote

$$-\sigma_\nu^D(t) = \xi_\nu(t), \qquad -\sigma_\nu^M(t) = \eta_\nu(t) \tag{13}$$

and note that, using (8), we have

$$\sigma_\nu^R(t) = \sigma_\nu(t) + \xi_\nu(t) + \eta_\nu(t). \tag{14}$$

We now substitute the equalities (13) and (14) in (9)–(12) and gather the resulting relations with (7) to obtain the contact condition (5).

## 3 Variational Analysis

In the study of Problem $\mathscr{P}$ we use standard notation for Lebesgue and Sobolev spaces. For the displacement field we use the space

$$V = \left\{ \mathbf{v} = (v_i) \in H^1(\Omega; \mathbb{R}^d) \mid \mathbf{v} = \mathbf{0} \text{ on } \Gamma_1 \right\}$$

where, here and below, for $\mathbf{v} \in H^1(\Omega; \mathbb{R}^d)$ we use the same symbol $\mathbf{v}$ for the trace of $\mathbf{v}$ on $\Gamma$. Since meas $(\Gamma_1) > 0$, it is known that $V$ is a Hilbert space with the inner product

$$(\mathbf{u}, \mathbf{v})_V = \int_\Omega \boldsymbol{\varepsilon}(\mathbf{u}) \cdot \boldsymbol{\varepsilon}(\mathbf{v})\, dx$$

and the associated norm $\|\cdot\|_V$. By the Sobolev trace theorem, we have

$$\|\mathbf{v}\|_{L^2(\Gamma_3;\mathbb{R}^d)} \le \|\gamma\|\, \|\mathbf{v}\|_V \quad \text{for all } \mathbf{v} \in V, \tag{15}$$

$\|\gamma\|$ being the norm of the trace operator $\gamma \colon V \to L^2(\Gamma_3; \mathbb{R}^d)$. We denote by $V^*$ the topological dual of $V$, and by $\langle \cdot, \cdot \rangle_{V^* \times V}$ the duality pairing of $V^*$ and $V$.

For the stress field we use the space $Q = L^2(\Omega; \mathbb{S}^d)$ which is a Hilbert space with the canonical inner product $(\cdot, \cdot)_Q$ and the associated norm $\|\cdot\|_Q$. We also denote by $\mathbf{Q}_\infty$ the space of fourth order tensor fields given by

$$\mathbf{Q}_\infty = \{\, \mathscr{E} = (\mathscr{E}_{ijkl}) \mid \mathscr{E}_{ijkl} = \mathscr{E}_{jikl} = \mathscr{E}_{klij} \in L^\infty(\Omega),\ \ 1 \le i, j, k, l \le d \,\}.$$

We note that $\mathbf{Q}_\infty$ is a real Banach space with the norm

$$\|\mathscr{E}\|_{\mathbf{Q}_\infty} = \sum_{0 \le i,j,k,l \le d} \|\mathscr{E}_{ijkl}\|_{L^\infty(\Omega)}.$$

In addition, a simple calculation shows that

$$\|\mathscr{E}\boldsymbol{\tau}\|_Q \le \|\mathscr{E}\|_{\mathbf{Q}_\infty} \|\boldsymbol{\tau}\|_Q \quad \text{for all } \mathscr{E} \in \mathbf{Q}_\infty,\ \boldsymbol{\tau} \in Q. \tag{16}$$

Finally, we use $\mathbb{N}$ for the set of positive integers and $\mathbb{R}_+$ for the set of nonnegative real numbers, i.e., $\mathbb{R}_+ = [0, +\infty)$. For a normed space $X$, we use the notation $C(\mathbb{R}_+; X)$ for the space of continuous functions defined on $\mathbb{R}_+$ with values in $X$. For a subset $K \subset X$ we still use the symbols $C(\mathbb{R}_+; K)$ for the set of continuous functions defined on $\mathbb{R}_+$ with values in $K$.

We now list the assumptions on the data. First, we assume that the elasticity operator $\mathscr{F} : \Omega \times \mathbb{S}^d \to \mathbb{S}^d$ satisfies the following properties.

$$\left\{\begin{array}{l}
\text{(a) There exists } L_{\mathscr{F}} > 0 \text{ such that for all } \boldsymbol{\varepsilon}_1, \boldsymbol{\varepsilon}_2 \in \mathbb{S}^d,\ \text{ a.e. } \mathbf{x} \in \Omega,\\
\qquad \|\mathscr{F}(\mathbf{x}, \boldsymbol{\varepsilon}_1) - \mathscr{F}(\mathbf{x}, \boldsymbol{\varepsilon}_2)\| \le L_{\mathscr{F}} \|\boldsymbol{\varepsilon}_1 - \boldsymbol{\varepsilon}_2\|;\\
\text{(b) There exists } m_{\mathscr{F}} > 0 \text{ such that for all } \boldsymbol{\varepsilon}_1, \boldsymbol{\varepsilon}_2 \in \mathbb{S}^d,\ \text{ a.e. } \mathbf{x} \in \Omega,\\
\qquad (\mathscr{F}(\mathbf{x}, \boldsymbol{\varepsilon}_1) - \mathscr{F}(\mathbf{x}, \boldsymbol{\varepsilon}_2)) \cdot (\boldsymbol{\varepsilon}_1 - \boldsymbol{\varepsilon}_2) \ge m_{\mathscr{F}} \|\boldsymbol{\varepsilon}_1 - \boldsymbol{\varepsilon}_2\|^2;\\
\text{(c) } \mathscr{F}(\cdot, \boldsymbol{\varepsilon}) \text{ is measurable on } \Omega \text{ for all } \boldsymbol{\varepsilon} \in \mathbb{S}^d;\\
\text{(d) } \mathscr{F}(\mathbf{x}, \mathbf{0}) = \mathbf{0} \text{ for a.e. } \mathbf{x} \in \Omega.
\end{array}\right. \tag{17}$$

The relaxation tensor $\mathscr{R}$ is such that

$$\mathscr{R} \in C(\mathbb{R}_+; \mathbf{Q}_\infty) \tag{18}$$

and, on the potential function $j_\nu : \Gamma_3 \times \mathbb{R} \to \mathbb{R}$, we assume

$$\left\{\begin{array}{l}
\text{(a) } j_\nu(\cdot, r) \text{ is measurable on } \Gamma_3 \text{ for all } r \in \mathbb{R} \text{ and there}\\
\qquad \text{exists } \overline{e} \in L^2(\Gamma_3) \text{ such that } j_\nu(\cdot, \overline{e}(\cdot)) \in L^1(\Gamma_3);\\
\text{(b) } j_\nu(\mathbf{x}, \cdot) \text{ is locally Lipschitz on } \mathbb{R} \text{ for a.e. } \mathbf{x} \in \Gamma_3;\\
\text{(c) } |\partial j_\nu(\mathbf{x}, r)| \le \overline{c}_0 + \overline{c}_1 |r| \text{ for a.e. } \mathbf{x} \in \Gamma_3,\\
\qquad \text{for all } r \in \mathbb{R} \text{ with } \overline{c}_0,\ \overline{c}_1 \ge 0;\\
\text{(d) } j_\nu^0(\mathbf{x}, r_1; r_2 - r_1) + j_\nu^0(\mathbf{x}, r_2; r_1 - r_2) \le \alpha_{j_\nu} |r_1 - r_2|^2\\
\qquad \text{for a.e. } \mathbf{x} \in \Gamma_3,\ \text{ all } r_1, r_2 \in \mathbb{R} \text{ with } \alpha_{j_\nu} \ge 0.
\end{array}\right. \tag{19}$$

Next, we assume that the penetration bound $g : \Gamma_3 \to \mathbb{R}$, the memory function $F_m : \Gamma_3 \times \mathbb{R} \to \mathbb{R}_+$ and the friction bound $F_b : \Gamma_3 \times \mathbb{R} \to \mathbb{R}$ are such that

$$g \in L^2(\Gamma_3), \quad g(\mathbf{x}) \geq 0 \quad \text{a.e. on } \Gamma_3. \tag{20}$$

$$\left\{ \begin{array}{l} \text{(a) There exists } L_{F_m} > 0 \text{ such that} \\ \qquad |F_m(\mathbf{x}, r_1) - F_m(\mathbf{x}, r_2)| \leq L_{F_m} |r_1 - r_2| \\ \qquad\quad \text{for all } r_1, r_2 \in \mathbb{R}, \text{ a.e. } \mathbf{x} \in \Gamma_3; \\ \text{(b) } F_m(\cdot, r) \text{ is measurable on } \Gamma_3 \text{ for all } r \in \mathbb{R}; \\ \text{(c) } x \mapsto F_m(\mathbf{x}, 0) \in L^2(\Gamma_3). \end{array} \right. \tag{21}$$

$$\left\{ \begin{array}{l} \text{(a) There exists } L_{F_b} > 0 \text{ such that} \\ \qquad |F_b(\mathbf{x}, r_1) - F_b(\mathbf{x}, r_2)| \leq L_{F_b} |r_1 - r_2| \\ \qquad\quad \text{for all } r_1, r_2 \in \mathbb{R}, \text{ a.e. } \mathbf{x} \in \Gamma_3; \\ \text{(b) } F_b(\cdot, r) \text{ is measurable on } \Gamma_3 \text{ for all } r \in \mathbb{R}; \\ \text{(c) } F_b(\mathbf{x}, r) = 0 \text{ for } r \leq 0, \ F_b(\mathbf{x}, r) > 0 \text{ for } r > 0, \text{ a.e. } \mathbf{x} \in \Gamma_3. \end{array} \right. \tag{22}$$

We also assume that the densities of body forces and surface tractions have the regularity

$$\mathbf{f}_0 \in C(\mathbb{R}_+; L^2(\Omega; \mathbb{R}^d)), \quad \mathbf{f}_2 \in C(\mathbb{R}_+; L^2(\Gamma_2; \mathbb{R}^d)) \tag{23}$$

and, finally, we assume the smallness condition

$$L_{F_b} \|\gamma\| + \alpha_{j_\nu} < m_{\mathscr{F}}. \tag{24}$$

We now introduce the set of the admissible displacement fields $U \subset V$ and the function $\mathbf{f}\colon \mathbb{R}_+ \to V^*$ defined by

$$U = \{ \mathbf{v} \in V \mid v_\nu \leq g \text{ on } \Gamma_3 \}, \tag{25}$$

$$\langle \mathbf{f}(t), \mathbf{v} \rangle_{V^* \times V} = (\mathbf{f}_0(t), \mathbf{v})_{L^2(\Omega;\mathbb{R}^d)} + (\mathbf{f}_2(t), \mathbf{v})_{L^2(\Gamma_2;\mathbb{R}^d)} \tag{26}$$
$$\text{for all } \mathbf{v} \in V, \ t \in \mathbb{R}_+.$$

Assume now that $(\mathbf{u}, \boldsymbol{\sigma})$ represents a couple of regular functions which satisfy (1)–(6) and let $t \in \mathbb{R}_+$, $\mathbf{v} \in U$. We perform an integration by parts, split the surface integral on three integrals on $\Gamma_1$, $\Gamma_2$ and $\Gamma_3$, and use the equalities (2)–(4) to deduce that

$$\int_\Omega \boldsymbol{\sigma}(t) \cdot (\boldsymbol{\varepsilon}(\mathbf{v}) - \boldsymbol{\varepsilon}(\mathbf{u}(t)))\, dx = \int_\Omega \mathbf{f}_0(t) \cdot (\mathbf{v} - \mathbf{u}(t))\, dx \tag{27}$$
$$+ \int_{\Gamma_2} \mathbf{f}_2(t) \cdot (\mathbf{v} - \mathbf{u}(t))\, d\Gamma + \int_{\Gamma_3} \sigma_\nu(t)(v_\nu - u_\nu(t))\, d\Gamma + \int_{\Gamma_3} \boldsymbol{\sigma}_\tau \cdot (\mathbf{v}_\tau - \mathbf{u}_\tau(t))\, d\Gamma.$$

Next, we use the contact boundary condition (5), the definition (25) and the definition of the Clarke subdifferential to obtain that

$$\int_{\Gamma_3} \sigma_\nu(t)(v_\nu - u_\nu(t))\,d\Gamma + \int_{\Gamma_3} F_m\Big(\int_0^t u_\nu^+(s)\,ds\Big)(v_\nu^+ - u_\nu^+(t))\,d\Gamma \quad (28)$$
$$+ \int_{\Gamma_3} j_\nu^0(u_\nu(t); v_\nu - u_\nu(t))\,d\Gamma \geq 0.$$

Note that here and below we use notation $j_\nu^0(r_1; r_2)$ for the generalized directional derivative of $j_\nu$ at $r_1$ in the direction $r_2$, see [1, 2] for details.

On the other hand, the friction law (6) yields

$$\int_{\Gamma_3} \boldsymbol{\sigma}_\tau(t)\cdot(\mathbf{v}_\tau - \mathbf{u}_\tau(t))\,d\Gamma + \int_{\Gamma_3} F_b(u_\nu(t))\big(\|\mathbf{v}_\tau\| - \|\mathbf{u}_\tau(t)\|\big)\,d\Gamma \geq 0. \quad (29)$$

We now combine equality (27) with inequalities (28), (29) to deduce that

$$\int_{\Omega} \boldsymbol{\sigma}(t)\cdot(\boldsymbol{\varepsilon}(\mathbf{v}) - \boldsymbol{\varepsilon}(\mathbf{u}(t)))\,dx + \int_{\Gamma_3} F_b(u_\nu(t))\,(\|\mathbf{v}_\tau\| - \|\mathbf{u}_\tau(t)\|)\,d\Gamma \quad (30)$$
$$+ \int_{\Gamma_3} F_m\Big(\int_0^t u_\nu^+(s)\,ds\Big)(v_\nu^+ - u_\nu^+(t))\,d\Gamma + \int_{\Gamma_3} j_\nu^0(u_\nu(t); v_\nu - u_\nu(t))\,d\Gamma$$
$$\geq \int_{\Omega} \mathbf{f}_0(t)\cdot(\mathbf{v} - \mathbf{u}(t))\,dx + \int_{\Gamma_2} \mathbf{f}_2(t)\cdot(\mathbf{v} - \mathbf{u}(t))\,d\Gamma$$

Finally, we substitute the consitutive law (1) in (30) and use notation (26) to obtain the following variational formulation of Problem $\mathscr{P}$, in terms of displacement.

PROBLEM $\mathscr{P}_V$. *Find a displacement field* $\mathbf{u}\colon \mathbb{R}_+ \to U$ *such that*

$$(\mathscr{F}(\boldsymbol{\varepsilon}(\mathbf{u}(t))), \boldsymbol{\varepsilon}(\mathbf{v}) - \boldsymbol{\varepsilon}(\mathbf{u}(t)))_Q + \Big(\int_0^t \mathscr{R}(t-s)(\boldsymbol{\varepsilon}(\mathbf{u}(s)))\,ds, \boldsymbol{\varepsilon}(\mathbf{v}) - \boldsymbol{\varepsilon}(\mathbf{u}(t))\Big)_Q \quad (31)$$
$$+ \int_{\Gamma_3} F_m\Big(\int_0^t u_\nu^+(s)\,ds\Big)\big(v_\nu^+ - u_\nu^+(t)\big)\,d\Gamma + \int_{\Gamma_3} F_b(u_\nu(t))\,(\|\mathbf{v}_\tau\| - \|\mathbf{u}_\tau(t)\|)\,d\Gamma$$
$$+ \int_{\Gamma_3} j_\nu^0(u_\nu(t); v_\nu - u_\nu(t))\,d\Gamma \geq \langle \mathbf{f}(t), \mathbf{v} - \mathbf{u}(t)\rangle_{V^*\times V} \quad \text{for all } \mathbf{v} \in U,\ t \in \mathbb{R}_+.$$

The unique solvability of Problem $\mathscr{P}_V$ is given by the following existence and uniqueness result, that we state here and prove in the next section.

**Theorem 8.1** *Assume that (17)–(24) hold. Then, Problem* $\mathscr{P}_V$ *has a unique solution* $\mathbf{u} \in C(\mathbb{R}_+; U)$.

We end this section with some remarks on the weak solvability of the contact problem $\mathscr{P}$.

First, a couple of functions $(\mathbf{u}, \boldsymbol{\sigma})$ defined on the positive real line $\mathbb{R}_+$ with values on the product space $V \times Q$ is called a weak solution to Problem $\mathscr{P}$ if $\mathbf{u}$ is a solution of the variational problem $\mathscr{P}_V$ and $\boldsymbol{\sigma}$ satisfies the constitutive law (1). We conclude that, under the assumption of Theorem 8.1, Problem $\mathscr{P}$ has a unique weak solution. Moreover, the solution has the regularity $\mathbf{u} \in C(\mathbb{R}_+; V)$ and $\boldsymbol{\sigma} \in C(\mathbb{R}_+; Q)$.

Next, recall that Theorem 8.1 provides the weak solvability of the contact problem $\mathscr{P}$ under the smallness assumption (24) involving the friction bound $F_b$, and the normal compliance potential $j_\nu$. The question whether this smallness assumption represents an intrinsic feature of our contact model or it is a limitation of our mathematical tools is left open and, clearly, deserves more investigation in the future.

Finally, note that the unknowns $\eta_\nu$ and $\xi_\nu$ of Problem $\mathscr{P}$ cannot be recovered since they cannot be computed when the solution $\mathbf{u}$ of Problem $\mathscr{P}$ is known. Actually, these unknowns represent interface forces and, as usual in solving contact problems with unilateral constraints, we do not have information neither on the uniqueness of these functions, nor on their regularity.

## 4 An Abstract Existence and Uniqueness Result

We present in this section an abstract result on history-dependent variational-hemivariational inequalities that we shall use to prove the unique solvability of Problem $\mathscr{P}_V$. For more details on the material presented in this section, we send the reader to [1, 2, 8, 10, 16].

Let $X$ be a reflexive Banach space and $Y$ be a normed space. We denote by $X^*$ the dual of $X$ and by $\langle \cdot, \cdot \rangle_{X^* \times X}$ the duality pairing of $X$ and $X^*$. Let $K$ be a subset of $X$ and $A\colon X \to X^*$, $\mathscr{S}\colon C(\mathbb{R}_+; X) \to C(\mathbb{R}_+; Y)$ be given operators. Consider also a function $\varphi\colon Y \times K \times K \to \mathbb{R}$, a locally Lipschitz function $j\colon X \to \mathbb{R}$ and a function $f\colon \mathbb{R}_+ \to X^*$. With these data we consider the problem of finding a function $u\colon \mathbb{R}_+ \to U$ such that, for each $t \in \mathbb{R}_+$, the following inequality holds

$$\begin{aligned} &\langle Au(t), v - u(t) \rangle + \varphi((\mathscr{S}u)(t), u(t), v) - \varphi((\mathscr{S}u)(t), u(t), u(t)) \\ &\quad + j^0(u(t); v - u(t)) \geq \langle f(t), v - u(t) \rangle \quad \text{for all } v \in K. \end{aligned} \tag{32}$$

In the study of (32), we assume the following hypotheses.

$$K \text{ is a nonempty, closed and convex subset of } X. \tag{33}$$

$$\left\{ \begin{array}{l} A\colon X \to X^* \text{ is an operator such that} \\ \text{(a) } A \text{ is pseudomonotone and there exist} \\ \quad \alpha_A > 0,\ \beta_A,\ \gamma_A \in \mathbb{R} \text{ and } u_0 \in K \text{ such that} \\ \quad\quad \langle Av, v - u_0 \rangle \geq \alpha_A \|v\|_X^2 - \beta_A \|v\|_X - \gamma_A \quad \text{for all } v \in X. \\ \text{(b) } A \text{ is strongly monotone, i.e., there exists } m_A > 0 \text{ such that} \\ \quad\quad \langle Av_1 - Av_2, v_1 - v_2 \rangle \geq m_A \|v_1 - v_2\|_X^2 \quad \text{for all } v_1, v_2 \in X. \end{array} \right. \tag{34}$$

$$\left\{\begin{array}{l}\varphi\colon Y \times K \times K \to \mathbb{R} \text{ is a function such that} \\ \text{(a) } \varphi(y,u,\cdot)\colon K \to \mathbb{R} \text{ is convex and l.s.c. on } K, \text{ for all } y \in Y,\ u \in K. \\ \text{(b) there exist } \alpha_\varphi,\ \beta_\varphi > 0 \text{ such that} \\ \qquad \varphi(y_1,u_1,v_2) - \varphi(y_1,u_1,v_1) + \varphi(y_2,u_2,v_1) - \varphi(y_2,u_2,v_2) \\ \qquad\quad \le \alpha_\varphi \|u_1 - u_2\|_X \|v_1 - v_2\|_X + \beta_\varphi \|y_1 - y_2\|_Y \|v_1 - v_2\|_X \\ \qquad\quad \text{for all } y_1,\ y_2 \in Y,\ u_1,\ u_2,\ v_1,\ v_2 \in K. \end{array}\right. \tag{35}$$

$$\left\{\begin{array}{l} j\colon X \to \mathbb{R} \text{ is a function such that} \\ \text{(a) } j \text{ is locally Lipschitz;} \\ \text{(b) } \|\partial j(v)\|_{X^*} \le c_0 + c_1 \|v\|_X \text{ for all } v \in X \text{ with } c_0, c_1 \ge 0. \\ \text{(c) there exists } \alpha_j > 0 \text{ such that} \\ \qquad j^0(v_1; v_2 - v_1) + j^0(v_2; v_1 - v_2) \le \alpha_j \|v_1 - v_2\|_X^2 \\ \qquad \text{for all } v_1, v_2 \in X. \end{array}\right. \tag{36}$$

$$\left\{\begin{array}{l} \text{For any } n \in \mathbb{N}, \text{ there exists } s_n > 0 \text{ such that} \\ \qquad \|(\mathscr{S}u_1)(t) - (\mathscr{S}u_2)(t)\|_Y \le s_n \displaystyle\int_0^t \|u_1(s) - u_2(s)\|_X\, ds \\ \text{for all } u_1,\ u_2 \in C(\mathbb{R}_+; X), \text{ for all } t \in [0,n]. \end{array}\right. \tag{37}$$

$$\alpha_\varphi + \alpha_j < m_A, \qquad \alpha_j < \alpha_A. \tag{38}$$

$$f \in C(\mathbb{R}_+; X^*). \tag{39}$$

Note that an operator $\mathscr{S}$ which satisfies condition (37) is called a *history-dependent operator*. Inequality (32) is governed both by the function $\varphi$ which is assumed to be convex with respect its second argument and by the function $j$ which is locally Lipschitz and could be nonconvex. Therefore, this inequality is a *variational-hemivariational inequality*. In addition, the function $\varphi$ in (32) depends on the operator $\mathscr{S}$, assumed to be history-dependent. For this reason, we refer to (32) as a *history-dependent variational-hemivariational inequality*. For this problem we have the following existence and uniqueness result.

**Theorem 8.2** *Let $X$ be a reflexive Banach space, $Y$ a normed space, and assume that (33)–(39) hold. Then, inequality (32) has a unique solution $u \in C(\mathbb{R}_+; K)$.*

The proof of Theorem 8.2 can be found in [16]. It is obtained by using arguments of elliptic variational-hemivariational inequalities and a fixed point result for history-dependent operators.

## 5 Proof of Theorem 8.1

We start by defining the operators $A\colon V \to V^*$, $\mathscr{S}\colon C(\mathbb{R}_+; V) \to C(\mathbb{R}_+; Q \times L^2(\Gamma_3))$ and the functions $\varphi\colon L^2(\Gamma_3) \times V \times V \to \mathbb{R}$ and $j\colon V \to \mathbb{R}$ by

$$\langle A\mathbf{u}, \mathbf{v}\rangle = \int_\Omega \mathscr{F}\boldsymbol{\varepsilon}(\mathbf{u}) \cdot \boldsymbol{\varepsilon}(\mathbf{u})\, dx \quad \text{for all } \mathbf{u}, \mathbf{v} \in V, \tag{40}$$

$$(\mathscr{S}\mathbf{u})(t) = \Big(\int_\Omega \mathscr{R}(t-s)\boldsymbol{\varepsilon}(\mathbf{u})\,ds,\ F_m\Big(\int_0^t u_\nu^+(s)\,ds\Big)\Big) \tag{41}$$
$$\text{for all } \mathbf{u} \in C(\mathbb{R}_+; V),\ t \in \mathbb{R}_+,$$
$$\varphi(\boldsymbol{\xi}, \mathbf{u}, \mathbf{v}) = (\boldsymbol{\xi}_1, \boldsymbol{\varepsilon}(\mathbf{v}))_Q + (\xi_2, v_\nu^+)_{L^2(\Gamma_3)} + (F_b(u_\nu), \|\mathbf{v}_\tau\|)_{L^2(\Gamma_3)} \tag{42}$$
$$\text{for all } \boldsymbol{\xi} = (\boldsymbol{\xi}_1, \xi_2) \in Q \times L^2(\Gamma_3),\ \mathbf{u},\ \mathbf{v} \in V,$$
$$j(\mathbf{v}) = \int_{\Gamma_3} j_\nu(v_\nu)\,d\Gamma \quad \text{for all } \mathbf{v} \in V. \tag{43}$$

Then, it is easy to see that Problem $\mathscr{P}_V$ is equivalent to the problem of finding a function $\mathbf{u}\colon \mathbb{R}_+ \to U$ such that for each $t \in \mathbb{R}_+$, the following inequality holds

$$\langle A\mathbf{u}(t), \mathbf{v} - \mathbf{u}(t)\rangle + \varphi((\mathscr{S}\mathbf{u})(t), \mathbf{u}(t), \mathbf{v}) - \varphi((\mathscr{S}\mathbf{u})(t), \mathbf{u}(t), \mathbf{u}(t)) \tag{44}$$
$$+j^0(\mathbf{u}(t); \mathbf{v} - \mathbf{u}(t)) \geq \langle \mathbf{f}(t), \mathbf{v} - \mathbf{u}(t)\rangle \quad \text{for all } \mathbf{v} \in U.$$

To solve this problem, we use Theorem 8.2 with $X = V$, $Y = L^2(\Gamma_3)$ and $K = U$ and, to this end, we check in what follows that assumptions (33)–(39) hold. We use arguments similar to those used in our previous works [8, 9, 14, 15] and, for this reason, we skip the details and we resume the proof as follows.

First, we note that assumption (20) and definition (25) imply (33). Next, a simple calculation based on the definition (40) of the operator $A$ and the properties (17) of the elasticity operator show that (34) holds with $m_A = \alpha_A = m_{\mathscr{F}}$. Moreover, using assumption (22) and the trace inequality (15), it is easy to see that the function $\varphi$ defined by (42) satisfies condition (35) with $\alpha_\varphi = L_{F_b}\|\gamma\|$.

On the other hand, assumption (19) on the function $j_\nu$ and definition (43) show that condition (36) holds with $\alpha_j = \alpha_{j_\nu}$. And, a simple calculation based on assumptions (18), (22) and inequality (16) imply that the operator (41) is a history-dependent operator, i.e., it satisfies condition (37). Now, keeping in mind that $m_A = \alpha_A = m_{\mathscr{F}}$, $\alpha_\varphi = L_{F_b}\|\gamma\|$ and $\alpha_j = \alpha_{j_\nu}$, we easily deduce that the smallness assumption (24) shows that conditions (38) hold, too. Finally, we note that regularity (23) on the densites of the body forces and tractions combined with definition (26) show that condition (39) is satisfied.

We are now in a position to use Theorem 8.2 to deduce the existence of a unique function $\mathbf{u} \in C(\mathbb{R}_+; U)$ such that (44) holds, for each $t \in \mathbb{R}_+$. And, using notation (40)–(43), we deduce that $\mathbf{u}$ is the unique solution to Problem $\mathscr{P}_V$ which concludes the proof. □

**Acknowledgements** Research supported by the National Science Center of Poland under the Maestro 3 Project No. DEC-2012/06/A/ST1/00262.

## References


1. Clarke, F.H.: Optimization and Nonsmooth Analysis. Wiley Interscience, New York (1983)
2. Denkowski, Z., Migórski, S., Papageorgiou, N.S.: An Introduction to Nonlinear Analysis: Theory. Kluwer Academic/Plenum Publishers, Boston, Dordrecht (2003)
3. Eck, C., Jarušek, J., Krbec, M.: Unilateral Contact Problems: Variational Methods and Existence Theorems, Pure and Applied Mathematics, vol. 270. Chapman/CRC Press, New York (2005)
4. Han, W., Sofonea, M.: Quasistatic Contact Problems in Viscoelasticity and Viscoplasticity, Studies in Advanced Mathematics, vol. 30. American Mathematical Society/RI-International Press, Providence, Somerville (2002)
5. Hlaváček, I., Haslinger, J., Nečas, J., Lovíšek, J.: Solution of Variational Inequalities in Mechanics. Springer-Verlag, New York (1988)
6. Kazmi, K., Barboteu, M., Han, W., Sofonea, M.: Numerical analysis of history-dependent quasivariational inequalities with applications in contact mechanics. Math. Model. Numer. Anal. M2AN **48**, 919–942 (2014)
7. Migórski, S., Ochal, A., Sofonea, M.: History-dependent subdifferential inclusions and hemivariational inequalities in contact mechanics. Nonlinear Anal. Real World Appl. **12**, 3384–3396 (2011)
8. Migórski, S., Ochal, A., Sofonea, M.: Nonlinear Inclusions and Hemivariational Inequalities, Models and Analysis of Contact Problems. Advances in Mechanics and Mathematics, vol. 26. Springer, New York (2013)
9. Migórski, S., Ochal, A., Sofonea, M.: History-dependent variational-hemivariational inequalities in contact mechanics. Nonlinear Anal. Real World Appl. **22**, 604–618 (2015)
10. Naniewicz, Z., Panagiotopoulos, P.D.: Mathematical Theory of Hemivariational Inequalities and Applications. Marcel Dekker Inc, New York (1995)
11. Panagiotopoulos, P.D.: Inequality Problems in Mechanics and Applications. Birkhäuser, Boston (1985)
12. Panagiotopoulos, P.D.: Hemivariational Inequalities: Applications in Mechanics and Engineering. Springer-Verlag, Berlin Heidelberg (1993)
13. Shillor, M., Sofonea, M., Telega, J.J.: Models and Analysis of Quasistatic Contact. Lecture Notes in Physics, vol. 655. Springer, Berlin Heidelberg (2004)
14. Sofonea, M., Han, W., Migórski, S.: Numerical analysis of history-dependent variational inequalities with applications to contact problems. Eur. J. Appl. Math. **26**, 427–452 (2015)
15. Sofonea, M., Matei, A.: History-dependent quasivariational inequalities arising in contact mechanics. Eur. J. Appl. Math. **22**, 471–491 (2011)
16. Sofonea, M., Matei, A.: Mathematical Models in Contact Mechanics, London Mathematical Society Lecture Note Series, vol. 398. Cambridge University Press, Cambridge (2012)
17. Sofonea, M., Migórski, S.: A class of history-dependent variational-hemivariational inequalities. Nonlinear Differ. Equ. Appl. **23**, 23 (2016) Art. 38 doi:10.1007/s00030-016-0391-0
18. Sofonea, M., Xiao, Y.: Fully history-dependent quasivariational inequalities in contact mechanics. Appl. Anal. **95**, 2464–2484 (2016)

# Monolithic Algorithm for Dynamic Fluid-Structure Interaction Problem

Cornel Marius Murea

**Abstract** We consider a numerical method for a fluid-structure interaction problem. Updated Lagrangian method is used for the structure and fluid equations are written in Arbitrary Lagrangian Eulerian coordinates. The global moving mesh for the fluid-structure domain is aligned with the fluid-structure interface. At each time step, we solve a monolithic system of unknowns velocity and pressure defined on the global mesh. The continuity of velocity at the interface is automatically satisfied, while the continuity of stress does not appear explicitly in the monolithic fluid-structure system. At each time step we solve only one linear system. Numerical results are presented.

## 1 Introduction

Fluid-structure interaction problem can be solved numerically using partitioned procedure or monolithic approaches. Partitioned procedure strategy consists in solving separately the fluid and structure sub-problems using iterative process as fixed-point iterations or Newton like methods. Monolithic methods solve the fluid-structure interaction problem as a single system of equations and, in many cases, the boundary conditions at the interface are included in the global system.

In this work we use a monolithic strategy with the particularity that we employ a global moving mesh for the fluid-structure domain and the interface is an "interior boundary" of the global mesh. Since we use continuous finite elements over the fluid-structure domain, the continuity of velocity at the interface is automatically satisfied. The continuity of stress at the fluid-structure interface does not appear explicitly in the monolithic fluid-structure system due to the action and reaction principle.

C.M. Murea (✉)
Laboratoire de Mathématiques, Informatique et Applications, Université de
Haute Alsace, 6 Rue des Frères Lumière, 68093 Mulhouse, France
e-mail: cornel.murea@uha.fr

© Springer Nature Singapore Pte Ltd. 2017
F. dell'Isola et al. (eds.), *Mathematical Modelling in Solid Mechanics*,
Advanced Structured Materials 69, DOI 10.1007/978-981-10-3764-1_9

## 2 Setting the Fluid-Structure Interaction Problem

We study a two dimensional fluid-structure interaction problem. We denote by $\Omega_0^S$ the initial structure domain and we assume that its boundary admits the decomposition $\partial\Omega_0^S = \Gamma_D \cup \Gamma_0$. We suppose that the initial structure domain is undeformed (stress-free). At the time instant $t$, the structure occupies the domain $\Omega_t^S$ bounded by $\partial\Omega_t^S = \Gamma_D \cup \Gamma_t$. On the boundary $\Gamma_D$, we impose zero displacements.

Let $D$ be a rectangle of boundary $\partial D = \Sigma_1 \cup \Sigma_2 \cup \Sigma_3 \cup \Sigma_4$, with $\Sigma_1$ the left, $\Sigma_2$ the bottom, $\Sigma_3$ the right and $\Sigma_4$ the top boundary, (see Fig. 1).

We assume that the structure is completely embedded into the fluid, therefore at the time instant $t$, the fluid occupies the domain $\Omega_t^F = D \setminus \overline{\Omega}_t^S$. The boundary $\partial\Omega_t^S$ is common of both domains.

We denote by $\mathbf{U}^S : \Omega_0^S \times [0, T] \to \mathbb{R}^2$ the displacement of the structure. A particle of the structure whose initial position was the point $\mathbf{X}$ will occupies the position $\mathbf{x} = \mathbf{X} + \mathbf{U}^S(\mathbf{X}, t)$ in the deformed domain $\Omega_t^S$.

We denote by $\mathbf{F}(\mathbf{X}, t) = \mathbf{I} + \nabla_{\mathbf{X}}\mathbf{U}^S(\mathbf{X}, t)$ the gradient of the deformation, where $\mathbf{I}$ is the unity matrix and we set $J(\mathbf{X}, t) = \det \mathbf{F}(\mathbf{X}, t)$.

The first and the second Piola–Kirchhoff stress tensors are denoted by $\boldsymbol{\Pi}$ and $\boldsymbol{\Sigma}$, respectively and the following equality holds $\boldsymbol{\Pi} = \mathbf{F}\boldsymbol{\Sigma}$. We suppose that the material of the structure is elastic, homogeneous, isotropic.

We have assumed that the fluid is governed by the Navier–Stokes equations. For each time instant $t \in [0, T]$, we denote the fluid velocity by $\mathbf{v}^F(t) = \left(v_1^F(t), v_2^F(t)\right)^T : \Omega_t^F \to \mathbb{R}^2$ and the fluid pressure by $p^F(t) : \Omega_t^F \to \mathbb{R}$. Let us remark that the fluid domain $\Omega_t^F$ depends on the position of the interface $\Gamma_t$, which is the image of $\Gamma_0$ via the map $\mathbf{X} \to \mathbf{X} + \mathbf{U}^S(\mathbf{X}, t)$.

Let $\varepsilon\left(\mathbf{v}^F\right) = \frac{1}{2}\left(\nabla\mathbf{v}^F + \left(\nabla\mathbf{v}^F\right)^T\right)$ be the fluid rate of strain tensor and let $\sigma^F = -p^F\mathbf{I} + 2\mu^S\varepsilon\left(\mathbf{v}^F\right)$ be the fluid stress tensor. In order to simplify the notation,

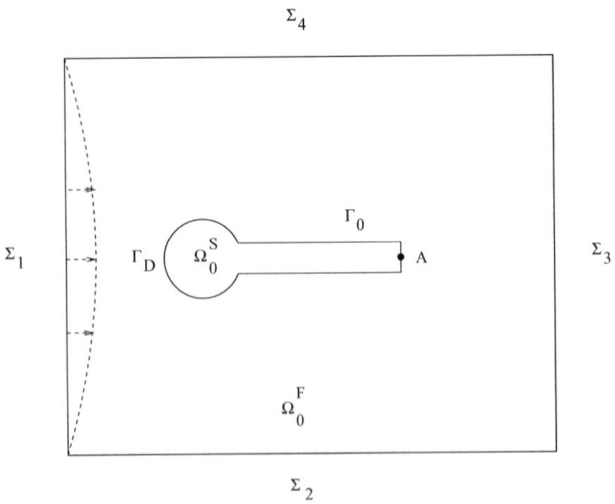


**Fig. 1** Geometrical configuration

we write $\nabla \mathbf{v}^F$ in place of $\nabla_\mathbf{x} \mathbf{v}^F$, when the gradients are computed with respect to the Eulerian coordinates $\mathbf{x}$.

The problem is to find the structure displacement $\mathbf{U}^S$, the fluid velocity $\mathbf{v}^F$ and the fluid pressure $p^F$ such that:

$$\rho_0^S(\mathbf{X}) \frac{\partial^2 \mathbf{U}^S}{\partial t^2}(\mathbf{X}, t) - \nabla_\mathbf{X} \cdot (\mathbf{F}\boldsymbol{\Sigma})(\mathbf{X}, t) = \rho_0^S(\mathbf{X})\,\mathbf{g} \quad \text{in } \Omega_0^S \times (0, T), \tag{1}$$

$$\mathbf{U}^S(\mathbf{X}, t) = 0 \quad \text{on } \Gamma_D \times (0, T), \tag{2}$$

$$\rho^F \left( \frac{\partial \mathbf{v}^F}{\partial t} + (\mathbf{v}^F \cdot \nabla)\mathbf{v}^F \right) - 2\mu^F \nabla \cdot \varepsilon\left(\mathbf{v}^F\right) + \nabla p^F = \rho^F \mathbf{g}, \tag{3}$$

$$\forall t \in (0, T), \forall \mathbf{x} \in \Omega_t^F,$$

$$\nabla \cdot \mathbf{v}^F = 0, \ \forall t \in (0, T), \forall \mathbf{x} \in \Omega_t^F, \tag{4}$$

$$\mathbf{v} = \mathbf{v}_{in} \text{ on } \Sigma_1 \times (0, T), \tag{5}$$

$$\sigma^F \mathbf{n}^F = \mathbf{h}_{out} \text{ on } \Sigma_3 \times (0, T), \tag{6}$$

$$\mathbf{v}^F = 0 \text{ on } \Sigma_2 \cup \Sigma_4 \cup \Gamma_D, \tag{7}$$

$$\mathbf{v}^F\left(\mathbf{X} + \mathbf{U}^S(\mathbf{X}, t), t\right) = \frac{\partial \mathbf{U}^S}{\partial t}(\mathbf{X}, t) \ \text{ on } \Gamma_0 \times (0, T), \tag{8}$$

$$\left(\sigma^F \mathbf{n}^F\right)_{(\mathbf{X}+\mathbf{U}^S(\mathbf{X},t),t)} = -(\mathbf{F}\boldsymbol{\Sigma})(\mathbf{X}, t)\,\mathbf{N}^S(\mathbf{X}) \ \text{ on } \Gamma_0 \times (0, T), \tag{9}$$

$$\mathbf{U}^S(\mathbf{X}, 0) = \mathbf{U}^{S,0}(\mathbf{X}) \ \text{ in } \Omega_0^S, \tag{10}$$

$$\frac{\partial \mathbf{U}^S}{\partial t}(\mathbf{X}, 0) = \mathbf{V}^{S,0}(\mathbf{X}) \ \text{ in } \Omega_0^S, \tag{11}$$

$$\mathbf{v}^F(\mathbf{X}, 0) = \mathbf{v}^{F,0}(\mathbf{X}) \ \text{ in } \Omega_0^F. \tag{12}$$

Here $\rho_0^S : \Omega_0^S \to \mathbb{R}$ is the initial mass density of the structure, $\mathbf{g}$ is the acceleration of gravity vector and it is assumed to be constant, $\mathbf{N}^S$ is the unit outer normal vector along the boundary $\partial\Omega_0^S$, $\rho^F > 0$ and $\mu^F > 0$ are constants and its represent the mass density and the viscosity of the fluid, respectively, $\mathbf{v}_{in}$ is the prescribed inflow velocity, $\mathbf{h}_{out}$ is prescribed outflow boundary stress, $\mathbf{n}^F$ is the unit outer normal vector along the boundary $\partial\Omega_t^F$.

For the structure Eqs. (1) and (2), we have used the Lagrangian coordinates, while for the fluid Eqs. (3)–(7) the Eulerian coordinates have been used. The Eqs. (8) and (9) represent the continuity of velocity and of stress at the interface, respectively. Initial conditions are given by (10)–(12). To conclude, the governing equations and conditions for fluid-structure interaction are (1)–(12).

## 3 Total Lagrangian Framework for the Structure Approximation

Let us introduce $\mathbf{V}^S$ the velocity of the structure in the Lagrangian coordinates. The Eq. (1) is equivalent to

$$\rho_0^S(\mathbf{X})\frac{\partial \mathbf{V}^S}{\partial t}(\mathbf{X},t) - \nabla_{\mathbf{X}}\cdot(\mathbf{F}\boldsymbol{\Sigma})(\mathbf{X},t) = \rho_0^S(\mathbf{X})\,\mathbf{g},\quad \text{in } \Omega_0^S\times(0,T) \tag{13}$$

$$\frac{\partial \mathbf{U}^S}{\partial t}(\mathbf{X},t) = \mathbf{V}^S(\mathbf{X},t),\quad \text{in } \Omega_0^S\times(0,T). \tag{14}$$

Let $N\in\mathbb{N}^*$ be the number of time steps and $\Delta t = T/N$ the time step. We set $t_n = n\Delta t$ for $n = 0,1,\ldots,N$. Let $\mathbf{V}^{S,n}(\mathbf{X})$ and $\mathbf{U}^{S,n}(\mathbf{X})$ be approximations of $\mathbf{V}^S(\mathbf{X},t_n)$ and $\mathbf{U}^S(\mathbf{X},t_n)$. We also use the notations

$$\mathbf{F}^n = \mathbf{I} + \nabla_{\mathbf{X}}\mathbf{U}^{S,n},\quad \boldsymbol{\Sigma}^n = \boldsymbol{\Sigma}(\mathbf{F}^n),\ n\ge 0.$$

The system (13) and (14) will be approached by the implicit Euler scheme

$$\rho_0^S(\mathbf{X})\frac{\mathbf{V}^{S,n+1}(\mathbf{X}) - \mathbf{V}^{S,n}(\mathbf{X})}{\Delta t} - \nabla_{\mathbf{X}}\cdot\left(\mathbf{F}^{n+1}\boldsymbol{\Sigma}^{n+1}\right)(\mathbf{X}) = \rho_0^S(\mathbf{X})\,\mathbf{g},\ \text{in } \Omega_0^S \tag{15}$$

$$\frac{\mathbf{U}^{S,n+1}(\mathbf{X}) - \mathbf{U}^{S,n}(\mathbf{X})}{\Delta t} = \mathbf{V}^{S,n+1}(\mathbf{X}),\ \text{in } \Omega_0^S \tag{16}$$

From (16), we get $\mathbf{F}^{n+1} = \mathbf{F}^n + \Delta t\nabla_{\mathbf{X}}\mathbf{V}^{S,n+1}$ and consequently, $\mathbf{F}^{n+1}$ and $\boldsymbol{\Sigma}^{n+1}$ depend on the velocity $\mathbf{V}^{S,n+1}$ but not in the displacement $\mathbf{U}^{S,n+1}$. In other words, we have eliminated the unknown displacement and we have now an equation of unknown $\mathbf{V}^{S,n+1}$.

The weak form of the Eq. (15) is as follows: find $\mathbf{V}^{S,n+1}:\Omega_0^S\to\mathbb{R}^2$, $\mathbf{V}^{S,n+1}=0$ on $\Gamma_D$, such that

$$\int_{\Omega_0^S}\rho_0^S\frac{\mathbf{V}^{S,n+1}-\mathbf{V}^{S,n}}{\Delta t}\cdot\mathbf{W}^S\,d\mathbf{X} + \int_{\Omega_0^S}\mathbf{F}^{n+1}\boldsymbol{\Sigma}^{n+1} : \nabla_{\mathbf{X}}\mathbf{W}^S\,d\mathbf{X}$$
$$= \int_{\Omega_0^S}\rho_0^S\mathbf{g}\cdot\mathbf{W}^S\,d\mathbf{X} + \int_{\Gamma_0}\mathbf{F}^{n+1}\boldsymbol{\Sigma}^{n+1}\mathbf{N}^S\cdot\mathbf{W}^S\,dS \tag{17}$$

for all $\mathbf{W}^S:\Omega_0^S\to\mathbb{R}^2$, $\mathbf{W}^S=0$ on $\Gamma_D$. Here we assume that the forces $\mathbf{F}^{n+1}\boldsymbol{\Sigma}^{n+1}\mathbf{N}^S$ on the interface $\Gamma_0$ are known.

# 4 Updated Lagrangian Framework for the Structure Approximation

We follow a similar approach that in [3], where the structure is a Neo–Hookean material. In the present paper, the structure is governed by the linear elasticity equations. We denote by $\Omega_n^S$ the image of $\Omega_0^S$ via the map $\mathbf{X} \to \mathbf{X} + \mathbf{U}^{S,n}(\mathbf{X})$ and we set $\widehat{\Omega}^S = \Omega_n^S$ the computational domain for the structure.

The map from $\Omega_0^S$ to $\Omega_{n+1}^S$ defined by $\mathbf{X} \to \mathbf{x} = \mathbf{X} + \mathbf{U}^{S,n+1}(\mathbf{X})$ is the composition of the map from $\Omega_0^S$ to $\widehat{\Omega}^S$ defined by $\mathbf{X} \to \widehat{\mathbf{x}} = \mathbf{X} + \mathbf{U}^{S,n}(\mathbf{X})$ with the map from $\widehat{\Omega}^S$ to $\Omega_{n+1}^S$ defined by

$$\widehat{\mathbf{x}} \to \mathbf{x} = \widehat{\mathbf{x}} + \mathbf{U}^{S,n+1}(\mathbf{X}) - \mathbf{U}^{S,n}(\mathbf{X}) = \widehat{\mathbf{x}} + \widehat{\mathbf{u}}(\widehat{\mathbf{x}}).$$

With the notations $\widehat{\mathbf{F}} = \mathbf{I} + \nabla_{\widehat{\mathbf{x}}}\widehat{\mathbf{u}}$ and $\widehat{J} = \det \widehat{\mathbf{F}}$, $J^n = \det \mathbf{F}^n$, we obtain

$$\mathbf{F}^{n+1}(\mathbf{X}) = \widehat{\mathbf{F}}(\widehat{\mathbf{x}})\,\mathbf{F}^n(\mathbf{X}), \quad J^{n+1}(\mathbf{X}) = \widehat{J}(\widehat{\mathbf{x}})\,J^n(\mathbf{X}). \tag{18}$$

The relation between the Cauchy stress tensor of the structure $\sigma^S$ and the second Piola–Kirchhoff stress tensor $\boldsymbol{\Sigma}$ is the following $\sigma^S(\mathbf{x}, t) = \left(\frac{1}{J}\mathbf{F}\boldsymbol{\Sigma}\mathbf{F}^T\right)(\mathbf{X}, t)$, where $\mathbf{x} = \mathbf{X} + \mathbf{U}^S(\mathbf{X}, t)$. The mass conservation assumption gives $\rho^S(\mathbf{x}, t) = \frac{\rho_0^S(\mathbf{X})}{J(\mathbf{X}, t)}$, where $\rho^S(\mathbf{x}, t)$ is the mass density of the structure in the Eulerian framework.

For the semi-discrete scheme, we use the notations

$$\sigma^{S,n+1}(\mathbf{x}) = \left(\frac{1}{J^{n+1}}\mathbf{F}^{\mathbf{n+1}}\boldsymbol{\Sigma}^{n+1}\left(\mathbf{F}^{n+1}\right)^T\right)(\mathbf{X}), \quad \mathbf{x} = \mathbf{X} + \mathbf{U}^{S,n+1}(\mathbf{X})$$

and $\rho^{S,n}(\widehat{\mathbf{x}}) = \frac{\rho_0^S(\mathbf{X})}{J^n(\mathbf{X})}$, $\widehat{\mathbf{x}} = \mathbf{X} + \mathbf{U}^{S,n}(\mathbf{X})$.

Let us introduce $\widehat{\mathbf{v}}^{S,n+1} : \widehat{\Omega}^S \to \mathbb{R}^2$ and $\mathbf{v}^{S,n} : \widehat{\Omega}^S \to \mathbb{R}^2$ defined by $\widehat{\mathbf{v}}^{S,n+1}(\widehat{\mathbf{x}}) = \mathbf{V}^{S,n+1}(\mathbf{X})$ and $\mathbf{v}^{S,n}(\widehat{\mathbf{x}}) = \mathbf{V}^{S,n}(\mathbf{X})$. Also, for $\mathbf{W}^S : \Omega_0^S \to \mathbb{R}^2$, we define $\widehat{\mathbf{w}}^S : \widehat{\Omega}^S \to \mathbb{R}^2$ and $\mathbf{w}^S : \Omega_{n+1}^S \to \mathbb{R}^2$ by $\widehat{\mathbf{w}}^S(\widehat{\mathbf{x}}) = \mathbf{w}^S(\mathbf{x}) = \mathbf{W}^S(\mathbf{X})$.

Now, we rewrite the Eq. (17) over the domain $\widehat{\Omega}^S$. For the first term of (17), we get

$$\int_{\Omega_0^S} \rho_0^S \frac{\mathbf{V}^{S,n+1} - \mathbf{V}^{S,n}}{\Delta t} \cdot \mathbf{W}^S \, d\mathbf{X} = \int_{\widehat{\Omega}^S} \rho^{S,n} \frac{\widehat{\mathbf{v}}^{S,n+1} - \mathbf{v}^{S,n}}{\Delta t} \cdot \widehat{\mathbf{w}}^S \, d\widehat{\mathbf{x}}$$

and, similarly,

$$\int_{\Omega_0^S} \rho_0^S \mathbf{g} \cdot \mathbf{W}^S \, d\mathbf{X} = \int_{\widehat{\Omega}^S} \rho^{S,n} \mathbf{g} \cdot \widehat{\mathbf{w}}^S \, d\widehat{\mathbf{x}}.$$

Using the identity $\left(\nabla \mathbf{w}^S(\mathbf{x})\right)\mathbf{F}^{n+1}(\mathbf{X}) = \nabla_{\mathbf{X}}\mathbf{W}^S(\mathbf{X})$ and the definition of $\sigma^{S,n+1}$, we get

$$\int_{\Omega_0^S} \mathbf{F}^{n+1}\boldsymbol{\Sigma}^{n+1} : \nabla_{\mathbf{X}}\mathbf{W}^S \, d\mathbf{X} = \int_{\Omega_{n+1}^S} \sigma^{S,n+1} : \nabla \mathbf{w}^S \, d\mathbf{x}.$$

Details about this kind of transformation could be found in [1], Chap. 1.2.

In order to write the above integral over the domain $\widehat{\Omega}^S$, let us introduce the tensor

$$\widehat{\boldsymbol{\Sigma}}\left(\widehat{\mathbf{x}}\right)=\widehat{J}\left(\widehat{\mathbf{x}}\right)\widehat{\mathbf{F}}^{-1}\left(\widehat{\mathbf{x}}\right)\sigma^{S,n+1}\left(\mathbf{x}\right)\widehat{\mathbf{F}}^{-T}\left(\widehat{\mathbf{x}}\right). \tag{19}$$

Since $\left(\nabla\mathbf{w}^S\left(\mathbf{x}\right)\right)\widehat{\mathbf{F}}\left(\widehat{\mathbf{x}}\right)=\nabla_{\widehat{\mathbf{x}}}\widehat{\mathbf{w}}^S\left(\widehat{\mathbf{x}}\right)$, see [1], Chap. 1.2 and taking into account (19), we get

$$\int_{\Omega_{n+1}^S}\sigma^{S,n+1}:\nabla\mathbf{w}^S\,d\mathbf{x}=\int_{\widehat{\Omega}^S}\widehat{\mathbf{F}}\widehat{\boldsymbol{\Sigma}}:\nabla_{\widehat{\mathbf{x}}}\widehat{\mathbf{w}}^S\,d\widehat{\mathbf{x}}.$$

Now, it is possible to present the updated Lagrangian version of (17). Knowing $\mathbf{U}^{S,n}:\Omega_0^S\to\mathbb{R}^2$, $\widehat{\Omega}^S=\Omega_n^S$ and $\mathbf{v}^{S,n}:\widehat{\Omega}^S\to\mathbb{R}^2$, we try to find $\widehat{\mathbf{v}}^{S,n+1}:\widehat{\Omega}^S\to\mathbb{R}^2$, $\widehat{\mathbf{v}}^{S,n+1}=0$ on $\Gamma_D$ such that

$$\begin{aligned}\int_{\widehat{\Omega}^S}\rho^{S,n}\frac{\widehat{\mathbf{v}}^{S,n+1}-\mathbf{v}^{S,n}}{\Delta t}\cdot\widehat{\mathbf{w}}^S\,d\widehat{\mathbf{x}}+\int_{\widehat{\Omega}^S}\widehat{\mathbf{F}}\widehat{\boldsymbol{\Sigma}}:\nabla_{\widehat{\mathbf{x}}}\widehat{\mathbf{w}}^S\,d\widehat{\mathbf{x}}&\\=\int_{\widehat{\Omega}^S}\rho^{S,n}\mathbf{g}\cdot\widehat{\mathbf{w}}^S\,d\widehat{\mathbf{x}}+\int_{\Gamma_0}\mathbf{F}^{n+1}\boldsymbol{\Sigma}^{n+1}\mathbf{N}^S\cdot\mathbf{W}^S\,dS&\end{aligned} \tag{20}$$

for all $\widehat{\mathbf{w}}^S:\widehat{\Omega}^S\to\mathbb{R}^2$, $\widehat{\mathbf{w}}^S=0$ on $\Gamma_D$. We recall that the forces $\mathbf{F}^{n+1}\boldsymbol{\Sigma}^{n+1}\mathbf{N}^S$ on the interface $\Gamma_0$ are assumed known.

Using the identity $\widehat{\mathbf{u}}\left(\widehat{\mathbf{x}}\right)=\mathbf{U}^{S,n+1}\left(\mathbf{X}\right)-\mathbf{U}^{S,n}\left(\mathbf{X}\right)=\Delta t\,\mathbf{V}^{S,n+1}\left(\mathbf{X}\right)=\Delta t\,\widehat{\mathbf{v}}^{S,n+1}\left(\widehat{\mathbf{x}}\right)$, we obtain

$$\widehat{\mathbf{F}}=\mathbf{I}+\Delta t\nabla_{\widehat{\mathbf{x}}}\widehat{\mathbf{v}}^{S,n+1}. \tag{21}$$

Moreover, using (18) and (19), it follows that

$$\begin{aligned}\widehat{\boldsymbol{\Sigma}}&=\widehat{J}\widehat{\mathbf{F}}^{-1}\sigma^{S,n+1}\widehat{\mathbf{F}}^{-T}=\widehat{J}\widehat{\mathbf{F}}^{-1}\frac{1}{J^{n+1}}\mathbf{F}^{n+1}\boldsymbol{\Sigma}^{n+1}\left(\mathbf{F}^{n+1}\right)^T\widehat{\mathbf{F}}^{-T}\\&=\frac{1}{J^n}\mathbf{F}^n\boldsymbol{\Sigma}^{n+1}\left(\mathbf{F}^n\right)^T.\end{aligned} \tag{22}$$

For the linear elastic material, we have

$$\boldsymbol{\Sigma}(\mathbf{U})=\lambda^S(\nabla_{\mathbf{X}}\cdot\mathbf{U})+\mu^S\left(\nabla_{\mathbf{X}}\mathbf{U}+(\nabla_{\mathbf{X}}\mathbf{U})^T\right)$$

where $\lambda^S$ and $\mu^S$ are the Lamé coefficients. Therefore,

$$\boldsymbol{\Sigma}^{n+1}=\boldsymbol{\Sigma}(\mathbf{U}^{S,n+1})=\boldsymbol{\Sigma}(\mathbf{U}^{S,n})+(\Delta t)\boldsymbol{\Sigma}(\mathbf{V}^{S,n+1})=\boldsymbol{\Sigma}^n+(\Delta t)\boldsymbol{\Sigma}(\mathbf{V}^{S,n+1}).$$

We introduce $\boldsymbol{\Sigma}_{\widehat{\mathbf{x}}}(\widehat{\mathbf{u}})=\lambda^S(\nabla_{\widehat{\mathbf{x}}}\cdot\widehat{\mathbf{u}})+\mu^S\left(\nabla_{\widehat{\mathbf{x}}}\widehat{\mathbf{u}}+(\nabla_{\widehat{\mathbf{x}}}\widehat{\mathbf{u}})^T\right)$ and $\boldsymbol{\Sigma}(\mathbf{V}^{S,n+1})$ could be approached by $\boldsymbol{\Sigma}_{\widehat{\mathbf{x}}}(\widehat{\mathbf{v}}^{S,n+1})$. We can approach the map $\widehat{\mathbf{v}}^{S,n+1}\to\widehat{\mathbf{F}}\widehat{\boldsymbol{\Sigma}}$ by the linear application

$$\widehat{\mathbf{F}}\widehat{\boldsymbol{\Sigma}} \approx \frac{1}{J^n}\mathbf{F}^n\boldsymbol{\Sigma}^n\left(\mathbf{F}^n\right)^T + \Delta t\nabla_{\widehat{\mathbf{x}}}\widehat{\mathbf{v}}^{S,n+1}\frac{1}{J^n}\mathbf{F}^n\boldsymbol{\Sigma}^n\left(\mathbf{F}^n\right)^T + \frac{\Delta t}{J^n}\mathbf{F}^n\boldsymbol{\Sigma}_{\widehat{\mathbf{x}}}(\widehat{\mathbf{v}}^{S,n+1})\left(\mathbf{F}^n\right)^T$$
$$= \sigma^{S,n} + \Delta t\nabla_{\widehat{\mathbf{x}}}\widehat{\mathbf{v}}^{S,n+1}\sigma^{S,n} + \frac{\Delta t}{J^n}\mathbf{F}^n\boldsymbol{\Sigma}_{\widehat{\mathbf{x}}}(\widehat{\mathbf{v}}^{S,n+1})\left(\mathbf{F}^n\right)^T.$$

We define $\widehat{\mathbf{u}}^{S,n}(\widehat{\mathbf{x}}) = \mathbf{U}^{S,n}(\mathbf{X})$ and, for the small deformations, we have $\mathbf{F}^n \approx \mathbf{I}$, $J^n \approx 1, \sigma^{S,n} \approx \boldsymbol{\Sigma}_{\widehat{\mathbf{x}}}(\widehat{\mathbf{u}}^{S,n+1})$. Finally, we replace the map $\widehat{\mathbf{v}}^{S,n+1} \rightarrow \widehat{\mathbf{F}}\widehat{\boldsymbol{\Sigma}}$ by the linear application

$$\widehat{\mathbf{L}}\left(\widehat{\mathbf{v}}^{S,n+1}\right) = \boldsymbol{\Sigma}_{\widehat{\mathbf{x}}}(\widehat{\mathbf{u}}^{S,n}) + (\Delta t)\boldsymbol{\Sigma}_{\widehat{\mathbf{x}}}(\widehat{\mathbf{v}}^{S,n+1}). \tag{23}$$

The linearized updated Lagrangian weak formulation of the structure is: knowing $\mathbf{U}^{S,n} : \Omega_0^S \rightarrow \mathbb{R}^2, \widehat{\Omega}^S = \Omega_n^S$ and $\mathbf{v}^{S,n} : \widehat{\Omega}^S \rightarrow \mathbb{R}^2$, find $\widehat{\mathbf{v}}^{S,n+1} : \widehat{\Omega}^S \rightarrow \mathbb{R}^2, \widehat{\mathbf{v}}^{S,n+1} = 0$ on $\Gamma_D$ such that

$$\int_{\widehat{\Omega}^S}\rho^{S,n}\frac{\widehat{\mathbf{v}}^{S,n+1} - \mathbf{v}^{S,n}}{\Delta t}\cdot\widehat{\mathbf{w}}^S\,d\widehat{\mathbf{x}} + \int_{\widehat{\Omega}^S}\widehat{\mathbf{L}}\left(\widehat{\mathbf{v}}^{S,n+1}\right) : \nabla_{\widehat{\mathbf{x}}}\widehat{\mathbf{w}}^S\,d\widehat{\mathbf{x}}$$
$$= \int_{\widehat{\Omega}^S}\rho^{S,n}\mathbf{g}\cdot\widehat{\mathbf{w}}^S\,d\widehat{\mathbf{x}} + \int_{\Gamma_0}\mathbf{F}^{n+1}\boldsymbol{\Sigma}^{n+1}\mathbf{N}^S\cdot\mathbf{W}^S\,dS \tag{24}$$

for all $\widehat{\mathbf{w}}^S : \widehat{\Omega}^S \rightarrow \mathbb{R}^2, \widehat{\mathbf{w}}^S = 0$ on $\Gamma_D$.

## 5 Monolithic Algorithm for the Fluid-Structure Equations

We have $\partial\Omega_n^S = \Gamma_D \cup \Gamma_n$, where $\Gamma_n$ is a approximation of the moving interface $\Gamma_{t_n}$, $\Omega_n^F = D \setminus \overline{\Omega}_n^S$ and let us introduce the global velocity, pressure and test function

$$\widehat{\mathbf{v}}^{n+1} : D \rightarrow \mathbb{R}^2, \quad \widehat{p}^{n+1} : D \rightarrow \mathbb{R}, \quad \widehat{\mathbf{w}} : D \rightarrow \mathbb{R}^2,$$

$$\widehat{\mathbf{v}}^{n+1} = \begin{cases}\widehat{\mathbf{v}}^{F,n+1} \text{ in } \Omega_n^F\\ \widehat{\mathbf{v}}^{S,n+1} \text{ in } \Omega_n^S\end{cases}, \quad \widehat{p}^{n+1} = \begin{cases}\widehat{p}^{F,n+1} \text{ in } \Omega_n^F\\ \widehat{p}^{S,n+1} \text{ in } \Omega_n^S\end{cases}, \quad \widehat{\mathbf{w}} = \begin{cases}\widehat{\mathbf{w}}^F \text{ in } \Omega_n^F\\ \widehat{\mathbf{w}}^S \text{ in } \Omega_n^S\end{cases}.$$

### Algorithm for fluid-structure interaction

**Time advancing scheme from $n$ to $n+1$**

We assume that we know the mesh $\mathscr{T}_h^n$, the velocity $\mathbf{v}^n$, the pressure $p^n$, and the mesh velocity $\boldsymbol{\vartheta}^n$.

**Step 1**: Solve the monolithic **linear** system and get the velocity $\widehat{\mathbf{v}}^{n+1} \in \left(H^1(D)\right)^2$, $\widehat{\mathbf{v}}^{n+1} = \mathbf{v}_{in}$ on $\Sigma_1$, $\widehat{\mathbf{v}}^{n+1} = 0$ on $\partial D \cup \Gamma_D$ and the pressure $\widehat{p}^{n+1} \in L^2(D)$, $\widehat{p}^{n+1} = 0$ in $\Omega_n^S$, such that:

$$\int_{\Omega_n^F} \rho^F \frac{\widehat{\mathbf{v}}^{n+1}}{\Delta t} \cdot \widehat{\mathbf{w}} d\widehat{\mathbf{x}} + \int_{\Omega_n^F} \rho^F \left( \left( \left( \mathbf{v}^n - \boldsymbol{\vartheta}^n \right) \cdot \nabla_{\widehat{\mathbf{x}}} \right) \widehat{\mathbf{v}}^{n+1} \right) \cdot \widehat{\mathbf{w}} d\widehat{\mathbf{x}}$$
$$- \int_{\Omega_n^F} \left( \nabla_{\widehat{\mathbf{x}}} \cdot \widehat{\mathbf{w}} \right) \widehat{p}^{n+1} d\widehat{\mathbf{x}} + \int_{\Omega_n^F} 2\mu^F \varepsilon \left( \widehat{\mathbf{v}}^{n+1} \right) : \varepsilon \left( \widehat{\mathbf{w}} \right) d\widehat{\mathbf{x}}$$
$$+ \int_{\Omega_n^S} \rho^{S,n} \frac{\widehat{\mathbf{v}}^{n+1}}{\Delta t} \cdot \widehat{\mathbf{w}} \, d\widehat{\mathbf{x}} + \int_{\Omega_n^S} \widehat{\mathbf{L}} \left( \widehat{\mathbf{v}}^{n+1} \right) : \nabla_{\widehat{\mathbf{x}}} \widehat{\mathbf{w}} \, d\widehat{\mathbf{x}}$$
$$= \int_{\Omega_n^F} \rho^F \frac{\mathbf{v}^n}{\Delta t} \cdot \widehat{\mathbf{w}} d\widehat{\mathbf{x}} + \int_{\Omega_n^F} \mathbf{f}^{F,n} \cdot \widehat{\mathbf{w}} \, d\widehat{\mathbf{x}} + \int_{\Sigma_3} \mathbf{h}_{out} \cdot \widehat{\mathbf{w}} \, d\widehat{\mathbf{x}}$$
$$+ \int_{\Omega_n^S} \rho^{S,n} \frac{\mathbf{v}^n}{\Delta t} \cdot \widehat{\mathbf{w}} \, d\widehat{\mathbf{x}} + \int_{\Omega_n^S} \rho^{S,n} \mathbf{g} \cdot \widehat{\mathbf{w}} \, d\widehat{\mathbf{x}}, \tag{25}$$
$$\int_{\Omega_n^F} \left( \nabla_{\widehat{\mathbf{x}}} \cdot \widehat{\mathbf{v}}^{n+1} \right) \widehat{q} \, d\widehat{\mathbf{x}} = 0, \tag{26}$$

for all $\widehat{\mathbf{w}} \in \left( H^1 (D) \right)^2$ $\widehat{\mathbf{w}} = 0$ on $\partial D \cup \Gamma_D$ and for all $\widehat{q} \in L^2 (D)$.

**Step 2**: Compute the mesh velocity such that $\widehat{\boldsymbol{\vartheta}}^{n+1} : D \to \mathbb{R}^2$

$$\begin{cases} \Delta_{\widehat{\mathbf{x}}} \widehat{\boldsymbol{\vartheta}}^{n+1} & = 0 \quad \text{in} \quad D, \\ \widehat{\boldsymbol{\vartheta}}^{n+1} & = 0 \quad \text{on} \quad \partial D \cup \Gamma_D, \\ \widehat{\boldsymbol{\vartheta}}^{n+1} & = \widehat{\mathbf{v}}^{n+1} \quad \text{on} \quad \Gamma_n. \end{cases} \tag{27}$$

We can replace in (27), the Laplacian by the linear elasticity operator in order to improve the quality of the mesh.

**Step 3**: Define the map $\mathbb{T}_n : \overline{D} \to \mathbb{R}^2$ by:

$$\mathbb{T}_n(\widehat{\mathbf{x}}) = \widehat{\mathbf{x}} + (\Delta t) \widehat{\boldsymbol{\vartheta}}^{n+1}(\widehat{\mathbf{x}}) \chi_{\Omega_n^F}(\widehat{\mathbf{x}}) + (\Delta t) \widehat{\mathbf{v}}^{n+1}(\widehat{\mathbf{x}}) \chi_{\Omega_n^S}(\widehat{\mathbf{x}})$$

where $\chi_{\Omega_n^F}$ and $\chi_{\Omega_n^S}$ are the characteristic functions of fluid and structure domains.

The new mesh is $\mathbb{T}_n(\mathscr{T}_h^n) = \mathscr{T}_h^{n+1}$.

**Step 4**: We define $\mathbf{v}^{n+1} : D \to \mathbb{R}^2$, $p^{n+1} : D \to \mathbb{R}$ and $\boldsymbol{\vartheta}^{n+1} : D \to \mathbb{R}^2$ by:

$$\mathbf{v}^{n+1}(\mathbf{x}) = \widehat{\mathbf{v}}^{n+1}(\widehat{\mathbf{x}}), \;\; p^{n+1}(\mathbf{x}) = \widehat{p}^{n+1}(\widehat{\mathbf{x}}), \;\; \boldsymbol{\vartheta}^{n+1}(\mathbf{x}) = \widehat{\boldsymbol{\vartheta}}^{n+1}(\widehat{\mathbf{x}})$$

for all $\widehat{\mathbf{x}} \in D$ and $\mathbf{x} = \mathbb{T}_n(\widehat{\mathbf{x}})$.

We solve the monolithic system (25) and (26) using globally continuous finite element for the velocity $\widehat{\mathbf{v}}^{n+1} \in \left( H^1 (D) \right)^2$ defined all over the fluid-structure global mesh. Then the both continuity conditions at the interface hold. For the global pressure $\widehat{p}^{n+1} \in L^2 (D)$, we have to impose $\widehat{p}^{n+1} = 0$ in $\Omega_n^S$. More precisely, we impose $\widehat{p}^{n+1} = 0$ at each node of the structure sub-domain excepting the nodes on the interface $\Gamma_n$.

This algorithm is similar to [4], where the Newmark method was employed for the structure, but the actual algorithm is not a particular case of the cited paper. In addition, the quality of the mesh is augmented in the actual version by solving the mesh velocity after the resolution of the monolithic linear system. Another improvement is that we use now the facility of `FreeFem++` to integrate over a sub-domain, which is faster that using the characteristic function.

# 6 Numerical Test. Flow Around a Flexible Thin Structure Attached to a Fixed Cylinder

We have tested the benchmark FSI3 from [5]. The numerical tests have been produced using *FreeFem++* (see [2]).

The structure is composed by a rectangular flexible beam attached to a fixed circle, see Fig. 1. The circle center is positioned at (0.2, 0.2) m measured from the left bottom corner of the channel. The circle has the radius $r = 0.5$ m and the rectangular beam is of length $\ell = 0.35$ m, thickness $h = 0.02$ m. The mass density is $\rho^S = 1000$ Kg/(m$^3$), the Young modulus is $E^S = 5.6 \times 10^6$ Pa and the Poisson's ratio is $\nu^S = 0.4$.

The channel has the length $L = 2.5$ m and the width $H = 0.41$ m. The fluid dynamic viscosity is $\mu^F = 1$ Kg/(ms) and the mass density is $\rho^F = 1000$ Kg/(m$^3$).

We denote by $\Sigma_1 = \{0\} \times [0, H]$, $\Sigma_3 = \{L\} \times [0, H]$ the left and the right vertical boundaries of the channel and by $\Sigma_2 = [0, L] \times \{0\}$, $\Sigma_4 = [0, L] \times \{H\}$ the bottom and the top boundaries, respectively.

We have used the boundary condition $\mathbf{v} = \mathbf{v}_{in}$ at the inflow $\Sigma_1$, where

$$\mathbf{v}_{in}(x_1, x_2, t) = \begin{cases} \left(1.5\,\overline{U}\frac{x_2(H-x_2)}{(H/2)^2}\frac{(1-\cos(\pi t/2))}{2},\ 0\right), & (x_1, x_2) \in \Sigma_1, 0 \le t \le 2 \\ \left(1.5\,\overline{U}\frac{x_2(H-x_2)}{(H/2)^2},\ 0\right), & (x_1, x_2) \in \Sigma_1, 2 \le t \le T = 8 \end{cases}$$

and $\overline{U} = 2$. At $\Sigma_2$, $\Sigma_4$, as well as on the boundary of the circle, we have imposed the no-slip boundary condition $\mathbf{v} = \mathbf{0}$. At the outflow $\Sigma_3$, we have imposed the traction free $\sigma^F(\mathbf{v}, p)\,\mathbf{n}^F = 0$. Initially, the fluid and the structure are at rest.

Using *FreeFem++* [2], it is possible to construct a global fluid-structure mesh with an "interior boundary" which is the fluid-structure interface. The global moving mesh for the fluid-structure domain is aligned with the fluid-structure interface and changes at at each time step. For the finite element approximation of the fluid-structure velocity, we have used the triangular finite element $\mathbb{P}_1 + bubble$ and we

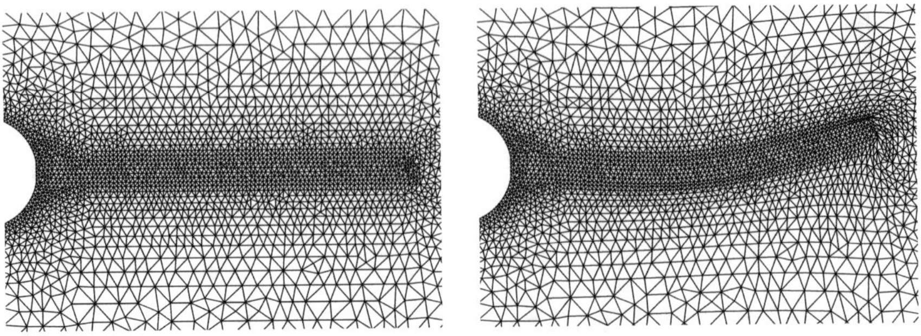

**Fig. 2** Details of the fluid-structure mesh of 9382 triangles and 4859 vertices at $t = 0$ and $t = 6.016$

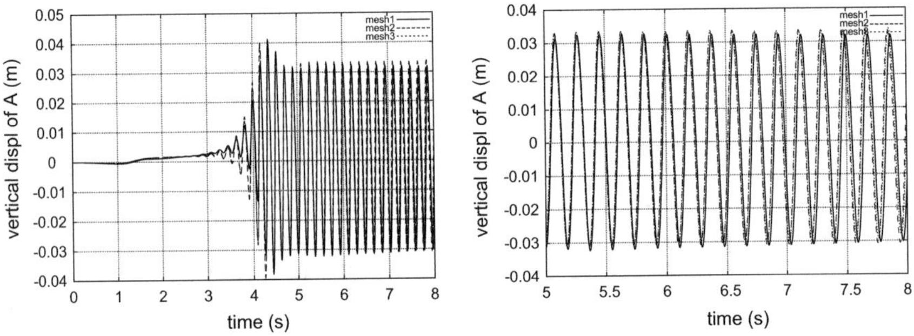


**Fig. 3** Time history of the vertical displacement of the point A in meters for $\Delta t = 0.002$ s using three different meshes and detail in the time interval 5–8 s

have employed for the pressure the finite element $\mathbb{P}_1$. The linear fluid-structure system is solved using the LU decomposition. First, we use three global meshes for the fluid-structure domain: *mesh1* of 9382 triangles and 4859 vertices, *mesh2* of 12532 triangles and 6464 vertices, *mesh3* of 19318 triangles and 9903 vertices, see Fig. 2, with the time step $\Delta t = 0.002$ s and the number of time steps $N = 4000$. After an initial transient period, the system settles into periodic oscillations, Fig. 3. The average frequency in the time interval [5, 8] is about 5.33 Hz. The results are similar to [5], where the reference amplitude of the periodic oscillations is 0.034, but the structure is a St. Venant–Kirchhoff material. The pressure in the structure domain has no physical signification and it is fixed to zero, Fig. 5. Also, we have used three time steps $\Delta t = 0.001$ s, $\Delta t = 0.002$ s and $\Delta t = 0.004$ s with the *mesh1*, see Fig. 4. We observe that the numerical behavior is more sensitive to the time step.

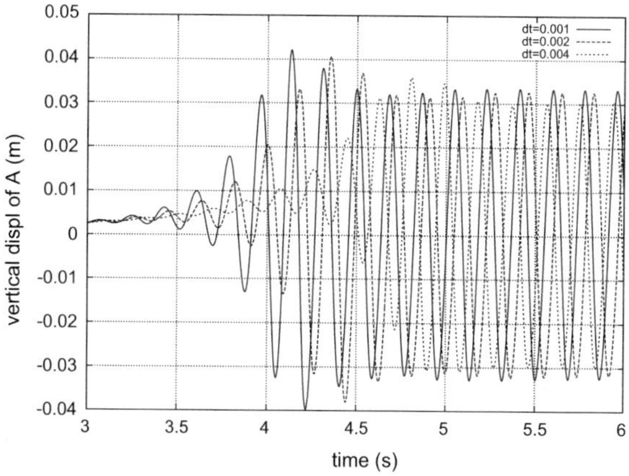


**Fig. 4** Time history of the vertical displacement of the point A in meters for $\Delta t = 0.001$ s, $\Delta t = 0.002$ s and $\Delta t = 0.004$ s in the time interval 3–6 s

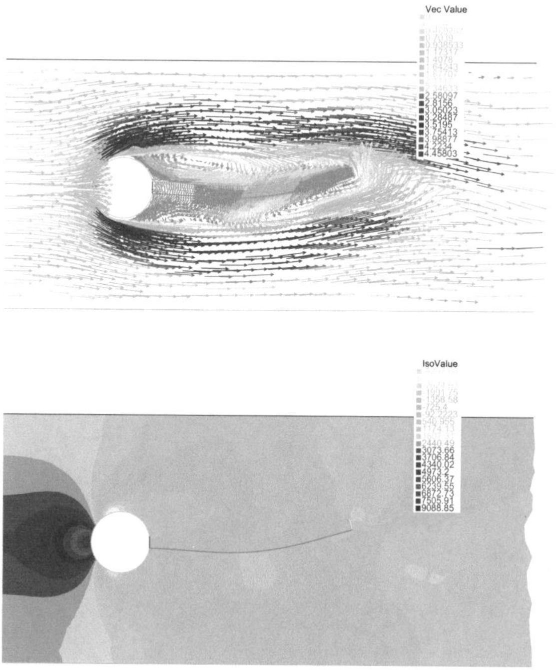


**Fig. 5** Velocity (*top*) and pressure (*bottom*) at $t = 6.016$

# 7 Conclusions

We have used global moving meshes for the fluid-structure domain aligned with the fluid-structure interface. We solved a linear monolithic system in which the unknowns are the velocity and the pressure, defined on the global mesh, at each time step. The continuity of velocity at the interface was automatically satisfied, since we used continuous finite elements.

## References

1. Ciarlet, P.G.: Élasticité tridimensionnelle, Masson (1986)
2. Hecht, F.: New development in FreeFem++. J. Numer. Math. **20**(3–4), 251–265 (2012). http://www.freefem.org
3. Murea, C.M., Sy, S.: Updated Lagrangian/Arbitrary Lagrangian Eulerian framework for interaction between a compressible Neo-Hookean structure and an incompressible fluid. Int. J. Numer. Meth. Eng. (2016). doi:10.1002/nme.5302
4. Sy, S., Murea, C.M.: Algorithm for solving fluid-structure interaction problem on a global moving mesh. Coupled Syst. Mech. Int. J. **1**(1), 99–113 (2012)
5. Turek, S., Hron, J.: Proposal for numerical benchmarking of fluid-structure interaction between an elastic object and laminar incompressible flow. In: Bungartz, H.-J., Schfer, M. (eds.) Fluid-Structure Interaction - Modelling, Simulation, Optimization. Lecture Notes in Computational Science and Engineering, vol. 53, pp. 371–385. Springer, Berlin (2006)

# Elastic-Plastic Rolling Contact Problems with Graded Materials and Heat Exchange

Andrzej Myśliński

**Abstract** This paper is concerned with the rolling contact problems including heat flow across a contact surface. The nonhomogeneous two-layer material model of the obstacle is assumed, i.e., mechanical and thermal properties of the obstacle coating material near its surface are dependent on the spatial variable. The elastic-plastic graded model of the coating layer rather than elastic one is assumed. A variational formulation of this dynamic contact phenomenon is derived in the framework of general thermo-elastic-viscoplastic material models. The displacements and temperatures of the bodies in contact are governed by the coupled system consisting of dynamic variational inequality and parabolic equation. The existence of solutions to this coupled boundary value problem is shown using monotonicity and fixed-point arguments. The rolling contact problem is discretized using the finite element method and numerically solved using the semi-smooth Newton method. Numerical results, including the distribution of contact stresses and temperature, are provided and discussed.

## 1 Introduction

Dynamic or quasistatic contact phenomena for elastic-plastic or viscoplastic materials with heat flow appear in many engineering problems [1, 5, 7, 25] and are intensively studied in literature (see references in [1–3, 8, 10, 12, 14, 17, 21, 22]). General model of thermo-elastic-viscoplastic material is characterized by a rate–type constitutive equation with internal variables [23] modeling their impact on the behavior of real bodies in contact under plastic deformation. The considered internal state variables include, among others, spatial display of dislocations, the work hardening of materials, the temperature or the damage field [3]. The existence of solutions to

A. Myśliński (✉)
Systems Research Institute, ul. Newelska 6, 01-447 Warsaw, Poland
e-mail: myslinsk@ibspan.waw.pl

A. Myśliński
Faculty of Manufacturing Engineering, Warsaw University of Technology, ul. Narbutta 85, 02-524 Warsaw, Poland

© Springer Nature Singapore Pte Ltd. 2017
F. dell'Isola et al. (eds.), *Mathematical Modelling in Solid Mechanics*,
Advanced Structured Materials 69, DOI 10.1007/978-981-10-3764-1_10

these contact problems is studied in monographs [11, 13] and papers [1, 3, 9, 12, 15, 16, 18, 19, 23, 24].

Rolling contact problems between a wheel and a rail or between a tyre and a road belong to the class of dynamic contact problems [1, 25]. Due to the appearance of the large displacements and stresses or the surface material plastification generated by the frictional heat flow they should be considered in the framework of elastic-plastic material models rather than elastic models only [25]. Moreover the application of the nonhomogeneous materials [6] leads to the reduction of maximal contact stress, the wear of the contacting surfaces or noise generated during the movement comparing to the homogeneous materials. The considered nonhomogeneous materials consist of two or many layers of functionally graded materials characterized by different mechanical or thermal properties depending on the spatial variables. Functionally graded materials [5, 21] are generally multiphase composites consisted mainly from a ceramic and a metal phase with continuously varying volume fractions. They exploit the heat, oxidation and corrosion resistance typical of ceramics, and the strength, ductility and toughness typical of metals. These material coatings are widely used in engineering structures where the contact stress is a major concern.

The paper is concerned with the analysis and numerical modeling of the rolling contact between a rigid wheel and an elasto-viscoplastic rail lying on a rigid foundation. The contact phenomenon includes also a heat generation and flow through the contact surface [6, 18]. The obstacle is assumed to be covered with functionally graded coating material which properties depending on the spatial variables according to the power law. In the paper the nonhomogeneous plastically graded model of the coating layer rather than elastic one as in [7, 20] is assumed. Based on plasticity theory [11, 12] the time-dependent model of this rolling contact problem is formulated in the paper. The elastic and plastic responses of the graded layer are approximated, respectively, by Hooke's law and the von Mises yield criterion with isotropic power law hardening [4, 12]. The rolling contact problem is governed by the coupled system consisting from elastoplastic and conductive equations. The existence of solutions to this coupled hyperbolic–parabolic boundary value problem is shown in the framework of general thermo-elastic-viscoplastic material models. The finite element method is used as a discretization method. It is well known that the application of the classical finite element method, where material properties are constant, to solution of problems with the functionally graded materials may lead to large numerical errors. A proper approach to solve such problems requires the application of the nonhomogeneous finite element method containing additional approximation functions in order to interpolate material properties at the level of each finite element. This idea is implemented in the framework of the graded [6] or multi-scale [6] finite element methods. The semi-smooth Newton method [10] is used to solve this discrete system numerically. The distribution of stresses including the normal and tangent contact stresses as well as the distribution of the temperature are numerically calculated and the provided results are discussed.

## 2 General Contact Problem

Consider a dynamic frictionless contact problem for a body occupying a bounded domain $\Omega \subset R^2$ with a Lipschitz continuous boundary $\Gamma$. This boundary is divided into three disjoint measurable parts $\Gamma_1$, $\Gamma_2$ and $\Gamma_3$. Assume $meas(\Gamma_1) > 0$. A body is assumed clamped along the boundary $\Gamma_1$, i.e., the displacement vanishes there. Along the boundary $\Gamma_3 \times (0, T)$ the body is assumed to be in contact with the foundation. The surface traction $f_2$ acts on the boundary $\Gamma_2 \times (0, T)$. The body is loaded by a volume force of density $f_1$ in $\Omega \times (0, T)$. The external heat source $q$ is applied in $\Omega \times (0, T)$. The body is assumed to undergo the coupled thermal as well as elastic-viscoplastic deformation with linear isotropic and kinematic hardening [2, 11]. Let us denote by $u = (u_1, u_2)$, $u = u(x, t)$, $x \in \Omega$, $t \in (0, T)$, $T > 0$ is given, and by $\theta = \theta(x, t)$ a displacement and a temperature of the body, respectively. Moreover $\sigma = \sigma(u) = \{\sigma_{ij}\}_{i=1}^2$ and $\varepsilon = \varepsilon(u) = \frac{1}{2}(u_{i,j} + u_{j,i})$, $i, j = 1, 2$, $u_{i,j} = \frac{\partial u_i}{\partial x_j}$ denote the Cauchy stress tensor and the linearized strain tensor, respectively. The divergence operator $div$ is defined as $div(\sigma) = \{\sigma_{ij,j}\}$, $i, j = 1, 2$, and $\sigma_{ij,j} = \frac{\partial \sigma_{ij}}{\partial x_j}$. The summation convention over repeated indices is used. The dot above the symbol denotes the derivative with respect the time variable, i.e., $\dot{u} = \frac{\partial u}{\partial t}$ and $\ddot{u} = \frac{\partial^2 u}{\partial t^2}$.

By $\nu$ we denote the unit outward normal vector to the boundary $\Gamma$. Normal and tangential components of the displacement field $u$ are denoted by $u_\nu = u \cdot \nu = u_i \cdot \nu_i$, $i = 1, 2$, and by $u_\tau = u - u_\nu \nu$, respectively. Similarly normal and tangential components of the stress field $\sigma$ are denoted by $\sigma_\nu = \sigma\nu \cdot \nu$ and by $\sigma_\tau = \sigma\nu - \sigma_\nu \nu$, respectively. Denote by $\mathbb{S}_2$ the space of second-order symmetric tensors on $R^2$. Moreover $Q = \{\tilde{q} = (\tilde{q}_{ij})_{2\times 2} : \tilde{q}_{ij} = \tilde{q}_{ji}, \tilde{q}_{ij} \in L^2(\Omega)\}$ and $Q_0 = \{\tilde{q} \in Q : \mathrm{tr}\, \tilde{q} = 0\}$ is closed subspace of $Q$. Additive small strain plasticity model is used [10–12] where $\varepsilon^p$ denotes the plastic part of the strain tensor. By $(\sigma, \chi) \in Q \times [L^2(\Omega)]^2$ and $(\varepsilon^p, \zeta) \in Q_0 \times [L^2(\Omega)]^2$ we denote the generalized stress and strain tensors [11]. For a given yield function $\varphi$ the set of admissible generalized stresses $K$ is defined by $K = \{(\sigma, \chi) : \varphi(\sigma, \chi) \leq 0\}$. Denote by $N_K$ a normal cone to the set $K$ at a point $(\sigma, \chi)$ and by $\phi$ the support function of the set $K$ called the dissipation function [11] as well as by $K_p = \mathrm{dom}\, \phi$.

Consider the following contact problem: find the stress field $\sigma : \Omega \times [0, T] \to \mathbb{S}_2$, the displacement $u : \Omega \times [0, T] \to R^2$, the internal field $(\varepsilon^p, \zeta) : \Omega \times [0, T] \to R^{2\times 2} \times R^2$ and the temperature $\theta : \Omega \times [0, T] \to R$ satisfying:

$$\sigma(t) = \mathscr{A}(\varepsilon(\dot{u}(t))) + \mathscr{E}(\varepsilon(u(t))) + \int_0^t \mathscr{G}(\sigma(s) - \mathscr{A}(\dot{\varepsilon}(\dot{u}(s)), \varepsilon(u(s)), \theta(s), \zeta(s))ds \text{ in } \Omega \text{ a.e. } t \in (0, T), \quad (1)$$

$$\rho\ddot{u} = div\sigma + f_1 \text{ in } \Omega \times (0, T), \quad (2)$$

$$(\dot{\varepsilon}^p, \dot{\zeta}) \in K_p \text{ and } \phi(q, \eta) - \phi(\dot{\varepsilon}^p, \dot{\zeta}) - \sigma(q - \dot{\varepsilon}^p) - \chi(\eta - \dot{\zeta}) \geq 0 \ \forall (q, \eta) \in K_p \text{ in } \Omega \times (0, T), \quad (3)$$

$$\rho\dot{\theta} - \Delta\theta = \psi(\sigma - \mathscr{A}(\varepsilon(\dot{u})), \varepsilon(u), \theta, \zeta) + g \text{ in } \Omega \times (0, T), \quad (4)$$

$$u = 0 \quad \text{on} \quad \Gamma_1 \times (0, T), \tag{5}$$
$$\sigma_\nu = f_2 \quad \text{on} \quad \Gamma_2 \times (0, T), \tag{6}$$
$$-\sigma_\nu = p_\nu(u_\nu) \quad \text{on} \quad \Gamma_3 \times (0, T), \tag{7}$$
$$-\sigma_\tau = p_\tau \quad \text{on} \quad \Gamma_3 \times (0, T), \tag{8}$$
$$\nabla\theta\nu + k_1\theta = \bar{g} \quad \text{on} \quad \Gamma_3 \times (0, T), \tag{9}$$
$$u(0) = u_0, \ \dot{u}(0) = u_1, \ \varepsilon^p(0) = \varepsilon_0^p, \ \zeta(0) = \zeta_0, \ \theta(0) = \theta_0. \tag{10}$$

Equation (1) represents the thermo-elastic-viscoplastic constitutive law with operators $\mathscr{A}$ and $\mathscr{E}$ governing the viscous and the elastic properties of the material as well as with nonlinear constitutive function $\mathscr{G}$ governing viscoplastic properties of the material. The motion of the body is governed by the equation (2) where $\rho$ denotes the material mass density. Inequality (3), equivalent to $(\dot{\varepsilon}^p, \dot{\zeta}) \in N_K(\sigma, \chi)$ [4, 11, 12, 14], describes the plastic flow with linear isotropic and kinematic hardening characterized by the hardening modulus $\mathbf{H}$ relating $\chi = -\mathbf{H}\zeta$. Heat flow is governed by the equation (4) where $\psi$ is a constitutive function representing the heat generated by the work of internal forces and $g$ is a given volume heat source. Displacement and stress boundary conditions are given by (5)–(6), respectively. Normal compliance condition with a given positive function $p_\nu$ is described by (7). In (8) tangential traction $p_\tau$ is a given function. Fourier type boundary condition for temperature is given in (9) with a given function $\bar{g}$ and constant $k_1 > 0$. Suitable regular initial data functions $u_0$, $u_1$, $\varepsilon_0^p$, $\zeta_0$, $\theta_0$ in (10) are assumed to be given. Before we formulate initial problem (1)–(10) in variational form let us introduce the following spaces and subspaces:

$$H = [L^2(\Omega)]^2 = \{u = \{u_i\}_{i=1}^2 : u_i \in L^2(\Omega)\}, \tag{11}$$
$$\mathscr{H} = \{\sigma = \{\sigma_{ij}\}_{i,j=1}^2 : \sigma_{ij} = \sigma_{ji} \in L^2(\Omega)\}, \tag{12}$$
$$H_1 = \{u \in H : \varepsilon(u) \in \mathscr{H}\}, \quad \mathscr{H}_1 = \{\sigma \in \mathscr{H} : div(\sigma) \in H\}, \tag{13}$$
$$V = H^1(\Omega), \quad \mathscr{V} = \{v \in H_1 : v = 0 \text{ on } \Gamma_1\}. \tag{14}$$

The spaces $H$, $\mathscr{H}$, $H_1$, $\mathscr{H}_1$, $V$ are endowed with the canonical inner products, i.e., $(\cdot, \cdot)_H = \int_\Omega u_i v_i dx$, $i = 1, 2$. The inner product on the space $\mathscr{V}$ is equal to $(u, v)_{\mathscr{V}} = (\varepsilon(u), \varepsilon(v))_{\mathscr{H}}$. Let $\mathscr{V}'$ and $V'$ denote dual spaces to the spaces $\mathscr{V}$ and $V$, respectively. Recall [13] identifying the spaces $H$ and $L^2(\Omega)$ with their dual spaces $H'$ and $L^2(\Omega)$ we have inclusions $\mathscr{V} \subset H \subset \mathscr{V}'$ and $V \subset L^2(\Omega) \subset V'$. The duality pairing between spaces $\mathscr{V}'$ and $\mathscr{V}$ is denoted by $< \cdot, \cdot >_{\mathscr{V}' \times \mathscr{V}}$. For every real space $X$ by $C(0, T; X)$ or $C^1(0, T; X)$ we denote the space of continuous or continuously differentiable functions from $[0, T]$ to $X$, respectively [12, 13, 16].

Let us introduce the following assumptions. The viscosity operator $\mathscr{A} : \Omega \times \mathbb{S}_2 \to \mathbb{S}_2$ satisfies:

$$\begin{cases} (a)\ \text{There exists a constant } L_1 > 0 \text{ such that} \\ \quad \mid \mathscr{A}(x,\varepsilon_1) - \mathscr{A}(x,\varepsilon_2) \mid \leq L_1 \mid \varepsilon_1 - \varepsilon_2 \mid \ \forall\, \varepsilon_1, \varepsilon_2 \in \mathbb{S}_2, \text{ a.e. } x \in \Omega. \\ (b)\ \text{There exists a constant } m_1 \text{ such that} \\ \quad (\mathscr{A}(x,\varepsilon_1) - \mathscr{A}(x,\varepsilon_2)) \geq m_1 \mid \varepsilon_1 - \varepsilon_2 \mid^2 \ \forall\, \varepsilon_1, \varepsilon_2 \in \mathbb{S}_2, \text{ a.e. } x \in \Omega. \\ (c)\ \text{The mapping } x \to \mathscr{A}(x,\varepsilon) \text{ is Lebesgue measurable on } \Omega \ \forall\, \varepsilon \in \mathbb{S}_2 \\ (d)\ \text{The mapping } x \to \mathscr{A}(x,0) \in \mathscr{H}. \end{cases} \tag{15}$$

The elasticity operator $\mathscr{E} : \Omega \times \mathbb{S}_2 \to \mathbb{S}_2$ satisfies:

$$\begin{cases} (a)\ \text{There exists a constant } L_2 > 0 \text{ such that} \\ \quad \mid \mathscr{E}(x,\varepsilon_1) - \mathscr{E}(x,\varepsilon_2) \mid \leq L_2 \mid \varepsilon_1 - \varepsilon_2 \mid \ \forall\, \varepsilon_1, \varepsilon_2 \in \mathbb{S}_2, \text{ a.e. } x \in \Omega. \\ (b)\ \text{The mapping } x \to \mathscr{E}(x,\varepsilon) \text{ is Lebesgue measurable on } \Omega \ \forall\, \varepsilon \in \mathbb{S}_2. \\ (c)\ \text{The mapping } x \to \mathscr{E}(x,0) \in \mathscr{H}. \end{cases} \tag{16}$$

The viscoplasticity operator $\mathscr{G} : \Omega \times \mathbb{S}_2 \times \mathbb{S}_2 \times R \times R \to \mathbb{S}_2$ is assumed to satisfy:

$$\begin{cases} (a)\ \text{There exists a constant } L_3 > 0 \text{ such that} \\ \quad \mid \mathscr{G}(x,\sigma_1,\varepsilon_1,\theta_1,\zeta_1) - \mathscr{G}(x,\sigma_2,\varepsilon_2,\theta_2,\zeta_2) \mid \leq L_3(\mid \sigma_1 - \sigma_2 \mid + \\ \quad \mid \varepsilon_1 - \varepsilon_2 \mid + \mid \theta_1 - \theta_2 \mid + \mid \zeta_1 - \zeta_2 \mid) \ \forall\, \sigma_1, \sigma_2 \in \mathbb{S}_2, \ \forall\, \varepsilon_1, \varepsilon_2 \in \mathbb{S}_2, \\ \quad \forall\, \theta_1, \theta_2 \in R, \ \forall\, \zeta_1, \zeta_2 \in R, \ \text{ a.e. } x \in \Omega. \\ (b)\ \text{The mapping } x \to \mathscr{G}(x,\sigma,\varepsilon,\theta,\zeta) \text{ is Lebesgue measurable on } \Omega \\ \quad \forall\, \sigma, \varepsilon \in \mathbb{S}_2, \theta, \zeta \in R. \\ (c)\ \text{The mapping } x \to \mathscr{G}(x,0,0,0,0) \in \mathscr{H}. \end{cases} \tag{17}$$

The dissipation function $\phi : \Omega \times \mathbb{S}_2 \times R^2 \to R$ as well as the set $K_p$ of the admissible states and the hardening modulus $\mathbf{H}$ satisfy

$$\begin{cases} (a)\ \phi \text{ is a proper, convex and lower semicontinuous function,} \\ (b)\ K_p \text{ is nonempty, closed and convex set in } L^2(\Omega; R^{2\times 2} \times R^2), \\ (c)\ \text{the hardening modulus } \mathbf{H} \text{ is symmetric, positive definite} \\ \quad \text{and linear operator from } R^2 \text{ into itself.} \end{cases} \tag{18}$$

The function $\psi : \Omega \times \mathbb{S}_2 \times \mathbb{S}_2 \times R \times R \to R$ satisfies:

$$\begin{cases} (a)\ \text{There exists a constant } L_5 > 0 \text{ such that} \\ \quad \mid \psi(x,\sigma_1,\varepsilon_1,\theta_1,\zeta_1) - \psi(x,\sigma_2,\varepsilon_2,\theta_2,\zeta_2) \mid \leq L_5(\mid \sigma_1 - \sigma_2 \mid + \\ \quad \mid \varepsilon_1 - \varepsilon_2 \mid + \mid \theta_1 - \theta_2 \mid + \mid \zeta_1 - \zeta_2 \mid) \ \forall \sigma_1, \sigma_2 \in \mathbb{S}_2, \ \forall\, \varepsilon_1, \varepsilon_2 \in \mathbb{S}_2, \\ \quad \forall\, \theta_1, \theta_2 \in R, \ \forall\, \zeta_1, \zeta_2 \in R, \ \text{ a.e. } x \in \Omega. \\ (b)\ \text{The mapping } x \to \psi(x,\sigma,\varepsilon,\theta,\zeta) \text{ is Lebesgue measurable on } \Omega \\ \quad \forall\, \sigma, \varepsilon \in \mathbb{S}_2, \theta, \zeta \in R. \\ (c)\ \text{The mapping } x \to \psi(x,0,0,0,0) \in \mathscr{H}. \end{cases} \tag{19}$$

The normal compliance function $p_\nu : \Gamma_3 \times R \to R_+$ is assumed to satisfy:

$$\begin{cases} (a) \text{ There exists a constant } L_6 > 0 \text{ such that} \\ \quad \mid p_\nu(x, z_1) - p_\nu(x, z_2) \mid \leq L_6 \mid z_1 - z_2 \mid \ \forall \ z_1, z_2 \in R, \text{ a.e. } x \in \Gamma_3. \\ (b) \text{ The mapping } x \to p_\nu(x, z) \text{ is Lebesgue measurable on } \Gamma_3 \ \forall z \in R. \\ (c) \text{ The mapping } x \to p_\nu(x, z) = 0 \text{ for any } z \leq 0, \text{ a.e. } x \in \Gamma_3. \end{cases} \tag{20}$$

We shall also assume:

$$f_1 \in L^2(0, T; H), \quad f_2 \in L^2(0, T; [L^2(\Gamma_2)]^2), \tag{21}$$

$$g \in L^2(0, T; L^2(\Omega)), \quad \bar{g} \in L^2(\Gamma_3), \quad k_1 > 0, \quad p_\tau \in L^\infty(\Gamma_3), \tag{22}$$

$$\rho \in L^\infty(\Omega), \quad (\varepsilon_0^p, \zeta_0) \in K_p, \quad u_0 \in \mathscr{V}, \quad u_1 \in H, \quad \theta_0 \in V. \tag{23}$$

Let us define the following bilinear and linear forms:

$$a_\theta : V \times V \to R, \quad a_\theta(\zeta, \xi) = \int_\Omega \rho \nabla\zeta \nabla\xi dx + k_1 \int_{\Gamma_3} \zeta\xi ds \tag{24}$$

$$f(t) \in L^2(0, T; \mathscr{V}'), \quad < f(t), \upsilon >_{\mathscr{V}' \times \mathscr{V}} = \int_\Omega f_1(t)\upsilon dx + \int_{\Gamma_2} f_2 \upsilon ds, \tag{25}$$

$$j_c : \mathscr{V} \times \mathscr{V} \to R, \quad j_c(u, \upsilon) = \int_{\Gamma_3} (p_\nu(u_\nu)\upsilon_\nu + p_\tau \upsilon_\tau) ds, \tag{26}$$

$$j_p : Q_0 \times H \to R, \quad j_p(q, \zeta) = \int_\Omega \phi(q, \zeta) dx. \tag{27}$$

Consider problem (1)–(10) in the variational form: find the stress field $\sigma : [0, T] \to \mathbb{S}_2$, the displacement field $u : [0, T] \to R$, the internal variable $(\varepsilon^p, \zeta) : [0, T] \to R^{2\times 2} \times R^2$, the temperature field $\theta : [0, T] \to R$, such that

$$\sigma(t) = \mathscr{A}(\varepsilon(\dot{u}(t))) + \mathscr{E}(\varepsilon(u(t))) + \int_0^t \mathscr{G}(\sigma(s) -$$

$$\mathscr{A}(\dot{\varepsilon}(\dot{u}(s))), \varepsilon(u(s)), \theta(s), \zeta(s)) ds \quad \text{in } \Omega \text{ a.e. } t \in (0, T), \tag{28}$$

$$< \rho\ddot{u}, \upsilon >_{\mathscr{V}' \times \mathscr{V}} + \int_\Omega \sigma(t)\varepsilon(\upsilon) dx + j_c(u, \upsilon) + j_p(q, \eta) -$$

$$j_p(\dot{\varepsilon}^p, \dot{\zeta}) \geq < f(t), \upsilon >_{\mathscr{V}' \times \mathscr{V}} \quad \forall (\upsilon, q, \eta) \in \mathscr{V} \times K_p \text{ a.e. } t \in (0, T), \tag{29}$$

$$< \rho\dot{\theta}, \upsilon >_{V' \times V} + a_\theta(\theta, \upsilon) = < \psi(\sigma(t) - \mathscr{A}(\varepsilon(\dot{u}(t)), \varepsilon(u(t)), \theta(t), \tag{30}$$

$$\zeta(t), \upsilon >_{V' \times V} + \int_\Omega g(t)\upsilon dx + \int_{\Gamma_3} \bar{g}\upsilon ds, \quad \forall \upsilon \in V, \text{ a.e. } t \in (0, T), \tag{31}$$

$$u(0) = u_0, \ \dot{u}(0) = u_1, \ \theta(0) = \theta_0, \ \varepsilon^p(0) = \varepsilon_0^p, \ \zeta(0) = \zeta_0. \tag{32}$$

The existence of a unique solution to contact problem (28)–(32) is shown in:

**Theorem 10.1** *Assume conditions (15)–(23) hold. There exists a unique solution* $(\sigma, u, \varepsilon^p, \zeta, \theta)$ *to the problem (28)–(32). Moreover*

$$u \in C^0(0,T;\mathscr{V}) \cap C^1(0,T;H), \quad \dot{u} \in L^2(0,T;\mathscr{V}), \quad \ddot{u} \in L^2(0,T;\mathscr{V}'), \tag{33}$$
$$\varepsilon^p \in L^2(0,T;V) \cap C^0(0,T;L^2(\Omega)), \quad \dot{\varepsilon}^p \in L^2(0,T;V'), \tag{34}$$
$$\zeta \in L^2(0,T;V) \cap C^0(0,T;L^2(\Omega)), \quad \dot{\zeta} \in L^2(0,T;V'), \tag{35}$$
$$\sigma \in L^2(0,T;\mathscr{H}), \quad \theta \in L^2(0,T;V) \cap C^0(0,T;L^2(\Omega)), \quad \dot{\theta} \in L^2(0,T;V'). \tag{36}$$

In order to prove Theorem 10.1 we need the following auxiliary results. For a given $\gamma \in L^2(0,T;\mathscr{V}')$ let us define the auxiliary problem $P_\gamma$: find the displacement field $u_\gamma : [0,T] \times \Omega \to R^2$ and the internal variable $(\varepsilon^p_\gamma, \zeta_\gamma) : [0,T] \times \Omega \to R^{2\times 2} \times R^2$, satisfying

$$< \rho\ddot{u}_\gamma, \upsilon >_{\mathscr{V}'\times\mathscr{V}} + \int_\Omega \mathscr{A}(\varepsilon(\dot{u}_\gamma(t)))\varepsilon(\upsilon)dx + j_p(q,\eta) - j_p(\dot{\varepsilon}^p\gamma, \dot{\zeta}_\gamma) + \tag{37}$$
$$< \gamma(t), \upsilon >_{\mathscr{V}'\times\mathscr{V}} \geq < f(t), \upsilon >_{\mathscr{V}'\times\mathscr{V}}, \quad \forall(\upsilon, q, \eta) \in \mathscr{V} \times K_p, \text{ a.e. } t \in (0,T),$$
$$\text{and} \quad u_\gamma(0) = u_0, \quad \dot{u}_\gamma(0) = u_1, \quad \varepsilon^p_\gamma(0) = \varepsilon^p_0, \quad \zeta_\gamma(0) = \zeta_0. \tag{38}$$

**Lemma 10.1** *For all $\gamma \in L^2(0,T;\mathscr{V}')$ there exists a unique solution $u_\gamma : [0,T] \times \Omega \to R^2$, $(\varepsilon^p_\gamma, \zeta_\gamma) : [0,T] \times \Omega \to R^{2\times 2} \times R$, to the auxiliary problem $P_\gamma$ satisfying (33)–(35).*

*Proof* From the assumption (15) it follows that operator $\mathscr{A}$ is bounded, semi-continuous and coercive on $\mathscr{V}$. Since $\gamma \in L^2(0,T;\mathscr{V}')$ and (23) holds by standard arguments concerning the parabolic inequalities it results the existence of $(u_\gamma, p_\gamma, \zeta_\gamma)$ satisfying (37)–(38). For details see [2, 3, 19]. □

Let $\alpha \in L^2(0,T;V')$ be given. Define the auxiliary problem $P_\alpha$: find the temperature $\theta_\alpha : [0,T] \times \Omega \to R$ satisfying

$$< \rho\dot{\theta}_\alpha, \upsilon >_{V'\times V} + a_\theta(\theta_\alpha, \upsilon) =$$
$$< \alpha, \upsilon >_{V'\times V} + \int_\Omega g\upsilon dx + \int_{\Gamma_3} \bar{g}\upsilon ds, \forall \upsilon \in V, \text{ a.e. } t \in (0,T), \tag{39}$$
$$\theta_\alpha(0) = \theta_0. \tag{40}$$

**Lemma 10.2** *For all $\alpha \in L^2(0,T;V')$ there exists a unique solution $\theta_\alpha : [0,T] \times \Omega \to R$ to the auxiliary problem $P_\alpha$ satisfying (36).*

*Proof* From Poincaré-Friedrich's inequality it follows that the bilinear form $a_\theta$ is $V$-elliptic. Hence by standard arguments the parabolic boundary value problem (39)–(40) possesses a unique solution $\theta_\alpha : [0,T] \times \Omega \to R$ satisfying (34). For details see [3, 19]. □

Let us consider the following auxiliary problem $P_{\gamma,\alpha}$: find the stress field $\sigma_{\gamma,\alpha} : [0,T] \times \Omega \to \mathbb{S}_2$ solving the equation:

$$\sigma_{\gamma,\alpha}(t) = \mathscr{E}(\varepsilon(u_\gamma)(t))) + \int_0^t \mathscr{G}(\sigma_{\gamma,\alpha}(s), \varepsilon(u_\gamma(s)), \theta_\alpha(s), \zeta_\gamma(s))ds, \quad \forall t \in [0, T]. \tag{41}$$

**Lemma 10.3** *There exists a unique solution* $\sigma_{\gamma,\alpha} : [0, T] \times \Omega \to \mathbb{S}_2$ *to the problem* $P_{\gamma,\alpha}$ *satisfying (36). Let for* $i = 1, 2$, $u_{\gamma_i}$, $\theta_{\alpha_i}$, $\zeta_{\gamma_i}$ *and* $\sigma_{\gamma_i,\alpha_i}$ *denote the solutions to problems* $P_{\gamma_i}$, $P_{\alpha_i}$ *and* $P_{\gamma_i,\alpha_i}$, *respectively. Then there exists constant* $C > 0$ *such that*

$$\| \sigma_{\gamma_1,\alpha_1}(t) - \sigma_{\gamma_2,\alpha_2}(t) \|^2_{\mathscr{H}} \le C(\| u_{\gamma_1}(t) - u_{\gamma_2}(t) \|^2_{\mathscr{V}} + \int_0^t (\| u_{\gamma_1}(s) - u_{\gamma_2}(s) \|^2_{\mathscr{V}} + \| \theta_{\alpha_1}(s) - \theta_{\alpha_2}(s) \|^2_V + \| \zeta_{\gamma_1}(s) - \zeta_{\gamma_2}(s) \|^2_V)ds). \tag{42}$$

*Proof* By $\Pi_{\gamma,\alpha} : L^2(0, T; \mathscr{H}) \to L^2(0, T; \mathscr{H})$ we denote the mapping

$$\Pi_{\gamma,\alpha}\sigma(t) = \mathscr{E}(\varepsilon(u_\gamma)(t))) + \int_0^t \mathscr{G}(\sigma_{\gamma,\alpha}(s), \varepsilon(u_\gamma(s)), \theta_\alpha(s), \zeta_\gamma(s))ds.$$

Assume $\sigma_i \in L^2(0, T; \mathscr{H})$, $i = 1, 2$, and $t^\star \in (0, T)$. From the assumption (17) and Hölder's inequality we obtain

$$\| \Pi_{\gamma,\alpha}\sigma_1(t^\star) - \Pi_{\gamma,\alpha}\sigma_2(t^\star) \|^2_{\mathscr{H}} \le L_3^2 T \int_0^{t^\star} \| \sigma_1(s) - \sigma_2(s) \|^2_{\mathscr{H}} . \tag{43}$$

Repeating this evaluation $k$ times and integrating on the time interval $(0, T)$ we obtain

$$\| \Pi^k_{\gamma,\alpha}\sigma_1 - \Pi_{\gamma,\alpha}\sigma_2 \|^2_{L^2(0,T;\mathscr{H})} \le \frac{L_3^{2k} T^{2k}}{k!} \| \sigma_1(s) - \sigma_2(s) \|^2_{L^2(0,T;\mathscr{H})} . \tag{44}$$

Hence for $k$ large enough operator $\Pi^k_{\gamma,\alpha}$ is a contraction on the space $L^2(0, T; \mathscr{H})$. By Banach fixed point theorem there exists a unique solution $\sigma_{\gamma,\alpha} \in L^2(0, T; \mathscr{H})$ to the equation $\Pi_{\gamma,\alpha}\sigma_{\gamma,\alpha} = \sigma_{\gamma,\alpha}$ which is also a unique solution to (41). Since for $i = 1, 2$ $u_{\gamma_i}$, $\theta_{\alpha_i}$, $\zeta_{\gamma_i}$ are solutions to problems (37)–(38), (39)–(40), respectively, applying Young's inequality and (15)–(17) we obtain (42). □

**Lemma 10.4** *Assume the mapping* $\Lambda : L^2(0, T; \mathscr{V}' \times V') \to L^2(0, T; \mathscr{V}' \times V')$ *is defined as follows:*

$$\Lambda(\gamma(t), \alpha(t)) = (\Lambda_0(\gamma(t), \alpha(t)), \Lambda_1(\gamma(t), \alpha(t))),$$

$$(\Lambda_0(\gamma(t), \alpha(t)), v) = (\mathscr{E}(\varepsilon(u_\gamma)(t)), \varepsilon(v))_{\mathscr{H}} + j_c(u_\gamma(t), v) + (\int_0^t \mathscr{G}(\sigma_{\gamma,\alpha}(s), \varepsilon(u_\gamma(s)), \theta_\alpha(s), \zeta_\gamma(s))ds, \varepsilon(v))_{\mathscr{H}} \quad \forall v \in \mathscr{V}, \tag{45}$$

$$\Lambda_1(\gamma(t), \alpha(t)) = \psi(\sigma_{\gamma,\alpha}(t), \varepsilon(u_\gamma(t)), \theta_\alpha(t), \zeta_\gamma(t)), \tag{46}$$

*The mapping $\Lambda$ has a fixed point $(\gamma^\star, \alpha^\star) \in L^2(0, T; \mathcal{V}' \times V')$.*

*Proof* Using assumptions (15)–(20) as well as Hölder's and Young's inequalities we show that

$$\| \Lambda(\gamma_1(t), \alpha_1(t)) - \Lambda(\gamma_2(t), \alpha_2(t)) \|^2_{\mathcal{V}' \times V'} \leq \\ C \| \gamma_1(t) - \gamma_2(t) \|^2_{\mathcal{V}'} + \| \alpha_1(t) - \alpha_2(t) \|^2_{V'} . \tag{47}$$

Reiterating this inequality k times results in

$$\| \Lambda^k(\gamma_1, \alpha_1) - \Lambda^k(\gamma_2, \alpha_2) \|^2_{\mathcal{V}' \times V'} \leq \\ \frac{C^k T^k}{k!} \| \gamma_1 - \gamma_2 \|^2_{L^2(0,T;\mathcal{V}')} + \| \alpha_1 - \alpha_2 \|^2_{L^2(0,T;V')} . \tag{48}$$

For $k$ large enough operator $\Lambda^k$ is a contraction on the space $L^2(0, T; \mathcal{V}' \times V')$. By Banach fixed point theorem it follows that $\Lambda$ possesses a unique fixed point $(\gamma^\star, \alpha^\star) \in L^2(0, T; \mathcal{V}' \times V')$. □

Using Lemmas 10.1–10.4 we prove Theorem 10.1.

*Proof (of Theorem 10.1)* Denote by $(\gamma^\star, \alpha^\star) \in L^2(0, T; \mathcal{V}' \times V')$ the fixed point of the operator $\Lambda$ defined by (45)–(46). Let

$$u = u_{\gamma^\star}, \quad \theta = \theta_{\alpha^\star}, \quad \varepsilon^p = \varepsilon^p_{\gamma^\star}, \quad \zeta = \zeta_{\gamma^\star}, \quad \sigma = \mathscr{A}\varepsilon(\dot{u}) + \sigma_{\gamma^\star, \alpha^\star}. \tag{49}$$

Setting in (41) $\gamma = \gamma^\star$, $\alpha = \alpha^\star$ and using (49) it results that (28) holds. From (37) with $\gamma = \gamma^\star$ and (49) we obtain

$$< \rho\ddot{u}, v >_{\mathcal{V}' \times \mathcal{V}} + \int_\Omega (\mathscr{A}\varepsilon(\dot{u})\varepsilon(v))dx + < \gamma^\star(t), v >_{\mathcal{V}' \times \mathcal{V}} + j_p(q, \eta) - \\ j_p(\dot{\varepsilon}^p, \dot{\zeta}) = < f(t), v >_{\mathcal{V}' \times \mathcal{V}} \quad \forall (v, q, \eta) \in \mathcal{V} \times K_p, \text{ a.e. } t \in (0, T). \tag{50}$$

From (45)–(46), (49) as well as

$$\Lambda_0(\gamma^\star, \alpha^\star) = \gamma^\star, \quad \Lambda_1(\gamma^\star, \alpha^\star) = \alpha^\star,$$

we obtain

$$< \gamma^\star, v >_{\mathcal{V}' \times \mathcal{V}} = \int_\Omega \mathscr{E}(\varepsilon(u(t)))\varepsilon(v)dx + j_c(u(t), v) + j_p(q, \eta) - \\ j_p(\dot{\varepsilon}^p, \dot{\zeta}) + \int_\Omega (\int_0^t \mathscr{G}(\sigma(s) - \mathscr{A}(\dot{\varepsilon}(\dot{u}(s))), \varepsilon(u(s)), \\ \theta(s), \zeta(s))ds)\varepsilon(v)dx \quad \forall (v, q, \eta) \in \mathcal{V} \times K_p, \tag{51}$$

$$\alpha^\star(t) = \psi(\sigma(t) - \mathscr{A}(\varepsilon(\dot{u}(t))), \varepsilon(u(t)), \theta(t), \zeta(t)). \tag{52}$$

Inserting (51) into (50), using (28) we obtain that (29) is satisfied. Setting $\alpha = \alpha^\star$ in (39) and using (49) as well as (52) we conclude that (30) is satisfied. From Lemmas 10.1–10.4 it results that (10) and (34)–(35) hold. From the uniqueness of solutions to problems (37)–(38), (39)–(40), (41) as well as from the uniqueness of the fixed point of the operator (45)–(46) follows the uniqueness of the solution to the problem (28)–(32). □

Many engineering problems including the rolling contact problems between a wheel and a rail or a tyre and a road are described by the system (28)–(32). In the next section wheel-rail rolling contact problem is investigated.

## 3 Rolling Contact Problem

Consider deformations of a rail lying on a rigid foundation (Fig. 1). The rail strip occupies domain $\Omega \in R^2$ with the boundary $\Gamma$ and has constant height $h$. The strip is assumed to consist of two elastic-plastic and thermally conductive layers denoted by $\Omega_c$ and $\Omega_s$ such that $\Omega = \Omega_c \cup \Omega_s$. Subdomains $\Omega_c$ and $\Omega_s$ denote the coating and substrate layers having thicknesses $h_c$ and $h_s$, respectively. Moreover $h_c < h_s$ and $h = h_c + h_s$. By $\Gamma_c$ and $\Gamma_s$ we denote the boundaries of $\Omega_c$ and $\Omega_s$ respectively. The boundary $\Gamma$ of strip $\Omega$ is assumed to consist of two parts $\Gamma_1$ and $\Gamma_3$ such that $\Gamma = \Gamma_1 \cup \Gamma_3$. Obviously the boundary $\Gamma$ is equal to sum of the boundaries $\Gamma_c$ and $\Gamma_s$ minus their intersection. A wheel having radius $r_0$, rotating speed $\omega$ and linear velocity *Ve* rolls along the upper surface $\Gamma_3$ of the strip $\Omega$ and is pressed in it. We assume that the length of the strip is much bigger than the radius of the wheel. The head and tail ends of the strip are clamped. No mass forces in the strip are assumed. The coating layer of the rail is assumed to consist of a steel and ceramic. The functionally graded material coating of rail is assumed to be processed in such a way

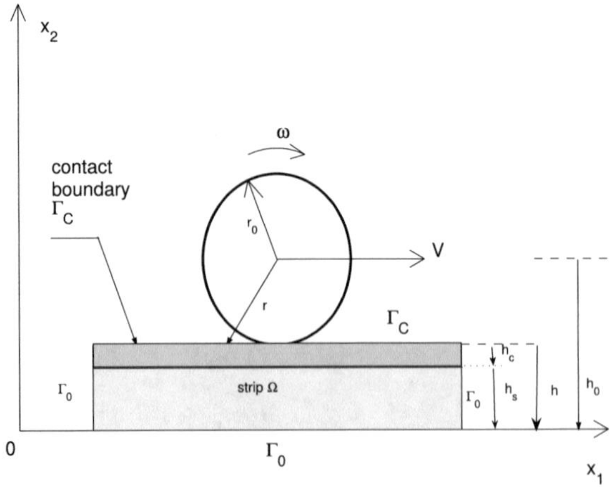


**Fig. 1** Wheel rolling over the two-layered rail

that the property grading is smooth. The models governing the material distribution in the graded layer have been reported in many papers [5, 21]. Here we use the power model in the form

$$P = P(x) = P_c + (P_s - P_c)\left(\frac{h_s + h_c - x_2}{h_c}\right)^n, \quad h_s \leq x_2 \leq h_s + h_c, \tag{53}$$

where $x = (x_1, x_2) \in \Omega$, $P = P(x)$, denote the material property dependent on spatial variable $x_2$ and $P_c$, $P_s$ are the coating and steel layers properties, respectively. Real number $n > 0$ denotes the non homogeneity parameter of the graded material.

## 3.1 Thermo-Elastoplastic Model

Let us denote as previously, by $u = (u_1, u_2)$, $u = u(x, t)$, $x \in \Omega$, $t \in (0, T)$, $T > 0$, a displacement of the strip and by $\theta = \theta(x, t)$ the absolute temperature of the strip. Since the rail consist of the coating layer $\Omega_c$ and the substrate layer $\Omega_s$ we denote by $u_c$ and $u_s$ as well as $\theta_c$ and $\theta_s$ the displacement and temperature of these layers, respectively. Therefore $u = u_c$ in $\Omega_c \times (0, T)$ and $u = u_s$ in $\Omega_s \times (0, T)$. The similar relation holds for $\theta$. Assume the wheel and the rail are brought into contact under the action of the moving wheel load. The displacement $u$ and temperature $\theta$ of the strip $\Omega$ are governed by the system of the following equations [20]:

$$\rho \ddot{u} = \mathrm{div}\sigma(u) - \tilde{\alpha}(3\tilde{\lambda} + 2\tilde{\gamma})\nabla\theta \quad \text{in } \Omega \times (0, T), \tag{54}$$

$$\rho c \dot{\theta} = \mathrm{div}(\bar{\kappa}\nabla\theta) \quad \text{in } \Omega \times (0, T). \tag{55}$$

Mass density $\rho \in L^\infty(\Omega)$, thermal expansion coefficient $\tilde{\alpha} \in L^\infty(\Omega)$, Lamé coefficients $\tilde{\lambda}$ and $\tilde{\gamma}$, thermal conductivity coefficient $\bar{\kappa} \in L^\infty(\Omega)$, heat capacity coefficient $c \in L^\infty(\Omega)$ [5] depend on $x_2$ in $\Omega_c \times (0, T)$ according to (53) and are assumed constant in $\Omega_s \times (0, T)$. Consider the stress-strain relation. The strain $\varepsilon(u)$ is additively decomposed [12] into its viscous elastic and plastic parts such that

$$\sigma(u) = A_1\varepsilon(\dot{u}) + A_2\varepsilon^e(u) = A_1\varepsilon(\dot{u}) + A_2(\varepsilon(u) - \varepsilon^p), \quad \mathrm{tr}\, \varepsilon^p = 0 \text{ in } \Omega, \tag{56}$$

The fourth order viscous and elasticity tensors $A_1 \in L^\infty(\Omega)$ and $A_2 \in L^\infty(\Omega)$, respectively, satisfy usual symmetry, boundedness and $H^1(\Omega)$ ellipticity conditions. Consider the plasticity conditions [10, 12, 14]. First denote by $\varsigma$ and $f^p$ relative deviatoric stress and the yield function, respectively, defined by [10, 12]:

$$\varsigma = \mathrm{dev}\sigma - \frac{1}{2}K\varepsilon^p, \quad \mathrm{dev}\sigma = \sigma - \frac{1}{2}\mathrm{tr}\, \sigma \; Id, \tag{57}$$

$$f^p(\varsigma, \beta) = \parallel \varsigma \parallel - Y^p(\beta), \quad Y^p(\beta) = \frac{1}{\sqrt{2}}(\sigma_0 + H\beta), \tag{58}$$

where $Id$ is the identity operator, the nonnegative constants $K$ and $H$ denote kinematic and isotropic hardening parameters, respectively, such that $\mathbf{H}$ takes the form of the diagonal $2 \times 2$ matrix with main diagonal entries $K$ and $H$ [12]. Moreover $\sigma_0 > 0$ denotes the yield stress and $\beta$ is the equivalent plastic strain such that the rate $\dot{\beta}$ is associated with the rate $\dot{\varepsilon}^p$. The plastic strain $\varepsilon^p \in Q_0$ and the relative deviatoric stress $\varsigma$ satisfy the flow rule in $\Omega \times [0, T]$

$$\dot{\varepsilon}^p = \begin{cases} \tilde{\gamma}^p \frac{\varsigma}{\|\varsigma\|} & \text{for } \| \varsigma \| > 0, \\ 0 & \text{for } \| \varsigma \| = 0, \end{cases} \tag{59}$$

with $\varepsilon^p(0) = \varepsilon^p(x, 0) = 0$. The yield function $f^p$ and the consistency parameter $\tilde{\gamma}^p$ satisfy the complementarity conditions in $\Omega \times [0, T]$

$$\tilde{\gamma}^p \geq 0, \quad f^p(\varsigma, \beta) \leq 0, \quad \tilde{\gamma}^p f^p(\varsigma, \beta) = 0. \tag{60}$$

The equivalent plastic strain $\beta$ satisfies the evolution law:

$$\dot{\beta} = \frac{1}{\sqrt{2}} \tilde{\gamma}^p \quad \text{and} \quad \beta(x, 0) = \beta(0) = 0. \tag{61}$$

From (61) it follows the plasticity law is associative, i.e., the inner variable $(z, \varepsilon^p)$ with $z = -H\beta$ satisfies

$$\dot{z} = \tilde{\gamma}^p \frac{\partial f^p}{\partial \beta} \quad \text{and} \quad \dot{\varepsilon}^p = \tilde{\gamma}^p \frac{\partial f^p}{\partial \varsigma}. \tag{62}$$

The contact conditions on the boundary $\Gamma_3 \times (0, T)$ take the form (7)–(8). Function $\bar{g} \in L^2(\Gamma_3)$ in the heat flow condition (9) is equal to $\bar{g} = \bar{\alpha}\mu(Ve)p_\nu$ where a constant $0 \leq \bar{\alpha} \leq 1$ represents the fraction of frictional heat flow rate entering the rail and $0 \leq \mu \leq 1$ is a suitable small constant denoting the friction coefficient. Moreover the following boundary and initial conditions are imposed:

$$u_c = u_s = 0, \quad \theta_c = \theta_s = \theta_g, \quad -\bar{\kappa}\frac{\partial \theta_c}{\partial \nu}(x, t) = 0, \quad \text{on } \Gamma_1 \times (0, T), \tag{63}$$

$$u_s = u_c, \quad \theta_s = \theta_c, \quad \text{on } \Gamma_c \cap \Gamma_s, \quad \theta_c(0, x) = \theta_s(0, x) = \theta_g \quad \text{in } \Omega_c \cup \Omega_s, \tag{64}$$

$$u_c(0) = \bar{u}_{0c} \quad \dot{u}_c(0) = \bar{u}_{1c} \quad \text{in } \Omega_c, \quad u_s(0) = \bar{u}_{0s} \quad \dot{u}_s(0) = \bar{u}_{1s} \quad \text{in } \Omega_s, \tag{65}$$

where for $j =$ "c" or $j =$ "s" $\bar{u}_{0j}, \bar{u}_{1j}$ and $\theta_g$ are given functions. Let us formulate the variational problem associated with the strong model (54)–(65). For the displacement and the temperature let introduce the closed subspaces $V_u = \{\upsilon \in [H^1(\Omega)]^2 : \upsilon = 0 \text{ on } \Gamma_1\}$ and $V_\theta = \{\lambda \in H^1(\Omega) : \lambda = \theta_g \text{ on } \Gamma_1\}$. The inner variables $(\varsigma, \beta)$ governing the plastic behavior of material belong to the subspace $V^p = Q_0 \times L^2(\Omega)$. The admissible set of plastic variables is denoted by $\Lambda_p = \{(\varsigma, \beta) \in V^p : f^p(\varsigma, \beta) \leq 0\}$. For linear yield function (58) this set is convex. Let us define the following forms: $\tilde{a}_u : V_u \times V_\theta \times V_u \to R$, $\tilde{a}_\theta : V_\theta \times V_\theta \to R$, $\tilde{j}_p : \Lambda_p \to R$ and $\tilde{j}_c : V_u \times V_u \to R$

given by

$$\tilde{a}_u(u,\theta,\upsilon) = \int_\Omega (A_1\varepsilon(\dot{u})\varepsilon(\upsilon) + A_2\varepsilon(u)\varepsilon(\upsilon) + \tilde{\alpha}(3\tilde{\lambda} + 2\tilde{\gamma})\nabla\theta\upsilon)dx, \quad (66)$$

$$\tilde{a}_\theta(\theta,\varphi) = \int_\Omega \bar{\kappa}\nabla\theta\nabla\varphi dx + k_1\int_{\Gamma_3}\theta\varphi ds, \quad (67)$$

$$\tilde{j}_p(\iota,\omega) = -\int_\Omega A_2\iota\omega dx, \qquad \tilde{j}_c(u,\upsilon) = \int_{\Gamma_3}(p_\nu(u_\nu)\upsilon_\nu + p_\tau\upsilon_\tau)ds. \quad (68)$$

The strong problem (54)–(65) has the following variational form: find $(u, \theta, (\varsigma, \beta)) \in V_u \times V_\theta \times \Lambda_p$ such that for each time $t \in [0, T]$ it holds:

$$< \rho\ddot{u}, \upsilon >_{V_u'\times V_u} + \tilde{a}_u(u,\theta,\upsilon) + \tilde{j}_p(\varepsilon^p, \varepsilon(\upsilon)) + \tilde{j}_c(u,\upsilon) = 0 \ \ \forall \upsilon \in V_u, \quad (69)$$

$$< c\rho\dot{\theta}, \varphi >_{V_\theta'\times V_\theta} + \tilde{a}_\theta(\theta,\varphi) - \int_{\Gamma_3}\bar{\kappa}\bar{g}\varphi ds = 0 \ \ \forall\varphi \in V_\theta, \quad (70)$$

$$-\tilde{j}_p((A_2)^{-1}((\iota - \varsigma), \dot{\varepsilon}^p) - \int_\Omega H\dot{\beta}(\varpi - \beta)dx \le 0 \ \ \forall(\iota,\varpi) \in \Lambda_p, \quad (71)$$

as well as (63)–(65). Remark, the coupled system consisting of the stress–strain relation (56), (69) and (71) combined in one inequality [12] as well as (70), is a particular representation of general system (28)–(32). Therefore by Theorem 10.1 follows the existence of a unique solution $(\sigma, u, \theta, (\varsigma, \beta)) \in L^2(0, T; \mathscr{H}) \times V_u \times V_\theta \times \Lambda_p$ to problem (56), (63)–(65), (69)–(71). Remark in [1, 8] are provided the existence results of a unique solution to quasistatic thermo-rigid plastic contact problem with nonlocal contact and friction conditions as well as to quasistatic thermo elastic-viscoplastic contact problem with bilateral contact, friction and wear, respectively.

## 3.2 Numerical Results

The discrete quasistatic system of equations for nonhomogeneous materials (69)–(71) with nonpenetration and friction conditions (see [6] for details) has been solved numerically using semi-smooth Newton method [10, 14]. Polygonal domain $\Omega$ occupied by the rail has a form $\Omega = \{(x_1, x_2) \in R^2 \ : \ x_1 \in (-2, 2), \ x_2 \in (0, 1)\}$. The triangular linear elements were used in elastoplastic and thermal models. The total number of nodes and elements are 1701 and 1600 respectively. Other data are as follows: the velocity $Ve = 10$ m/s, radius of the wheel $r_0 = 0.46$ m, the friction coefficient $\mu$ is equal to 0.45. The penetration of the wheel is taken as $\delta = 0.25 \cdot 10^{-3}$ m. Functions $\bar{u}_{0j}$ and $\bar{u}_{1j}$, for $j =$ "c" or $j =$ "s", in (57)–(58) are selected as equal to 0. Ambient temperature $\theta_g$ is equal to 20°C. Time step size $\Delta t = 2 \cdot 10^{-4}$. Figures 2 and 3 show, that plastic pressure is almost flat at the central part of the contact zone in both directions $x_1$ and $x_2$, i.e. where the yield criterion is satisfied. The graded layer can reduce the values of the normal contact stress and the maximal temperature

Fig. 2 Normal contact traction distribution **a** elastic case **b** elastoplastic case along $x_1$ ($x_2 = 1$)

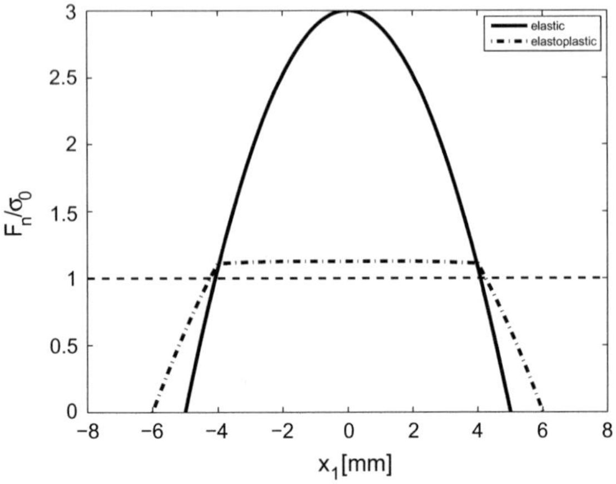


Fig. 3 Normal contact traction distribution **a** elastic case **b** elastoplastic along $x_2$ ($x_1 = 0$)

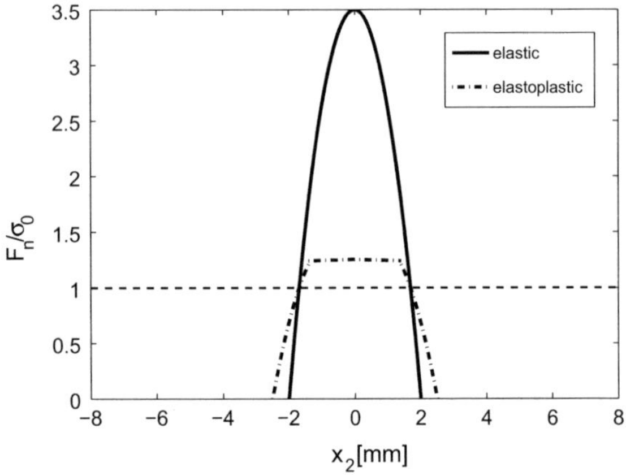


in the contact zone. Figure 4 compares the normal contact pressure distributions on the contact surface for different values of the nonhomogeneity gradient $n$. In all cases the maximum contact pressure occurs at the center of the contact zone. As $n$ increases and the coated surface becomes more flexible the maximal contact pressure value is slightly increasing at a cost of shrinking the contact zone length. The obtained maximal values are lower than in pure homogeneous case. In elastoplastic model the contact zone is generally greater than in elastic model. The evolution during time of plastic and contact active sets is shown in Fig. 5. The number of finite elements expressing plastic behavior is increasing in time to reach maximum and than stabilizes on lower level. The number of contact nodes is increasing during the computations to reach maximum in the last iteration. Figures 6 and 7 display the distribution of temperature in the strip in normal and tangential directions in the contact zone. Maximal value of the temperature appears at in the middle of the contact area and in adjacent area. As $n$ increases the temperature maximal

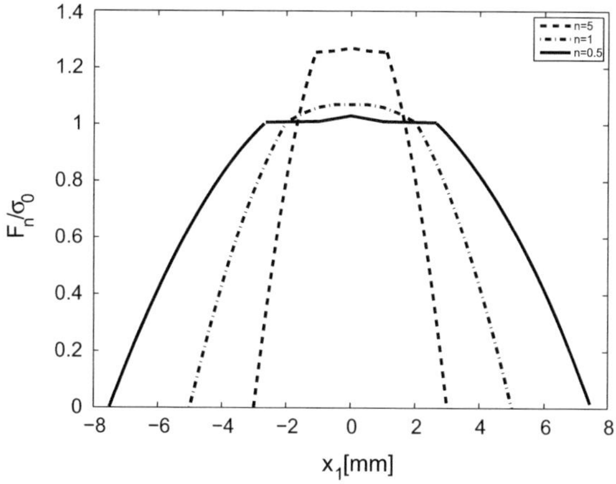


**Fig. 4** Normal contact distribution for different nonhomogeneity parameters $n$

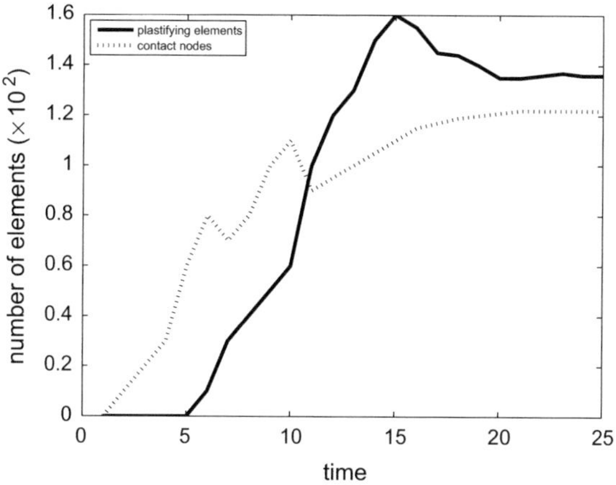


**Fig. 5** Evolution of number of plastifying elements and contact active nodes during time

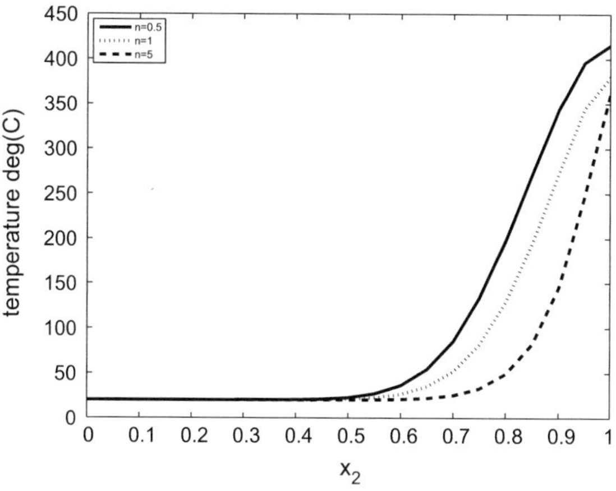


**Fig. 6** Rail temperature distribution along $x_2$ direction at $x_1 = 0$

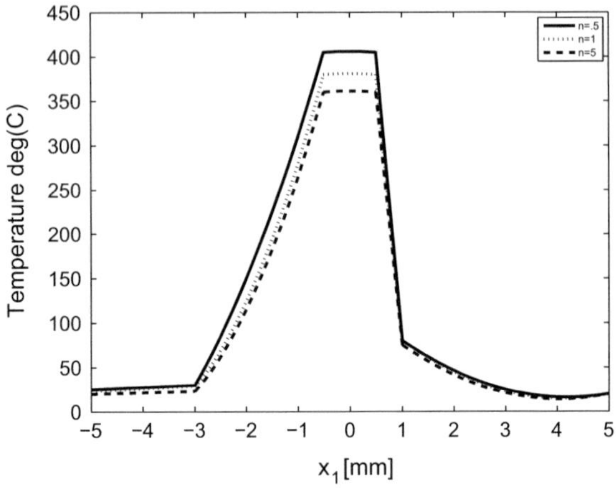


**Fig. 7** Temperature distribution along $x_1$ direction on the contact interface at $x_2 = 1$

value decreases. In front of the moving wheel the temperature rapidly decays to the ambient temperature of the surrounding medium while behind of the wheel the decay of temperature is relatively mild. As nonhomogeneity gradient $n$ increases from 0.5 to 5 the temperature maximal value decreases. The obtained temperature maximal values are lower than in pure homogeneous material case, i.e., $n = 0$.

## References

1. Angelov, T.: Variational analysis of thermo-mechanically coupled steady-state rolling problem. Appl. Math. Mech. **34**(11), 1361–1372 (2013)
2. Bartels, S., Mielke, A., Roubiček, T.: Quasistatic small strain plasticity in the limit of vanishing hardening and its numerical approximation. SIAM J. Numer. Anal. **50**(2), 951–976 (2012)
3. Boutechebak, S., Ahmed, A.A.: Analysis of a dynamic thermo-elastic-viscoplastic contact problem. Electron. J. Qual. Theory Differ. Equ. **71**, 1–17 (2013)
4. Carstensen, C.: Numerical analysis of the primal problem of elastoplasticity with hardening. Numer. Math. **82**, 577–597 (1999)
5. Choi, I.S., Dao, M., Suresh, S.: Mechanics of indentation of plastically graded materials I: analysis. J. Mech. Phys. Solids **56**, 157–171 (2008)
6. Chudzikiewicz, A., Myśliński, A.: Rolling contact problems for plastically graded materials. In: Kleiber, M., Burczyński, T., Wilde, K., Górski, J., Winkielmann, K., Smakosz, Ł. (eds.) Advances in Mechanics. Theoretical, Computational and Interdisciplinary Issues, pp. 423–428. CRC Press/Balkema, EH Leiden, The Netherlands (2016)
7. Chudzikiewicz, A., Myśliński, A.: Thermoelastic wheel-rail contact problem with elastic graded materials. Wear **271**(1–2), 417–425 (2011)
8. Djabi, A., Merouani, A., Aissaoui, A.: A frictional contact problem with wear involving elastic-viscoplastic materials with damage and thermal effects. Electron. J. Qual. Theory Differ. Equ. **27**, 1–18 (2015)
9. Eck, C., Jarušek, J., Sofonea, M.: A dynamic elastic-viscoplastic unilateral contact problem with normal damped response and Coulomb friction. Eur. J. Appl. Math. **21**, 229–251 (2010)
10. Hager, C., Wohlmuth, B.I.: Nonlinear complementarity functions for plasticity problems with frictional contact. Comput. Methods Appl. Mech. Eng. **198**, 3411–3427 (2009)

11. Han, W., Reddy B.D.: Plasticity: mathematical theory and numerical analysis. 2nd edn. Springer, New York (2013)
12. Han, W., Reddy, B.D.: Computational plasticity: the variational basis and numerical analysis. Comput. Mech. Adv. **2**, 283–400 (1995)
13. Han, W., Sofonea, M.: Quasistatic Contact Problems in Viscoelasticity and Viscoplasticity. American Mathematical Society and International Press of Boston Inc, Studies in Advanced Mathematics **30** (2002)
14. Hintermüller, M., Rösel, S.: A duality based path following semismooth Newton method for elasto-plastic contact problems. J. Comput. Appl. Math. **292**, 150–173 (2016)
15. Jakabčin, L.: Existence of solutions to an elastoviscoplastic model with kinematic hardening and r-Laplacian fracture approximation. Math. Model. Numer. Anal. **50**(2), 455–473 (2016)
16. Jarušek, J., Sofonea, M.: On the solvability of dynamic elastic-visco-plastic contact problems with adhesion. Ann. Acad. Romanian Sci. Ser. Math. Appl. **1**(2), 191–214 (2009)
17. Johnson, C.: On plasticity with hardening. J. Math. Anal. Appl. **62**, 325–336 (1978)
18. Krejči, P., Petrov, A.: Elasto-plastic contact problems with heat exchange. Nonlinear Anal. Real World Appl. **22**, 551–567 (2015)
19. Merouani, A., Messelmi, F.: Dynamic evolution of damage in elastic-thermo-viscoplastic materials. Electron. J. Differ. Equ. **129**, 1–15 (2010)
20. Myśliński, A.: Thermoelastic rolling contact problems for multilayer materials. Nonlinear Anal. Real World Appl. **22**, 619–631 (2015)
21. Serkan, D.: Contact mechanics of graded materials. Analysis using singular integrated equations. In: Multiscale and Functionally Graded Materials 2006, vol. 973, pp. 820–825. AIP Conference Proceedings (2008)
22. Shillor, M., Sofonea, M., Telega, J.J.: Models and analysis of quasistatic contact: variational methods. Springer, Berlin (2004)
23. Sofonea, M.: Quasistatic processes for elastic-viscoplastic materials with internal state variable. Annales Scientifiques de l'Université de Clermont. Mathématiques **94**(25), 47–60 (1989)
24. Sofonea, M., Pătrulescu, F., Farcaş, A.: A viscoplastic contact problem with normal compliance, unilateral constraint and memory term. Appl. Math. Optim. **69**, 175–198 (2014)
25. Wu, L., Wen, Z., Li, W., Jin, X.: Thermo-elastic-plastic finite element analysis of wheel-rail sliding contact. Wear **271**, 437–443 (2011)

# New Thoughts in Nonlinear Elasticity Theory via Hencky's Logarithmic Strain Tensor

**Patrizio Neff, Robert J. Martin and Bernhard Eidel**

**Abstract** We consider the two logarithmic strain measures $\omega_{\rm iso} = \|\operatorname{dev}_n \log U\|$ and $\omega_{\rm vol} = |\operatorname{tr}(\log U)|$, which are isotropic invariants of the Hencky strain tensor $\log U = \log(F^T F)$, and show that they can be uniquely characterized by purely geometric methods based on the geodesic distance on the general linear group $\operatorname{GL}(n)$. Here, $F$ is the deformation gradient, $U = \sqrt{F^T F}$ is the right Biot-stretch tensor, log denotes the principal matrix logarithm, $\| \,.\, \|$ is the Frobenius matrix norm, tr is the trace operator and $\operatorname{dev}_n X = X - \frac{1}{n}\operatorname{tr}(X) \cdot \mathbb{1}$ is the $n$-dimensional deviator of $X \in \mathbb{R}^{n\times n}$. This characterization identifies the Hencky (or true) strain tensor as the natural nonlinear extension of the linear (infinitesimal) strain tensor $\varepsilon = \operatorname{sym} \nabla u$, which is the symmetric part of the displacement gradient $\nabla u$, and reveals a close geometric relation between the classical quadratic isotropic energy potential in linear elasticity and the geometrically nonlinear quadratic isotropic Hencky energy. Our deduction involves a new fundamental logarithmic minimization property of the orthogonal polar factor $R$, where $F = RU$ is the polar decomposition of $F$.

## 1 Strain and Strain Measures in Nonlinear Elasticity

The concept of *strain* is of fundamental importance in elasticity theory. In linearized elasticity, one assumes that the Cauchy stress tensor $\sigma$ is a linear function of the symmetric infinitesimal strain tensor

---

P. Neff (✉)
Head of Chair for Nonlinear Analysis and Modelling, Fakultät Für Mathematik, Universität Duisburg-Essen, Campus Essen, Thea-Leymann Straße 9, 45141 Essen, Germany
e-mail: patrizio.neff@uni-due.de

R.J. Martin
Lehrstuhl Für Nichtlineare Analysis und Modellierung, Fakultät für Mathematik, Universität Duisburg-Essen, Thea-Leymann Str. 9, 45127 Essen, Germany
e-mail: robert.martin@uni-due.de

B. Eidel
Chair of Computational Mechanics, Universität Siegen, Paul-Bonatz-Straße 9-11, 57068 Siegen, Germany
e-mail: bernhard.eidel@uni-siegen.de



F. dell'Isola et al. (eds.), *Mathematical Modelling in Solid Mechanics*, Advanced Structured Materials 69, DOI 10.1007/978-981-10-3764-1_11

$$\varepsilon = \operatorname{sym} \nabla u = \operatorname{sym}(\nabla \varphi - \mathbb{1}) = \operatorname{sym}(F - \mathbb{1}) \,,$$

where $\varphi\colon \Omega \to \mathbb{R}^n$ is the deformation of an elastic body with a given reference configuration $\Omega \subset \mathbb{R}^n$, $\varphi(x) = x + u(x)$ with the displacement $u$, $F = \nabla\varphi$ is the deformation gradient, $\operatorname{sym} \nabla u = \frac{1}{2}(\nabla u + (\nabla u)^T)$ is the symmetric part of the displacement gradient $\nabla u$ and $\mathbb{1} \in \mathrm{GL}^+(n)$ is the identity tensor in the group of invertible tensors with positive determinant. In geometrically nonlinear elasticity models, it is no longer necessary to postulate a linear connection between some stress and some strain. However, nonlinear strain tensors are often used in order to simplify the stress response function, and many constitutive laws are expressed in terms of linear relations between certain strains and stresses [2, 3, 6].

There are different definitions of what exactly the term "strain" encompasses: while Truesdell and Toupin [42, p. 268] consider "any uniquely invertible isotropic second order tensor function of [the right Cauchy-Green deformation tensor] $C = F^T F$" to be a strain tensor, it is commonly assumed [20, p. 230] (cf. [5, 21, 22, 36]) that a (material or Lagrangian[1]) strain takes the form of a *primary matrix function* of the right Biot-stretch tensor $U = \sqrt{F^T F}$ of the deformation gradient $F \in \mathrm{GL}^+(n)$, i.e. an isotropic tensor function $E\colon \mathrm{Sym}^+(n) \to \mathrm{Sym}(n)$ from the set of positive definite tensors to the set of symmetric tensors of the form

$$E(U) = \sum_{i=1}^{n} \mathrm{e}(\lambda_i) \cdot e_i \otimes e_i \quad \text{for} \quad U = \sum_{i=1}^{n} \lambda_i \cdot e_i \otimes e_i \tag{1}$$

with a *scale function* $\mathrm{e}\colon (0,\infty) \to \mathbb{R}$, where $\otimes$ denotes the tensor product, $\lambda_i$ are the eigenvalues and $e_i$ are the corresponding eigenvectors of $U$.

The general idea underlying these definitions is clear: strain is a measure of deformation (i.e. the change in form and size) of a body with respect to a chosen (arbitrary) reference configuration. Furthermore, the strain of the deformation gradient $F \in \mathrm{GL}^+(n)$ should correspond only to the *non-rotational* part of $F$. In particular, the strain must vanish if and only if $F$ is a pure rotation, i.e. if and only if $F \in \mathrm{SO}(n)$, where $\mathrm{SO}(n) = \{Q \in \mathrm{GL}(n) \mid Q^T Q = \mathbb{1},\ \det Q = 1\}$ denotes the special orthogonal group. This ensures that the only strain-free deformations are rigid body movements [33].

In contrast to *strain* or *strain tensor*, we use the term ***strain measure*** to refer to a nonnegative real-valued function $\omega\colon \mathrm{GL}^+(n) \to [0,\infty)$ depending on the deformation gradient which vanishes if and only if $F$ is a pure rotation, i.e. $\omega(F) = 0$ if and only if $F \in \mathrm{SO}(n)$.

In the following we consider the question of what strain measures are appropriate for the theory of nonlinear isotropic elasticity. Since, by our definition, a strain measure attains zero if and only if $F \in \mathrm{SO}(n)$, a simple geometric approach is to consider a *distance function* on the group $\mathrm{GL}^+(n)$ of admissible deformation gradients, i.e. a function $\mathrm{dist}\colon \mathrm{GL}^+(n) \times \mathrm{GL}^+(n) \to [0,\infty)$ with $\mathrm{dist}(A,B) = \mathrm{dist}(B,A)$

[1]Similarly, a *spatial* or *Eulerian* strain tensor $\widehat{E}(V)$ depends on the left Biot-stretch tensor $V = \sqrt{FF^T}$ (cf. [14]).

which satisfies the triangle inequality and vanishes if and only if its arguments are identical. Such a distance function induces a "natural" strain measure on $\mathrm{GL}^+(n)$ by means of the distance to the special orthogonal group $\mathrm{SO}(n)$:

$$\omega(F) := \operatorname{dist}(F, \mathrm{SO}(n)) := \inf_{Q \in \mathrm{SO}(n)} \operatorname{dist}(F, Q) . \tag{2}$$

In this way, the search for an appropriate strain measure reduces to the task of finding a *natural, intrinsic distance function* on $\mathrm{GL}^+(n)$.

# 2 Euclidean Strain Measures

## 2.1 The Euclidean Strain Measure in Linear Isotropic Elasticity

An approach similar to the definition of strain measures via distance functions on $\mathrm{GL}^+(n)$, as stated in Eq. (2), can be employed in linearized elasticity theory: let $\varphi(x) = x + u(x)$ with the displacement $u$. Then the *infinitesimal strain measure* may be obtained by taking the distance of the displacement gradient $\nabla u \in \mathbb{R}^{n\times n}$ to the set of *linearized rotations* $\mathfrak{so}(n) = \{A \in \mathbb{R}^{n\times n} : A^T = -A\}$, which is the vector space of skew symmetric matrices. An obvious choice for a distance measure on the linear space $\mathbb{R}^{n\times n} \cong \mathbb{R}^{n^2}$ of $n \times n$-matrices is the *Euclidean distance* induced by the canonical Frobenius norm

$$\|X\| = \sqrt{\operatorname{tr}(X^T X)} = \sqrt{\sum_{i,j=1}^{n} X_{ij}^2} .$$

We use the more general weighted norm defined by

$$\|X\|^2_{\mu,\mu_c,\kappa} = \mu \, \|\operatorname{dev}_n \operatorname{sym} X\|^2 + \mu_c \, \|\operatorname{skew} X\|^2 + \frac{\kappa}{2} \, [\operatorname{tr}(X)]^2 \quad \mu, \mu_c, \kappa > 0 , \tag{3}$$

which separately weights the *deviatoric* (or *trace free*) *symmetric part* $\operatorname{dev}_n \operatorname{sym} X = \operatorname{sym} X - \frac{1}{n} \operatorname{tr}(\operatorname{sym} X) \cdot \mathbb{1}$, the *spherical part* $\frac{1}{n} \operatorname{tr}(X) \cdot \mathbb{1}$, and the *skew symmetric part* $\operatorname{skew} X = \frac{1}{2}(X - X^T)$ of $X$; note that $\|X\|_{\mu,\mu_c,\kappa} = \|X\|$ for $\mu = \mu_c = 1, \kappa = \frac{2}{n}$, and that $\| \, . \, \|_{\mu,\mu_c,\kappa}$ is induced by the inner product

$$\langle X, Y \rangle_{\mu,\mu_c,\kappa} = \mu \, \langle \operatorname{dev}_n \operatorname{sym} X, \operatorname{dev}_n \operatorname{sym} Y \rangle + \mu_c \, \langle \operatorname{skew} X, \operatorname{skew} Y \rangle + \tfrac{\kappa}{2} \operatorname{tr}(X) \operatorname{tr}(Y) \tag{4}$$

on $\mathbb{R}^{n\times n}$, where $\langle X, Y\rangle = \operatorname{tr}(X^T Y)$ denotes the canonical inner product. In fact, every isotropic inner product on $\mathbb{R}^{n\times n}$, i.e. every inner product $\langle\cdot,\cdot\rangle_{\text{iso}}$ with

$$\langle Q^T X Q,\ Q^T Y Q\rangle_{\text{iso}} = \langle X, Y\rangle_{\text{iso}}$$

for all $X, Y \in \mathbb{R}^{n\times n}$ and all $Q \in \mathrm{O}(n)$, is of the form (4), cf. [11]. The suggestive choice of variables $\mu$ and $\kappa$, which represent the *shear modulus* and the *bulk modulus*, respectively, will prove to be justified later on. The remaining parameter $\mu_c$ will be called the *spin modulus*.

Of course, the element of best approximation in $\mathfrak{so}(n)$ to $\nabla u$ with respect to the weighted Euclidean distance $\operatorname{dist}_{\text{Euclid},\mu,\mu_c,\kappa}(X, Y) = \|X - Y\|_{\mu,\mu_c,\kappa}$ is given by the associated orthogonal projection of $\nabla u$ to $\mathfrak{so}(n)$. Since $\mathfrak{so}(n)$ and the space $\operatorname{Sym}(n)$ of symmetric matrices are orthogonal with respect to $\langle\cdot,\cdot\rangle_{\mu,\mu_c,\kappa}$, this projection is given by the *continuum rotation*, i.e. the skew symmetric part $\operatorname{skew}\nabla u = \frac{1}{2}(\nabla u - (\nabla u)^T)$ of $\nabla u$, the axial vector of which is $\operatorname{curl} u$. Thus the distance is

$$\begin{aligned}\operatorname{dist}_{\text{Euclid},\mu,\mu_c,\kappa}(\nabla u, \mathfrak{so}(n)) :&= \inf_{A\in\mathfrak{so}(n)} \|\nabla u - A\|_{\mu,\mu_c,\kappa}\\ &= \|\nabla u - \operatorname{skew}\nabla u\|_{\mu,\mu_c,\kappa} = \|\operatorname{sym}\nabla u\|_{\mu,\mu_c,\kappa}\,. \end{aligned} \tag{5}$$

We therefore find

$$\begin{aligned}\operatorname{dist}^2_{\text{Euclid},\mu,\mu_c,\kappa}(\nabla u, \mathfrak{so}(n)) &= \|\operatorname{sym}\nabla u\|^2_{\mu,\mu_c,\kappa}\\ &= \mu\,\|\operatorname{dev}_n \operatorname{sym}\nabla u\|^2 + \frac{\kappa}{2}\,[\operatorname{tr}(\operatorname{sym}\nabla u)]^2\\ &= \mu\,\|\operatorname{dev}_n \varepsilon\|^2 + \frac{\kappa}{2}\,[\operatorname{tr}(\varepsilon)]^2 = W_{\text{lin}}(\nabla u)\end{aligned}$$

for the linear strain tensor $\varepsilon = \operatorname{sym}\nabla u$, which is the quadratic isotropic elastic energy.

## 2.2 *The Euclidean Strain Measure in Nonlinear Isotropic Elasticity*

In order to obtain a strain measure in the geometrically nonlinear case, we must compute the distance

$$\operatorname{dist}(\nabla\varphi, \mathrm{SO}(n)) = \operatorname{dist}(F, \mathrm{SO}(n)) = \inf_{Q\in\mathrm{SO}(n)} \operatorname{dist}(F, Q)$$

of the deformation gradient $F = \nabla\varphi \in \mathrm{GL}^+(n)$ to the actual set of pure rotations $\mathrm{SO}(n) \subset \mathrm{GL}^+(n)$. It is therefore necessary to choose a distance function on $\mathrm{GL}^+(n)$;

an obvious choice is the restriction of the Euclidean distance on $\mathbb{R}^{n\times n}$ to $\mathrm{GL}^+(n)$. For the canonical Frobenius norm $\|\,.\,\|$, the Euclidean distance between $F, P \in \mathrm{GL}^+(n)$ is

$$\mathrm{dist}_{\mathrm{Euclid}}(F, P) = \|F - P\| = \sqrt{\mathrm{tr}[(F-P)^T(F-P)]}\,.$$

Now let $Q \in \mathrm{SO}(n)$. Since $\|\,.\,\|$ is orthogonally invariant, i.e. $\|\widehat{Q}X\| = \|X\widehat{Q}\| = \|X\|$ for all $X \in \mathbb{R}^{n\times n}$, $\widehat{Q} \in \mathrm{O}(n)$, we find

$$\mathrm{dist}_{\mathrm{Euclid}}(F, Q) = \|F - Q\| = \|Q^T(F - Q)\| = \|Q^T F - \mathbb{1}\|\,. \tag{6}$$

Thus the computation of the strain measure induced by the Euclidean distance on $\mathrm{GL}^+(n)$ reduces to the *matrix nearness problem* [19]

$$\mathrm{dist}_{\mathrm{Euclid}}(F, \mathrm{SO}(n)) = \inf_{Q\in\mathrm{SO}(n)} \|F - Q\| = \min_{Q\in\mathrm{SO}(n)} \|Q^T F - \mathbb{1}\|\,.$$

By a well-known optimality result discovered by Giuseppe Grioli [15] (cf. [9, 16, 27, 32]), also called "Grioli's Theorem" by Truesdell and Toupin [42, p. 290], this minimum is attained for the orthogonal polar factor $R$.

**Theorem 1** (Grioli's Theorem [15, 32, 42]) *Let* $F \in \mathrm{GL}^+(n)$. *Then*

$$\min_{Q\in\mathrm{SO}(n)} \|Q^T F - \mathbb{1}\| = \|R^T F - \mathbb{1}\| = \|\sqrt{F^T F} - \mathbb{1}\| = \|U - \mathbb{1}\|\,,$$

*where* $F = R\,U$ *is the polar decomposition of* $F$ *with* $R = \mathrm{polar}(F) \in \mathrm{SO}(n)$ *and* $U = \sqrt{F^T F} \in \mathrm{Sym}^+(n)$. *The minimum is uniquely attained at the orthogonal polar factor* $R$.

Thus for nonlinear elasticity, the restriction of the Euclidean distance to $\mathrm{GL}^+(n)$ yields the strain measure

$$\mathrm{dist}_{\mathrm{Euclid}}(F, \mathrm{SO}(n)) = \|U - \mathbb{1}\|\,.$$

In analogy to the linear case, we obtain

$$\mathrm{dist}^2_{\mathrm{Euclid}}(F, \mathrm{SO}(n)) = \|U - \mathbb{1}\|^2 = \|E_{1/2}\|^2\,, \tag{7}$$

where $E_{1/2} = U - \mathbb{1}$ is the Biot strain tensor. Note the similarity between this expression and the *Saint-Venant–Kirchhoff* energy [24]

$$\|E_1\|^2_{\mu,\mu_c,\kappa} = \mu\,\|\mathrm{dev}_3\,E_1\|^2 + \frac{\kappa}{2}\,[\mathrm{tr}(E_1)]^2\,, \tag{8}$$

where $E_1 = \frac{1}{2}(C - \mathbb{1}) = \frac{1}{2}(U^2 - \mathbb{1})$ is the Green–Lagrangian strain.

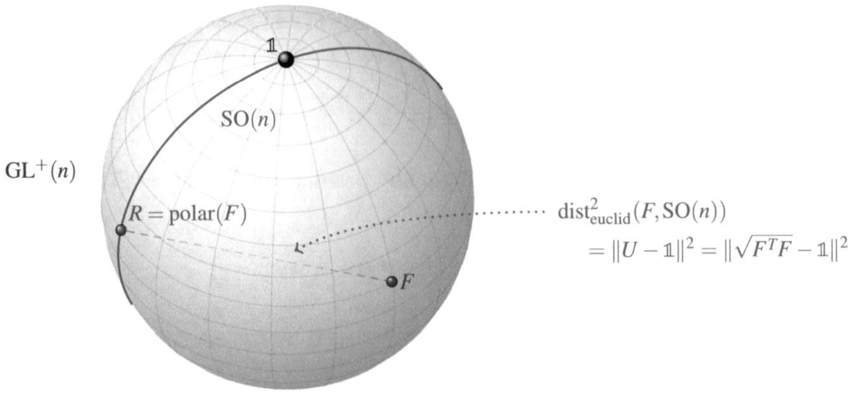


**Fig. 1** The Euclidean distance as an extrinsic measure on $\text{GL}^+(n)$; note that the representation of the manifold $\text{GL}^+(n)$ as a sphere only serves to demonstrate the necessity of an intrinsic distance measure and does not reflect the actual topological properties of $\text{GL}^+(n)$

However, the resulting strain measure $\omega(U) = \text{dist}_{\text{Euclid}}(F, \text{SO}(n)) = \|U - \mathbb{1}\|$ does not truly seem appropriate for finite elasticity theory: for $U \to 0$ we find $\|U - \mathbb{1}\| \to \|\mathbb{1}\| = \sqrt{n} < \infty$, thus singular deformations do not necessarily correspond to an infinite measure $\omega$. Furthermore, the above computations are not compatible with the weighted norm introduced in Sect. 2.1: in general [12, 13, 31],

$$\min_{Q \in \text{SO}(n)} \|F - Q\|^2_{\mu,\mu_c,\kappa} \neq \min_{Q \in \text{SO}(n)} \|Q^T F - \mathbb{1}\|^2_{\mu,\mu_c,\kappa} \neq \|\sqrt{F^T F} - \mathbb{1}\|^2_{\mu,\mu_c,\kappa}, \tag{9}$$

thus the Euclidean distance of $F$ to $\text{SO}(n)$ with respect to $\|\,.\,\|_{\mu,\mu_c,\kappa}$ does not equal $\|\sqrt{F^T F} - \mathbb{1}\|_{\mu,\mu_c,\kappa}$ in general. In these cases, the element of best approximation is not the orthogonal polar factor $R = \text{polar}(F)$.

We also observe that the Euclidean distance is not an *intrinsic* distance measure on $\text{GL}^+(n)$: in general, $A - B \notin \text{GL}^+(n)$ for $A, B \in \text{GL}^+(n)$, hence the term $\|A - B\|$ depends on the underlying linear structure of $\mathbb{R}^{n \times n}$.

Most importantly, because $\text{GL}^+(n)$ is not convex, the straight line $\{A + t\,(B - A)\,|\,t \in [0, 1]\}$ connecting $A$ and $B$ is not necessarily contained in $\text{GL}^+(n)$, which shows that the characterization of the Euclidean distance as the length of a shortest connecting curve is also not possible in a way intrinsic to $\text{GL}^+(n)$, as the intuitive sketches in Figs. 1 and 2 indicate.

These issues amply demonstrate that the Euclidean distance can only be regarded as an *extrinsic* distance measure on the general linear group. We therefore need to expand our view to allow for a more appropriate, truly *intrinsic* distance measure on $\text{GL}^+(n)$.

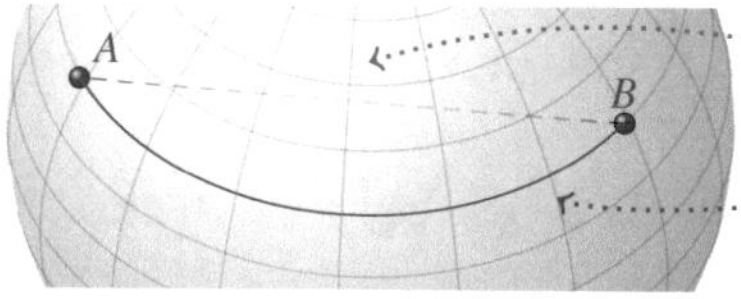


**Fig. 2** The geodesic (intrinsic) distance compared to the Euclidean (extrinsic) distance

# 3 The Riemannian Strain Measure in Nonlinear Isotropic Elasticity

## *3.1* $\mathrm{GL}^+(n)$ *as a Riemannian Manifold*

In order to find an intrinsic distance function on $\mathrm{GL}^+(n)$ that alleviates the drawbacks of the Euclidean distance, we endow $\mathrm{GL}(n)$ with a *Riemannian metric* $g$, which is defined by an inner product $g_A\colon T_A\,\mathrm{GL}(n)\times T_A\,\mathrm{GL}(n)\to\mathbb{R}$ on each tangent space $T_A\,\mathrm{GL}(n)$, $A\in\mathrm{GL}(n)$. Then the length of a sufficiently smooth curve $\gamma\colon[0,1]\to\mathrm{GL}(n)$ is given by $L(\gamma)=\int_0^1\sqrt{g_{\gamma(t)}(\dot\gamma(t),\dot\gamma(t))}\,\mathrm{d}t$, where $\dot\gamma(t)=\frac{\mathrm{d}}{\mathrm{d}t}\gamma(t)$, and the *geodesic distance* (cf. Fig. 2) between $A,B\in\mathrm{GL}^+(n)$ is defined as the infimum over the lengths of all (twice continuously differentiable) curves connecting $A$ to $B$:

$$\mathrm{dist}_{\mathrm{geod}}(A,B)=\inf\{L(\gamma)\mid\gamma\in C^2([0,1];\mathrm{GL}^+(n)),\ \gamma(0)=A,\ \gamma(1)=B\}\,.$$

Our search for an appropriate strain measure is thereby reduced to the task of finding an appropriate Riemannian metric on $\mathrm{GL}(n)$. Although it might appear as an obvious choice, the metric $\breve g$ with

$$\breve g_A(X,Y):=\langle X,Y\rangle\quad\text{for all } A\in\mathrm{GL}^+(n),\ X,Y\in\mathbb{R}^{n\times n}\tag{10}$$

provides no improvement over the already discussed Euclidean distance on $\mathrm{GL}^+(n)$: since the length of a curve $\gamma$ with respect to $\breve g$ is its classical (Euclidean) length, the shortest connecting curves with respect to $\breve g$ are straight lines of the form $t\mapsto A+t(B-A)$ with $A,B\in\mathrm{GL}^+(n)$. Locally, the geodesic distance induced by $\breve g$ is therefore equal to the Euclidean distance, and thus many of the shortcomings of the Euclidean distance apply to the geodesic distance induced by $\breve g$ as well.

In order to find a more viable Riemannian metric $g$ on $\mathrm{GL}(n)$, we consider the mechanical interpretation of the induced geodesic distance $\mathrm{dist}_{\mathrm{geod}}$: while our focus lies on the strain measure induced by $g$, that is the geodesic distance of the deformation gradient $F$ to the special orthogonal group $\mathrm{SO}(n)$, the distance $\mathrm{dist}_{\mathrm{geod}}(F_1,F_2)$ between two deformation gradients $F_1,F_2$ can also be motivated directly as a *measure of difference* between two linear (or *homogeneous*) deformations $F_1,F_2$ of the same body $\Omega$. More generally, we can define a difference measure between two inhomogeneous deformations $\varphi_1,\varphi_2:\Omega\subset\mathbb{R}^n\to\mathbb{R}^n$ via

$$\operatorname{dist}(\varphi_1, \varphi_2) := \int_\Omega \operatorname{dist}_{\text{geod}}(\nabla\varphi_1(x), \nabla\varphi_2(x))\,\mathrm{d}x \tag{11}$$

under suitable regularity conditions for $\varphi_1, \varphi_2$ (e.g. if $\varphi_1, \varphi_2$ are sufficiently smooth with $\det \nabla\varphi_i > 0$ up to the boundary).

In order to find an appropriate Riemannian metric $g$ on $\operatorname{GL}(n)$, we must discuss the required properties of this "difference measure". First, the requirements of objectivity (left-invariance) and isotropy (right-invariance) suggest that the metric $g$ should be *bi*-$\operatorname{O}(n)$-*invariant*, i.e. satisfy

$$\underbrace{g_{QA}(QX, QY) = \overbrace{g_A(X, Y)}_{\text{objectivity}} = g_{AQ}(XQ, YQ)}^{\text{isotropy}} \tag{12}$$

for all $Q \in \operatorname{O}(n)$, $A \in \operatorname{GL}(n)$ and $X, Y \in T_A \operatorname{GL}(n)$, to ensure that $\operatorname{dist}_{\text{geod}}(A, B) = \operatorname{dist}_{\text{geod}}(Q\,A, Q\,B) = \operatorname{dist}_{\text{geod}}(A\,Q, B\,Q)$.

However, these requirements do not sufficiently determine a specific Riemannian metric. For example, (12) is satisfied by the metric $\breve{g}$ defined in (10) as well as by the metric $\breve{\breve{g}}$ with $\breve{\breve{g}}_A(X, Y) = \langle A^T X, A^T Y\rangle$. In order to rule out unsuitable metrics, we need to impose further restrictions on $g$. If we consider the distance measure $\operatorname{dist}(\varphi_1, \varphi_2)$ between two deformations $\varphi_1, \varphi_2$ introduced in (11), a number of further invariances can be motivated: if we require that the distance is not changed by the superposition of a homogeneous deformation, i.e. that

$$\operatorname{dist}(B \cdot \varphi_1, B \cdot \varphi_2) = \operatorname{dist}(\varphi_1, \varphi_2)$$

for all constant $B \in \operatorname{GL}(n)$, then $g$ must be *left*-$\operatorname{GL}(n)$-*invariant*, i.e.

$$g_{BA}(B\,X, B\,Y) = g_A(X, Y) \tag{13}$$

for all $A, B \in \operatorname{GL}(n)$ and $X, Y \in T_A \operatorname{GL}(n)$.

It can easily be shown [26] that a Riemannian metric $g$ is left-$\operatorname{GL}(n)$-invariant as well as right-$\operatorname{O}(n)$-invariant if and only if $g$ is of the form

$$g_A(X, Y) = \langle A^{-1}X, A^{-1}Y\rangle_{\mu,\mu_c,\kappa}\,, \tag{14}$$

where $\langle\cdot,\cdot\rangle_{\mu,\mu_c,\kappa}$ is the fixed inner product on the tangent space $\mathfrak{gl}(n) = T_{\mathbb{1}} \operatorname{GL}(n) = \mathbb{R}^{n\times n}$ at the identity with

$$\langle X, Y\rangle_{\mu,\mu_c,\kappa} = \mu\,\langle \operatorname{dev}_n \operatorname{sym} X, \operatorname{dev}_n \operatorname{sym} Y\rangle + \mu_c\langle \operatorname{skew} X, \operatorname{skew} Y\rangle + \tfrac{\kappa}{2}\operatorname{tr}(X)\operatorname{tr}(Y)$$

for constant positive parameters $\mu, \mu_c, \kappa > 0$, and where $\langle X, Y\rangle = \operatorname{tr}(X^T Y)$ denotes the canonical inner product on $\mathfrak{gl}(n) = \mathbb{R}^{n\times n}$. In the following, we will always assume that $\operatorname{GL}(n)$ is endowed with a Riemannian metric of the form (14) unless indicated otherwise.

In order to find the geodesic distance

$$\operatorname{dist}_{\text{geod}}(F, \operatorname{SO}(n)) = \inf_{Q\in\operatorname{SO}(n)} \operatorname{dist}_{\text{geod}}(F, Q)$$

of $F \in \operatorname{GL}^+(n)$ to $\operatorname{SO}(n)$, we need to consider the *geodesic curves* on $\operatorname{GL}^+(n)$. It has been shown [1, 17, 26, 28] that every geodesic on $\operatorname{GL}^+(n)$ with respect to the left-$\operatorname{GL}(n)$-invariant Riemannian metric (14) is of the form

$$\gamma_F^\xi(t) = F\,\exp(t(\operatorname{sym}\xi - \tfrac{\mu_c}{\mu}\operatorname{skew}\xi))\,\exp(t(1+\tfrac{\mu_c}{\mu})\operatorname{skew}\xi) \qquad (15)$$

with $F \in \operatorname{GL}^+(n)$ and some $\xi \in \mathfrak{gl}(n)$, where exp denotes the matrix exponential. Since the geodesic curves are defined globally, $\operatorname{GL}^+(n)$ is *geodesically complete* with respect to the metric $g$. We can therefore apply the Hopf-Rinow theorem [23, 26] to find that for all $F, P \in \operatorname{GL}^+(n)$ there exists a *length minimizing geodesic* $\gamma_F^\xi$ connecting $F$ and $P$. Without loss of generality, we can assume that $\gamma_F^\xi$ is defined on the interval $[0, 1]$. Then the end points of $\gamma_F^\xi$ are

$$\gamma_F^\xi(0) = F \quad \text{and} \quad P = \gamma_F^\xi(1) = F\,\exp(\operatorname{sym}\xi - \tfrac{\mu_c}{\mu}\operatorname{skew}\xi)\,\exp((1+\tfrac{\mu_c}{\mu})\operatorname{skew}\xi),$$

and the length of the geodesic $\gamma_F^\xi$ starting in $F$ with initial tangent $F\xi \in T_F\operatorname{GL}^+(n)$ (cf. (15) and Fig. 3) is given by [26]

$$L(\gamma_F^\xi) = \|\xi\|_{\mu,\mu_c,\kappa}\,.$$

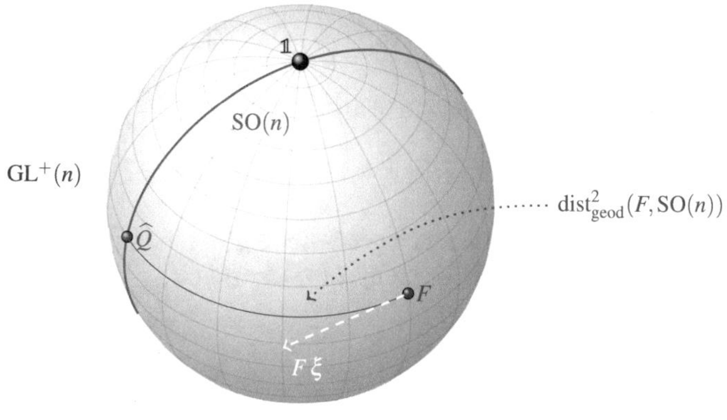


**Fig. 3** The geodesic (intrinsic) distance to $\operatorname{SO}(n)$; neither the element $\widehat{Q}$ of best approximation nor the initial tangent $F\xi \in T_F\operatorname{GL}^+(n)$ of the connecting geodesic is known beforehand

The geodesic distance between $F$ and $P$ can therefore be characterized as

$$\operatorname{dist}_{\text{geod}}(F, P) = \min\{\|\xi\|_{\mu,\mu_c,\kappa} \mid \xi \in \mathfrak{gl}(n) : \gamma_F^\xi(1) = P\},$$

that is the minimum of $\|\xi\|_{\mu,\mu_c,\kappa}$ over all $\xi \in \mathfrak{gl}(n)$ which connect $F$ and $P$, i.e. satisfy

$$\exp(\operatorname{sym}\xi - \tfrac{\mu_c}{\mu}\operatorname{skew}\xi)\,\exp((1+\tfrac{\mu_c}{\mu})\operatorname{skew}\xi) = F^{-1}P\,. \tag{16}$$

Although some numerical computations have been employed [43] to approximate the geodesic distance in the special case of the canonical left-$\mathrm{GL}(n)$-invariant metric, i.e. for $\mu = \mu_c = 1$, $\kappa = \frac{2}{n}$, there is no known closed-form solution to the highly nonlinear system (16) in terms of $\xi$ for given $F, P \in \mathrm{GL}^+(n)$ and thus no known method of directly computing $\operatorname{dist}_{\text{geod}}(F, P)$ in the general case exists. However, this parametrization of the geodesic curves will still allow us to obtain a lower bound on the distance of $F$ to $\mathrm{SO}(n)$.

## *3.2 The Geodesic Distance to* SO($n$)

Having defined the geodesic distance on $\mathrm{GL}^+(n)$, we can now consider the geodesic strain measure, i.e. the geodesic distance of the deformation gradient $F$ to $\mathrm{SO}(n)$:

$$\operatorname{dist}_{\text{geod}}(F, \mathrm{SO}(n)) = \inf_{Q\in\mathrm{SO}(n)} \operatorname{dist}_{\text{geod}}(F, Q)\,. \tag{17}$$

Now, let $F = R\,U$ denote the polar decomposition of $F$ with $U \in \mathrm{Sym}^+(n)$ and $R \in \mathrm{SO}(n)$. In order to establish a simple upper bound on the geodesic distance $\operatorname{dist}_{\text{geod}}(F, \mathrm{SO}(n))$, we construct a particular curve $\gamma_R$ connecting $F$ to its orthogonal factor $R \in \mathrm{SO}(n)$ and compute its length $L(\gamma_R)$. For

$$\gamma_R(t) := R\,\exp((1-t)\,\log U)\,,$$

where $\log U \in \mathrm{Sym}(n)$ is the principal matrix logarithm of $U$, we find

$$\gamma_R(0) = R\,\exp(\log U) = R\,U = F \qquad \text{and} \qquad \gamma_R(1) = R\,\exp(0) = R \in \mathrm{SO}(n)\,.$$

It is easy to confirm that $\gamma_R$ is in fact a geodesic as given in (15) with $\xi = \log U \in \mathrm{Sym}(n)$, thus the length of $\gamma_R$ is given by $L(\gamma_R) = \|\log U\|_{\mu,\mu_c,\kappa}$. We can thereby establish the *upper bound*

$$\operatorname{dist}^2_{\text{geod}}(F, \mathrm{SO}(n)) = \inf_{Q\in \mathrm{SO}(n)} \operatorname{dist}^2_{\text{geod}}(F, Q) \leq \operatorname{dist}^2_{\text{geod}}(F, R) \tag{18}$$

$$\leq L^2(\gamma_R) = \|\log U\|^2_{\mu,\mu_c,\kappa} = \mu\,\|\operatorname{dev}_n \log U\|^2 + \frac{\kappa}{2}\,[\operatorname{tr}(\log U)]^2 \tag{19}$$

for the geodesic distance of $F$ to $\mathrm{SO}(n)$.

Our task in the remainder of this section is to show that the right hand side of inequality (19) is *also a lower bound* for the (squared) geodesic strain measure, i.e. that, altogether,

$$\operatorname{dist}^2_{\text{geod}}(F, \mathrm{SO}(n)) = \mu\,\|\operatorname{dev}_n \log U\|^2 + \frac{\kappa}{2}\,[\operatorname{tr}(\log U)]^2 .$$

However, while the orthogonal polar factor $R$ is the element of best approximation in the Euclidean case (for $\mu = \mu_c = 1, \kappa = \frac{2}{n}$) due to Grioli's Theorem, it is not clear whether $R$ is indeed the element in $\mathrm{SO}(n)$ with the shortest geodesic distance to $F$ (and thus whether equality holds in (18)). Furthermore, it is not even immediately obvious that the geodesic distance between $F$ and $R$ is actually given by the right hand side of (19), since a shorter connecting geodesic might exist (and hence inequality might hold in (19)).

Nonetheless, the following fundamental logarithmic minimization property of the orthogonal polar factor, combined with the computations in Sect. 3.1, allows us to show that (19) is indeed also a lower bound for $\operatorname{dist}_{\text{geod}}(F, \mathrm{SO}(n))$.

**Proposition 2** *Let $F = R\sqrt{F^TF}$ be the polar decomposition of $F \in \mathrm{GL}^+(n)$ with $R \in \mathrm{SO}(n)$ and let $\|\,.\,\|$ denote the Frobenius norm on $\mathbb{R}^{n\times n}$. Then*

$$\inf_{Q\in \mathrm{SO}(n)} \|\operatorname{sym} \operatorname{Log}(Q^TF)\| = \|\operatorname{sym} \log(R^TF)\| = \|\log \sqrt{F^TF}\|\,,$$

*where*

$$\inf_{Q\in \mathrm{SO}(n)} \|\operatorname{sym} \operatorname{Log}(Q^TF)\| := \inf_{Q\in \mathrm{SO}(n)} \inf\{\|\operatorname{sym} X\| \mid X \in \mathbb{R}^{n\times n}\,,\ \exp(X) = Q^TF\}$$

*is defined as the infimum of $\|\operatorname{sym}\,.\,\|$ over "all real matrix logarithms" of $Q^TF$.*

Proposition 2, which can be seen as the natural logarithmic analogue of Grioli's Theorem (cf. Sect. 2.2), was first shown for dimensions $n = 2, 3$ by Neff et al. [35] using the so-called sum-of-squared-logarithms inequality [7, 8, 10, 37]. A generalization to all unitarily invariant norms and complex logarithms for arbitrary dimension was given by Lankeit, Neff and Nakatsukasa [25]. We also require the following corollary involving the weighted Frobenius norm, which is not orthogonally invariant.

**Corollary 3** *Let*

$$\|X\|^2_{\mu,\mu_c,\kappa} = \mu\,\|\operatorname{dev}_n \operatorname{sym} X\|^2 + \mu_c\,\|\operatorname{skew} X\|^2 + \frac{\kappa}{2}\,[\operatorname{tr}(X)]^2\,, \qquad \mu, \mu_c, \kappa > 0\,,$$

*for all* $X \in \mathbb{R}^{n\times n}$, *where* $\|\,.\,\|$ *is the Frobenius matrix norm. Then*

$$\inf_{Q\in \mathrm{SO}(n)} \|\operatorname{sym} \operatorname{Log}(Q^T F)\|_{\mu,\mu_c,\kappa} = \|\log \sqrt{F^T F}\|_{\mu,\mu_c,\kappa}\,.$$

We are now ready to prove our main result.

**Theorem 4** *Let* $g$ *be the left-*$\mathrm{GL}(n)$*-invariant, right-*$\mathrm{O}(n)$*-invariant Riemannian metric on* $\mathrm{GL}(n)$ *defined by*

$$g_A(X,Y) = \langle A^{-1}X, A^{-1}Y\rangle_{\mu,\mu_c,\kappa}\,, \qquad \mu, \mu_c, \kappa > 0\,,$$

*for* $A \in \mathrm{GL}(n)$ *and* $X, Y \in \mathbb{R}^{n\times n}$, *where*

$$\langle X, Y\rangle_{\mu,\mu_c,\kappa} = \mu\,\langle \operatorname{dev}_n \operatorname{sym} X, \operatorname{dev}_n \operatorname{sym} Y\rangle + \mu_c \langle \operatorname{skew} X, \operatorname{skew} Y\rangle + \tfrac{\kappa}{2} \operatorname{tr}(X)\operatorname{tr}(Y)\,. \tag{20}$$

*Then for all* $F \in \mathrm{GL}^+(n)$, *the geodesic distance of* $F$ *to the special orthogonal group* $\mathrm{SO}(n)$ *induced by* $g$ *is given by*

$$\operatorname{dist}^2_{\mathrm{geod}}(F, \mathrm{SO}(n)) = \mu\,\|\operatorname{dev}_n \log U\|^2 + \frac{\kappa}{2}\,[\operatorname{tr}(\log U)]^2\,, \tag{21}$$

*where* $\log$ *is the principal matrix logarithm,* $\operatorname{tr}(X) = \sum_{i=1}^n X_{i,i}$ *denotes the trace and* $\operatorname{dev}_n X = X - \frac{1}{n}\operatorname{tr}(X)\cdot \mathbb{1}$ *is the n-dimensional deviatoric part of* $X \in \mathbb{R}^{n\times n}$. *In particular, the geodesic distance does not depend on the* spin modulus $\mu_c$.

*Remark 5* It can also be shown [30] that the orthogonal factor $R \in \mathrm{SO}(n)$ of the polar decomposition $F = R\,U$ is the unique element of best approximation in $\mathrm{SO}(n)$, i.e. that for $Q \in \mathrm{SO}(n)$, $\operatorname{dist}_{\mathrm{geod}}(F, \mathrm{SO}(n)) = \operatorname{dist}_{\mathrm{geod}}(F, Q)$ if and only if $Q = R$.

*Proof* (*of Theorem* 4) Let $F \in \mathrm{GL}^+(n)$ and $\widehat{Q} \in \mathrm{SO}(n)$. Then according to our previous considerations (cf. Sect. 3.1) there exists $\xi \in \mathfrak{gl}(n)$ with

$$\exp\left(\operatorname{sym}\,\xi - \tfrac{\mu_c}{\mu}\operatorname{skew}\,\xi\right)\,\exp\left(\left(1 + \tfrac{\mu_c}{\mu}\right)\operatorname{skew}\,\xi\right) = F^{-1}\widehat{Q} \tag{22}$$

and

$$\|\xi\|_{\mu,\mu_c,\kappa} = \mathrm{dist}_{\mathrm{geod}}(F, \widehat{Q})\,. \tag{23}$$

In order to find a lower estimate on $\|\xi\|_{\mu,\mu_c,\kappa}$ (and thus on $\mathrm{dist}_{\mathrm{geod}}(F, \widehat{Q})$), we compute

$$\begin{aligned}
& \exp\left(\mathrm{sym}\ \xi - \tfrac{\mu_c}{\mu}\mathrm{skew}\ \xi\right) \exp\left(\left(1+\tfrac{\mu_c}{\mu}\right)\mathrm{skew}\ \xi\right) = F^{-1}\widehat{Q} \\
\Longrightarrow \quad & \exp\left(\left(1+\tfrac{\mu_c}{\mu}\right)\mathrm{skew}\ \xi\right)^{-1} \exp\left(\mathrm{sym}\ \xi - \tfrac{\mu_c}{\mu}\mathrm{skew}\ \xi\right)^{-1} = \widehat{Q}^T F \\
\Longrightarrow \quad & \exp\left(-\,\mathrm{sym}\ \xi + \tfrac{\mu_c}{\mu}\mathrm{skew}\ \xi\right) = \underbrace{\exp(\,(1+\tfrac{\mu_c}{\mu})\mathrm{skew}\ \xi\,)}_{\in \mathfrak{so}(n)}\widehat{Q}^T F\,.
\end{aligned}$$

Since $\exp(W) \in \mathrm{SO}(n)$ for all skew symmetric $W \in \mathfrak{so}(n)$, we find

$$\exp(\underbrace{-\,\mathrm{sym}\ \xi + \tfrac{\mu_c}{\mu}\mathrm{skew}\ \xi}_{=:Y}) = Q_\xi^T F \tag{24}$$

with $Q_\xi = \widehat{Q}\exp(-(1+\frac{\mu_c}{\mu})\mathrm{skew}\ \xi\,) \in \mathrm{SO}(n)$; note that $\mathrm{sym}\,Y = -\,\mathrm{sym}\ \xi$. According to (24), $Y = -\,\mathrm{sym}\ \xi + \frac{\mu_c}{\mu}\mathrm{skew}\ \xi$ is "a logarithm"[2] of $Q_\xi^T F$. The weighted Frobenius norm of the symmetric part of $Y = -\,\mathrm{sym}\ \xi + \frac{\mu_c}{\mu}\mathrm{skew}\ \xi$ is therefore bounded below by the infimum of $\|\mathrm{sym}\,X\|_{\mu,\mu_c,\kappa}$ over "all logarithms" $X$ of $Q_\xi^T F$:

$$\begin{aligned}
\|\mathrm{sym}\,\xi\|_{\mu,\mu_c,\kappa} &= \|\mathrm{sym}\,Y\|_{\mu,\mu_c,\kappa} \\
&\overset{(24)}{\geq} \inf\{\|\mathrm{sym}\,X\|_{\mu,\mu_c,\kappa} \,|\, X \in \mathbb{R}^{n\times n},\ \exp(X) = Q_\xi^T F\} \\
&\geq \inf_{Q\in\mathrm{SO}(n)} \inf\{\|\mathrm{sym}\,X\|_{\mu,\mu_c,\kappa} \,|\, X \in \mathbb{R}^{n\times n},\ \exp(X) = Q^T F\} \\
&= \inf_{Q\in\mathrm{SO}(n)} \|\mathrm{sym}\,\mathrm{Log}(Q^T F)\|_{\mu,\mu_c,\kappa}\,.
\end{aligned} \tag{25}$$

We can now apply Corollary 3 to find

$$\begin{aligned}
\mathrm{dist}^2_{\mathrm{geod}}(F, \widehat{Q}) = \|\xi\|^2_{\mu,\mu_c,\kappa} &= \mu\,\|\mathrm{dev}_n\,\mathrm{sym}\ \xi\|^2 + \mu_c\,\|\mathrm{skew}\ \xi\|^2 + \frac{\kappa}{2}\,[\mathrm{tr}(\mathrm{sym}\ \xi)]^2 \\
&\geq \mu\,\|\mathrm{dev}_n\,\mathrm{sym}\ \xi\|^2 + \frac{\kappa}{2}\,[\mathrm{tr}(\mathrm{sym}\ \xi)]^2 \qquad (26) \\
&= \|\mathrm{sym}\ \xi\|^2_{\mu,\mu_c,\kappa} \\
&\overset{(25)}{\geq} \inf_{Q\in\mathrm{SO}(n)} \|\mathrm{sym}\,\mathrm{Log}(Q^T F)\|^2_{\mu,\mu_c,\kappa} \\
&\overset{\text{Corollary 3}}{=} \mu\,\|\log\sqrt{F^T F}\|^2_{\mu,\mu_c,\kappa} \\
&= \mu\,\|\mathrm{dev}_n\,\log U\|^2 + \frac{\kappa}{2}\,[\mathrm{tr}(\log U)]^2
\end{aligned}$$

[2]Loosely speaking, we use the term "a logarithm of $A \in \mathrm{GL}^+(n)$" to denote any (real) solution $X$ of the matrix equation $\exp X = A$.

for $U = \sqrt{F^T F}$. Since this inequality is independent of $\widehat{Q}$ and holds for all $\widehat{Q} \in \mathrm{SO}(n)$, we obtain the desired lower bound

$$\mathrm{dist}^2_{\mathrm{geod}}(F, \mathrm{SO}(n)) = \inf_{\widehat{Q} \in \mathrm{SO}(n)} \mathrm{dist}^2_{\mathrm{geod}}(F, \widehat{Q}) \geq \mu \, \|\mathrm{dev}_n \log U\|^2 + \frac{\kappa}{2} \, [\mathrm{tr}(\log U)]^2$$

on the geodesic distance of $F$ to $\mathrm{SO}(n)$. Together with the upper bound already established in (19), we finally find

$$\mathrm{dist}^2_{\mathrm{geod}}(F, \mathrm{SO}(n)) = \mathrm{dist}^2_{\mathrm{geod}}(F, R) = \mu \, \|\mathrm{dev}_n \log U\|^2 + \frac{\kappa}{2} \, [\mathrm{tr}(\log U)]^2 . \qquad \square$$

According to Theorem 4, the squared geodesic distance between $F$ and $\mathrm{SO}(n)$ with respect to any left-$\mathrm{GL}(n)$-invariant, right-$\mathrm{O}(n)$-invariant Riemannian metric on $\mathrm{GL}(n)$ is the *isotropic quadratic Hencky energy*

$$W_{\mathrm{H}}(F) = \mu \, \|\mathrm{dev}_n \log U\|^2 + \frac{\kappa}{2} \, [\mathrm{tr}(\log U)]^2 ,$$

where the parameters $\mu, \kappa > 0$ represent the shear modulus and the bulk modulus, respectively (cf. Fig. 4). The Hencky energy function was introduced in 1929 by H. Hencky [18], who derived it from geometrical considerations as well: his deduction was based on a set of axioms including a law of superposition for the stress response function [29], an approach previously employed by G. F. Becker [4, 34] in 1893 and later followed in a more general context by H. Richter [39], cf. [38, 40, 41].

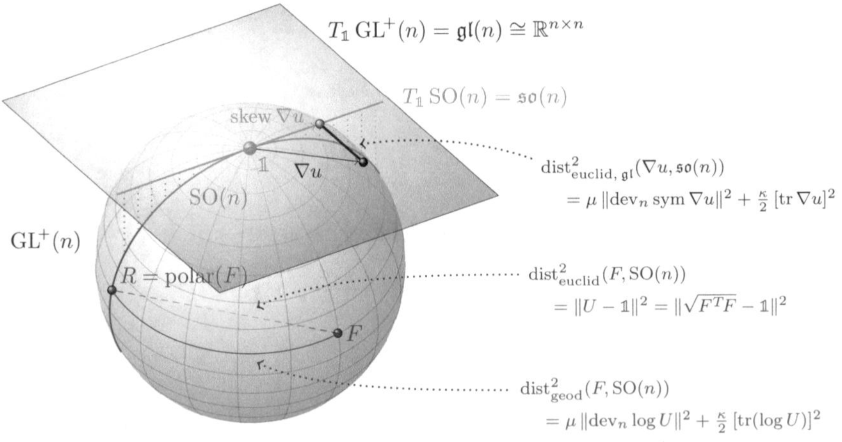


**Fig. 4** The isotropic Hencky energy of $F$ measures the geodesic distance between $F$ and $\mathrm{SO}(n)$. The linear Euclidean strain measure is obtained via linearization of the tangent space $\mathfrak{gl}(n)$ at $\mathbb{1}$

## References

1. Andruchow, E., Larotonda, G., Recht, L., Varela, A.: The left invariant metric in the general linear group. J. Geom. Phys. **86**, 241–257 (2014)
2. Batra, R.C.: Linear constitutive relations in isotropic finite elasticity. J. Elast. **51**(3), 243–245 (1998)
3. Batra, R.C.: Comparison of results from four linear constitutive relations in isotropic finite elasticity. Int. J. Non-Linear Mech. **36**(3), 421–432 (2001)
4. Becker, G.F.: The finite elastic stress-strain function. Am. J. Sci. **46**, 337–356 (1893). https://www.uni-due.de/imperia/md/content/mathematik/ag_neff/becker_latex_new1893.pdf
5. Bertram, A.: Elasticity and Plasticity of Large Deformations. Springer, Heidelberg (2008)
6. Bertram, A., Böhlke, T., Šilhavý, M.: On the rank 1 convexity of stored energy functions of physically linear stress-strain relations. J. Elast. **86**(3), 235–243 (2007)
7. Bîrsan, M., Neff, P., Lankeit, J.: Sum of squared logarithms - an inequality relating positive definite matrices and their matrix logarithm. J. Inequalities Appl. **2013**(1), 1–16 (2013). doi:10.1186/1029-242X-2013-168
8. Borisov, L., Neff, P., Sra, S., Thiel, C.: The sum of squared logarithms inequality in arbitrary dimensions. to appear in Linear Algebra Appl. (2015). arXiv:1508.04039
9. Bouby, C., Fortuné, D., Pietraszkiewicz, W., Vallée, C.: Direct determination of the rotation in the polar decomposition of the deformation gradient by maximizing a Rayleigh quotient. Zeitschrift für Angewandte Mathematik und Mechanik **85**(3), 155–162 (2005)
10. Dannan, F.M., Neff, P., Thiel, C.: On the sum of squared logarithms inequality and related inequalities. to appear in JMI J. Math. Inequalities (2014). arXiv:1411.1290
11. De Boor, C.: A naive proof of the representation theorem for isotropic, linear asymmetric stress-strain relations. J. Elast. **15**(2), 225–227 (1985). doi:10.1007/BF00041995
12. Fischle, A., Neff, P.: The geometrically nonlinear cosserat micropolar shear-stretch energy. Part I: A general parameter reduction formula and energy-minimizing microrotations in 2d. to appear in Zeitschrift für Angewandte Mathematik und Mechanik (2015). arXiv:1507.05480
13. Fischle, A., Neff, P.: The geometrically nonlinear cosserat micropolar shear-stretch energy. part ii: Non-classical energy-minimizing microrotations in 3d and their computational validation. Submitted (2015). arXiv:1509.06236
14. Fosdick, R.L., Wineman, A.S.: On general measures of deformation. Acta Mech. **6**(4), 275–295 (1968)
15. Grioli, G.: Una proprieta di minimo nella cinematica delle deformazioni finite. Bollettino dell'Unione Matematica Italiana **2**, 252–255 (1940)
16. Grioli, G.: Mathematical Theory of Elastic Equilibrium (recent results). Ergebnisse der angewandten Mathematik, vol. 7. Springer, Heidelberg (1962)
17. Hackl, K., Mielke, A., Mittenhuber, D.: Dissipation distances in multiplicative elastoplasticity. In: Wendland, W.L., Efendiev, M. (eds.) Analysis and Simulation of Multifield Problems, pp. 87–100. Springer, Heidelberg (2003)
18. Hencky, H.: Welche Umstände bedingen die Verfestigung bei der bildsamen Verformung von festen isotropen Körpern? Zeitschrift für Physik **55**, 145–155 (1929). www.uni-due.de/imperia/md/content/mathematik/ag_neff/hencky1929.pdf
19. Higham, N.J.: Matrix Nearness Problems and Applications. University of Manchester, Department of Mathematics, Manchester (1988)
20. Hill, R.: On constitutive inequalities for simple materials - I. J. Mech. Phys. Solids **11**, 229–242 (1968)
21. Hill, R.: Constitutive inequalities for isotropic elastic solids under finite strain. Proc. R. Soc. A Math. Phys. Sci. **314**, 457–472 (1970)
22. Hill, R.: Aspects of invariance in solid mechanics. Adv. Appl. Mech. **18**, 1–75 (1978)
23. Hopf, H., Rinow, W.: Über den Begriff der vollständigen differentialgeometrischen Fläche. Commentarii Mathematici Helvetici **3**(1), 209–225 (1931)

24. Kirchhoff, G.R.: Über die Gleichungen des Gleichgewichtes eines elastischen Körpers bei nicht unendlich kleinen Verschiebungen seiner Theile. Sitzungsberichte der Mathematisch-Naturwissenschaftlichen Classe der Kaiserlichen Akademie der Wissenschaften in Wien **IX** (1852)
25. Lankeit, J., Neff, P., Nakatsukasa, Y.: The minimization of matrix logarithms: on a fundamental property of the unitary polar factor. Linear Algebra Appl. **449**, 28–42 (2014). doi:10.1016/j.laa.2014.02.012
26. Martin, R.J., Neff, P.: Minimal geodesics on gl(n) for left-invariant, right-o(n)-invariant riemannian metrics. to appear in The J. Geom. Mech. (2014). arXiv:1409.7849
27. Martins, L.C., Podio-Guidugli, P.: A variational approach to the polar decomposition theorem. Rendiconti delle sedute dell'Accademia nazionale dei Lincei **66**(6), 487–493 (1979)
28. Mielke, A.: Finite elastoplasticity, Lie groups and geodesics on SL(d). In: Newton, P., Holmes, P., Weinstein, A. (eds.) Geometry, Mechanics, and Dynamics - Volume in Honor of the 60th Birthday of J.E. Marsden, pp. 61–90. Springer, New York (2002)
29. Neff, P., Eidel, B., Martin, R.J.: The axiomatic deduction of the quadratic Hencky strain energy by Heinrich Hencky (a new translation of Hencky's original German articles). (2014). arXiv:1402.4027
30. Neff, P., Eidel, B., Martin, R.J.: Geometry of logarithmic strain measures in solid mechanics. Archive for Rational Mechanics and Analysis (2016). doi:10.1007/s00205-016-1007-x. arXiv:1505.02203
31. Neff, P., Fischle, A., Münch, I.: Symmetric Cauchy-stresses do not imply symmetric Biot-strains in weak formulations of isotropic hyperelasticity with rotational degrees of freedom. Acta Mech. **197**, 19–30 (2008)
32. Neff, P., Lankeit, J., Madeo, A.: On Grioli's minimum property and its relation to Cauchy's polar decomposition. Int. J. Eng. Sci. **80**, 209–217 (2014)
33. Neff, P., Münch, I.: Curl bounds Grad on SO(3). ESAIM: Control Optim. Calc. Var. **14**(1), 148–159 (2008)
34. Neff, P., Münch, I., Martin, R.J.: Rediscovering G. F. Becker's early axiomatic deduction of a multiaxial nonlinear stress-strain relation based on logarithmic strain. to appear in Math. Mech. Solids (2014). doi:10.1177/1081286514542296. arXiv:1403.4675
35. Neff, P., Nakatsukasa, Y., Fischle, A.: A logarithmic minimization property of the unitary polar factor in the spectral and frobenius norms. SIAM J. Matrix Anal. Appl. **35**(3), 1132–1154 (2014). doi:10.1137/130909949
36. Norris, A.N.: Higher derivatives and the inverse derivative of a tensor-valued function of a tensor. Q. Appl. Math. **66**, 725–741 (2008)
37. Pompe, W., Neff, P.: On the generalised sum of squared logarithms inequality. J. Inequalities Appl. **2015**(1), 1–17 (2015). doi:10.1186/s13660-015-0623-6
38. Richter, H.: Das isotrope Elastizitätsgesetz. Zeitschrift für Angewandte Mathematik und Mechanik **28**(7/8), 205–209 (1948). https://www.uni-due.de/imperia/md/content/mathematik/ag_neff/richter_isotrop_log.pdf
39. Richter, H.: Verzerrungstensor, Verzerrungsdeviator und Spannungstensor bei endlichen Formänderungen. Zeitschrift für Angewandte Mathematik und Mechanik **29**(3), 65–75 (1949). https://www.uni-due.de/imperia/md/content/mathematik/ag_neff/richter_deviator_log.pdf
40. Richter, H.: Zum Logarithmus einer Matrix. Archiv der Mathematik **2**(5), 360–363 (1949). doi:10.1007/BF02036865. https://www.uni-due.de/imperia/md/content/mathematik/ag_neff/richter_log.pdf
41. Richter, H.: Zur Elastizitätstheorie endlicher Verformungen. Mathematische Nachrichten **8**(1), 65–73 (1952)
42. Truesdell, C., Toupin, R.: The classical field theories. In: Flügge, S. (ed.) Handbuch der Physik, vol. III/1. Springer, Heidelberg (1960)
43. Zacur, E., Bossa, M., Olmos, S.: Multivariate tensor-based morphometry with a right-invariant Riemannian distance on $GL^+(n)$. J. Math. Imaging Vis. **50**, 19–31 (2014)

# Continuum Physics with Violations of the Second Law of Thermodynamics

Martin Ostoja-Starzewski

**Abstract** As dictated by the modern statistical physics, the second law is to be replaced by the fluctuation theorem on very small length and/or time scales. This means that the deterministic continuum thermomechanics must be generalized to a stochastic theory allowing randomly spontaneous violations of the Clausius–Duhem inequality to take place anywhere in the material domain. This paper outlines possible extensions of stochastic continuum thermomechanics in coupled field problems: (i) thermoviscous fluids, (ii) thermo-elastodynamics, and (iii) poromechanics with dissipation within the skeleton, the fluid, and the temperature field. Linear dissipative processes are being considered, with the thermodynamic orthogonality providing the average constitutive response and the fluctuation theorem providing the violations of the second law of thermodynamics. Special attention is paid to the fact that one can develop hyperbolic theories (i.e. free of the paradox of infinite speeds of signal transmission) while working with the Fourier-type conduction for which the fluctuation theorem has already been developed.

## 1 Violations of the Second Law of Thermodynamics in Heat Conduction and Viscous Flow

In continuum thermomechanics (e.g. [1]), the second fundamental law may be written in terms of the reversible ($s^{*(r)}$) and irreversible ($s^{*(ir)}$) parts of entropy production rate ($\dot{s}$)

$$\dot{s} = s^{*(r)} + s^{*(ir)} \quad with \ \ s^{*(r)} = -\left(\frac{q_i}{\theta}\right)_{,i} \quad and \ \ s^{*(ir)} \geq 0. \tag{1}$$

M. Ostoja-Starzewski (✉)
Department of Mechanical Science & Engineering, Institute for Condensed Matter Theory,
University of Illinois at Urbana-Champaign, Urbana, IL 61801, USA
e-mail: martinos@illinois.edu

M. Ostoja-Starzewski
Department of Mechanical Science & Engineering, Beckman Institute,
University of Illinois at Urbana-Champaign, Urbana, IL 61801, USA

© Springer Nature Singapore Pte Ltd. 2017
F. dell'Isola et al. (eds.), *Mathematical Modelling in Solid Mechanics*,
Advanced Structured Materials 69, DOI 10.1007/978-981-10-3764-1_12

Here $\theta$ is the absolute temperature while $q_i$ is the heat flux [Throughout we interchangeably use the subscript ($f_{i...}$) and the symbolic ($\boldsymbol{f}$) notations for tensors, as the need arises; an overdot means the material time derivative.]. The inequality in $(1)_3$ is assumed to hold instantaneously, i.e. for all $t$.

In contemporary statistical physics (e.g. [2, 3]) the second law is replaced by the fluctuation theorem which gives the relative probability of observing processes that have positive ($A$) and negative ($-A$) total dissipation in non-equilibrium systems:

$$\frac{\mathsf{P}\,(\phi_t = A)}{\mathsf{P}\,(\phi_t = -A)} = e^{At}. \tag{2}$$

Here $\phi_t$ is the total dissipation for a trajectory $\boldsymbol{\Gamma} \equiv \{q_1, p_1, ..., q_N, p_N\}$ of $N$ particles originating at $\boldsymbol{\Gamma}\,(0)$ and evolving for a time $t$:

$$\phi_t\,(\boldsymbol{\Gamma}\,(0)) = \int_0^t \phi\,(\boldsymbol{\Gamma}\,(s))\,ds. \tag{3}$$

The integral in (3) involves an instantaneous dissipation function:

$$\phi\,(\boldsymbol{\Gamma}\,(t)) = \frac{d\phi_t\,(\boldsymbol{\Gamma}\,(0))}{dt}. \tag{4}$$

The second law of thermodynamics is recovered upon ensemble averaging, time averaging, or upscaling.

Note for future reference that the dissipation function $\phi$, albeit on a coarser length scale, is also employed in continuum thermomechanics to describe the dissipative part of constitutive behavior of an elementary volume $dV$ (or a corresponding elementary mass $dm$). That function is taken as a functional $\phi\,(\boldsymbol{V})$ over the space of velocities $\boldsymbol{V}$ (or as a functional $\phi\,(\boldsymbol{Y})$ over the space of dissipative forces $\boldsymbol{Y}$), such that its value equals the instantaneous irreversible entropy production:

$$\phi\,(\boldsymbol{V}) = s^{*(ir)}. \tag{5}$$

Effectively, the functional $\phi\,(\boldsymbol{V})$ or $\phi\,(\boldsymbol{Y})$ is employed to derive the constitutive laws of continua. As is well known, one of the simplest models of continuum physics is the linear (Fourier) heat conduction, whereby the functional becomes a quadratic form and the inequality $(1)_3$ implies the positive definiteness of the thermal conductivity tensor.

In view of (2) above, the dissipation function is a stochastic (not deterministic) quantity which possibly and spontaneously takes negative values, so that the positive-definiteness does not absolutely hold. Therefore, we write (5) as

$$\boldsymbol{Y}\,(\omega)\cdot\boldsymbol{V} = \phi\,(\boldsymbol{V}, \omega) = s^{*(ir)}, \quad \omega \in \Omega, \tag{6}$$

where $\boldsymbol{Y}(\omega)$ are the dissipative forces conjugate to $\boldsymbol{V}$. Given that $\Omega$ is the set of possible outcomes, the argument $\omega$ indicates that $\phi(\boldsymbol{V}, \omega)$ is a stochastic functional, while $\boldsymbol{Y}(\omega)$ is a random quantity for a non-random (prescribed) velocity $\boldsymbol{V}$. An analogous picture holds for $\boldsymbol{Y}$ being prescribed and $\boldsymbol{V}$ the random outcome. It is tacitly assumed that $\Omega$ is equipped with a $\sigma$-algebra of observable events $\mathscr{A}$ and a probability measure $\mathsf{P}$ defined on $(\Omega, \mathscr{A})$.

The fluctuation theorem as expressed by (1) states that (i) positive dissipation is exponentially more likely to be observed than negative dissipation, and (ii) ensemble averaging of $\phi_t$ leads to

$$\langle \phi_t \mid \mathscr{F}_t \rangle \geq 0. \tag{7}$$

Here $\mid \mathscr{F}_t$ indicates the conditioning on the past and is discussed below, while $\langle f \rangle := \int f \, d\mathsf{P}$. Thus, the entropy production rate is non-negative on average. In view of the random fluctuations, $\phi_t$ is a stochastic process with a specific type of memory effect: a submartingale [4]. Treating time as a continuous parameter, we have

$$\langle \phi_{t+dt} \mid \text{past history} \rangle \geq \phi_t. \tag{8}$$

Next, we recall the Doob–Meyer decomposition to write $\phi_t$ as a sum of a martingale $(M)$ and a "drift" process $(G)$:

$$\phi_t(\boldsymbol{V}, \omega) = M + G \quad and \quad \phi(\boldsymbol{V}, \omega) = \dot{M} + \dot{G}. \tag{9}$$

Thus, $M \neq 0$ reflects the fluctuations of entropy production about the zero level $\langle s^{(i)} \rangle = 0$. The four different cases depending on whether $M = 0$ or $M \neq 0$ and $G = 0$ or $G > 0$ have been discussed in [4]. Overall, deterministic continuum mechanics is smoothly recovered as the time and/or spatial scale increases (so that $M \to 0$) or via ensemble averaging.

Note that one might also work with a discrete time formulation, making the mathematical analysis simpler.

There are three types of phenomena in classical physics where the fluctuation theorem is applicable: viscous [2], thermal [3], and electrical [2]. Here we concern ourselves with the first two, so that, contact with continuum thermomechanics is made by writing the scalar product $\boldsymbol{Y} \cdot \boldsymbol{V}$ as one involving the intrinsic mechanical dissipation (which includes the viscous effects) and thermal dissipation in spatial (Eulerian) description:

$$\phi(\boldsymbol{V}, \omega) = \phi_{th}(\boldsymbol{V}_1, \omega) + \phi_{mech}(\boldsymbol{V}_2, \omega), \quad \boldsymbol{V} \equiv (\boldsymbol{V}_1, \boldsymbol{V}_2) = \left( \frac{-\nabla\theta}{\theta}, \boldsymbol{d} \right). \tag{10}$$

Thus, the generalized velocity vector $\boldsymbol{V}$ is made up of two parts: the temperature gradient divided by the temperature $-\nabla\theta/\theta$ and the deformation rate $\boldsymbol{d}$. The reason we take the former as the argument of $\phi_{th}$ is that the fluctuation theorem for heat flow was derived for controllable temperature differences [3], with the heat flux being the stochastic outcome. Analogously, the fluctuation theorem for Couette and Poiseuille flows was derived for controllable velocities [3], with the Cauchy stress being the stochastic outcome. Thus, the dissipative force corresponding to $\boldsymbol{V}$ is made up of the heat flux and the dissipative stress

$$\boldsymbol{Y} \equiv (\boldsymbol{Y}_1, \boldsymbol{Y}_2) = \left(\boldsymbol{q}, \boldsymbol{\sigma}^{(d)}\right). \tag{11}$$

It is largely a matter of convenience whether $-\nabla\theta/\theta$ or $\boldsymbol{q}$ should be taken as a velocity or a dissipative force. In the section on thermoviscous fluids we work with the setup outlined above, while in the section on inviscid thermoelastic solids we invert the roles of $-\nabla\theta/\theta$ and $\boldsymbol{q}$.

There are two basic possibilities here:

- both processes in (10) may independently exhibit spontaneous random violations of the second law;
- both processes in (10) are coupled implying that the thermal and viscous violations of the second law are coupled, for which the relevant statistical physics has not yet been studied.

In what follows, we shall consider the first possibility above focusing on: (i) thermoviscous fluids with parabolic or hyperbolic type heat conduction, (ii) thermoelasticity with parabolic or hyperbolic type heat conduction, and (iii) poromechanics with dissipation within the skeleton, the fluid, and the temperature field. The reason we consider parabolic or hyperbolic cases is that the statistical physics has established the spontaneous violations of the Fourier type law [3], but a hyperbolic heat conduction in fluids and solids can still be modeled if a relaxation time in the entropy constitutive law is introduced. The theoretical developments below hinge on the fact that the balance laws apply irrespective of the conventional second law being obeyed or not. At the same time, we are interested in formulating models which are hyperelastic and hyperdissipative in ensemble average sense (or, for long time averages), thereby extending such class beyond the deterministic media fully obeying the second law [5, 6].

## 2 Random Fields

One of the key problems in constitutive modeling in continuum mechanics concerns the finding of a solution of (6), i.e. determining a constitutive relation linking $\boldsymbol{Y}$ with $\boldsymbol{V}$. The most effective and popular approach is based on a thermodynamic orthogonality [1] which also provides a stepping-stone to more complex models in continuum thermodynamics. To this end, take $\phi$ as a functional of $\boldsymbol{V}$, and obtain $\boldsymbol{Y}$ as its gradient in the velocity space:

$$\boldsymbol{Y} = \lambda \nabla_{\boldsymbol{V}} \phi \geq 0, \quad \lambda = \left(\boldsymbol{V} \cdot \nabla_{\boldsymbol{V}} \phi\right)^{-1} \phi. \tag{12}$$

The meaning of (11) is that, provided the dissipative force $\boldsymbol{Y}$ is prescribed, the actual velocity maximizes the dissipation rate $l^{(d)} = \boldsymbol{Y} \cdot \boldsymbol{V}$ subject to the side condition $\phi(\boldsymbol{V}) = \boldsymbol{Y} \cdot \boldsymbol{V} = l^{(d)} \geq 0$.

Replacing the deterministic picture by a stochastic one, the internal energy density $u$ and the entropy $s$ are real-valued random fields over the material ($\mathscr{D}$) and time ($T$) domains:

$$u : \mathscr{D} \times T \times \Omega \to \mathbb{R}, \quad s : \mathscr{D} \times T \times \Omega \to \mathbb{R}, \tag{13}$$

where we consider the heat conduction problem in a rigid (undeformable) conductor. The randomness disappears as the time and/or spatial scales become large: the field quantities simplify to deterministic functions of a homogeneous continuum.

Considering, say, the thermal dissipation in (10), we have

$$\phi_{th}\left(\frac{-\nabla\theta}{\theta}, \omega\right) = -q_k \frac{\theta_{,k}}{\theta} \equiv -\boldsymbol{q} \cdot \frac{\nabla\theta}{\theta}. \tag{14}$$

Given the stochastic violations of second law,

$$\phi_{th}(\boldsymbol{q}, \omega) = \dot{G}(\boldsymbol{q}) + \dot{M}(\boldsymbol{q}, \omega), \tag{15}$$

which, for the linear Fourier-type conductivity, becomes more explicit with

$$\dot{G}(\boldsymbol{q}) = \frac{1}{\theta} q_i \kappa_{ij} q_j, \ \dot{M}(\boldsymbol{q}, \omega) = \frac{1}{\theta} q_i \mathscr{M}_{ij}(\omega) q_j. \tag{16}$$

Here $\dot{G}(\boldsymbol{q})$ involves the thermal conductivity $\kappa_{ij}$ which is positive definite, and $\dot{M}(\boldsymbol{q}, \omega) = dM(\boldsymbol{q}, \omega)/dt$, with $M$ being the martingale modeling the random fluctuations according to (2). Clearly, the randomness residing in $M(\boldsymbol{d}, \omega)$ allows the total thermal conductivity $\kappa_{ij} + \mathscr{M}_{ij}$ to become negative since $\mathscr{M}_{ij}$ is not required to be positive definite, thus signifying the violations of the second law. More specifically, $\mathscr{M}_{ij} : \mathscr{V} \to \mathscr{V}$ (where $\mathscr{V}$ is a real vector space) is a second-order rank 2 tensor random field (e.g. [7, 8])

$$\mathscr{M}_{ij} : \mathscr{D} \times \Omega \to \mathscr{V}^2. \tag{17}$$

In view of the Gaussian character of nanoscale fluctuations, $\mathscr{M}_{ij}$ is a Gaussian tensor random field.

The same approach as in (15)–(17) may be used to introduce fluctuations in mechanical dissipation $\phi_{th}(\boldsymbol{d}, \omega)$ having spontaneously negative viscous responses.

## 3 Thermoviscous Fluid with a Thermal Relaxation Time

The internal energy $u$ is taken as a function of the strain $\varepsilon_{ij}$ and the entropy $s$:

$$u = u(\varepsilon_{ij}, s). \tag{18}$$

Switching from $u$ to $\psi = u - \theta s$ by a Legendre transformation, we find

$$\psi = \psi(\varepsilon_{ij}, \theta) \quad and \quad s = s(\theta). \tag{19}$$

The first fundamental law (energy balance) is

$$\rho \dot{u} = \sigma_{ij} d_{ij} - q_{i,i} , \tag{20}$$

where the first term on the right is the specific power of deformation. The free energy function $\psi$ (taken per unit mass), assuming no elastic response but the presence of a relaxation time $t_0$, is

$$\rho\psi(\varepsilon_{ij}, \theta) = \rho\psi_0 - \rho s_0 \vartheta - \frac{C_E}{2\theta_0}\vartheta^2 - \frac{C_E}{\theta_0} t_0 \vartheta \dot{\vartheta}, \quad \vartheta = \theta - \theta_0. \tag{21}$$

Here $\psi_0$ and $s_0$ are the free energy and entropy in the reference state, $\mu$ is the shear elastic modulus, $C_E$ is the specific heat at constant strain, and $\vartheta = \theta - \theta_0$ is the temperature difference from the reference temperature $\theta_0$. The last term on the right hand side is taken by analogy to the thermoelasticity with two relaxation times in the next section, so as to retain the Fourier-type heat conduction, but to obey the hyperbolic (finite speed) heat propagation. Also note that $\psi$ does not depend on the strain $\varepsilon_{ij}$, so the resulting fluid does not poses any elasticity. In this section, an overdot denotes a material derivative $\mathscr{D}/\mathscr{D}t$ for absolute tensors (like the temperature gradient and deformation rate) and an Oldroyd derivative for tensor densities (like the heat flux and stress tensor).

The free energy being a potential for quasi-conservative stresses $\sigma_{ij}^{(q)}$ and the entropy $s$, we find

$$\begin{aligned} \sigma_{ij}^{(q)} &= \rho \frac{\partial \psi}{\partial \varepsilon_{ij}} = 0, \\ s = -\rho \frac{\partial \psi}{\partial \theta} &= \frac{C_E}{\theta_0}\vartheta + \frac{C_E}{\theta_0} t_0 \dot{\vartheta}. \end{aligned} \tag{22}$$

The relation $(22)_2$ is immediately identified as the constitutive equation for entropy. In view of the fluid's incompressibility, $\sigma_{ij}^{(q)}$ is taken as the deviatoric part of the quasi-conservative stress tensor, the corresponding spherical part being zero. As always in TIV [1],

$$\sigma_{ij} = \sigma_{ij}^{(q)} + \sigma_{ij}^{(d)}, \quad \beta_{ij}^{(q)} = -\beta_{ij}^{(d)}, \tag{23}$$

where $\sigma_{ij}^{(d)}$ is the dissipative stress. Also, $\beta_{ij}^{(q)}$ is the internal quasi-conservative stress and $\beta_{ij}^{(d)}$ is the internal dissipative stress, the first being conjugate to the internal variable $\alpha_{ij}$ and the second one to its rate $\dot{\alpha}_{ij}$. The fluid under consideration has no elastic response, so $\alpha_{ij} \equiv 0$.

The above Ansatz leads to the Clausius–Duhem inequality in the form

$$\rho\theta s^{*(ir)} = -\frac{q_i\theta,_i}{\theta} + \sigma_{ij}^{(d)} d_{ij} \geq 0, \tag{24}$$

where $\sigma_{ij}^{(d)}$ is the dissipative stress, which is now equal to the total stress $\sigma_{ij}$ in view of $(22)_1$ and $(23)_1$. As discussed in the first two sections, the inequality (24) may spontaneously be violated.

Consistent with (10), we take the specific (per unit mass) dissipation $\phi$ as:

$$\rho\theta s^{*(ir)} = \phi\left(-\frac{\theta,_i}{\theta}, d_{ij}\right). \tag{25}$$

Clearly, the inequality (24) may be stated in terms of the scalar product (6): $\boldsymbol{Y} \cdot \boldsymbol{V} \geq 0$.

Next, for the entropy production rate we adopt the dissipation functional in the space of velocity $\boldsymbol{V} = (-\theta^{-1}\nabla\theta, \boldsymbol{d})$:

$$\rho\theta s^{*(ir)} \equiv \rho\phi(\boldsymbol{V}) = \frac{\kappa}{\theta}\theta,_i\theta,_i + H d_{ij} d_{ij}, \tag{26}$$

where $\kappa$ is the Fourier conductivity and $H$ is the fluid viscosity. This is seen as a special case of (10) with both processes being effectively compound [1]. By thermodynamic orthogonality, (26) yields

$$\begin{aligned} -q_i &= \frac{1}{2}\rho\frac{\partial\phi}{\partial\theta,_i} = \kappa\theta,_i, \\ \sigma_{ij} = \sigma_{ij}^{(d)} &= \frac{1}{2}\rho\frac{\partial\phi}{\partial d_{ij}} = H d_{ij}. \end{aligned} \tag{27}$$

Collecting the three parts of the constitutive law: mechanical, Fourier law, and entropy:

$$\begin{aligned} q_i &= -\kappa\theta,_i, \\ \sigma_{ij} &= H d_{ij}, \\ s &= \frac{C_E}{\theta_0}\left(\vartheta + t_0\dot{\vartheta}\right), \end{aligned} \tag{28}$$

which shows that, while the Fourier-type law holds, there is a relaxation effect involved in the entropy. As a result, there also are violations according to [3], while the heat is conducted with finite speeds – i.e. not infinite speeds as would be the case with $t_0 = 0$. In other words, instead of having a parabolic (diffusion) equation for temperature, we have (by application of the energy balance (20) and the entropy-temperature relation $(28)_3$)

$$\kappa \vartheta_{,ii} = \rho C_E \left( \dot{\vartheta} + t_0 \ddot{\vartheta} \right). \tag{29}$$

Here we have also used the approximation of small temperature fluctuations. In effect, $\theta$ (just like $\vartheta$) is governed by the telegraph (damped hyperbolic) Eq. (29), whose limiting case (for $t_0 \to 0$) is the conventional (parabolic) heat conduction equation.

## 4 Thermoelasticity with Two Relaxation Times

Conventional thermo-elastodynamics is hyperbolic in elastic response and parabolic in heat conduction. The standard way to obtain a purely hyperbolic (and still linear) thermo-elastodynamics is to replace the Fourier law by the Maxwell-Cattaneo law [9]. However, a fluctuation theorem for the latter type of thermal response does not (yet) exist and we need to work with a Fourier-type law. Thus, one may proceed by using the *theory of thermoelasticity with two relaxation times* [10, 11]. While the original derivation of that reference had used the free energy functional only, one may proceed by using a different free energy functional along with a dissipation functional. The approach is similar to that in the preceding section, although we take the heat flux and its rate as the argument of $\phi$. First, we adopt the internal energy $u$ as a function of the infinitesimal elastic strain $\varepsilon_{ij}$ and the entropy $s$

$$u = u(\varepsilon_{ij}, s), \tag{30}$$

along with the (specific, per unit mass) dissipation functional $\phi$ as a function of the strain rate $\dot{\varepsilon}_{ij}$, the heat flux, and its rate

$$\phi = \phi(\dot{\varepsilon}_{ij}, q_i). \tag{31}$$

By the Legendre transformation $\psi = u - \theta s$, we now obtain

$$\dot{\psi} = \dot{u} - \dot{\theta} s - \theta \dot{s}, \tag{32}$$

whereas by the postulate of hyperelasticity for quasi-conservative stress and the entropy

$$\sigma_{ij}^{(q)} = \rho \frac{\partial \psi}{\partial \varepsilon_{ij}} \quad and \quad s = -\rho \frac{\partial \psi}{\partial \theta}. \tag{33}$$

Noting the balance of energy (20), (33) becomes

$$\rho\theta\dot{s} = \sigma_{ij}^{(d)}\dot{\varepsilon}_{ij} - q_{i,i}\,, \tag{34}$$

which, in view of (1), yields the Clausius–Duhem in standard form:

$$\rho\theta s^{*(ir)} = -\frac{q_i\theta_{,i}}{\theta} + \sigma_{ij}^{(d)}\dot{\varepsilon}_{ij} \geq 0. \tag{35}$$

Now, adopt the free energy (with $\vartheta = \theta - \theta_0$ as before)

$$\psi = \psi(\varepsilon_{ij}, \theta) = \frac{1}{2}\varepsilon_{ij}C_{ijkl}\varepsilon_{kl} + M_{ij}\varepsilon_{ij}\vartheta - \frac{C_E}{2\theta_0}\vartheta^2 - \frac{C_E}{\theta_0}t_0\vartheta\dot{\vartheta}, \tag{36}$$

so that

$$\sigma_{ij}^{(q)} = C_{ijkl}\varepsilon_{kl} + M_{ij}\vartheta \quad and \quad s = -M_{ij}\varepsilon_{ij} + \frac{C_E}{\theta_0}\vartheta + \frac{C_E}{\theta_0}t_0\dot{\vartheta}\,. \tag{37}$$

Also, adopt the dissipation function (this time in the space of heat flux and strain rate)

$$\phi(q_i, \dot{\varepsilon}_{ij}) = \rho\theta s^{*(ir)} = \frac{\lambda_{ij}}{\theta}q_iq_j + t_1M_{ij}\dot{\varepsilon}_{ij}\dot{\vartheta}, \tag{38}$$

so that, by treating both processes as compound [1],

$$-\frac{\theta_{,i}}{\theta} = \frac{1}{2}\frac{\partial\phi}{\partial q_i} = \frac{\lambda_{ij}}{\theta}q_j \quad and \quad \sigma_{ij}^{(d)} = \frac{\partial\phi}{\partial\dot{\varepsilon}_{ij}} = t_1M_{ij}\dot{\vartheta}. \tag{39}$$

On account of (23), we obtain

$$\begin{aligned} \sigma_{ij} = \sigma_{ij}^{(q)} + \sigma_{ij}^{(d)} &= C_{ijkl}\varepsilon_{kl} + M_{ij}(\vartheta + t_1\dot{\vartheta}), \\ \theta_0 s &= -\theta_0M_{ij}\varepsilon_{ij} + C_E(\vartheta + t_0\dot{\vartheta}), \\ q_i &= -k_{ij}\vartheta_{,j}\,, \end{aligned} \tag{40}$$

where, again (recalling Sect. 2), $\kappa_{ij} + \mathscr{M}_{ij}$ is a random field of the Fourier thermal conductivity in space-time with spontaneous violations of positive-definiteness property. Note that $k_{ij}$ in $(40)_3$ equals $\kappa_{ij} + \mathscr{M}_{ij}$, anisotropy being possible because we are now dealing with a solid, not fluid. It is well known that, (40) lead to coupled and hyperbolic-type equations for the $(u_i, \vartheta)$ pair

$$\begin{aligned} (C_{ijkl}u_{k,l})_{,j} - \rho\ddot{u}_i + [M_{ij}(\vartheta + t_1\dot{\vartheta})]_{,j} &= -b_i, \\ (k_{ij}\vartheta_{,j})_{,i} - C_E(\dot{\vartheta} + t_0\ddot{\vartheta}) + \theta_0M_{ij}\dot{u}_{i,j} &= -r. \end{aligned} \tag{41}$$

Observe:

(i) The constitutive relations (40) are the same as those of the Green-Lindsay theory, but their derivation is based on treating the Fourier-type heat conduction as a purely dissipative process, and thus as a process described by the dissipation function rather than by the free energy function in [10, 11].

(ii) The inequalities $t_1 \geq t_0 \geq 0$ have to hold. By setting $t_1 = t_0 = 0$, we obtain the classical thermoelasticity. Also, one may only consider the limit $t_0 \to 0$, so that $(41)_2$ reduces to the conventional heat conduction equation.

(iii) Transient phenomena (such as wavefronts), if occurring on very short length scales, are expected to deviate from the hyperbolic thermo-elastodynamics obeying the second law [11].

## 5 Violations of Second Law in Poromechanics

The preceding considerations apply to physics of porous media, in the sense that:

- the nanoscale dimensions of the porous channel network are nanoscale;
- the viscous fluid flow (Poiseuille and Couette type) in the channels violates the second law;
- the temperature field in the fluid occupying the channels violates the second law.

As a reference, in classical poromechanics obeying the second law [12], the Clausius–Duhem inequality is written in terms of irreversible entropy production $S^{*(ir)}$ $(= \rho s^{*(ir)})$ taking the form

$$\theta S^{*(ir)} = \theta S^{*(ir)}_{(th)} + \theta S^{*(ir)}_{(fluid)} + \theta S^{*(ir)}_{(skeleton)} \geq 0, \tag{42}$$

where three possible contributions to dissipation are identified:

1. thermal dissipation: $S^{*(ir)}_{(th)}$;
2. fluid dissipation: $S^{*(ir)}_{(fluid)}$;
3. skeleton dissipation: $S^{*(ir)}_{(skeleton)}$.

Conventionally, each of these contributions to dissipation is assumed to satisfy its own second law inequality. It now follows that, in case of poromechanics describing phenomena on very small space and time scales, the spontaneous violations of the second law can occur in either one or two or three processes, and these can be modeled according to what has been presented in the preceding sections.

## 6 Conclusions

The recent works [4, 13, 14] investigated extensions of continuum thermomechanics to account for spontaneous, random violations of the second law that become relevant on very small length and/or time scales... although in cholesteric liquids the time of such a violation may be up to 3 s. As dictated by modern statistical physics, the second law is then to be replaced by the fluctuation theorem. The particular phenomena and aspects included: Newtonian fluids with either parabolic or hyperbolic heat conduction, random field models including spatial fractal and Hurst effects, acceleration wavefront of nanoscale thickness, Lyapunov function for the heat field, random fluctuations of the microrotation field in a viscous micropolar fluid, Couette flow, and permeability of a medium with nanoscale pores.

This paper outlines possible extensions of stochastic continuum thermomechanics in coupled field problems: (i) thermoviscous fluids, (ii) thermoelastic solids, and (iii) poromechanics with dissipation within the skeleton, the fluid, and the temperature field. Special attention is paid to the fact that one can develop hyperbolic theories (i.e. free of the paradox of infinite speeds of signal transmission) while working with the Fourier-type heat conduction for which the fluctuation theorem has already been developed.

There are various directions in which this research may further be developed, of which we list two. On one hand, the details and extensions of what has been considered here need to be worked. On the other hand, one can start from the so-called Crooks Fluctuation Theorem (CFT) in statistical mechanics [15, 16] that relates the work done on a system during a non-equilibrium transformation to the free energy difference between the final and the initial state of the transformation. In general, the CFT says that, if the dynamics of the system satisfies microscopic reversibility, then the forward space-time trajectory $\boldsymbol{\Gamma}(t)$ is exponentially more likely than the time-reversed trajectory $\tilde{\boldsymbol{\Gamma}}(t)$, given that it produces entropy $s^{*(ir)}$,

$$\frac{\mathsf{P}\left[\boldsymbol{\Gamma}(t)\right]}{\mathsf{P}\left[\tilde{\boldsymbol{\Gamma}}(t)\right]} = e^{\sigma^{(ir)}(\boldsymbol{\Gamma})}. \tag{43}$$

Here $\sigma^{(ir)}(\boldsymbol{\Gamma})$ is the microscopic version of $\Delta S^{(ir)} = \Delta S - Q/\theta = (W - \Delta\Psi)/\theta$ written for the macroscopic system, whereby we also recall (1), the first law of thermodynamics, and the classical relation $\Psi = U - S\theta$, see [17]. The latter reference reviews this and many related issues as well as the fact that the CFT implies the so-called Jarzynski equality [18, 19]

$$\left\langle e^{-W/k_B\theta}\right\rangle = e^{-\Delta\Psi/k_B\theta}, \tag{44}$$

where $k_B$ is the Boltzmann constant and $\theta$ is the initial temperature of the system in the reservoir. One step in the direction of extending the phenomenological non-equilibrium thermodynamics to account for that equality has been taken in [20].

**Acknowledgements** This material is based upon work partially supported by the NSF under grants CMMI-1462749 and IIP-1362146 (I/UCRC on Novel High Voltage/Temperature Materials and Structures).

## References


1. Ziegler, H.: An Introduction to Thermomechanics. North-Holland (1983)
2. Evans, D.J., Searles, D.J.: The fluctuation theorem. Adv. Phys. **51**(7), 1529–1585 (2002)
3. Searles, D.J., Evans, D.J.: Fluctuation theorem for heat flow. Int. J. Thermophys. **22**(1), 123–134 (2001)
4. Ostoja-Starzewski, M., Malyarenko, A.: Continuum mechanics beyond the second law of thermodynamics. Proc. R. Soc. A **470**, 20140531 (2014)
5. Goddard, J.D.: Edelen's dissipation potentials and the visco-plasticity of particulate media. Acta Mech. **225**, 2239–2259 (2014)
6. Goddard, J.D.: Dissipation potentials for reaction-diffusion systems. Ind. Eng. Chem. Res. **54**(16), 4078–4083 (2015)
7. Malyarenko, A., Ostoja-Starzewski, M.: Statistically isotropic tensor random fields: Correlation structures. Math. Mech. Complet. Syst. (MEMOCS) **2**(2), 209–231 (2014)
8. Malyarenko, A., Ostoja-Starzewski, M.: Spectral expansions of homogeneous and isotropic tensor-valued random fields. J. Appl. Math. Phys. (ZAMP) **67**(3), paper 59 (2016)
9. Lord, H.W., Shulman, Y.: A generalized dynamical theory of thermoelasticity. J. Mech. Phys. Solids **15**, 299–309 (1967)
10. Green, A.E., Lindsay, K.A.: Thermoelasticity. J. Elast. **2**(1), 1–7 (1972)
11. Ignaczak, M. Ostoja-Starzewski, M.: Thermoelasticity with Finite Wave Speeds. Oxford University Press, Oxford (2010)
12. Coussy, O.: Poromechanics. Wiley, New Jersey (2004)
13. Ostoja-Starzewski, M.: Second law violations, continuum mechanics, and permeability. Continuum Mech. Thermodyn. **28**, 489–501 (2016). Erratum: doi:10.1007/s00161-016-0534-x
14. Ostoja-Starzewski, M., Raghavan, B.: Continuum mechanics versus violations of the second law of thermodynamics. J. Therm. Stresses **39**(6), 734–749 (2016)
15. Crooks, G.E.: Nonequilibrium measurements of free energy differences for microscopically reversible Markovian systems. J. Stat. Phys. **90**(5/6), 1481–1487 (1998)
16. Crooks, G.E.: Entropy production fluctuation theorem and the nonequilibrium work relation for free energy differences. Phys. Rev. E **60**(3), 2721–2726 (1999)
17. Jarzynski, C.: Equalities and inequalities: Irreversibility and the second law of thermodynamics at the nanoscale. Annu. Rev. Condens. Matter Phys. **2**, 329–351 (2011)
18. Jarzynski, C.: Nonequilibrium equality for free-energy differences. Phys. Rev. Lett. **78**(14), 2690–2693 (1997)
19. Jarzynski, C.: Equilibrium free-energy differences from nonequilibrium measurements: a master-equation approach. Phys. Rev. E **56**(5), 5018–5035 (1997)
20. Muschik, W.: Non-equilibrium equilibrium thermodynamics and stochasticity, a phenomenological look on Jarzynski's equality. Continuum Mech. Thermodyn. **28**(6), 1887–1903 (2016). doi:https://arxiv.org/abs/1603.02135

# An Inverse Method to Get Further Analytical Solutions for a Class of Metamaterials Aimed to Validate Numerical Integrations

**Luca Placidi, Emilio Barchiesi and Antonio Battista**

**Abstract** We consider an isotropic second gradient elastic two-dimensional solid. Besides, we relax the isotropic hypothesis and consider a D4 orthotropic material. The reason for this last choice is that such anisotropy is the most general for pantographic structures, which exhibit attracting mechanical properties. In this paper we analyze the role of the external body double force $m^{ext}$ on the partial differential equations and we subsequently revisit some analytical solutions that have been considered in the literature for identification purposes. The revisited analytical solutions will be employed as well for identification purposes in a further contribution.

## 1 Introduction

Design of metamaterials (see [25, 35] for recent review papers) is nowadays a very important challenge, and new possibility are available due to an enormously increased capacity of big data analysis. In the present introduction we want to frame this problem in the existing literature and offer some prospects on potentially useful tools.

L. Placidi (✉)
Faculty of Engineering, International Telematic University Uninettuno,
C.so Vittorio Emanuele II, 39, 00186 Rome, Italy
e-mail: luca.placidi@uninettunounirsity.net

E. Barchiesi
Dipartimento di Ingegneria Meccanica e Aerospaziale, Universitá di
Roma La Sapienza, Via Eudossiana 18, 00184 Rome, Italy
e-mail: emilo.barchiesi@uniroma1.it

A. Battista
Laboratory of Science for Environmental Engineering, Université de
La Rochelle, La Rochelle, France
e-mail: antonio.battista@univ-lr.fr



F. dell'Isola et al. (eds.), *Mathematical Modelling in Solid Mechanics*,
Advanced Structured Materials 69, DOI 10.1007/978-981-10-3764-1_13

Due to the aforementioned computational power, discrete models are becoming increasingly capable of capturing all the important features of continuum mechalics systems (see [2, 14, 41, 56, 60] for some recent numerical and theoretical result). Besides, metamaterials are designed to perform with expected mechanical properties, and the microstructure can be very complicated and difficult to manage from a numerical point of view [17, 20, 40, 59] due to the high computational demand of structures, especially if one wants to include description of impact-like behaviors [15, 19, 63], instabilities [46, 53, 54, 61, 62, 64, 66] and/or surface effects [3, 26] and damage or plastic behaviour [24, 41, 42]. In fact, the size of the microstructure (for nano-sized objects see for instance [4–7]) can be designed in such a way that the number of cells is very high and the effective geometry of the resulting body is complex, see e.g. the geometry of pantographic structures [33, 36, 67] and of truss structures [1, 69]. The numerical simulation of a 3D body with such a geometry goes usually beyond the standard numerical capabilities if one employs standard 3D Cahcy elasticity models. Thus, the necessity to find new models that are able to deal with complex microstructures from a continuum point of view, via e.g. an homogenization criterion [21, 37, 42, 60], is strongly felt by the scientific community since the time of Piola [31, 34], when numerical simulations could not be performed with the aid of computers. Higher order gradient continua [12, 32] fulfill the above mentioned characteristics and second gradient [29, 30, 47, 55, 65, 68, 75] 2D elastic materials, in general anisotropic [13, 28, 45, 73, 74, 77], is the topic of this contribution. This research field can be seen as a specialization of the well-established field of micromorphic/microstructured continua [48–52, 70], in which the kinematical descriptors added to the Cauchy model are independent of the classic ones. In particular, here we consider the role of distributed double force [27] in some cases in which analytical solutions are possible. In general, analytical solutions are very difficult to achieve, and in some repsects less importnant than in the past due to the new numerical tools; however, they still play the very important role of benchmark in many cases, even including difficult/pathological mechanical systems [16]. In these models new numerical schemes [8, 9, 22, 43, 71] must be conceived with finite elements, and particularly suitable for higher order continua are those allowing a higher degree of continuity [18, 23, 44]. In a further contribution we will use such analytical solutions for purposes of identification [56, 72, 76].

## 2 Formulation of the Problem

### 2.1 *Definition of the Deformation Energy Functional*

$\mathcal{B}$ is a 2-dimensional body that is considered in the reference configuration, where $X$ are the coordinates of its points. $U\,(G, \nabla G)$ is the internal energy density functional that is a function of the deformation matrix $G = \left(F^T F - I\right)/2$ and of its gradient $\nabla G$. Here, $F = \nabla \chi$, where $\chi$ is the placement function, $F^T$ is the transpose of $F$, and $\nabla$ is the gradient operator. The energy functional $\mathscr{E}(u(X))$ depends on the displacement $u = \chi - X$ and possesses two contributions: the internal and the external energies,

$$\mathscr{E}(u(X)) = \int_{\mathscr{B}} \left[U(G, \nabla G) - b^{ext} \cdot u - m^{ext} \cdot \nabla u\right] dA \tag{1}$$
$$- \int_{\partial\mathscr{B}} \left[t^{ext} \cdot u + \tau^{ext} \cdot [(\nabla u)\, n]\right] ds - \int_{[\partial\partial\mathscr{B}]} f^{ext} \cdot u$$

where $n$ is the unit external normal and the dot $\cdot$ indicates the scalar product between vectors or tensors. $b^{ext}$ and $m^{ext}$ are (per unit area) the external bulk force and double force, respectively; $t^{ext}$ and $\tau^{ext}$ are (per unit length) the external force and double force; $f^{ext}$ is the external concentrated force, that is applied on the vertices $[\partial\partial\mathscr{B}]$. In other words, the last integral is the sum of the external works made by the concentrated forces applied at the vertices. Besides,

$$\partial\mathscr{B} = \bigcup_{c=1}^{m} \Sigma_c, \qquad [\partial\partial\mathscr{B}] = \bigcup_{c=1}^{m} \mathscr{V}_c.$$

Thus, the boundary $\partial\mathscr{B}$ is the union of $m$ regular parts $\Sigma_c$ (with $c = 1, \ldots, m$) and the so-called boundary of the boundary$[\partial\partial\mathscr{B}]$ is the union of the corresponding $m$ vertex-points $\mathscr{V}_c$ (with $c = 1, \ldots, m$) with coordinates $X^c$. Finally, for the sake of simplicity, we make explicit that line and vertex-integrals of a generic field $g(X)$ are

$$\int_{\partial\mathscr{B}} g(X)\, ds = \sum_{c=1}^{m} \int_{\Sigma_c} g(X)\, ds, \quad \int_{[\partial\partial\mathscr{B}]} g(X) = \sum_{c=1}^{m} g(X^c). \tag{2}$$

It is worth to be noted that the new contribution of this paper is in the tensor $m^{ext}$, that represents (per unit area) the external body double force. This term will be crucial for the identification that will be analyzed in another paper.

## 2.2 Formulation of the Variational Principle

Assuming that first variation of (1) vanishes (see [57]) yields:

$$\delta\mathscr{E} = -\int_{\mathscr{B}} \delta u_i \left[\left(S_{ij} - T_{ijh,h}\right)_{,j} + b_i^{ext} - m_{ij,j}^{ext}\right] dA$$
$$+ \int_{\partial\mathscr{B}} \left[\delta u_i \left(t_i - t_i^{ext} - m_{ij}^{ext} n_j\right) + \delta u_{i,j} n_j \left(\tau_i - \tau_i^{ext}\right)\right] ds$$
$$+ \int_{[\partial\partial\mathscr{B}]} \delta u_i \left(f_i - f_i^{ext}\right). \tag{3}$$

For the sake of simplicity, we skip to index notations (derivative with respect to $X_j$, that is the $j$-th component of the position $X$, is indicated by the subscript $j$ after comma) and the following positions have been used,

$$t_i = \left(S_{ij} - T_{ijh,h}\right) n_j - P_{ka}\left(T_{ihj} P_{ah} n_j\right)_{,k} \tag{4}$$

$$\tau_i = T_{ijk} n_j n_k \tag{5}$$

$$f_i = T_{ihj} V_{hj} \tag{6}$$

$P$ is the tangential projector operator ($P_{ij} = \delta_{ij} - n_i n_j$), $V$ is the vertex operator

$$V_{hj} = \nu_h^l n_j^l + \nu_h^r n_j^r,$$

where superscripts $l$ and $r$ refer (roughly speaking, left and right), respectively, to one and to the other sides that define a certain vertex-point $\mathscr{V}_c$; $\nu$ is the external tangent unit vector. Stress and hyper stress tensors are,

$$S_{ij} = \frac{\partial U}{\partial G_{ij}}, \qquad T_{ijh} = \frac{\partial U}{\partial G_{ij,h}}. \tag{7}$$

It is worth to be noted from the first addend of (3) that the external part $t^{ext} + m^{ext} n$ of the dual of the virtual displacement is not equal to the external force per unit length $t^{ext}$.

## 2.3 Balance of Forces and Moments

Partial differential equations that govern the deformation process have been derived assuming the arbitrariness of the displacement variation $\delta u_i$ inside the body. The balance of forces and moments, in the present formulation, are obtained by considering the subset of admissible motions constituted by the particular case of rigid motion, which in our case is a superposition of (i) a rigid translation $u_i^0$ and (ii) a rotation, e.g., around the origin and of an arbitrary angle $\theta$,

$$u_i = u_i^0 + \theta \varepsilon_{ijh} \delta_{3j} X_h = u_i^0 - \theta \delta_{1i} X_2 + \theta \delta_{2i} X_1 \implies$$
$$\delta u_i = \delta u_i^0 - \delta\theta \left(\delta_{1i} X_2 - \delta_{2i} X_1\right). \tag{8}$$

With this assumption we have from (11) that $U = 0$, from (4)–(6) $t = 0$, $\tau = 0$ and $f = 0$, respectively, while the variation of the deformation energy functional is

$$0 = -\delta\mathscr{E} = \int_{\mathscr{B}} \delta u_i b_i^{ext} + \int_{\partial\mathscr{B}} \left[\delta u_i t_i^{ext} + \delta u_{i,j} n_j \tau_i^{ext}\right] + \int_{[\partial\partial\mathscr{B}]} \delta u_i f_i^{ext}. \tag{9}$$

Inserting the right-hand side of (8) into the (9) yields

$$0 = -\delta\mathscr{E} = \delta u_i^0 \left\{ \int_{\mathscr{B}} b_i^{ext} + \int_{\partial\mathscr{B}} t_i^{ext} + \int_{[\partial\partial\mathscr{B}]} f_i^{ext} \right\}$$
$$-\delta\theta \left( \int_{\mathscr{B}} X_2 b_1^{ext} - X_1 b_2^{ext} + \int_{\partial\mathscr{B}} \left[ X_2 t_1^{ext} - X_1 t_2^{ext} + n_2 \tau_1^{ext} - n_1 \tau_2^{ext} \right] \right.$$
$$\left. + \int_{[\partial\partial\mathscr{B}]} X_2 f_1^{ext} - X_1 f_2^{ext} \right).$$

Thus, for an arbitrary pure translation ($\delta\theta = 0$) we have the so called balance of forces,

$$\int_{\mathscr{B}} b_\alpha^{ext} + \sum_{c=1}^{m} \int_{\Sigma_c} t_\alpha^{ext} + \sum_{c=1}^{m} f_\alpha^{ext} \left( X_i^c \right) = 0, \tag{10}$$

and for an arbitrary pure rotation ($\delta u_\alpha^0 = 0$) we have the so called balance of moments,

$$\int_{\mathscr{B}} X_2 b_1^{ext} - X_1 b_2^{ext} + \sum_{c=1}^{m} \int_{\Sigma_c} \left[ X_2 t_1^{ext} - X_1 t_2^{ext} + n_2 \tau_1^{ext} - n_1 \tau_2^{ext} \right]$$
$$+ \sum_{c=1}^{m} \left( X_2^c f_1^{ext} - X_1^c f_2^{ext} \right) = 0$$

where we have used the definitions given in Eq. (2).

## 2.4 The Deformation Energy Functional for 2D Isotropic and for D4 Anisotropic Linear Second Gradient Elasticity

In [47, 58] a general form of the density of the deformation energy functional $U$ of a linear isotropic and in [10, 11, 13, 56] for $D4$ anisotropic second gradient elastic material is given. For the isotropic case it is already proved that

$$U\left(G_{ij}, G_{ij,h}\right) = \tilde{U}\left(u_i\right) = (\lambda + 2\mu)\left(u_{1,1}^2 + u_{2,2}^2\right) + \mu\left(u_{1,2}^2 + u_{2,1}^2\right) \tag{11}$$
$$+2\lambda u_{1,1} u_{2,2} + 2\mu u_{1,2} u_{2,1} + \frac{1}{2} A \left(u_{1,22}^2 + u_{2,11}^2\right) + \frac{1}{2} B \left(u_{1,11}^2 + u_{2,22}^2\right)$$
$$+C\left(u_{1,12}^2 + u_{2,12}^2\right) + 2D\left(u_{1,11} u_{2,12} + u_{2,22} u_{1,12}\right)$$
$$+\frac{1}{2}(A + B - 2C)\left(u_{1,11} u_{1,22} + u_{2,11} u_{2,22}\right)$$
$$+(B - A - 2D)\left(u_{1,12} u_{2,11} + u_{1,22} u_{2,12}\right),$$

where $\lambda$ and $\mu$ are the Lamé's coefficients (see also [38, 39]) and $A$, $B$, $C$ and $D$ are 4 second gradient constitutive parameters.

In the general linear anisotropic case we have

$$U\left(G,\nabla G\right)=\hat{U}\left(\varepsilon,\eta\right)=\frac{1}{2}C_{IJ}\varepsilon_I\varepsilon_J+\frac{1}{2}A_{\alpha\beta}\eta_\alpha\eta_\beta \tag{12}$$

where the indices $I$ and $J$ vary from 1 to 3, the indices $\alpha$ and $\beta$ vary from 1 to 6, $\varepsilon_I$ is the $I$-th component of the column-vector $\varepsilon$

$$\varepsilon=\begin{pmatrix} G_{11}\\ G_{22}\\ \sqrt{2}G_{12}\end{pmatrix}, \tag{13}$$

$\eta_\alpha$ is the $\alpha$-th component of the column-vector $\eta$

$$\eta=\begin{pmatrix} G_{11,1}\\ G_{22,1}\\ \sqrt{2}G_{12,2}\\ G_{22,2}\\ G_{11,2}\\ \sqrt{2}G_{12,1}\end{pmatrix}, \tag{14}$$

$C_{IJ}$ is the $IJ$-th component of the $3\times3$ matrix $C$, that for isotropic case is

$$C^{ISO}=\begin{pmatrix} c_{11} & c_{12} & 0\\ c_{12} & c_{11} & 0\\ 0 & 0 & c_{11}-c_{12}\end{pmatrix} \tag{15}$$

and for $D_4$ anisotropic case is

$$C^{D_4}=\begin{pmatrix} c_{11} & c_{12} & 0\\ c_{12} & c_{11} & 0\\ 0 & 0 & c_{33}\end{pmatrix}, \tag{16}$$

$A_{\alpha\beta}$ is the $\alpha\beta$-th component of the matrix $A$, that for isotropic case is

$$A^{ISO}=\begin{pmatrix} a_{11} & a_{12} & \frac{a_{11}-a_{22}}{\sqrt{2}}-a_{23} & 0 & 0 & 0\\ a_{12} & a_{22} & a_{23} & 0 & 0 & 0\\ \frac{a_{11}-a_{22}}{\sqrt{2}}-a_{23} & a_{23} & \frac{a_{11}+a_{22}}{2}-a_{12} & 0 & 0 & 0\\ 0 & 0 & 0 & a_{11} & a_{12} & \frac{a_{11}-a_{22}}{\sqrt{2}}-a_{23}\\ 0 & 0 & 0 & a_{12} & a_{22} & a_{23}\\ 0 & 0 & 0 & \frac{a_{11}-a_{22}}{\sqrt{2}}-a_{23} & a_{23} & \frac{a_{11}+a_{22}}{2}-a_{12}\end{pmatrix} \tag{17}$$

and for $D_4$ anisotropic case is

$$A^{D_4} = \begin{pmatrix} a_{11} & a_{12} & a_{13} & 0 & 0 & 0 \\ a_{12} & a_{22} & a_{23} & 0 & 0 & 0 \\ a_{13} & a_{23} & a_{33} & 0 & 0 & 0 \\ 0 & 0 & 0 & a_{11} & a_{12} & a_{13} \\ 0 & 0 & 0 & a_{12} & a_{22} & a_{23} \\ 0 & 0 & 0 & a_{13} & a_{23} & a_{33} \end{pmatrix}, \tag{18}$$

With (11), or with the insertion of (15) and (17) into (12) the system of partial differential equations that can be extrapolated by the first line of (3) is calculated for the isotropic linear case,

$$\begin{aligned} & u_{1,11} (\lambda + 2\mu) + u_{1,22}\mu + u_{2,12} (\lambda + \mu) = \\ & = u_{1,1111}B + u_{1,2222}A + u_{1,1122} (A + B) \\ & + \left(u_{2,1222} + u_{2,1112}\right) (B - A) - b_1^{ext} + m_{11,1}^{ext} + m_{12,2}^{ext}. \end{aligned} \tag{19}$$

We remark that an interchange of the indices 1 and 2 in the displacement field $u_i$ and in its derivatives, in the external force per unit area $b_i^{ext}$ and in the external double force per unit area $m_{ij}^{ext}$ and in its derivatives (not in the indices of the constitutive coefficients of Eqs. (15) and (17)) gives the second partial differential equation of such a system. The system of PDEs for anisotropic $D_4$ elastic second gradient materials has been deduced again by the first line of (3) and by insertion of (16) and (18) into (12). Here, it is made explicit the first partial differential equation,

$$\begin{aligned} & c_{11}u_{1,11} + \frac{1}{2}c_{33} \left(u_{1,22} + u_{2,12}\right) + c_{12}u_{2,12} = \\ & = a_{11}u_{1,1111} + \sqrt{2} \left(a_{13} + a_{23}\right) \left(\frac{1}{2}u_{2,1222} + \frac{1}{2}u_{2,1112} + u_{1,1122}\right) \\ & + a_{22}u_{1,1122} + a_{12} \left(u_{2,1222} + u_{2,1112}\right) + a_{33} \left(u_{1,2222} + u_{1,1122} + u_{2,1222} + u_{2,1112}\right) \\ & - b_1^{ext} + m_{11,1}^{ext} + m_{12,2}^{ext} \end{aligned} \tag{20}$$

and again we remark that an interchange of the indices 1 and 2 in the displacement field $u$ and in its derivatives, in the external force per unit area $b^{ext}$ and in the external double force per unit area $m^{ext}$ and in its derivatives in (20) (not in the indices of the constitutive coefficients of Eqs. (16) and (18)), because of the symmetry $D_4$, gives the other partial differential equation of such a system.

## 3 The Case of a Rectangle

In this Section we define the case of a body of rectangular shape. The reason of this choice is twofold. First, all boundaries are straight. This means that the external normals do not depend on the space-coordinate $X$, and therefore, see e.g. Eq. (4), the boundary conditions are simplified. Second, the presence of vertices implies an

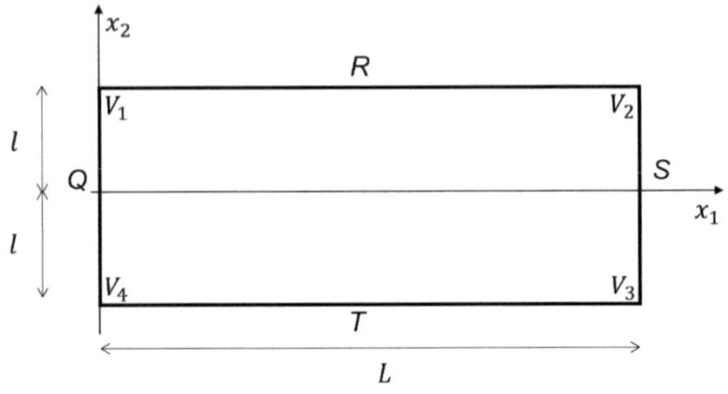


**Fig. 1** Nomenclature of the 2-dimensional body $\mathscr{B}$

increasing number of possible coefficient identification. The reason is that vertex-boundary conditions, as we will see, must be considered.

## 3.1 The General Framework of Straight Lines

In Fig. 1 the scheme of a rectangle is represented. Side-names are $Q$, $R$, $S$ and $T$ and vertex-names $\mathscr{V}_1$, $\mathscr{V}_2$, $\mathscr{V}_3$ and $\mathscr{V}_4$. In these hypotheses from (4) we have

$$\begin{aligned} t_i = \left(S_{ij} - T_{ijh,h}\right) n_j - P_{ka}\left(T_{ihj}P_{ah}n_j\right)_{,k} = \left(S_{ij} - T_{ijh,h}\right) n_j - P_{ka}\left(T_{ihj,k}P_{ah}n_j\right) \\ = \left(S_{ij} - T_{ijh,h}\right) n_j - P_{kh}\left(T_{ihj,k}n_j\right) = \left(S_{ij} - T_{ijh,h}\right) n_j - \left(\delta_{hk} - n_h n_k\right)\left(T_{ihj,k}n_j\right) \\ = \left(S_{ij} - T_{ijh,h}\right) n_j - \left(\delta_{hk}\right)\left(T_{ihj,k}n_j\right) + \left(n_h n_k\right)\left(T_{ihj,k}n_j\right) \\ = \left(S_{ij} - T_{ijh,h}\right) n_j - \left(T_{ihj,h}n_j\right) + \left(n_h n_k\right)\left(T_{ihj,k}n_j\right) \\ = S_{ij}n_j - \left(T_{ijh,h} + T_{ihj,h}\right) n_j + T_{ihj,k}n_h n_k n_j \end{aligned}$$

so that from (4)–(6) are simplified,

$$t_\alpha = S_{\alpha j}n_j - \left(T_{\alpha jh,h} + T_{\alpha hj,h}\right) n_j + T_{\alpha hj,k}n_h n_k n_j, \tag{21}$$

$$\tau_\alpha = T_{\alpha jk}n_j n_k, \quad f_\alpha = T_{\alpha ij}V_{ij}, \tag{22}$$

and we have an explicit form for $t_1$, $t_2$, $\tau_1$ and $\tau_2$ for both isotropic and anisotropic case for each of the four sides of the rectangle (next subsection) and for $f_1$ and $f_2$ for each of the four vertices (subsequent subsection).

### 3.1.1 Characterization of Sides

The characterization of side $S$ is done by setting $n_i = \delta_{i1}$. Thus, from (21) with $\alpha = 1, 2$, and from $(22)_1$ with $\alpha = 1, 2$, we have

$$t_1 = t_1^S = S_{11} - T_{112,2} - T_{111,1} - T_{121,2} \tag{23}$$
$$t_2 = t_2^S = S_{21} - T_{211,1} - T_{212,2} - T_{221,2}, \tag{24}$$
$$\tau_1 = \tau_1^S = T_{111}, \tag{25}$$
$$\tau_2 = \tau_2^S = T_{211}. \tag{26}$$

The characterization of side $Q$ is done by setting $n_i = -\delta_{i1}$. Thus, from (21) with $\alpha = 1, 2$, and from $(22)_1$ with $\alpha = 1, 2$, we have

$$t_1 = t_1^Q = -S_{11} + T_{112,2} + T_{111,1} + T_{121,2} \tag{27}$$
$$t_2 = t_2^Q = -S_{21} + T_{211,1} + T_{212,2} + T_{221,2}, \tag{28}$$
$$\tau_1 = \tau_1^Q = T_{111} \tag{29}$$
$$\tau_2 = \tau_2^Q = T_{211}. \tag{30}$$

We remark that, because of symmetry reasons, $t_1^Q$ in (27) and $t_2^Q$ in (28) are the opposite of $t_1^S$ in (23) and of $t_2^S$ in (24), respectively, and that $\tau_1^Q$ in (29) and $\tau_2^Q$ in (30) are the same of $\tau_1^S$ in (25) and of $\tau_2^S$ in (26), respectively.

The characterization of side $R$ is done by setting $n_i = \delta_{i2}$. Thus, from (21) with $\alpha = 1, 2$, and from $(22)_1$ with $\alpha = 1, 2$, we have

$$t_1 = t_1^R = S_{12} - T_{121,1} - T_{112,1} - T_{122,2}, \tag{31}$$
$$t_2 = t_2^R = S_{22} - T_{221,1} - T_{212,1} - T_{222,2}, \tag{32}$$
$$\tau_1 = \tau_1^R = T_{122}, \tag{33}$$
$$\tau_2 = \tau_2^R = T_{222}. \tag{34}$$

We remark that, because of symmetry reasons, $t_1^R$ in (31) and $t_2^R$ in (32) are the same of $t_2^S$ in (24) and of $t_1^S$ in (23), respectively, by changing the indices 1 and 2. Besides, because of symmetry reasons, $\tau_1^R$ in (33) and $\tau_2^R$ in (34) are the same of $\tau_2^S$ in (25) and of $\tau_1^S$ in (26), respectively, by changing the indices 1 and 2.

Finally, the characterization of side $T$ is done by setting $n_i = -\delta_{i2}$. Thus, from (21) with $\alpha = 1, 2$, and from (22) with $\alpha = 1, 2$, we have

$$t_1 = t_1^T = -S_{12} + T_{121,1} + T_{112,1} + T_{122,2}, \tag{35}$$
$$t_2 = t_2^T = -S_{22} + T_{221,1} + T_{212,1} + T_{222,2}, \tag{36}$$
$$\tau_1 = \tau_1^T = T_{122}, \tag{37}$$
$$\tau_2 = \tau_2^T = T_{222}. \tag{38}$$

We remark that, because of symmetry reasons, $t_1^T$ in (35) and $t_2^T$ in (36) are the opposite of $t_1^R$ in (31) and of $t_2^R$ in (32), respectively, and that $\tau_1^T$ in (37) and $\tau_2^T$ in (38) are the same of $\tau_1^R$ in (33) and of $\tau_2^R$ in (34), respectively.

Characterization of sides is concluded by insertion of (12) into (7) with the restrictions due to the given anisotropy.

### 3.2 Characterization of Vertices

The last term of (3) is reduced, because of $(2)_2$, to

$$\int_{[\partial\partial\mathscr{B}]} \delta u_\alpha \left(f_\alpha - f_\alpha^{ext}\right) = \\ = \left[\delta u_\alpha \left(T_{\alpha ij} V_{ij} - f_\alpha^{ext}\right)\right]_{\mathscr{V}_1} + \left[\delta u_\alpha \left(T_{\alpha ij} V_{ij} - f_\alpha^{ext}\right)\right]_{\mathscr{V}_2} \\ + \left[\delta u_\alpha \left(T_{\alpha ij} V_{ij} - f_\alpha^{ext}\right)\right]_{\mathscr{V}_3} + \left[\delta u_\alpha \left(T_{\alpha ij} V_{ij} - f_\alpha^{ext}\right)\right]_{\mathscr{V}_4}. \tag{39}$$

For vertex $\mathscr{V}_1$ the side $Q$ has $n_j = -\delta_{1j}$ and $\nu_i = \delta_{i2}$ and the side $R$ has $n_j = \delta_{2j}$ and $\nu_i = -\delta_{i1}$,

$$\left(V_{ij}\right)_{\mathscr{V}_1} = \nu_i^l n_j^l + \nu_i^r n_j^r = -\delta_{i2}\delta_{1j} - \delta_{i1}\delta_{2j}.$$

For vertex $\mathscr{V}_2$ the side $R$ has $n_j = \delta_{2j}$ and $\nu_i = \delta_{i1}$ and the side $S$ has $n_j = \delta_{1j}$ and $\nu_i = \delta_{i2}$,

$$\left(V_{ij}\right)_{\mathscr{V}_2} = \nu_i^l n_j^l + \nu_i^r n_j^r = \delta_{i1}\delta_{j2} + \delta_{i2}\delta_{1j} = -\left(V_{ij}\right)_{\mathscr{V}_1}.$$

For vertex $\mathscr{V}_3$ the side $S$ has $n_j = \delta_{1j}$ and $\nu_i = -\delta_{i2}$ and the side $T$ has $n_j = -\delta_{2j}$ and $\nu_i = \delta_{i1}$,

$$\left(V_{ij}\right)_{\mathscr{V}_3} = \nu_i^l n_j^l + \nu_i^r n_j^r = -\delta_{i2}\delta_{1j} - \delta_{i1}\delta_{2j} = \left(V_{ij}\right)_{\mathscr{V}_1}.$$

For vertex $\mathscr{V}_4$ the side $T$ has $n_j = -\delta_{2j}$ and $\nu_i = -\delta_{i1}$ and the side $Q$ has $n_j = -\delta_{1j}$ and $\nu_i = -\delta_{i2}$

$$\left(V_{ij}\right)_{\mathscr{V}_4} = \nu_i^l n_j^l + \nu_i^r n_j^r = \delta_{i1}\delta_{j2} + \delta_{i2}\delta_{1j} = -\left(V_{ij}\right)_{\mathscr{V}_1}.$$

Thus, finally, (39) yields

$$\int_{[\partial\partial\mathscr{B}]} \delta u_\alpha \left(f_\alpha - f_\alpha^{ext}\right) = \left[\delta u_\alpha \left(-T_{\alpha 21} - T_{\alpha 12} - f_\alpha^{ext}\right)\right]_{\mathscr{V}_1} \\ + \left[\delta u_\alpha \left(T_{\alpha 12} + T_{\alpha 21} - f_\alpha^{ext}\right)\right]_{\mathscr{V}_2} \\ + \left[\delta u_\alpha \left(-T_{\alpha 21} - T_{\alpha 12} - f_\alpha^{ext}\right)\right]_{\mathscr{V}_3} + \left[\delta u_\alpha \left(T_{\alpha 12} + T_{\alpha 21} - f_\alpha^{ext}\right)\right]_{\mathscr{V}_4}, \tag{40}$$

where $T_{\alpha 12} + T_{\alpha 21}$, in terms of the displacement field, is for $\alpha = 1$

$$T_{112} + T_{121} = 2Cu_{1,12} + (B - A - 2D)\, u_{2,11} + 2Du_{2,22}, \tag{41}$$

for the isotropic case and

$$T_{112} + T_{121} = \left( a_{22} + \sqrt{2}a_{23} + \frac{1}{2}a_{33} \right) u_{1,12}$$
$$+ \left( \frac{\sqrt{2}}{2}a_{23} + \frac{1}{2}a_{33} \right) u_{2,11} + \left( a_{12} + \frac{\sqrt{2}}{2}a_{13} \right) u_{2,22}, \quad (42)$$

for the $D_4$ anisotropic case. We again remark, because of the symmetry $D_4$ (or because of isotropy), an interchange of the indices of the displacement field $u$ and in its derivatives gives the other combination $T_{\alpha 12} + T_{\alpha 21}$ for $\alpha = 2$.

## 4 Analytical Solutions

In the following we revisit some analytical solutions for the above introduced second-gradient 2d continuum model which were first considered in [58]. Differently from the present paper (see Eq. (1)), in [58] no bulk double force per unit area was considered and, hence, from the very beginning $m^{ext}$ could be regarded as set to zero. In [58], for each imposed analytical solution, a compatible external bulk distributed action $b^{ext}$, i.e. satisfying partial differential equations resulting from the stationarity of the energy functional after having plugged the imposed analytical solution in terms of the displacement field, was chosen in view of the coefficients identification step. This resulted in the coefficient $c_{33}$, appearing in (16), to be vanishing in [56]. Hence, in the present extended theory we shall revisit the analytical solutions considered in [58] by choosing compatible combinations of the external distributed actions $b^{ext}$ and $m^{ext}$ (being this time possibly non-zero) in view of an identification step which will lead in a further contribution to a non-vanishing coefficient $c_{33}$. The material in this Section will be therefore instrumental for another paper dealing with the identification of coefficients $a_{11}, a_{12}, a_{13}, a_{22}, a_{23}, a_{33}, c_{11}, c_{12}, c_{33}$ in an analogous fashion to the one of [56].

### *4.1 Analytical Solutions for the Isotropic Case*

#### 4.1.1 Isotropic Heavy Sheet

The rectangle of Fig. 1 is now considered heavy (an heavy sheet) and hanged by the top side $R$. The word heavy corresponds to a weight loading, i.e. a constant distributed force in the vertical direction and directed downwards. Besides, the top-side of the rectangle can not displace vertically and both the vertical left- and right-hand sides can not displace horizontally. Let us consider the following solution, i.e. the exact solution for the first gradient case, in terms of the displacement field,

$$u_1 = 0, \qquad u_2 = \frac{\rho g \left(X_2 - l\right)\left(3l + X_2\right)}{2\left(\lambda + 2\mu\right)}, \tag{43}$$

The two partial differential equations from (19) are satisfied with

$$-b_1^{ext} + m_{11,1}^{ext} + m_{12,2}^{ext} = 0 \quad -b_2^{ext} + m_{21,1}^{ext} + m_{22,2}^{ext} = \rho g \tag{44}$$

that are compatible with a zero external double force per unit area

$$m_{ij}^{ext} = 0 \quad \forall i, j = 1, 2 \tag{45}$$

and an external force per unit area,

$$b_1^{ext} = 0, \quad b_2^{ext} = -\rho g, \tag{46}$$

that is due to the weight.

Let us remark that we could conceive different combinations of the external distributed actions $b^{ext}$ and $m^{ext}$ that are compatible with the imposed solution in (52).

### 4.1.2 Bending for Isotropic Case

Let us take into account the following displacement field,

$$u_1 = \frac{3M^{ext}\left(\lambda + 2\mu\right)X_1 X_2}{8l^3\mu\left(\lambda + \mu\right)}, \qquad u_2 = -\frac{3M^{ext}\left[\lambda X_2^2 + \left(\lambda + 2\mu\right)X_1^2\right]}{16l^3\mu\left(\lambda + \mu\right)}, \tag{47}$$

that is the exact bending solution for the first gradient case. The two partial differential equations from (19) are satisfied with

$$-b_1^{ext} + m_{11,1}^{ext} + m_{12,2}^{ext} = 0 \quad -b_2^{ext} + m_{21,1}^{ext} + m_{22,2}^{ext} = 0 \tag{48}$$

that are compatible with a zero external double force per unit area (45) and a zero external force per unit area,

$$b_1^{ext} = 0, \quad b_2^{ext} = 0. \tag{49}$$

### 4.1.3 Flexure for Isotropic Case

Let us take into account the following displacement field,

$$u_1 = -\frac{QX_2\left[(\lambda+2\mu)\left(3X_1^2 - X_2^2 - 6LX_1\right) + 2(\lambda+\mu)\left(6l^2 - X_2^2\right)\right]}{16l^3\mu(\lambda+\mu)}, \tag{50}$$

$$u_2 = -\frac{Q\left[(3L - X_1)(\lambda+2\mu)X_1^2 + 3(L - X_1)\lambda X_2^2\right]}{16l^3\mu(\lambda+\mu)}, \tag{51}$$

that is the exact flexure solution for the first gradient case. The two partial differential equations from (19) are satisfied with (48), that are compatible with a zero external double force per unit area (45) and with a zero external force per unit area (49).

## 4.2 Analytical Solutions for the $D_4$ anisotropic case

### 4.2.1 D4 Anisotropic Heavy Sheet

The heavy sheet case of Sect. 4.1.1 is now considered for the $D_4$ anisotropic case. The following displacement field is considered,

$$u_1 = 0, \qquad u_2 = \frac{\rho g (X_2 - l)(3l + X_2)}{2c_{11}}, \tag{52}$$

that is again the exact solution for the first gradient ($D_4$ anisotropy) case. The two partial differential equations from (20) are satisfied with (44) that are compatible with a zero external double force per unit area (45) and an external force per unit area (46), that is due to the weight.

Therefore, the two partial differential equations from (20) are satisfied with

$$b_1^{ext} - m_{11,1}^{ext} - m_{12,2}^{ext} = 0, \quad \rho g + b_2^{ext} - m_{21,1}^{ext} - m_{22,2}^{ext} = 0 \tag{53}$$

that are compatible, among other choices, with both a zero external double force per unit area (45) and with the following external force per unit area,

$$b_1^{ext} = 0, \quad b_2^{ext} = -\rho g,$$

that is the choice of [56].

### 4.2.2 D4 Anisotropic Non-conventional Bending

Let us take into account the following displacement field,

$$u_1 = 0, \qquad u_2 = -\frac{aX_1^2}{2}, \tag{54}$$

that represent a non-conventional bending field. Let us now explain why do we call such kind of deformation a non-conventional bending. It is true, in fact, that, for each horizontal material lines of the domain, the bending deformation that is achieved is conventional for that line. However, at the $2D$ domain-scale the deformation is completely different. For example the direction of each vertical material lines remains invariant, i.e. the vertical fibers do not rotate at all, and does not follow the rotation of the horizontal fibers as in the classical conventional bending case, where both horizontal and vertical fibers remain orthogonal to each other.

Therefore, the two partial differential equations from (20) are satisfied with

$$b_1^{ext} - m_{11,1}^{ext} - m_{12,2}^{ext} = 0 \quad b_2^{ext} - m_{21,1}^{ext} - m_{22,2}^{ext} = \frac{a}{2}c_{33} \tag{55}$$

that are compatible, among other choices, with both a zero external double force per unit area (45) and with the following external force per unit area,

$$b_1^{ext} = 0, \quad b_2^{ext} = -ac_{33},$$

that is the choice of [56] or with a non-zero non-homogeneous external double force per unit area

$$m_{11}^{ext} = 0, \quad m_{12}^{ext} = -\frac{a}{2}c_{33}X_1, \quad m_{21}^{ext} = -\frac{a}{2}c_{33}X_1, \quad m_{22}^{ext} = 0, \tag{56}$$

and with a zero external force per unit area (49). In the next paper we will see that this choice is the most convenient for certain identification purposes.

## 4.3 D4 Anisotropic Trapezoidal Case

Let us take into account the following displacement field

$$u_1 = 0, \qquad u_2 = bX_1X_2. \tag{57}$$

The two partial differential equations from (20) are satisfied with

$$-b_1^{ext} + m_{11,1}^{ext} + m_{12,2}^{ext} = b\left(c_{12} + \frac{1}{2}c_{33}\right) \quad -b_2^{ext} + m_{21,1}^{ext} + m_{22,2}^{ext} = 0 \tag{58}$$

that are compatible, e.g., with a zero external force per unit area (49) and with the following non-zero non-homogeneous external double force per unit area,

$$m_{11}^{ext} = 0, \quad m_{12}^{ext} = b\left(c_{12} + \frac{1}{2}c_{33}\right)X_2, \quad m_{21}^{ext} = b\left(c_{12} + \frac{1}{2}c_{33}\right)X_2, \quad m_{22}^{ext} = 0.$$

## 5 Conclusion

An inverse method has been used to obtain analytical solutions that are not present in the literature. Such solutions have been obtained for a class of metamaterials, i.e. for isotropic and for anisotropic D4 second gradient elastic 2D bodies. They will be used not only to get new results in terms of identifications but also to validate numerical integrations. Specifically, in this paper, the role of external bulk distributed double force has been analyzed.

## References

1. Alibert, J.-J., Seppecher, P., dell'Isola, F.: Truss modular beams with deformation energy depending on higher displacement gradients. Math. Mech. Solids **8**(1), 51–73 (2003)
2. Alibert, J.-J., Della, A.: Corte, Second-gradient continua as homogenized limit of pantographic microstructured plates: a rigorous proof. Zeitschrift fur angewandte Mathematik und Physik **66**(5), 2855–2870 (2015)
3. Altenbach, H., Eremeev, V.A., Morozov, N.F.: On equations of the linear theory of shells with surface stresses taken into account. Mech. Solids **45**(3), 331–342 (2010)
4. Aminpour, H., Rizzi, N., Salerno, G.: A one-dimensional beam model for single-wall carbon nano tube column buckling. Civil-Comp Proc. **106** (2014)
5. Aminpour, H., Rizzi, N.: A one-dimensional continuum with microstructure for single-wall carbon nanotubes bifurcation analysis. Math. Mech. Solids **21**(2), 168–181 (2016)
6. Aminpour, H., Rizzi, N.: On the continuum modelling of carbon nano tubes. Civil-Comp Proc. **108** (2015)
7. Aminpour, H., Rizzi, N.: On the modelling of carbon nano tubes as generalized continua. Adv. Struct. Mater. **42**(1), 15–35 (2016)
8. Assante, D., C. Cesarano, C.: Simple semi-analytical expression of the lightning base current in the frequency-domain. J. Eng. Sci. Technol. Rev. **7**(2), 1–6 (2014)
9. Atluri, S.N., Cazzani, A.: Rotations in computational solid mechanics. Arch. Comput. Methods Eng. **2**(1), 49–138 (1995)
10. Auffray, N., Bouchet, R., Brechet, Y.: Derivation of anisotropic matrix for bi-dimensional strain-gradient elasticity behavior. Int. J. Solids Struct. **46**(2), 440–454 (2009)
11. Auffray, N., Kolev, B., Petitot, M.: On anisotropic polynomial relations for the elasticity tensor. J. Elast. **115**(1), 77–103 (2014)
12. Auffray, N., dell'Isola, F., Eremeyev, V.A., Madeo, A., Rosi, G.: Analytical continuum mechanics à la Hamilton-Piola least action principle for second gradient continua and capillary fluids. Math. Mech. Solids **20**(4), 375–417 (2015)
13. Auffray, N., Dirrenberger, J., Rosi, G.: A complete description of bi-dimensional anisotropic strain-gradient elasticity, Submitted to International Journal of Solids and Structures (2017, in press). doi:10.1016/j.ijsolstr.2015.04.036
14. Baraldi, D., Reccia, E., Cazzani, A., Cecchi, A.: Comparative analysis of numerical discrete and finite element models: the case of in-plane loaded periodic brickwork. Compos.: Mech. Comput. Appl. **4**(4), 319–344 (2013)
15. Bersani, A.M., Giorgio, I., Tomassetti, G.: Buckling of an elastic hemispherical shell with an obstacle. Contin. Mech. Thermodyn. **25**(2–4), 443–467 (2013)
16. Bersani, A.M., Della Corte, A., Piccardo, G., Rizzi, N.L.: An explicit solution for the dynamics of a taut string of finite length carrying a traveling mass: the subsonic case. Zeitschrift für angewandte Mathematik und Physik **67**(4), 108 (2016)

17. Bilotta, A., Turco, E.: A numerical study on the solution of the Cauchy problem in elasticity. Int. J. Solids Struct. **46**(25–26), 4451–4477 (2009)
18. Bilotta, A., Formica, G., Turco, E.: Performance of a high-continuity finite element in three-dimensional elasticity. Int. J. Numer. Methods Biomed. Eng. **26**(9), 1155–1175 (2010)
19. Carcaterra, A., Roveri, N.: Energy distribution in impulsively excited structures. Shock Vib. **19**(5), 1143–1163 (2012)
20. Cazzani, A., Ruge, P.: Numerical aspects of coupling strongly frequency-dependent soil-foundation models with structural finite elements in the time-domain. Soil Dyn. Earthq. Eng. **37**, 56–72 (2012)
21. Cecchi, A., Rizzi, N.L.: Heterogeneous elastic solids: a mixed homogenization-rigidification technique. Int. J. Solids Struct. **38**(1), 29–36 (2001)
22. Cesarano, C., Assante, D.: A note on generalized Bessel functions. Int. J. Math. Models Methods Appl. Sci. **8**(1), 38–42 (2014)
23. Cuomo, M., Contrafatto, L., Greco, L.: A variational model based on isogeometric interpolation for the analysis of cracked bodies. Int. J. Eng. Sci. **80**, 173–188 (2014)
24. D'Annibale, F., Luongo, A.: A damage constitutive model for sliding friction coupled to wear. Contin. Mech. Thermodyn. **25**(2–4), 503–522 (2013)
25. Del Vescovo, D., Giorgio, I.: Dynamic problems for metamaterials: Review of existing models and ideas for further research. Int. J. Eng. Sci. **80**, 153–172 (2014)
26. Dell'Isola, F., Rotoli, G.: Validity of Laplace formula and dependence of surface tension on curvature in second gradient fluids. Mech. Res. Commun. **22**(5), 485–490 (1995)
27. Dell'Isola, F., Seppecher, P.: The relationship between edge contact forces, double forces and interstitial working allowed by the principle of virtual power. Comptes Rendus de l'Academie de Sciences, Serie IIb: Mecanique, Physique, Chimie, Astronomie **321**, 303–308 (1995)
28. Dell'Isola, F., Steigmann, D.: A two-dimensional gradient-elasticity theory for woven fabrics. J. Elast. **118**(1), 113–125 (2015)
29. Dell'Isola, F., Gouin, H., Seppecher, P.: Radius and surface tension of microscopic bubbles by second gradient theory. Comptes Rendus de l'Academie de Sciences - Serie IIb: Mecanique, Physique, Chimie, Astronomie **320**(6), 211–216 (1995)
30. Dell'Isola, F., Gouin, H., Rotoli, G.: Nucleation of spherical shell-like interfaces by second gradient theory: numerical simulations. Eur. J. Mech. B/Fluids **15**(4), 545–568 (1996)
31. dell'Isola, F., Andreaus, U., Placidi, L.: At the origins and in the vanguard of peri-dynamics, non-local and higher gradient continuum mechanics. An underestimated and still topical contribution of Gabrio Piola. Mech. Math. Solids (MMS) **20**, 887–928 (2015)
32. Dell'Isola, F., Seppecher, P., Della Corte, A.: The postulations à la D'Alembert and à la Cauchy for higher gradient continuum theories are equivalent: a review of existing results. Proc. R. Soc. Lond. A **471**(2183), 20150415 (2015)
33. Dell'Isola, F., Della Corte, A., Giorgio, I., Scerrato, D.: Pantographic 2D sheets: Discussion of some numerical investigations and potential applications. Int. J. Non-Linear Mech. (2015)
34. Dell'Isola, F., Della Corte, A., Esposito, R., Russo, L.: Some cases of unrecognized transmission of scientific knowledge: from antiquity to gabrio piola's peridynamics and generalized continuum theories. In: Generalized Continua as Models for Classical and Advanced Materials, pp. 77–128. Springer International Publishing (2016)
35. Dell'Isola, F., Steigmann, D., Della Corte, A.: Synthesis of fibrous complex structures: designing microstructure to deliver targeted macroscale response. Appl. Mech. Rev. **67**(6), 21 p. (2016)
36. Dell'Isola, F., Della Corte, A. Greco, L., Luongo, A.: Plane bias extension test for a continuum with two inextensible families of fibers: a variational treatment with Lagrange Multipliers and a perturbation solution. Int. J. Solids Struct. (2017, in press). doi:10.1016/j.ijsolstr.2015.08.029
37. Dos Reis, F., Ganghoffer, J.F.: Construction of micropolar continua from the asymptotic homogenization of beam lattices. Comput. Struct. **112–113**, 354–363 (2012)
38. Federico, S., Grillo, A., Wittum, G.: Considerations on incompressibility in linear elasticity. Nuovo Cimento C **32C**, 81–87 (2009)

39. Federico, S., Grillo, A., Imatani, S.: The linear elasticity tensor of incompressible materials. Math. Mech. Solids (2014, in press). doi:10.1177/1081286514550576
40. Garusi, E., Tralli, A., Cazzani, A.: An unsymmetric stress formulation for reissner-mindlin plates: A simple and locking-free rectangular element. Int. J. Comput. Eng. Sci. **5**(3), 589–618 (2004)
41. Goda, I., Assidi, M., Belouettar, S., Ganghoffer, J.F.: A micropolar anisotropic constitutive model of cancellous bone from discrete homogenization. J. Mech. Behav. Biomed. Mater. **16**(1), 87–108 (2012)
42. Goda, I., Assidi, M., Ganghoffer, J.F.: A 3D elastic micropolar model of vertebral trabecular bone from lattice homogenization of the bone microstructure. Biomech. Model. Mechanobiol. **13**(1), 53–83 (2014)
43. Greco, L., Cuomo, M.: Consistent tangent operator for an exact Kirchhoff rod model. Contin. Mech. Thermodyn. 1–7 (2014, in press)
44. Greco, L., Cuomo, M.: An implicit G1 multi patch B-spline interpolation for Kirchhoff-Love space rod. Comput. Methods Appl. Mech. Eng. **269**, 173–197 (2014)
45. Indelicato, G., Albano, A.: Symmetry properties of the elastic energy of a woven fabric with bending and twisting resistance. J. Elast. **94**(1), 33–54 (2009)
46. Luongo, A., D'Annibale, F.: Bifurcation analysis of damped visco-elastic planar beams under simultaneous gravitational and follower forces. Int. J. Mod. Phys. B **26**(25), Article number 1246015 (2012)
47. Mindlin, R.D.: Micro-structure in Linear Elasticity, Department of Civil Engineering Columbia University New York 27, New York (1964)
48. Misra, A., Huang, S.: Micromechanical stress-displacement model for rough interfaces: effect of asperity contact orientation on closure and shear behavior. Int. J. Solids Struct. **49**(1), 111–120 (2012)
49. Misra, A., Poorsolhjouy, P.: Micro-macro scale instability in 2D regular granular assemblies. Contin. Mech. Thermodyn. **27**(1–2), 63–82 (2013)
50. Misra, A., Singh, V.: Nonlinear granular micromechanics model for multi-axial rate-dependent behavior. Int. J. Solids Struct. **51**(13), 2272–2282 (2014)
51. Misra, A., Singh, V.: Thermomechanics-based nonlinear rate-dependent coupled damage-plasticity granular micromechanics model. Contin. Mech. Thermodyn. **27**(4), 787–817 (2015)
52. Misra, A., Parthasarathy, R., Singh, V., Spencer, P.: Micro-poromechanics model of fluid-saturated chemically active fibrous media. ZAMM Zeitschrift fur Angewandte Mathematik und Mechanik **95**(2), 215–234 (2015)
53. Nguyen, C.H., Freda, A., Solari, G., Tubino, F.: Aeroelastic instability and wind-excited response of complex lighting poles and antenna masts. Eng. Struct. **85**, 264–276 (2015)
54. Piccardo, G., Pagnini, L.C., Tubino, F.: Some research perspectives in galloping phenomena: critical conditions and post-critical behavior. Contin. Mech. Thermodyn. **27**(1–2), 261–285 (2014)
55. Pideri, C., Pierre Seppecher, P.: A second gradient material resulting from the homogenization of an heterogeneous linear elastic medium. Contin. Mech. Thermodyn. **9**(5), 241–257 (1997)
56. Placidi, L., Andreaus, U., Giorgio, I.: Identification of two-dimensional pantographic structure via a linear D4 orthotropic second gradient elastic model. J. Eng. Math. ISSN: 0022-0833 (2017, in press). doi:10.1007/s10665-016-9856-8
57. Placidi, L., El Dhaba, A.R.: Semi-inverse method á la Saint-Venant for two-dimensional linear isotropic homogeneous second gradient elasticity. Mech. Math. Solids (2017, in press). ISSN: 1081-2865
58. Placidi, L., Andreaus, U., Della Corte, A., Lekszycki, T.: Gedanken experiments for the determination of two-dimensional linear second gradient elasticity coefficients. Zeitschrift für angewandte Mathematik und Physik **66**, 3699–3725 (2015)
59. Presta, F., Hendy, C.R., Turco, E.: Numerical validation of simplified theories for design rules of transversely stiffened plate girders. Struct. Eng. **86**(21), 37–46 (2008)
60. Rahali, Y., Giorgio, I., Ganghoffer, J.F., dell'Isola, F.: Homogenization á la Piola produces second gradient continuum models for linear pantographic lattices. Int. J. Eng. Sci. **97**, 148–172 (2015)

61. Rizzi, N.L., Varano, V.: On the postbuckling analysis of thin-walled frames. In: Proceedings of the 13th International Conference on Civil, Structural and Environmental Engineering Computing. 2011, 14p 13th International Conference on Civil, Structural and Environmental Engineering Computing, CC 2011; Chania, Crete; Greece; 6 September 2011 through 9 September 2011; Code 89029
62. Rizzi, N.L., Varano, V.: The effects of warping on the postbuckling behaviour of thin-walled structures. Thin-Walled Struct. **49**(9), 1091–1097 (2011)
63. Roveri, N., Carcaterra, A., Akay, A.: Vibration absorption using non-dissipative complex attachments with impacts and parametric stiffness. J. Acoust. Soc. Am. **126**(5), 2306–2314 (2009)
64. Ruta, G.C., Varano, V., Pignataro, M., Rizzi, N.L.: A beam model for the flexural–torsional buckling of thin-walled members with some applications. Thin-Walled Struct. **46**(7), 816–822 (2008)
65. Sansour, C., Skatulla, S.: A strain gradient generalized continuum approach for modelling elastic scale effects. Comput. Methods Appl. Mech. Eng. **198**(15), 1401–1412 (2009)
66. Scerrato, D., Giorgio, I., Rizzi, N.L.: Three-dimensional instabilities of pantographic sheets with parabolic lattices: numerical investigations. Zeitschrift fur Angewandte Mathematik und Physik, 67(3), Article number 53, 1 June (2016)
67. Scerrato, D. Zhurba Eremeeva, I.A., Lekszycki, T., Rizzi, N.L.: On the effect of shear stiffness on the plane deformation of linear second gradient pantographic sheets. ZAMM. Zeitschrift fur Angewandte Mathematik und Mechanik. (2016). doi:10.1002/zamm.201600066
68. Selvadurai, A.P.S.: Plane strain problems in second-order elasticity theory. Int. J. Non-Linear Mech. **8**(6), 551–563 (1973)
69. Seppecher, P., Alibert J.-J., Dell'Isola, F.: Linear elastic trusses leading to continua with exotic mechanical interactions. J. Phys.: Conf. Seri. **319**(1), 13 p. (2011)
70. Serpieri, R., Della Corte, A., Travascio, F., Rosati, L.: Variational theories of two-phase continuum poroelastic mixtures: a short survey. In: Generalized Continua as Models for Classical and Advanced Materials, pp. 377–394. Springer International Publishing (2016)
71. Solari, G., Pagnini, L.C., Piccardo, G.: A numerical algorithm for the aerodynamic identification of structures. J. Wind Eng. Ind. Aerodyn. **69–71**, 719–730 (1997)
72. Solari, G., Pagnini, L.C., Piccardo, G.: A numerical algorithm for the aerodynamic identification of structures. J. Wind Eng. Ind. Aerodyn. **69–71**, 719–730 (1997)
73. Steigmann, D.J.: Linear theory for the bending and extension of a thin, residually stressed, fiber-reinforced lamina. Int. J. Eng. Sci. **47**(11–12), 1367–1378 (2009)
74. Steigmann, D.J., Dell'Isola, F.: Mechanical response of fabric sheets to three-dimensional bending, twisting, and stretching. Acta Mechanica Sinica/Lixue Xuebao **31**(3), 373–382 (2015)
75. Terravecchia, S., Panzeca, T., Polizzotto, C.: Strain gradient elasticity within the symmetric BEM formulation. Fract. Struct. Integr. **29**, 61–73 (2014)
76. Turco, E.: Identification of axial forces on statically indeterminate pin-jointed trusses by a nondestructive mechanical test. Open Civil Eng. J. **7**(1), 50–57 (2013)
77. Walpole, L.J.: Fourth-rank tensors of the thirty-two crystal classes: multiplication tables. Proc. R. Soc. Lond. Ser. A **391**, 149–179 (1984)

# Identification of Two-Dimensional Pantographic Structures with a Linear D4 Orthotropic Second Gradient Elastic Model Accounting for External Bulk Double Forces

**Luca Placidi, Emilio Barchiesi and Alessandro Della Corte**

**Abstract** The present paper deals with the identification of the nine constitutive parameters appearing in the strain energy density of a linear elastic second gradient D4 orthotropic two-dimensional continuum model accounting for an external bulk double force $m^{ext}$. The aim is to specialize the model for the description of pantographic fabrics, which show such a kind of anisotropy. Analytical solutions for model problems, which are here referred to as the heavy sheet, the non-conventional bending and the trapezoidal cases are recalled from a previous paper and further elaborated in order to perform *gedanken* experiments. We completely characterize the set of nine constitutive parameters in terms of the materials the fibers are made of (i.e. of the Young's modulus of the fiber materials), of their cross section (i.e. of the area and of the moment of inertia of the fiber cross sections), of the internal rotational spring positioned at each intersection point between the two families of fibers and of the pitch, i.e. the distance between adjacent pivots. Finally, the remarkable form of the strain energy, derived in terms of the displacement field, is shortly discussed.

## 1 Introduction

Generalized continuum theories represent nowadays one of the most promising research fields in continuum mechanics. Sound theoretical results are already provided in the literature (see [8, 18, 21, 35, 38, 41, 49] for general results in higher gradient theory, [1, 22, 40, 42, 44–46] for pantographic and fibrous materials and

L. Placidi (✉)
Faculty of Engineering, International Telematic University Uninettuno,
Rome, Italy
e-mail: luca.placidi@uninettunounirsity.net

E. Barchiesi · A. Della Corte
Dipartimento di Ingegneria Meccanica E Aerospaziale,
Università di Roma La Sapienza, Via Eudossiana 18, 00184 Rome, Italy
e-mail: emilo.barchiesi@uniroma1.it

A. Della Corte
e-mail: alessandro.dellacorte@uniroma1.it



F. dell'Isola et al. (eds.), *Mathematical Modelling in Solid Mechanics*,
Advanced Structured Materials 69, DOI 10.1007/978-981-10-3764-1_14

[7, 19, 20] for second gradient fluids) together with interesting results concerning the more general field of microstructured/micromorphic continua (see for instance [28, 29, 32, 34, 47, 48], and in particular [2–5] for application to carbon nanotube and [6, 23, 25, 26] for biological application; a general introduction for the particular case of micropolar continua is [33]). In this framework pantographic structures represent, in the opinion of the authors, an ideal starting point for the theoretical understanding of generalized continua. In fact, they represent the simplest possible 2D continuum in wich second gradient effects naturally arise from the geometry of the microstructure. Moreover, availability of technologies such as 3D printing and other computer-aided techniques, allows a ready experimental comparison that reveals very fruitful in boosting the modeling process for pantografic sheets and for microstructured continua in general. In order to have this strict interaction between theoretical and experimental research properly working, suitable numerical tools should be provided since, with the exception of some particular cases (such as the examples presented in this work), it is in general impossible to provide closed-form solutions. The numerical tools employed are often borrowed from well-established techniques based mainly on Finite Element Method [9–12, 24, 39, 43], but sometimes high-regularity elements ensuring suitably strong continuity conditions needed by higher gradient models are required (e.g. [13–16, 27]).

The work is organized as follows. In Sect. 2 we recall the derivation of governing PDEs, obtained in [50] for a homogeneous D4 anisotropic two-dimensional second gradient elastic material. In this case, explicit form of stress and hyper stress components have been derived in terms both of the strain (and of its gradient) and of the displacement field (i.e., of first, of second and of third derivatives of its components). Thus, the stress divergence and the hyper stress double divergence vectors have been derived, so that the final form of the PDEs, that were first exposed in [50], have been re-obtained. In Sect. 3, the analytical solutions of the heavy sheet, of the non-conventional bending and of the trapezoidal cases, that have been shown in [50], will be analyzed. In particular, the strain, the strain-gradient, the stress and the hyper stress components will be derived. Besides, contact force and double force, for each of the four sides of the rectangular body, will be given, as well as the contact vertex forces for each of its four vertexes. In Sect. 4 the pantographic case has been analyzed, generalizing the results of [37] to the case where in the pantographic sheet springs are present instead of perfect hinges. In particular, we will show how the analytical solutions of the heavy sheet, of the non-conventional bending and of the trapezoidal cases, that have been obtained in [37], can be achieved with the presence of these internal rotational springs by applying a system of external couples, able to annihilate the effects of the internal rotational springs. Such a system of external couples will be identified with out-of-diagonal components of the external bulk double forces. Finally, the identification between the pantograph with internal rotational springs (the micro-model) and the continuous homogeneous D4 anisotropic two-dimensional elastic second gradient rectangular body (the macro-model) will give an explicit identification of the nine constitutive parameters of the macro-model in terms of constitutive characteristics of the micro-model. In Sect. 5 a simple form of the internal energy, in terms of the displacement field and of its first and second gradient components, will be shown. This form includes not only the contributions of both families of fibers, for axial and for bending deformations of the micro-beams, but also of the internal rotational springs. In Sect. 6 some conclusions will be driven.

Finally, a linguistic remark: in this paper (as well as in [36]) we use the expression "*gedanken* experiment" in order to describe the type of reasoning that allows us to find the constitutive parameters we search for. Since this linguistic choice has been argued sometimes, we would like to point out the following. According to the *Stanford Encyclopedia of Philosophy* [51], a *gedanken* experiment is a thought experiment, i.e. a device of the imagination used to investigate the nature of things. *Gedanken* experiments are used for diverse reasons in a variety of areas. The experiments should be imagined and schematically described, regardless of any possibility of practical realization. As model cases one can consider the well known experiments on the concept of simultaneity in Special Relativity or the EPR experiment by Einstein, Podolsky and Rosen (see for instance [52]). Our line of reasoning, though of course simpler, is nonetheless exactly of this type, and therefore we stick to our previous choice.

## 2 Outline of the Model

In this Section we briefly recall the main facts about the linear elastic second gradient D4 orthotropic two-dimensional continuum model employed in this paper, which can be found in more detail in [50].

$\mathscr{B}$ is a 2-dimensional body that is considered in the reference configuration, where $X$ denotes the coordinates of its material points. $U\left(G, \nabla G\right)$ is the internal energy density functional, that is a function of the deformation matrix $G = \left(F^T F - I\right)/2$ and of its gradient $\nabla G$. Here, $F = \nabla \chi$, where $\chi$ is the placement function, $F^T$ is the transpose of $F$, and $\nabla$ is the gradient operator. The energy functional $\mathscr{E}\left(u\left(\cdot\right)\right)$ depends on the displacement $u = \chi - X$ and includes two contributions: the internal and the external energies,

$$\mathscr{E}\left(u\left(X\right)\right) = \int_{\mathscr{B}} \left[U\left(G, \nabla G\right) - b^{ext} \cdot u - m^{ext} \cdot \nabla u\right] dA \tag{1}$$
$$- \int_{\partial\mathscr{B}} \left[t^{ext} \cdot u + \tau^{ext} \cdot \left[\left(\nabla u\right) n\right]\right] ds - \int_{[\partial\partial\mathscr{B}]} f^{ext} \cdot u.$$

Here, $n$ is the unit external normal and the dot the scalar product between vectors or tensors. $b^{ext}$ and $m^{ext}$ are (per unit area) the external body force and double force, respectively; $t^{ext}$ and $\tau^{ext}$ are (per unit length) the external force and double force; $f^{ext}$ is the external concentrated force, that is applied on the vertices $[\partial\partial\mathscr{B}]$. The boundary $\partial\mathscr{B}$ is assumed to be the union of $m$ regular parts $\Sigma_c$ (with $c = 1, \ldots, m$) and the so-called boundary of the boundary $[\partial\partial\mathscr{B}]$ is assumed to be the union of the corresponding $m$ vertex-points $\mathscr{V}_c$ (with $c = 1, \ldots, m$) with coordinates $X^c$.

It is worth noting that the new contribution of [50] with respect to [36] is in the tensor $m^{ext}$. This term will be crucial for the identification that we are going to perform.

It is well-established that in the general second gradient linear case we have

$$U(G, \nabla G) = \hat{U}(\varepsilon, \eta) = \frac{1}{2} C_{IJ} \varepsilon_I \varepsilon_J + \frac{1}{2} A_{\alpha\beta} \eta_\alpha \eta_\beta \tag{2}$$

where the indices $I$ and $J$ go from 1 to 3 and the indices $\alpha$ and $\beta$ go from 1 to 6. In [50], Eqs. (14), (16), (17), (19), $\varepsilon$, $\eta$, $C$ and $A$ are reported component-wise, respectively.

In [50], the computation of the minimum of $\mathscr{E}$ is performed and (see Eq. (21) therein) the system of PDEs for anisotropic $D_4$ elastic second gradient materials has been deduced in terms of the displacement field.

We now shall show some explicit computations which are instrumental for the identification procedure. Referring to the notation introduced in [50], Eqs. (5)–(8), the stress components read:

$$S_{11} = \frac{\partial U}{\partial G_{11}} = \frac{\partial U}{\partial \varepsilon_1} = C_{1I}\varepsilon_I = C_{11}\varepsilon_1 + C_{12}\varepsilon_2 + C_{13}\varepsilon_3 = c_{11}G_{11} + c_{12}G_{22}, \tag{3}$$

$$\begin{aligned} S_{12} = S_{21} &= \frac{1}{2}\frac{\partial U}{\partial G_{12}} = \frac{\sqrt{2}}{2}\frac{\partial U}{\partial \varepsilon_3} \\ &= \frac{\sqrt{2}}{2} C_{3I}\varepsilon_I = \frac{\sqrt{2}}{2} C_{31}\varepsilon_1 + \frac{\sqrt{2}}{2} C_{32}\varepsilon_2 + \frac{\sqrt{2}}{2} C_{33}\varepsilon_3 = c_{33}G_{12}, \end{aligned} \tag{4}$$

$$S_{22} = \frac{\partial U}{\partial G_{22}} = \frac{\partial U}{\partial \varepsilon_2} = C_{2I}\varepsilon_I = C_{21}\varepsilon_1 + C_{22}\varepsilon_2 + C_{23}\varepsilon_3 = c_{12}G_{11} + c_{11}G_{22}. \tag{5}$$

The hyperstress components read:

$$T_{111} = \frac{\partial U}{\partial G_{11,1}} = \frac{\partial U}{\partial \eta_1} = A_{1\alpha}\eta_\alpha = a_{11}G_{11,1} + a_{12}G_{22,1} + \sqrt{2}a_{13}G_{12,2}, \tag{6}$$

$$T_{112} = \frac{\partial U}{\partial G_{11,2}} = \frac{\partial U}{\partial \eta_5} = A_{5\alpha}\eta_\alpha = a_{12}G_{22,2} + a_{22}G_{11,2} + \sqrt{2}a_{23}G_{12,1}, \tag{7}$$

$$\begin{aligned} T_{121} = T_{211} &= \frac{1}{2}\frac{\partial U}{\partial G_{12,1}} = \frac{\sqrt{2}}{2}\frac{\partial U}{\partial \eta_6} \\ &= \frac{\sqrt{2}}{2} A_{6\alpha}\eta_\alpha = \frac{\sqrt{2}}{2} a_{13}G_{22,2} + \frac{\sqrt{2}}{2} a_{23}G_{11,2} + a_{33}G_{12,1}, \end{aligned} \tag{8}$$

$$\begin{aligned} T_{122} = T_{212} &= \frac{1}{2}\frac{\partial U}{\partial G_{12,2}} = \frac{\sqrt{2}}{2}\frac{\partial U}{\partial \eta_3} \\ &= \frac{\sqrt{2}}{2} A_{3\alpha}\eta_\alpha = \frac{\sqrt{2}}{2} a_{13}G_{11,1} + \frac{\sqrt{2}}{2} a_{23}G_{22,1} + a_{33}G_{12,2}, \end{aligned} \tag{9}$$

$$T_{221} = \frac{\partial U}{\partial G_{22,1}} = \frac{\partial U}{\partial \eta_2} = A_{2\alpha}\eta_\alpha = a_{12}G_{11,1} + a_{22}G_{22,1} + \sqrt{2}a_{23}G_{12,2}, \quad (10)$$

$$T_{222} = \frac{\partial U}{\partial G_{22,2}} = \frac{\partial U}{\partial \eta_4} = A_{4\alpha}\eta_\alpha = a_{11}G_{22,2} + a_{12}G_{11,2} + \sqrt{2}a_{13}G_{12,1}. \quad (11)$$

It is worth to be noted that, by replacing the components of the strain and of the strain gradient tensors

$$G_{11} = u_{1,1}, \quad G_{12} = G_{21} = \frac{1}{2}\left(u_{1,2} + u_{2,1}\right), \quad G_{22} = u_{2,2}, \quad (12)$$

$$G_{11,1} = u_{1,11}, \quad G_{11,2} = u_{1,12}, \quad G_{22,1} = u_{2,12}, \quad G_{22,2} = u_{2,22}, \quad (13)$$

$$G_{12,1} = G_{21,1} = \frac{1}{2}\left(u_{1,12} + u_{2,11}\right), \quad G_{12,2} = G_{21,2} = \frac{1}{2}\left(u_{1,22} + u_{2,12}\right), \quad (14)$$

into (3)–(11), we obtain, respectively, the stress and the hyperstress components in terms of the displacement fields,

$$S_{11} = c_{11}u_{1,1} + c_{12}u_{2,2}, \quad (15)$$

$$S_{12} = S_{21} = \frac{1}{2}c_{33}\left(u_{1,2} + u_{2,1}\right), \quad (16)$$

$$S_{22} = c_{11}u_{2,2} + c_{12}u_{1,1}, \quad (17)$$

$$T_{111} = a_{11}u_{1,11} + a_{12}u_{2,12} + \frac{a_{13}}{\sqrt{2}}\left(u_{1,22} + u_{2,12}\right), \quad (18)$$

$$T_{112} = a_{12}u_{2,22} + a_{22}u_{1,12} + \frac{a_{23}}{\sqrt{2}}\left(u_{1,12} + u_{2,11}\right), \quad (19)$$

$$T_{121} = T_{211} = \frac{\sqrt{2}}{2}a_{13}u_{2,22} + \frac{\sqrt{2}}{2}a_{23}u_{1,12} + \frac{1}{2}a_{33}\left(u_{1,12} + u_{2,11}\right), \quad (20)$$

$$T_{122} = T_{212} = \frac{\sqrt{2}}{2}a_{13}u_{1,11} + \frac{\sqrt{2}}{2}a_{23}u_{2,12} + \frac{1}{2}a_{33}\left(u_{1,22} + u_{2,12}\right), \quad (21)$$

$$T_{221} = a_{12}u_{1,11} + a_{22}u_{2,12} + \frac{a_{23}}{\sqrt{2}}\left(u_{1,22} + u_{2,12}\right), \quad (22)$$

$$T_{222} = a_{11}u_{2,22} + a_{12}u_{1,12} + \frac{a_{13}}{\sqrt{2}}\left(u_{1,12} + u_{2,11}\right), \quad (23)$$

so that the first component of stress divergence in terms of the displacement field is

$$S_{11,1} + S_{12,2} = c_{11}u_{1,11} + c_{12}u_{2,12} + \frac{1}{2}c_{33}\left(u_{1,22} + u_{2,12}\right). \quad (24)$$

Besides, keeping in mind that

$$T_{111,11} = a_{11}u_{1,1111} + a_{12}u_{2,1112} + \frac{a_{13}}{\sqrt{2}}\left(u_{1,1122} + u_{2,1112}\right),$$
$$T_{112,21} = a_{12}u_{2,1222} + a_{22}u_{1,1122} + \frac{a_{23}}{\sqrt{2}}\left(u_{1,1122} + u_{2,1112}\right),$$
$$T_{121,12} = \frac{\sqrt{2}}{2}a_{13}u_{2,1222} + \frac{\sqrt{2}}{2}a_{23}u_{1,1122} + \frac{1}{2}a_{33}\left(u_{1,1122} + u_{2,1112}\right),$$
$$T_{122,22} = \frac{\sqrt{2}}{2}a_{13}u_{1,1122} + \frac{\sqrt{2}}{2}a_{23}u_{2,1222} + \frac{1}{2}a_{33}\left(u_{1,2222} + u_{2,1222}\right),$$

the first component of the double divergence of the hyper-stress in terms of the displacement field is

$$\begin{aligned} T_{1jh,hj} &= a_{11}u_{1,1111} + a_{12}\left(u_{2,1112} + u_{2,1222}\right) \\ &+\sqrt{2}\left(a_{13} + a_{23}\right)\left(u_{1,1122} + \frac{1}{2}u_{2,1112} + \frac{1}{2}u_{2,1222}\right) \\ &+a_{22}u_{1,1122} + a_{33}\left(u_{1,1122} + u_{2,1112} + u_{1,2222} + u_{2,1222}\right). \end{aligned} \tag{25}$$

For the sake of brevity, from now on, whenever we will refer to an equation ($n$) in [50] we shall use the notation ($n$-P1). With this new notation it is now evident that (21-P1) is given by insertion of (24), and (25) into the integral argument of (3-P1).

## 3 Analytical Solutions

In this Section we shall elaborate on the top of analytical solutions proposed in [50] and give more details about the choice of compatible combinations of the external distributed bulk actions $b^{ext}$ and $m^{ext}$.

Since in [50] the case of a body of rectangular geometry is considered, in order to elaborate some of the analytical solutions proposed therein to perform identification through *gedanken* experiments, we shall do the same. In Fig. 1 the scheme of the rectangle considered in [50] is represented.

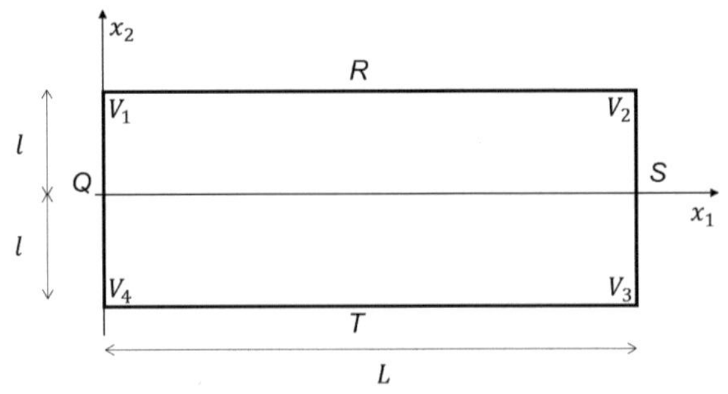


**Fig. 1** Nomenclature of the 2-dimensional body $\mathscr{B}$

## 3.1 Heavy Sheet

The following displacement field is considered,

$$u_1 = 0, \qquad u_2 = \frac{\rho g \left(X_2 - l\right)\left(3l + X_2\right)}{2c_{11}}. \tag{26}$$

At the bottom the vertical displacement is

$$u_2\left(X_1, X_2 = -l\right) = \frac{\rho g \left(-l - l\right)\left(3l - l\right)}{2c_{11}} = -2\frac{\rho g l^2}{c_{11}}.$$

We have by derivation of (26)

$$u_{1,1} = u_{1,2} = u_{2,1} = 0, \quad u_{2,2} = \frac{\rho g \left(X_2 + l\right)}{c_{11}}. \tag{27}$$

The strain components are by definition

$$G_{11} = 0, \quad G_{12} = G_{21} = 0, \quad G_{22} = \frac{\rho g \left(X_2 + l\right)}{c_{11}} \tag{28}$$

so that the strain gradient components are

$$G_{11,1} = G_{11,2} = G_{12,1} = G_{12,2} = G_{21,1} = G_{21,2} = G_{22,1} = 0, \quad G_{22,2} = \frac{\rho g}{c_{11}}. \tag{29}$$

Thus, from (15)–(17), the stress components are

$$S_{11} = c_{11}G_{11} + c_{12}G_{22} = c_{12}\frac{\rho g \left(X_2 + l\right)}{c_{11}}, \tag{30}$$

$$S_{12} = S_{21} = 2c_{33}G_{12} = 0, \tag{31}$$

$$S_{22} = c_{12}G_{11} + c_{11}G_{22} = \rho g \left(X_2 + l\right), \tag{32}$$

and, from (18)–(23), the hyperstress components are

$$T_{111} = 0, \tag{33}$$

$$T_{112} = a_{12}\frac{\rho g}{c_{11}}, \tag{34}$$

$$T_{121} = T_{211} = \frac{\sqrt{2}}{2}a_{13}\frac{\rho g}{c_{11}}, \tag{35}$$

$$T_{122} = T_{212} = 0, \tag{36}$$

$$T_{221} = 0, \tag{37}$$

$$T_{222} = a_{11}\frac{\rho g}{c_{11}}. \tag{38}$$

From (4-P1), at the boundary sides with unit normal $n$ we have the relations between, per unit length, the contact force $t$ and double force $\tau$ and the external force $t^{ext}$ and double force $\tau^{ext}$,

$$t_i = t_i^{ext} + m_{ij}^{ext} n_j, \qquad \tau_i = \tau_i^{ext}. \tag{39}$$

Insertion of (24-P1)-(27-P1) into (39) and use of (30)–(38) gives for sides $S$, $Q$, $R$ and $T$, respectively,

$$t_1 = t_1^S = c_{12} \frac{\rho g \left(X_2 + l\right)}{c_{11}} = t_1^{ext,S} + m_{11}^{ext} = t_1^{ext,S}, \tag{40}$$

$$t_2 = t_2^S = 0 = t_2^{ext,S} + m_{21}^{ext} = t_2^{ext,S}, \tag{41}$$

$$\tau_1 = \tau_1^S = 0 = \tau_1^{S,ext}, \tag{42}$$

$$\tau_2 = \tau_2^S = \frac{\sqrt{2}}{2} a_{13} \frac{\rho g}{c_{11}} = \tau_2^{S,ext}. \tag{43}$$

$$t_1 = t_1^Q = -c_{12} \frac{\rho g \left(X_2 + l\right)}{c_{11}} = t_1^{ext,Q} - m_{11}^{ext} = t_1^{ext,Q}, \tag{44}$$

$$t_2 = t_2^Q = 0 = t_2^{ext,Q} - m_{21}^{ext} = t_2^{ext,Q}, \tag{45}$$

$$\tau_1 = \tau_1^Q = 0 = \tau_1^{Q,ext}, \tag{46}$$

$$\tau_2 = \tau_2^Q = \frac{\sqrt{2}}{2} a_{13} \frac{\rho g}{c_{11}} = \tau_2^{Q,ext}. \tag{47}$$

$$t_1 = t_1^R = 0 = t_1^{ext,R} + m_{12}^{ext} = t_1^{ext,R}, \tag{48}$$

$$t_2 = t_2^R = \rho g \left(X_2 + l\right) = 2\rho g l + m_{22}^{ext} = t_2^{ext,R} + m_{22}^{ext} = t_2^{ext,R}, \tag{49}$$

$$\tau_1 = \tau_1^R = 0 = \tau_1^{R,ext}, \tag{50}$$

$$\tau_2 = \tau_2^R = a_{11} \frac{\rho g}{c_{11}} = \tau_2^{R,ext}. \tag{51}$$

$$t_1 = t_1^T = 0 = t_1^{ext,T} - m_{12}^{ext} = t_1^{ext,T}, \tag{52}$$

$$t_2 = t_2^T = -\rho g \left(X_2 + l\right) = -\rho g \left(-l + l\right) = 0 = t_2^{ext,T} - m_{22}^{ext} = t_2^{ext,T}, \tag{53}$$

$$\tau_1 = \tau_1^T = 0 = \tau_1^{T,ext}, \tag{54}$$

$$\tau_2 = \tau_2^T = a_{11} \frac{\rho g}{c_{11}} = \tau_2^{T,ext}. \tag{55}$$

At the vertices, from (42-P1) and (27), we have that the relevant combination of the hyperstress for characterization of vertex-forces are

$$T_{112} + T_{121} = \left( a_{12} + \frac{\sqrt{2}}{2} a_{13} \right) \frac{\rho g}{c_{11}}, \quad T_{212} + T_{221} = 0$$

so that, from (41-P1),

$$\left(f_1^{ext}\right)_{\mathscr{V}_i} = -\left(a_{12} + \frac{\sqrt{2}}{2}a_{13}\right)\frac{\rho g}{c_{11}}, \quad i = 1, 3 \tag{56}$$

$$\left(f_1^{ext}\right)_{\mathscr{V}_i} = \left(a_{12} + \frac{\sqrt{2}}{2}a_{13}\right)\frac{\rho g}{c_{11}}, \quad i = 2, 4 \tag{57}$$

$$\left(f_2^{ext}\right)_{\mathscr{V}_i} = 0, \quad i = 1, 2, 3, 4. \tag{58}$$

## 3.2 Non-conventional Bending

Let us take into account the following displacement field,

$$u_1 = 0, \qquad u_2 = -\frac{aX_1^2}{2}. \tag{59}$$

The maximum displacement (at side $S$) is

$$u_2\left(X_1 = L, X_2\right) = -\frac{aL^2}{2}.$$

We have by derivation of (59)

$$u_{1,1} = u_{1,2} = u_{2,2} = 0, \quad u_{2,1} = -aX_1. \tag{60}$$

The strain components are by definition

$$G_{11} = 0, \quad G_{12} = G_{21} = -\frac{1}{2}aX_1, \quad G_{22} = 0, \tag{61}$$

so that the strain gradient components are

$$G_{11,1} = G_{11,2} = G_{12,2} = G_{21,2} = G_{22,1} = G_{22,2} = 0, \quad G_{12,1} = G_{21,1} = -\frac{1}{2}a. \tag{62}$$

Thus, from (15)–(17), the stress components are

$$S_{11} = c_{11}G_{11} + c_{12}G_{22} = 0, \tag{63}$$

$$S_{12} = S_{21} = c_{33}G_{12} = -\frac{a}{2}X_1c_{33}, \tag{64}$$

$$S_{22} = c_{12}G_{11} + c_{11}G_{22} = 0, \tag{65}$$

and, from (18)–(23), the hyperstress components are

$$T_{111} = a_{11}G_{11,1} + a_{12}G_{22,1} + \sqrt{2}a_{13}G_{12,2} = 0, \tag{66}$$

$$T_{112} = a_{12}G_{22,2} + a_{22}G_{11,2} + \sqrt{2}a_{23}G_{12,1} = -\frac{\sqrt{2}}{2}a_{23}a, \tag{67}$$

$$T_{121} = T_{211} = \frac{\sqrt{2}}{2}a_{13}G_{22,2} + \frac{\sqrt{2}}{2}a_{23}G_{11,2} + a_{33}G_{12,1} = -\frac{1}{2}a_{33}a, \tag{68}$$

$$T_{122} = T_{212} = \frac{\sqrt{2}}{2}a_{13}G_{11,1} + \frac{\sqrt{2}}{2}a_{23}G_{22,1} + a_{33}G_{12,2} = 0, \tag{69}$$

$$T_{221} = a_{12}G_{11,1} + a_{22}G_{22,1} + \sqrt{2}a_{23}G_{12,2} = 0, \tag{70}$$

$$T_{222} = a_{11}G_{22,2} + a_{12}G_{11,2} + \sqrt{2}a_{13}G_{12,1} = -\frac{\sqrt{2}}{2}a_{13}a. \tag{71}$$

From (4-P1), at the boundary sides with unit normal $n$ we have the relations between, per unit length, the contact force $t$ and double force $\tau$ and the external force $t^{ext}$ and double force $\tau^{ext}$,

$$t_i = t_i^{ext} + m_{ij}^{ext} n_j, \qquad \tau_i = \tau_i^{ext}. \tag{72}$$

Insertion of (24-P1)-(27-P1) into (72) and use of (63)–(71) gives for sides $S$, $Q$, $R$ and $T$, respectively,

$$t_1 = t_1^S = 0 = t_1^{ext,S} + m_{11}^{ext} = t_1^{ext,S} = 0, \tag{73}$$

$$t_2 = t_2^S = -\frac{a}{2}X_1 c_{33} = t_2^{ext,S} + m_{21}^{ext} = t_2^{ext,S} - \frac{a}{2}c_{33}X_1 \Rightarrow t_2^{ext,S} = 0, \tag{74}$$

$$\tau_1 = \tau_1^S = 0 = \tau_1^{S,ext}, \tag{75}$$

$$\tau_2 = \tau_2^S = -\frac{a}{2}a_{33} = \tau_2^{S,ext}. \tag{76}$$

$$t_1 = t_1^Q = 0 = t_1^{ext,Q} - m_{11}^{ext} = t_1^{ext,Q} = 0, \tag{77}$$

$$t_2 = t_2^Q = \frac{a}{2}X_1 c_{33} = t_2^{ext,Q} - m_{21}^{ext} = t_2^{ext,Q} + \frac{a}{2}c_{33}X_1 \Rightarrow t_2^{ext,Q} = 0, \tag{78}$$

$$\tau_1 = \tau_1^Q = 0 = \tau_1^{Q,ext}, \tag{79}$$

$$\tau_2 = \tau_2^Q = -\frac{a}{2}a_{33} = \tau_2^{Q,ext}. \tag{80}$$

$$t_1 = t_1^R = -\frac{a}{2}X_1 c_{33} = t_1^{ext,R} + m_{12}^{ext} = t_1^{ext,R} - \frac{a}{2}X_1 c_{33} = 0 \Rightarrow t_1^{ext,R} = 0, \tag{81}$$

$$t_2 = t_2^R = 0 = t_2^{ext,R} + m_{22}^{ext} = t_2^{ext,R} = 0, \tag{82}$$

$$\tau_1 = \tau_1^R = 0 = \tau_1^{R,ext}, \tag{83}$$

$$\tau_2 = \tau_2^R = -\frac{\sqrt{2}}{2}a_{13}a = \tau_2^{R,ext}. \tag{84}$$

$$t_1 = t_1^T = \frac{a}{2} X_1 c_{33} = t_1^{ext,T} - m_{12}^{ext} = t_1^{ext,T} + \frac{a}{2} X_1 c_{33} \Rightarrow t_1^{ext,T} = 0, \tag{85}$$

$$t_2 = t_2^T = 0 = t_2^{ext,T} - m_{22}^{ext} = t_2^{ext,T}, \tag{86}$$

$$\tau_1 = \tau_1^T = 0 = \tau_1^{T,ext}, \tag{87}$$

$$\tau_2 = \tau_2^T = -\frac{\sqrt{2}}{2} a_{13} a = \tau_2^{T,ext}. \tag{88}$$

From (42-P1) and (60), at the vertices we have that the relevant combinations of hyperstress for characterization of vertex-forces are

$$T_{112} + T_{121} = -a\left(\frac{\sqrt{2}}{2} a_{23} + \frac{1}{2} a_{33}\right), \quad T_{212} + T_{221} = 0$$

so that, from (41-P1),

$$\left(f_1^{ext}\right)_{\mathscr{V}_i} = a\left(\frac{\sqrt{2}}{2} a_{23} + \frac{1}{2} a_{33}\right), \quad i = 1, 3 \tag{89}$$

$$\left(f_1^{ext}\right)_{\mathscr{V}_i} = -a\left(\frac{\sqrt{2}}{2} a_{23} + \frac{1}{2} a_{33}\right), \quad i = 2, 4 \tag{90}$$

$$\left(f_2^{ext}\right)_{\mathscr{V}_i} = 0, \quad i = 1, 2, 3, 4. \tag{91}$$

It is worth to be noted that the total external moment $M_S^{ext}$ on side $S$ is only due to the double force $\tau_2^{ext,S}$ of (76)

$$M_S^{ext} = \int_{-l}^{l} \tau_2^{ext,S} ds = -a_{33} \frac{a}{2} \int_{-l}^{l} ds = -a_{33} a l \tag{92}$$

that provides an interpretation of the parameter $a$ first introduced in (59), i.e.,

$$a = -\frac{M_S^{ext}}{a_{33} l}. \tag{93}$$

## 3.3 *Trapezoidal Case*

Let us take into account the following displacement field

$$u_1 = 0, \qquad u_2 = b X_1 X_2. \tag{94}$$

We have by derivation of (94)

$$u_{1,1} = u_{1,2} = 0, \quad u_{2,1} = bX_2 \quad u_{2,2} = bX_1. \tag{95}$$

The strain components are by definition

$$G_{11} = 0, \quad G_{12} = G_{21} = \frac{1}{2}bX_2, \quad G_{22} = bX_1 \tag{96}$$

so that the strain gradient components are

$$G_{11,1} = G_{11,2} = G_{12,1} = G_{21,1} = G_{22,2} = 0, \quad G_{12,2} = G_{21,2} = \frac{1}{2}b \quad G_{22,1} = b. \tag{97}$$

Thus, from (15)–(17), the stress components are

$$S_{11} = c_{11}G_{11} + c_{12}G_{22} = bc_{12}X_1, \tag{98}$$
$$S_{12} = S_{21} = c_{33}G_{12} = \frac{1}{2}bc_{33}X_2, \tag{99}$$
$$S_{22} = c_{12}G_{11} + c_{11}G_{22} = bc_{11}X_1, \tag{100}$$

and, from (18)–(23), the hyperstress components are

$$T_{111} = a_{11}G_{11,1} + a_{12}G_{22,1} + \sqrt{2}a_{13}G_{12,2} = b\left(a_{12} + \frac{\sqrt{2}}{2}a_{13}\right), \tag{101}$$

$$T_{112} = a_{12}G_{22,2} + a_{22}G_{11,2} + \sqrt{2}a_{23}G_{12,1} = 0, \tag{102}$$
$$T_{121} = T_{211} = \frac{1}{2}\sqrt{2}a_{13}G_{22,2} + \frac{1}{2}\sqrt{2}a_{23}G_{11,2} + a_{33}G_{12,1} = 0, \tag{103}$$
$$T_{122} = T_{212} = \frac{1}{2}\sqrt{2}a_{13}G_{11,1} + \frac{1}{2}\sqrt{2}a_{23}G_{22,1} + a_{33}G_{12,2}$$
$$= \frac{1}{2}b\left(\sqrt{2}a_{23} + a_{33}\right), \tag{104}$$

$$T_{221} = a_{12}G_{11,1} + a_{22}G_{22,1} + \sqrt{2}a_{23}G_{12,2} = b\left(a_{22} + \frac{\sqrt{2}}{2}a_{23}\right), \tag{105}$$

$$T_{222} = a_{11}G_{22,2} + a_{12}G_{11,2} + \sqrt{2}a_{13}G_{12,1} = 0. \tag{106}$$

Insertion of (24-P1)-(27-P1) into (39) and use of (63)–(71) gives for sides $S$, $Q$, $R$ and $T$, respectively,

$$t_1 = t_1^S = c_{12}bX_1 = c_{12}bL = t_1^{ext,S} + m_{11}^{ext} = t_1^{ext,S}, \tag{107}$$
$$t_2 = t_2^S = \frac{1}{2}bc_{33}X_2 = t_2^{ext,S} + m_{21}^{ext} = t_2^{ext,S} + b\left(c_{12} + \frac{1}{2}c_{33}\right)X_2$$
$$\Rightarrow t_2^{ext,S} = -bc_{12}X_2, \tag{108}$$

$$\tau_1 = \tau_1^S = b\left(a_{12} + \frac{\sqrt{2}}{2}a_{13}\right) = \tau_1^{S,ext}, \tag{109}$$

$$\tau_2 = \tau_2^S = 0 = \tau_2^{S,ext}. \tag{110}$$

$$t_1 = t_1^Q = -bc_{12}X_1 = 0 = t_1^{ext,Q} - m_{11}^{ext} = t_1^{ext,Q}, \tag{111}$$

$$t_2 = t_2^Q = -\frac{1}{2}bc_{33}X_2 = t_2^{ext,Q} - m_{21}^{ext} = t_2^{ext,Q} - b\left(c_{12} + \frac{1}{2}c_{33}\right)X_2$$

$$\Rightarrow t_2^{ext,Q} = bc_{12}X_2, \tag{112}$$

$$\tau_1 = \tau_1^Q = b\left(a_{12} + \frac{\sqrt{2}}{2}a_{13}\right) = \tau_1^{Q,ext}, \tag{113}$$

$$\tau_2 = \tau_2^Q = 0 = \tau_2^{Q,ext}. \tag{114}$$

$$t_1 = t_1^R = \frac{1}{2}bc_{33}X_2 = t_1^{ext,R} + m_{12}^{ext} = t_1^{ext,R} + b\left(c_{12} + \frac{1}{2}c_{33}\right)X_2 = 0$$

$$\Rightarrow t_1^{ext,R} = -bc_{12}X_2 = -blc_{12}, \tag{115}$$

$$t_2 = t_2^R = bc_{11}X_1 = t_2^{ext,R} + m_{22}^{ext} = t_2^{ext,R}, \tag{116}$$

$$\tau_1 = \tau_1^R = \frac{b}{2}\left(\sqrt{2}a_{23} + a_{33}\right) = \tau_1^{R,ext}, \tag{117}$$

$$\tau_2 = \tau_2^R = 0 = \tau_2^{R,ext}. \tag{118}$$

$$t_1 = t_1^T = -\frac{1}{2}bc_{33}X_2 = t_1^{ext,T} - m_{12}^{ext} = t_1^{ext,T} - b\left(c_{12} + \frac{1}{2}c_{33}\right)X_2$$

$$\Rightarrow \quad t_1^{ext,T} = -bc_{12}l, \tag{119}$$

$$t_2 = t_2^T = 0 = t_2^{ext,T} - m_{22}^{ext} = t_2^{ext,T}, \tag{120}$$

$$\tau_1 = \tau_1^T = \frac{b}{2}\left(\sqrt{2}a_{23} + a_{33}\right) = \tau_1^{T,ext}, \tag{121}$$

$$\tau_2 = \tau_2^T = 0 = \tau_2^{T,ext}. \tag{122}$$

From (42-P1) and (60), at the vertices we have that the relevant combination of the hyperstresses for characterization of vertex-forces are

$$T_{112} + T_{121} = 0, \quad T_{212} + T_{221} = b\left(a_{22} + \sqrt{2}a_{23} + \frac{1}{2}a_{33}\right)$$

so that, from (41-P1),

$$\left(f_1^{ext}\right)_{\mathcal{V}_i} = 0, \quad i = 1, 2, 3, 4 \tag{123}$$

$$\left(f_2^{ext}\right)_{\mathcal{V}_i} = -b\left(a_{22} + \sqrt{2}a_{23} + \frac{1}{2}a_{33}\right), \quad i = 1, 3 \tag{124}$$

$$\left(f_2^{ext}\right)_{\mathcal{V}_i} = b\left(a_{22} + \sqrt{2}a_{23} + \frac{1}{2}a_{33}\right), \quad i = 2, 4. \tag{125}$$

# 4 Pantographic Case

Let us assume that the two families of fibers in the pantographic structure, modelled as Euler beams in the micro-model, are aligned with the axes of the frame of reference. A series of intuitive considerations are done in this section. In other words a set of *gedanken* experiments is conceived for the purpose of parameters identification.

## 4.1 Heavy Sheet

We can prove that the vertical displacement of side $T$ is, in the micro-model, that of the free-side of a cantilever extensional beam of length $2l$, with axial rigidity $E_m A_m$ and with a distributed axial load $b_N = b_N^v + b_N^h$ that is due to its own weight $b_N^v = -\rho_m A_m g$ and to the weight of the horizontal beams $b_N^h = b_N^v$

$$u_2\left(X_1, X_2 = -l\right) = -\frac{2\rho g l^2}{c_{11}} = -\frac{4\rho_m g l^2}{E_m}, \tag{126}$$

where $g$ is gravity acceleration, $\rho_m$ is the mass per unit volume of the micro beams and $E_m$ is their Young modulus. Besides, equating the mass of the continuous macro model ($\rho L 2l$) with the one of the micro model ($\rho_m A_m L \frac{2l}{d_m} + \rho_m A_m 2l \frac{L}{d_m} = 2Ll\frac{2\rho_m A_m}{d_m}$) yields the relation

$$\rho = \frac{2\rho_m A_m}{d_m}, \tag{127}$$

where $A_m$ is the cross section area of each micro beam and $d_m$ is the distance between two adjacent families of micro beams. From (126) and (127) we have

$$c_{11} = E_m \frac{2\rho g l^2}{4\rho_m g l^2} = E_m \frac{2\rho_m A_m}{d_m}\frac{1}{2\rho_m} = \frac{E_m A_m}{d_m}. \tag{128}$$

Further considerations based upon the heavy sheet configuration can be done. First of all, the natural absence of the Poisson effect in this configuration makes the horizontal force per unit length for the vertical sides, that are given from (40). This and (128) give

$$t_1^{ext,S} = \frac{\rho g\,(l + X_2)}{c_{11}} c_{12} = 0, \quad \Rightarrow \quad c_{12} = 0. \tag{129}$$

The natural absence of double force in vertical sides gives from (43) and (128)

$$\tau_2^{ext,S} = \sqrt{2}\frac{a_{13}\rho g}{c_{11}} = 0, \quad \Rightarrow \quad a_{13} = 0. \tag{130}$$

The natural absence of double force in horizontal sides gives from (51) and (128)

$$\tau_2^{R,ext} = \frac{\rho g a_{11}}{c_{11}} = 0, \quad \Rightarrow \quad a_{11} = 0. \tag{131}$$

Moreover, in the same heavy sheet configuration, the natural absence of vertex forces gives from (56) and (128)

$$T_{112} + T_{121} = \frac{\left(a_{12} + \sqrt{2}a_{13}\right)\rho g}{2c_{11}} = 0, \quad \Rightarrow \quad a_{12} + \sqrt{2}a_{13} = 0, \tag{132}$$

that yields with (130),

$$a_{12} = 0. \tag{133}$$

## 4.2 Non-conventional Bending

We set an equivalence of such a case with a pantograph composed of a number (i.e. $\frac{2l}{d_m}$) of horizontal beam that are bent due to an external couple $M_m$ on the right-hand side of each beam, so that the total external moment $M_S^{ext}$ on side $S$ is related to $M_m$,

$$M_S^{ext} = -\frac{2l}{d_m} M_m, \tag{134}$$

and the vertical displacement of side $S$ is

$$u_2\,(X_1 = L, X_2) = -\frac{aL^2}{2} = -\frac{M_m L^2}{2E_m I_m}, \tag{135}$$

where $I_m$ is the moment of inertia of the micro-beams. Such an equivalence works nicely in the case of absence of internal rotational springs at the place of each internal hinges.

In the case where the internal rotational springs are present, we need to apply, in the micro-model, two moments at the position of each internal rotational springs. One moment $M_{ext}^{H}$ on the horizontal fibers and the other moment $M_{ext}^{V}$ on the vertical fibers. Such moments have the role to annihilate the effects of the moments, $M_{rs}^{H\rightarrow V}$ and $M_{rs}^{V\rightarrow H}$, that are due to the internal rotational springs and, therefore, are proportional to the relative angle $(\theta_H - \theta_V)$ between the horizontal ($\theta_H$ is the rotation of the horizontal beam) and the vertical $\theta_V$ ($\theta_V$ is the rotation of the vertical beam) beams. Let us take into account the horizontal beam. The internal rotational springs gives positive moment $M_{rs}^{V\rightarrow H}$ on the beam

$$M_{rs}^{V\rightarrow H} = -k_r\left(\theta_H - \theta_V\right)$$

because in this non-conventional bending case we have

$$\theta_V = 0, \qquad \theta_H = u_{2,1} = -aX_1,$$

that gives

$$M_{rs}^{V\rightarrow H} = k_r a X_1.$$

The same internal rotational spring gives an opposite moment $M_{rs}^{H\rightarrow V}$ on the vertical beam,

$$M_{rs}^{H\rightarrow V} = -k_r\left(\theta_V - \theta_H\right) = -k_r a X_1.$$

In order to annihilate the effects of the internal rotational springs, the external moments $M_{ext}^{H}$ and $M_{ext}^{V}$ need to be the opposite of those given by the rotational springs, i.e.,

$$M_{ext}^{H} = -M_{rs}^{V\rightarrow H} = -k_r a X_1, \qquad M_{ext}^{V} = -M_{rs}^{H\rightarrow V} = k_r a X_1.$$

Thus, two moments have been applied at the position of each internal hinge-rotational-spring in the micro-model. In the macro-model, such positions are distributed on the two-dimensional body. Therefore, we need to apply distributed external moments that are modelled by a distributed external double force $m^{ext}$. In particular only the out-of-diagonal components $m_{12}^{ext}$ and $m_{21}^{ext}$ have the role of distributed external couples. Since $m_{12}^{ext}$ does work on the component $u_{1,2}$ of the displacement gradient, then $m_{12}^{ext}$ is interpreted as the negative distributed couple on the vertical beams

$$m_{12}^{ext} = -M_{ext}^{V} = -k_r a X_1.$$

Because of (57-P1)

$$m_{12}^{ext} = -\frac{a}{2} c_{33} X_1, \quad \Rightarrow \quad c_{33} = 2k_r. \tag{136}$$

Equations (93), (134) and (135) give

$$a = -\frac{M_S^{ext}}{a_{33} l}.$$

$$a_{33} = -\frac{M_S^{ext}}{al} = \frac{2l}{d_m} M_m \frac{1}{al} = \frac{2l}{d_m} \frac{aL^2 2E_m I_m}{2L^2} \frac{1}{al} = 2\frac{E_m I_m}{d_m}. \quad (137)$$

Moreover, in the non-conventional bending case, from (89), the natural absence of vertex forces gives,

$$T_{112} + T_{121} = \frac{a}{2}\left(a_{33} + \sqrt{2}a_{23}\right) = 0, \quad \Rightarrow \quad a_{33} + \sqrt{2}a_{23} = 0. \quad (138)$$

## 4.3 Trapezoidal Case

From (124), assuming zero wedge forces, in the trapezoidal case we have

$$T_{212} + T_{221} = a_{22} + \sqrt{2}a_{23} + \frac{1}{2}a_{33} = 0 \Rightarrow 2a_{22} + 2\sqrt{2}a_{23} + a_{33} = 0. \quad (139)$$

## 4.4 Summary

Equations (128)–(132), (136)–(139) completely characterize the orthotropic material. In particular the two constitutive matrices are represented as follows

$$\mathbf{C} = \frac{E_m A_m}{d_m}\begin{pmatrix} 1 & 0 & 0 \\ 0 & 1 & 0 \\ 0 & 0 & 0 \end{pmatrix} + 2k_r \begin{pmatrix} 0 & 0 & 0 \\ 0 & 0 & 0 \\ 0 & 0 & 1 \end{pmatrix} = \begin{pmatrix} \frac{E_m A_m}{d_m} & 0 & 0 \\ 0 & \frac{E_m A_m}{d_m} & 0 \\ 0 & 0 & 2k_r \end{pmatrix}, \quad (140)$$

$$\mathbf{A} = \frac{E_m I_m}{d_m}\begin{pmatrix} 0 & 0 & 0 & 0 & 0 & 0 \\ 0 & 1 & -\sqrt{2} & 0 & 0 & 0 \\ 0 & -\sqrt{2} & 2 & 0 & 0 & 0 \\ 0 & 0 & 0 & 0 & 0 & 0 \\ 0 & 0 & 0 & 0 & 1 & -\sqrt{2} \\ 0 & 0 & 0 & 0 & -\sqrt{2} & 2 \end{pmatrix}. \quad (141)$$

### 4.5 *About the Redundancy of Equations Coming from gedanken experiments*

The same result in (136) could be achieved by identifying the component $m_{21}^{ext}$ that does work on the component $u_{2,1}$of the displacement gradient. Thus, $m_{21}^{ext}$ is interpreted as the positive distributed couple on the horizontal beams

$$m_{21}^{ext} = M_{ext}^{H} = -k_r a X_1,$$

that, with (57-P1), gives the same identification (136).

We remark that the identification in (129) could also be achieved, in the trapezoidal case, assuming zero horizontal force on the vertical sides and that the one in (130) could also be achieved in the non-conventional bending case, from (84), assuming zero double force at horizontal sides.

We remark that the identification in (133) could also be achieved in the trapezoidal case, from (113) and with (132), assuming zero horizontal double force at vertical sides.

We remark that the identification in (136) could also be achieved in the non-conventional bending case, from $(77)_1$ or from $(73)_1$, assuming zero horizontal force per unit length at vertical sides or, from (81), by assuming zero horizontal force per unit length at horizontal sides. Alternatively it could be achieved in the trapezoidal case, from $(78)_2$ and $(74)_2$, by assuming zero vertical force per unit length at vertical sides or, from (81), by assuming zero horizontal force per unit length at horizontal sides.

We finally remark that the identification in (138) could also be achieved in the trapezoidal case, from (117), assuming zero horizontal double force at horizontal sides.

Further redundant identification formulas could be derived in the trapezoidal case that, for the sake of simplicity, are not made explicit in this paper.

## 5 Some Remarks on the Internal Energy

It is interesting to recognize that the internal energy (2) can now be computed with (14-P1), (15-P1) and with the definition, in the linear case, of the deformation matrix $G$ and of its gradient $\nabla G$,

$$U(G, \nabla G) = \frac{1}{2}\frac{E_m A_m}{d_m}\left(G_{11}^2 + G_{22}^2\right) + \frac{1}{2}2k_r 2\left(G_{12}\right)^2$$
$$+\frac{1}{2}\frac{E_m I_m}{d_m}\left[G_{22,1}\left(G_{22,1} - 2G_{12,2}\right) + 2G_{12,2}\left(-G_{22,1} + 2G_{12,2}\right)\right]$$
$$+\frac{1}{2}\frac{E_m I_m}{d_m}\left[G_{11,2}\left(G_{11,2} - 2G_{12,1}\right) + 2G_{12,1}\left(-G_{11,2} + 2G_{12,1}\right)\right]$$

or, in terms of the displacement field,

$$U\left(G,\nabla G\right)=\frac{1}{2}k_r\left(u_{1,2}+u_{2,1}\right)^2+\frac{E_m A_m}{2d_m}\left(u_{1,1}^2+u_{2,2}^2\right)+\frac{E_m I_m}{2d_m}\left(u_{1,22}^2+u_{2,11}^2\right). \tag{142}$$

Expression (142) is a useful form of the energy, as in it the contributions of both families of fibers, for axial and for bending deformations of the micro-beams, and also of the internal rotational springs, appear explicitly.

## 6 Conclusion

In the same fashion of [37], in this paper we have identified the whole set of nine parameters characterizing a homogeneous linear second gradient D4 orthotropic continuum model accounting for external distributed bulk double forces, developed in [50] and which can be considered an extension of the model presented in [36]. Analytical solutions proposed in [50] were employed in order to perform an identification in terms of the Young's modulus of the fibers' material, of their area, of the moment of inertia of their cross sections, of the rigidity of the internal rotational spring and of the step. A suitable form of the strain energy, that closely resembles the contributions of both families of fibers, for axial and for bending deformations of the micro-beams, and also of the internal rotational springs, has been derived.

The results can of course be generalized in various ways. For instance, interesting progresses in this line of investigation may involve multi-physics coupling between mechanical and electric effects (a case in which anisotropy plays of course a relevant role [53]), as well as more general beam models for the fibers [54–56]. The extension of the results to nonlinear second gradient continua is of course far from trivial and may require substantial theoretical progresses.

## References


1. Alibert, J.-J., Seppecher, P., Dell'Isola, F.: Truss modular beams with deformation energy depending on higher displacement gradients. Math. Mech. Solids **8**(1), 51–73 (2003)
2. Aminpour, H., Rizzi, N.: A one-dimensional continuum with microstructure for single-wall carbon nanotubes bifurcation analysis. Math. Mech. Solids **21**(2), 168–181 (2016)
3. Aminpour, H., Rizzi, N.: On the continuum modelling of carbon nano tubes. Civil-Comp Proceedings, vol. 08 (2015)
4. Aminpour, H., Rizzi, N.: On the modelling of carbon nano tubes as generalized continua. Adv. Struct. Mater. **42**(1), 15–35 (2016)
5. Aminpour, H., Rizzi, N., Salerno, G.: A one-dimensional beam model for single-wall carbon nano tube column buckling. In: Civil-Comp Proceedings, vol. 106 (2014)
6. Giorgio, I., Andreaus, U., Scerrato, D., dell'Isola, F.: A visco-poroelastic model of functional adaptation in bones reconstructed with bio-resorbable materials. Biomech. Model. Mechanobiol. **15**(5), 1325–1343 (2016)

7. Auffray, N., dell'Isola, F., Eremeyev, V.A., Madeo, A., Rosi, G.: Analytical continuum mechanics à la Hamilton-Piola least action principle for second gradient continua and capillary fluids. Math. Mech. Solids **20**(4), 375–417 (2015)
8. Auffray, N., Dirrenberger, J., Rosi, G.: A complete description of bi-dimensional anisotropic strain-gradient elasticity. Int. J. Solids. Struct. **69–70**, 195–206 (2015). doi:10.1016/j.ijsolstr.2015.04.036
9. Baraldi, D., Reccia, E., Cazzani, A., Cecchi, A.: Comparative analysis of numerical discrete and finite element models: the case of in-plane loaded periodic brickwork. Comp. Mech. Comput. Appl. **4**(4), 319–344 (2013)
10. Bilotta, A., Formica, G., Turco, E.: Performance of a high-continuity finite element in three-dimensional elasticity. Int. J. Numer. Methods Biomed. Eng. **26**(9), 1155–1175 (2010)
11. Bilotta, A., Turco, E.: A numerical study on the solution of the Cauchy problem in elasticity. Int. J. Solids Struct. **46**(25–26), 4451–4477 (2009)
12. Cazzani, A., Ruge, P.: Numerical aspects of coupling strongly frequency-dependent soil-foundation models with structural finite elements in the time-domain. Soil Dyn. Earthq. Eng. **37**, 56–72 (2012)
13. Hughes, T.J., Cottrell, J.A., Bazilevs, Y.: Isogeometric analysis: CAD, finite elements, NURBS, exact geometry and mesh refinement. Comput. Methods Appl. Mech. Eng. **194**(39), 4135–4195 (2005)
14. Cazzani, A., Malagù, M., & Turco, E.: Isogeometric analysis of plane-curved beams. Math. Mech. Solids (2014). doi:10.1177/1081286514531265
15. Greco, L., Cuomo, M.: B-Spline interpolation of Kirchhoff-Love space rods. Comput. Methods Appl. Mech. Eng. **256**, 251–269 (2013)
16. Cuomo, M., Contrafatto, L., Greco, L.: A variational model based on isogeometric interpolation for the analysis of cracked bodies. Int. J. Eng. Sci. **80**, 173–188 (2014)
17. Del Vescovo, D., Giorgio, I.: Dynamic problems for metamaterials: review of existing models and ideas for further research. Int. J. Eng. Sci. **80**, 153–172 (2014)
18. Dell'Isola, F., Andreaus, U. and Placidi, L.: At the origins and in the vanguard of peri-dynamics, non-local and higher gradient continuum mechanics. An underestimated and still topical contribution of Gabrio Piola, Mechanics and Mathematics of Solids (MMS), vol. 20, p. 887–928 (2015)
19. Dell'Isola, F., Gouin, H., Seppecher, P.: Radius and surface tension of microscopic bubbles by second gradient theory. Comptes Rendus de l'Academie de Sciences - Serie IIb: Mecanique, Physique, Chimie, Astronomie **320**(6), 211–216 (1995)
20. Dell'Isola, F.G., Rotoli, G.: Validity of Laplace formula and dependence of surface tension on curvature in second gradient fluids. Mech. Res. Commun. **22**(5), 485–490 (1995)
21. Dell'Isola, F., Seppecher, P.: The relationship between edge contact forces, double forces and interstitial working allowed by the principle of virtual power. Comptes Rendus de l'Academie de Sciences, Serie IIb: Mecanique, Physique, Chimie, Astronomie **321**, 303–308 (1995)
22. Dell'Isola, F., Steigmann, D.: A two-dimensional gradient-elasticity theory for woven fabrics. J. Elast. **118**(1), 113–125 (2015)
23. Dos Reis, F., Ganghoffer, J.F.: Construction of micropolar continua from the asymptotic homogenization of beam lattices. Comput. Struct. **112–113**, 354–363 (2012)
24. Garusi, E., Tralli, A., Cazzani, A.: An unsymmetric stress formulation for reissner-mindlin plates: a simple and locking-free rectangular element. Int. J. Comput. Eng. Sci. **5**(3), 589–618 (2004)
25. Goda, I., Assidi, M., Belouettar, S., Ganghoffer, J.F.: A micropolar anisotropic constitutive model of cancellous bone from discrete homogenization. J. Mech. Behav. Biomed. Mater. **16**(1), 87–108 (2012)
26. Goda, I., Assidi, M., Ganghoffer, J.F.: A 3D elastic micropolar model of vertebral trabecular bone from lattice homogenization of the bone microstructure. Biomech. Model. Mechanobiol. **13**(1), 53–83 (2014)
27. Greco, L., Cuomo, M.: An implicit G1 multi patch B-spline interpolation for Kirchhoff-Love space rod. Comput. Methods Appl. Mech. Eng. **269**, 173–197 (2014)

28. Mindlin, R.D.: Micro-structure in Linear Elasticity, Department of Civil Engineering, vol. 27. Columbia University New York, New York (1964)
29. Misra, A., Huang, S.: Micromechanical stress-displacement model for rough interfaces: effect of asperity contact orientation on closure and shear behavior. Int. J. Solids Struct. **49**(1), 111–120 (2012)
30. Misra, A., Parthasarathy, R., Singh, V., Spencer, P.: Micro-poromechanics model of fluid-saturated chemically active fibrous media. ZAMM Zeitschrift fur Angewandte Mathematik und Mechanik **95**(2), 215–234 (2015)
31. Misra, A., Poorsolhjouy, P.: Micro-macro scale instability in 2D regular granular assemblies. Contin. Mech. Thermodyn. **27**(1–2), 63–82 (2013)
32. Altenbach, J., Altenbach, H., Eremeyev, V.A.: On generalized Cosserat-type theories of plates and shells: a short review and bibliography. Arch. Appl. Mech. **80**(1), 73–92 (2010)
33. Eremeyev, V.A., Lebedev, L.P., Altenbach, H.: Foundations of Micropolar Mechanics. Springer Science & Business Media (2012)
34. Misra, A., Singh, V.: Nonlinear granular micromechanics model for multi-axial rate-dependent behavior, 2014. Int. J. Solids Struct. **51**(13), 2272–2282 (2014)
35. Pideri, Catherine, Seppecher, P.: A second gradient material resulting from the homogenization of an heterogeneous linear elastic medium. Contin. Mech. Thermodyn. **9**(5), 241–257 (1997)
36. Placidi, L., Andreaus, U., Della Corte, A., Lekszycki, T.: Gedanken experiments for the determination of two-dimensional linear second gradient elasticity coefficients. Zeitschrift für angewandte Mathematik und Physik **66**, 3699–3725 (2015)
37. Placidi L., Andreaus U., Giorgio I.: Identification of two-dimensional pantographic structure via a linear D4 orthotropic second gradient elastic model. J. Eng. Math. ISSN: 0022-0833 (2017) doi:10.1007/s10665-016-9856-8
38. Sansour, C., Skatulla, S.: A strain gradient generalized continuum approach for modelling elastic scale effects. Comput. Methods Appl. Mech. Eng. **198**(15), 1401–1412 (2009)
39. Scerrato, D., Giorgio, I., Rizzi, N.L.: Three-dimensional instabilities of pantographic sheets with parabolic lattices: numerical investigations. Zeitschrift fur Angewandte Mathematik und Physik, vol. 67(3), Article number 53 (2016)
40. Scerrato, D., Zhurba Eremeeva, I.A., Lekszycki, T., Rizzi, N.L.: On the effect of shear stiffness on the plane deformation of linear second gradient pantographic sheets. ZAMM Zeitschrift fur Angewandte Mathematik und Mechanik, vol. 96, pp. 1268–1279 (2016). doi:10.1002/zamm.201600066
41. Selvadurai, A.P.S.: Plane strain problems in second-order elasticity theory. Int. J. Non-Linear Mech. **8**(6), 551–563 (1973)
42. Seppecher, P., Alibert, J.-J., Dell'Isola, F.: Linear elastic trusses leading to continua with exotic mechanical interactions, J. Phys. Conf. Ser. vol. 319(1), 13 p (2011)
43. Presta, F., Hendy, C.R., Turco, E.: Numerical validation of simplified theories for design rules of transversely stiffened plate girders. Struct. Eng. **86**(21), 37–46 (2008)
44. Rahali, Y., Giorgio, I., Ganghoffer, J.F., dell'Isola, F.: Homogenization á la Piola produces second gradient continuum models for linear pantographic lattices. Int. J. Eng. Sci. **97**, 148–172 (2015)
45. Steigmann, D.J.: Linear theory for the bending and extension of a thin, residually stressed, fiber-reinforced lamina. Int. J. Eng. Sci. **47**(11–12), 1367–1378 (2009)
46. Steigmann, D.J., dell'Isola, F.: Mechanical response of fabric sheets to three-dimensional bending, twisting, and stretching. Acta Mechanica Sinica/Lixue Xuebao **31**(3), 373–382 (2015)
47. Yang, Y., Ching, W.Y., Misra, A.: Higher-order continuum theory applied to fracture simulation of nanoscale intergranular glassy film. J. Nanomech. Micromech. **1**(2), 60–71 (2011)
48. Yang, Y., Misra, A.: Micromechanics based second gradient continuum theory for shear band modeling in cohesive granular materials following damage elasticity. Int. J. Solids Struct. **49**(18), 2500–2514 (2012)
49. Auffray, N., Bouchet, R., Brechet, Y.: Derivation of anisotropic matrix for bi-dimensional strain-gradient elasticity behavior. Int. J. Solids Struct. **46**(2), 440–454 (2009)

50. Placidi, L., Barchiesi, E., Battista, A., An inverse method to get further analytical solutions for a class of metamaterials aimed to validate numerical integrations, Proceedings of the ETAMM2016 conference EMERGING TRENDS IN APPLIED MATHEMATICS AND MECHANICS, May 30 - June 3, 2016, Perpignan, France
51. Nodelman, U., Allen, C., Perry, J.: Stanford encyclopedia of philosophy (2003)
52. Cohen, M.: Simultaneity and Einstein's Gedankenexperiment. Philosophy **64**(249), 391–396 (1989)
53. Abo-el-nour, N., Hamdan, A.M., Almarashi, A.A., and Battista, A.: The mathematical modeling for bulk acoustic wave propagation velocities in transversely isotropic piezoelectric materials. Mathematics and Mechanics of Solids (2015). doi:10.1177/1081286515613333
54. Silvestre, N., Camotim, D.: Second-order generalised beam theory for arbitrary orthotropic materials. Thin-Walled Struct. **40**(9), 791–820 (2002)
55. Piccardo, G., Ranzi, G., Luongo, A.: A complete dynamic approach to the generalized beam theory cross-section analysis including extension and shear modes. Math. Mech. Solids **19**(8), 900–924 (2014)
56. Piccardo, G., Ranzi, G., Luongo, A.: A direct approach for the evaluation of the conventional modes within the GBT formulation. Thin-Walled Struct. **74**, 133–145 (2014)

# Models of Debonding Caused by Vibrations, Heat and Humidity

**Meir Shillor**

**Abstract** This paper describes a 3D model for the process of debonding of two adhesively bonded rectangular components or solids caused by mechanical vibrations, temperature variations and moisture or humidity. These issues are very common in many parts of the world in which systems, such as automotive systems that have parts that are glued together, have to operate in adverse conditions of large variations in temperature and very high humidity. The model consists of a coupled system of dynamic equations for the displacements of the two components, the evolution equations for the temperature in each body and the evolution inclusions for the bonding field and the moisture field in the thin layer of glue on the contact surface. The solids may be either thermoelastic or thermoviscoelastic, and the viscosity may be of the short-memory or long-memory types. Then, this work presents 1D variations of the model. In particular, a model in which each of the components is described both as a rod and a beam so as to capture the tangential and vertical motions, which affect the strength of the bonding field. These models raise many interesting questions: theoretical, computational and experimental that are described in some detail. In particular, the 1D models may be used for parameter identification purposes, especially of the debonding source function.

## 1 Introduction

The usual method of joining metallic structures, especially plates, is by welding. However, this process does not work well when the plates are made of different materials and thicknesses, and certainly not when the components are not made of metal. In such cases, one often resort to various processes that involve adhesive bonding. Indeed, adhesively bonded plates can be found in many industrial settings, in particular in aviation and automotive systems.

M. Shillor (✉)
Department of Mathematics and Statistics, Oakland University,
Rochester, MI 48309, USA
e-mail: shillor@oakland.edu



F. dell'Isola et al. (eds.), *Mathematical Modelling in Solid Mechanics*,
Advanced Structured Materials 69, DOI 10.1007/978-981-10-3764-1_15

"Adhesively bonded composite joints have been widely used in various structural and mechanical applications due to their relative strength-to-weight ratio which leads to weight reduction as well as the energy savings. Joints are often the weakest link in a design, ... There are many advantages of using adhesives; any joint geometry, material combinations, and sizes can be accommodated. ... and they are fatigue and corrosion resistant as they slow or stop fatigue crack growth" [21].

The topic is obviously of considerable importance in engineering, as it relates to the reliability and safety of many systems. This work is motivated by the experimental and modeling results obtained very recently by Prof. S.A. Nassar and his students, [16, 17, 21], especially on the role played by humidity and temperature variations in the debonding processes. For additional information on their motivation and results we refer the reader to these articles and the references therein. We also note the reviews [2, 25] and the many engineering references in these articles concerning adhesive joining of solid components and structures.

Adhesion in mathematical literature is usually associated with models for contact, and as the Mathematical Theory of Contact Mechanics (MTCM) is reaching maturity, the number of results concerning adhesive contact is growing rapidly. Basic mathematical models and results on adhesive contact can be found in [7, 10, 12, 18, 22, 23] and the many references therein. More specialized results can be found in [1, 3, 4, 8, 9] where various specific settings and models were studied.

Here, we propose a rather general 3D model for the process of debonding of two rectangular solids or bodies that are glued on a flat portion of their common boundary. We assume that the adhesive layer is very thin, and so the debonding process takes place on the flat boundary.

We assume that the debonding process is caused by the mechanical vibrations of the tractions in the layer, the humidity or fraction of the moisture content and by temperature. The evolution of the adhesive is described by introducing the bonding field $\beta$ that measures the density fraction of active bonds on the surface, so that when $\beta = 1$ at a point all the bonds are active, when $\beta = 0$ all the bonds are severed, and when $0 < \beta < 1$ it is the fraction of the bonds is active (for further details see, e.g., [7, 22, 23]). To retain this interpretation, we introduce a differential inclusion that describes the evolution of the bonding field. Similarly, the evolution of humidity is described by a differential inclusion. The two bodies are assumed to be thermoelastic or thermoviscoelastic with general linear constitutive relations. Thus, the 3D model consists of a nonlinear coupled system of the dynamic equations for the displacements, the heat equations for the temperature and the two differential inclusions for the bonding field and the humidity. The latter two hold on the contact boundary.

The 'classical' formulation of the model is given in Model 1 in Sect. 2. Following the model, in Sect. 2.1, we pose a number of mathematical, computational and parameter identification open questions the resolution of which has both theoretical and applied interest. It is seen that these topics warrant further study. We note that related dynamic contact problems for thermoviscoelastic material were studied in [5, 6] (see also the references therein).

Next, there is considerable interest in a simpler 1D setting, where the model is more transparent and easier to analyze and simulate. The model for the debonding process in 1D is given in Model 2, Sect. 3. There, to capture the horizontal stress and the vertical shear, we model the system as two rod-beam systems. The model consists of a coupled system of two equations of motion of the rods and two of the beams, a parabolic differential inclusion for the evolution of the bonding field and a parabolic inclusion for the humidity or moisture function. The latter two hold on the common interval. The system is currently under study, [14].

Following the model, in Sect. 3.1, we again pose a number of mathematical, computational and parameter identification open questions. The main long-term interest is the possible identification of the form of the debonding source function, which controls the debonding processes. Moreover, there is considerable interest in obtaining a version of the system by using the ideas and tools in [19, 24] about asymptotic derivation of Model 2 from Model 1 when the relevant dimensions tend to zero, thus reducing the 3D setting into a 2D setting. Preliminary results have been recently attempted in [20].

Then, in Sect. 4 we simplify Model 2 by considering only two rods (disregarding vertical motion) and two differential inclusions for the bonding and humidity fields.

Finally, we mention is passing that all three models can be extended by allowing randomness in some of the system coefficients. Thus making them more applicable. Preliminary results in this direction can be found in [11, 13].

## 2 The 3D Model

We describe a 3D model for the process of debonding of two bonded rectangular solid bodies caused by mechanical vibrations, temperature variations and humidity or moisture. The assumption that the solid bodies are rectangular is for convenience only, what matters is that they are joined over a flat surface. This can be generalized to any Lipschitz surface, but we don't pursue it here.

The solid components are assumed to be thermoelastic or thermoviscoelastic, and are glued over a flat portion of their common surface. The glue is assumed to form a very thin layer, so we do not describe it as an intermediate body with volume, but just a layer on the boundary occupying a 2D portion of the common surface. The glue undergoes deterioration, debonding, caused by the vibrations and temperature variations in the solids, and diffusion of moisture in the glue layer.

A more complex model in which the adhesive is not very thin, so it occupies a volume and thus forming a third body, may be of interest in the future. We comment on this issue shortly at the end of the section.

The setting of the process, depicted in Fig. 1, is as follows. The upper body occupies the volume $\Omega_1 = (0, l_2) \times (0, l_{12}) \times (0, l_{13})$, the lower body occupies $\Omega_2 = (l_1, 1) \times (0, l_{12}) \times (-l_{23}, 0)$, and the adhesive occupies the common portion of the surface $\Sigma = (l_1, l_2) \times (0, l_{12}) \times (0)$. For the sake of simplicity, we assume that the widths (in the $y$-direction) are the same, otherwise the notation becomes a bit more cumbersome. Moreover, all the distances are scaled so that the length of the system (in the $x$-direction) is $L = 1$ and the system is oriented as in the figure.

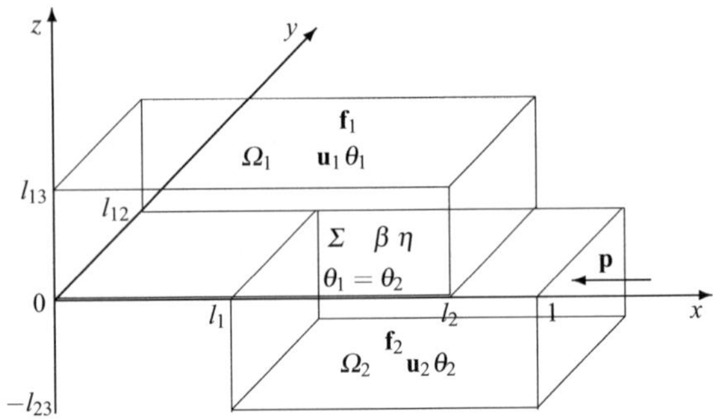


**Fig. 1** The 3D setting: the solids occupy $\Omega_1$ and $\Omega_2$ where the displacements $\mathbf{u}_1, \mathbf{u}_2$ and the temperatures $\theta_1, \theta_2$ are defined, respectively; the adhesive occupies the common surface $\Sigma$ where the bonding function $\beta$ and the moisture function $\eta$ are defined

The upper solid is attached to a rigid support on the face $x = 0$ and is free on the rest of the boundary, except on $\Sigma$. A given traction $\mathbf{p}$ acts on the face $x = 1$ of the second solid and the rest of its surface is free, except on $\Sigma$. We assume, for the sake of completeness, that volume forces $\mathbf{f}_1$ and $\mathbf{f}_2$ act on the two bodies, respectively.

In the following sections we are interested in the limit problems obtained when the cross sections are small, i.e., when $l_{12}, l_{13}, l_{23} \to 0$, which lead us to the rod-beam models.

We now introduce the rest of the notation. The indices $i, j, k, l$ have values $\{1, 2, 3\}$ and refer to the coordinates, while $\alpha = 1, 2$ refers to the top or bottom solids. For the sake of simplicity, we describe only the thermoviscoelastic case, and note that the thermoelastic case is obtained be setting the viscosity coefficients to be zero.

We denote by $\theta_\alpha$ the temperature in solid $\alpha$ and assume that, since the adhesive layer is thin, $\theta_1 = \theta_2 = \theta_\Sigma$ on the contact surface $\Sigma$. The evolution of the temperature is given by the usual thermoelastic equations, i.e., we assume that the diffusion of moisture and debonding are associated with negligible energy changes.

We let $\mathbf{u}_\alpha = (u_{\alpha 1}, u_{\alpha 2}, u_{\alpha 3})$ denote the displacements and $\mathbf{v}_\alpha = (v_{\alpha 1}, v_{\alpha 2}, v_{\alpha 3})$ the velocity vectors, defined on $\Omega_\alpha \times [0, T]$, the linearized strains $\varepsilon_\alpha = (\varepsilon_{\alpha ij})$ are defined as $\varepsilon_{\alpha ij} = (\partial u_{\alpha i}/\partial x_j + \partial u_{\alpha j}/\partial x_i)/2$, the stress tensors are $\sigma_\alpha$, and the bodies are assumed to be made of thermoviscoelastic materials with constitutive relations

$$\sigma_\alpha = A_\alpha \varepsilon_\alpha(\mathbf{u}_\alpha) + B_\alpha \varepsilon_\alpha(\mathbf{v}_\alpha) - M_\alpha \theta_\alpha. \tag{1}$$

Here, $A_\alpha = (a_{\alpha ijkl})$ and $B_\alpha = (b_{\alpha ijkl})$ are the fourth order *elasticity* and *viscosity* tensors, with the usual symmetry properties, i.e.,

$$a_{\alpha ijkl} = a_{\alpha jikl} = a_{\alpha klij}, \quad b_{\alpha ijkl} = b_{\alpha jikl} = b_{\alpha klij},$$

defined on $\Omega_\alpha$, and $M_\alpha = (m_{\alpha ij})$ is the scaled second order tensor of thermal expansion, with symmetry $m_{\alpha ij} = m_{\alpha ji}$. All the coefficients may depend on the position $(x, y, z)$. Moreover, we assume that the viscosity is of the short-term memory type. We comment on long-term memory or history-dependent viscosity below.

We recall that in an isotropic thermoelastic material

$$\sigma_{\alpha ij} = \lambda_{\alpha 1}\varepsilon_{\alpha kk}\delta_{ij} + 2\lambda_{\alpha 2}\varepsilon_{\alpha ij} - m_\alpha \theta_\alpha,$$

where $\delta_{ij}$ is the Kronecker symbol (the identity matrix in 3D), a repeated roman subscript $i, j, k, l$ indicates summation and $\lambda_{\alpha 1}$ and $\lambda_{\alpha 2}$ are the Lamé coefficients of the materials. Thus, $A_\alpha$ has only two distinct coefficients and $M_\alpha$ has only one thermal expansion coefficient, and when the material is also homogeneous the coefficients are constants. We assume that the contribution of the viscosity to the energy transport is negligible.

In an isotropic material, one can describe viscosity with one or two viscosity coefficients, namely

$$\sigma_{\alpha ij} = \lambda_1 \varepsilon_{\alpha kk}\delta_{ij} + 2\lambda_2 \varepsilon_{\alpha ij} + 2\nu_2 \dot{\varepsilon}_{\alpha ij} - m_\alpha \theta_\alpha,$$

or

$$\sigma_{\alpha ij} = \lambda_1 \varepsilon_{\alpha kk}\delta_{ij} + 2\lambda_2 \varepsilon_{\alpha ij} + \nu_1 \dot{\varepsilon}_{\alpha kk}\delta_{ij} + 2\nu_2 \dot{\varepsilon}_{\alpha ij} - m_\alpha \theta_\alpha,$$

where the dot represents the partial time derivative and $\nu_1$ and $\nu_2$ are the viscosity coefficients.

If one wishes to use long-term memory viscosity one has to replace the viscosity term in (1) with

$$\int_0^t b_{\alpha ijkl}(t-s)\varepsilon_{\alpha ij}(s)\, ds.$$

Here, the coefficients $b_{\alpha ijkl}(t)$ are assumed to be given and with the necessary properties to make the integrals meaningful.

Next, we denote by $\Gamma_0 = \partial\Omega_1 \cap \{x = 0\}$ the face of the first body where it is clamped, and by $\Gamma_1 = \partial\Omega_2 \cap \{x = 1\}$ the face of the second body where the traction $\mathbf{p}$ acts.

We let $\beta = \beta(x, y, t)$ denote the *bonding field* that is defined on $\Sigma$, and measures the fraction of active bonds, and so we must have

$$0 \le \beta(x, y, t) \le 1 \tag{2}$$

on $\Sigma$. When $\beta(x, y, t) = 1$ all the bonds are active near the point and the adhesive strength is maximal, when $\beta(x, y, t) = 0$ all the bonds are severed and the surfaces are not bonded, and $0 < \beta(x, y, t) < 1$ represents the fraction of active bonds.

We denote by $\eta = \eta(x, y, t)$ the *moisture field*, which is also defined on $\Sigma$ and measures the fraction of water content in the layer, and so we must have

$$0 \le \eta(x, y, t) \le 1 \tag{3}$$

on $\Sigma$. Similarly, when $\eta(x, y, t) = 1$ the adhesive is saturated with water near the point, when $\eta(x, y, t) = 0$ there is no water, and when $0 < \eta(x, y, t) < 1$ it measures the moisture fraction. For the sake of simplicity, we assume that the saturation level is constant and does not depend on the temperature, strain or the bonding field.

To proceed, we consider the processes on the contact surface $\Sigma$. To that end, we let $\mathbf{n}_\alpha = (-1)^\alpha \mathbf{n}$ be the unit outward normals from $\Omega_\alpha$ on $\Sigma$, where $\mathbf{n} = (0, 0, 1)$, and then the normal components of the displacements are given by

$$u_{\alpha n} = \mathbf{u}_\alpha \cdot \mathbf{n}_\alpha = (-1)^\alpha u_{\alpha 3},$$

the tangential displacements are

$$\mathbf{u}_{\alpha\tau} = \mathbf{u}_\alpha - u_{\alpha n}\mathbf{n}_\alpha = (u_{\alpha 1}, u_{\alpha 2}, 0),$$

and similarly for the velocities.

We note that if we do not allow the interpenetration of the surfaces, we must impose the *Signorini contact condition* that the lower solid is always below the upper solid on $\Sigma$, so that $u_{23} \leq u_{13}$, i.e.,

$$u_{2n} + u_{1n} \leq 0. \tag{4}$$

We note that the + sign is a result of the definitions of the normals $\mathbf{n}_\alpha = (-1)^\alpha$ $(0, 0, 1)$. Below, we use the notion of subdifferential to impose this condition.

Next, the normal components and tangential parts of the stresses are given by

$$\sigma_{\alpha n} = \mathbf{n}_\alpha \cdot \sigma_\alpha \cdot \mathbf{n}_\alpha, \quad \sigma_{\alpha\tau} = \sigma_\alpha \cdot \mathbf{n}_\alpha - \sigma_{\alpha n}\mathbf{n}_\alpha.$$

We turn to the dynamics of the adhesive layer on $\Sigma$, which is our main topic and as was noted above, depend on the tractions, temperature $\theta_\Sigma$ and moisture $\eta$. The common *normal adhesive traction* on $\Sigma$ is given by

$$\sigma_n = \sigma_{1n} = -\sigma_{2n} = -\beta K_n(u_{2n} + u_{1n}),$$

where $K_n$ is the normal stiffness of the glue (when fully bonded), which is tensile by (4) since $\sigma_n \geq 0$. Here, we assume that the deterioration in the normal stiffness is linear in $\beta$ and in the normal strain $u_{2n} + u_{1n}$. Other, more complex ways to describe the normal traction, are possible, but are not warranted at this time.

The *shear tractions* are given by

$$\sigma_{1\tau} = -\sigma_{2\tau} = \beta K_\tau(\mathbf{u}_{2\tau} - \mathbf{u}_{1\tau}),$$

where $K_\tau$ is the tangential stiffness of the adhesive layer, when fully bonded. Again, the deterioration of the tangential shear depends linearly on $\beta$ and the tangential shear.

We assume that the adhesive cannot mend or rebond, so that bonds that are severed or broken cannot be reestablished, i.e., the process is irreversible. The debonding process depends on the mechanical state of the surface, the moisture and the bonding field itself. Moreover, we assume that the bonding strength of neighboring elements affects the debonding at a point, which we describe by adding a diffusion term to the debonding process. The evolution of the adhesive field $\beta$, which measures the fraction of the density of active bonds, is assumed to depend on the shear stress and on the moisture, and the rate of debonding is assumed to be given by the nonlinear diffusion equation

$$\beta_t - \nabla \cdot (\kappa_\beta \nabla \beta) = -\Phi(\mathbf{u}_{2\tau} - \mathbf{u}_{1\tau}, u_{2n} + u_{1n}, \beta, \eta),$$

on $\Sigma$. Here, $\kappa_\beta$ is the diffusion rate function that may depend on the tractions, the temperature and humidity, and the debonding rate function $\Phi$ depends on the normal and tangential strains, the moisture and the bonding field itself. Eventually, this function must be deduced from experiments. However, experiments that deal with processes on contacting surfaces are notoriously difficult. Below, we assume, ad hoc, a simplified source function. However, for mathematical purposes all we need is that $\Phi$ is Lipschitz in all its arguments and nonnegative since we do not allow for the mending of broken bonds.

Next, we note that in addition we need to enforce condition (2). To that end we introduce the indicator function $I_{[0,1]}(x)$ which has the value 0 when $0 \le x \le 1$ and the value $+\infty$ elsewhere. The subdifferential of $I_{[0,1]}(x)$ is the set-valued function

$$\partial I_{[0,1]}(x) = \begin{cases} (-\infty, 0] & x = 0, \\ 0 & 0 < x < 1, \\ [0, \infty) & x = 1, \\ \emptyset & \text{otherwise.} \end{cases} \tag{5}$$

We use the subdifferential term to guarantee that $0 \le \beta \le 1$. Thus, the evolution equation for $\beta$ is the set inclusion

$$\beta_t - \nabla \cdot (\kappa_\beta \nabla \beta) + \Phi(\mathbf{u}_{2\tau} - \mathbf{u}_{1\tau}, u_{2n} + u_{1n}, \theta_\Sigma, \beta, \eta) \in -\partial I_{[0,1]}(\beta). \tag{6}$$

The subdifferential guarantees that $0 \le \beta \le 1$ as follows. When $0 < \beta < 1$ the subdifferential has the value 0 and we have the usual parabolic equation. When $\beta = 0$ the subdifferential has the exact negative value $\zeta \in (-\infty, 0]$ so that $-\zeta$ prevents $\beta$ from becoming negative, and when $\beta = 1$ the subdifferential has the exact value $\zeta \in [0, \infty)$ so that $-\zeta$ prevents $\beta$ from exceeding one.

We assume that there is no flux of $\beta$ on the boundary $\partial\Sigma$, denoting by $\mathbf{n}_\Sigma$ the outward normal to $\Sigma$ on $\partial\Sigma$, we have that the boundary condition for $\beta$ is $\partial\beta/\partial n_\Sigma = \mathbf{n}_\Sigma \cdot \nabla\beta = 0$.

Next, we consider the diffusion of moisture or water in the adhesive layer, driven from the boundary $\partial\Sigma$. Since we use $\eta$ as the fraction of water content, to guarantee (3), we use the subdifferential again, and we model the process as

$$\eta_t - \nabla(D\nabla\eta) \in -\partial I_{[0,1]}(\eta), \quad \text{on } \Sigma. \tag{7}$$

The explanation is as above. We assume that the diffusion coefficient depends on the tractions, temperature, moisture and the bonding field, $D = D(\mathbf{u}_{2\tau} - \mathbf{u}_{1\tau}, u_{2n} + u_{1n}, \theta_\Sigma, \beta, \eta)$. The moisture on $\partial\Sigma$ is a known function $\eta_{\partial\Sigma}$, which in applications is the humidity fraction of the environment.

Finally, we describe the heat diffusion in the two solids. We assume that there are no volume heat sources and so it is modeled with the linearized equation

$$\rho_\alpha c_\alpha \theta_{\alpha t} - \nabla \cdot (\kappa_\alpha \nabla \theta_\alpha) = -m_\alpha(\theta + \Theta_0)\nabla \cdot \mathbf{v}_\alpha. \tag{8}$$

Here, $\rho_\alpha$ is the density; $c_\alpha$ is the heat capacity per unit mass; $\kappa_\alpha$ is the coefficient of heat conduction, which may depend on the bonding and moisture fields; and $\Theta_0$ is the reference temperature from which we measure $\theta$, and is chosen as the ambient constant temperature. The term on the right-hand side represents the work done by the thermal stresses. We note that the viscosity term contributes a nonlinear term in $\mathbf{v}^2$ which we have neglected, to keep the equations linear, see for details [5, 6].

The temperature is prescribed on $\Gamma_0$ as $\theta_1 = \theta_{cl}$, heat exchange condition with the environment, which is at temperature $\Theta_0$, holds on the rest of the boundaries except $\Sigma$ with heat exchange coefficients $h_\alpha$, i.e.,

$$\kappa_\alpha \frac{\partial \theta_\alpha}{\partial n_\alpha} = h_\alpha \theta_\alpha,$$

and on $\Sigma$ we assume continuity of the temperature and the heat flux, that is $\theta_\Sigma = \theta_1 = \theta_2$ and $\kappa_1 \partial\theta_1/\partial n_1 = -\kappa_2 \partial\theta_2/\partial n_2$.

We next enforce the Signorini condition (4) since we do not allow the interpenetration of the solids into each other over $\Sigma$. To that end, we introduce the indicator function $I_{(-\infty,0]}(x)$ that has the value 0 when $-\infty < x \le 0$ and the value $+\infty$ when $0 < x$. The subdifferential of $I_{(-\infty,0]}(x)$ is the set-valued function

$$\partial I_{(-\infty,0]}(x) = \begin{cases} 0 & -\infty < x \le 0, \\ [0, \infty) & x = 0, \\ \emptyset & \text{otherwise.} \end{cases} \tag{9}$$

We use the subdifferential term in the normal stress condition on $\Sigma$ to guarantee that $u_{2n} + u_{1n} \le 0$, i.e., (4). Then, we let

$$\sigma_{1n} = -\sigma_{2n} \in -\beta K_n(u_{2n} + u_{1n}) - \partial I_{(-\infty,0]}(u_{2n} + u_{1n}),$$

which means that the subdifferential is zero as long as $u_{2n} + u_{1n} < 0$ and when $u_{2n} + u_{1n} = 0$ it generates the exact amount of 'resistance' (making $\sigma_{1n} < 0$, compressive and $\sigma_{2n}$, tensile) that prevent $u_{2n} + u_{1n} > 0$.

Next, to simplify a bit the notation, we let

$$\Omega_{1T} = \Omega_1 \times [0, T], \qquad \Omega_{2T} = \Omega_2 \times [0, T], \qquad \Sigma_T = \Sigma \times [0, T].$$

Collecting the equations and conditions above leads to the following model for the *irreversible adhesive debonding of two bonded bodies due to moisture, heat and contact stresses.*

**Model 1** Find the functions $\mathbf{u}_1, \mathbf{v}_1 = \mathbf{u}_{1t} : \Omega_{1T} \to \mathbb{R}^3$, $\mathbf{u}_2, \mathbf{v}_2 = \mathbf{u}_{2t} : \Omega_{2T} \to \mathbb{R}^3$, $\theta_1 : \Omega_1 \times [0, T] \to \mathbb{R}$, $\theta_2 : \Omega_2 \times [0, T] \to \mathbb{R}$, and $\beta, \eta : \Sigma_T \to \mathbb{R}$, such that ($\alpha = 1, 2$),

$$\mathbf{u}_{\alpha tt} - \nabla \cdot \sigma_\alpha = 0 \quad \text{in } \Omega_{\alpha T}, \tag{10}$$

$$c_\alpha \theta_{\alpha t} - \nabla \cdot (\kappa_\alpha \nabla \theta_\alpha) = -m_\alpha(\theta + \Theta_0) \nabla \cdot \mathbf{u}_{\alpha t} \quad \text{in } \Omega_{\alpha T}, \tag{11}$$

$$\beta_t - \nabla \cdot (\kappa_\beta \nabla \beta) + \Phi(\mathbf{u}_2 - \mathbf{u}_1, \theta_\Sigma, \beta, \eta) \in -\partial I_{[0,1]}(\beta) \quad \text{in } \Sigma_T, \tag{12}$$

$$\eta_t - \nabla(D \nabla \eta) \in \partial I_{[0,1]}(\eta) \quad \text{in } \Sigma_T, \tag{13}$$

where the stresses are

$$\sigma_\alpha = A_\alpha \varepsilon_\alpha(\mathbf{u}_\alpha) + B_\alpha \varepsilon_\alpha(\mathbf{v}_\alpha) - M_\alpha \theta_\alpha \quad \text{in } \Omega_{\alpha T}; \tag{14}$$

together with the boundary conditions, for $0 \le t \le T$,

$$\mathbf{u}_1 = 0, \quad \theta_1 = \theta_{cl} \quad \text{on } \Gamma_0, \tag{15}$$

$$\sigma_1 = 0, \quad \kappa_1 \frac{\partial \theta_1}{\partial n_1} = h_1 \theta_1 \quad \text{on } \partial\Omega_1 - (\Gamma_0 \cup \Sigma), \tag{16}$$

$$\kappa_2 \frac{\partial \theta_2}{\partial n_2} = h_2 \theta_2 \quad \text{on } \partial\Omega_2 - \Sigma, \tag{17}$$

$$\sigma_{1n} = -\sigma_{2n} \in -\beta K_n(u_{2n} + u_{1n}) - \partial I_{(-\infty,0]}(u_{2n} + u_{1n}), \quad \text{on } \Sigma, \tag{18}$$

$$\sigma_{1\tau} = -\sigma_{2\tau} = \beta K_\tau(\mathbf{u}_{2\tau} - \mathbf{u}_{1\tau}) \quad \text{on } \Sigma, \tag{19}$$

$$\theta_\Sigma = \theta_1 = \theta_2, \qquad \kappa_1 \frac{\partial \theta_1}{\partial n_1} = \kappa_2 \frac{\partial \theta_2}{\partial n_2} \quad \text{on } \Sigma, \tag{20}$$

$$\sigma_2 = 0 \quad \text{on } \partial\Omega_2 - (\Gamma_1 \cup \Sigma), \tag{21}$$

$$\sigma_2 = \mathbf{p} \quad \text{on } \Gamma_1, \tag{22}$$

$$\partial\beta/\partial n_\Sigma = 0 \quad \text{on } \partial\Sigma, \tag{23}$$

$$\eta = \eta_\Gamma \quad \text{on } \partial\Sigma; \tag{24}$$

and the initial conditions

$$\mathbf{u}_\alpha = \mathbf{u}_{\alpha 0}, \quad \mathbf{v}_\alpha = \mathbf{v}_{\alpha 0}, \quad \theta_\alpha = \theta_{\alpha 0} \quad \text{on } \Omega_\alpha, \tag{25}$$

$$\beta = \beta_0, \quad \eta = \eta_0 \quad \text{on } \Sigma. \tag{26}$$

Here, $\mathbf{u}_{\alpha 0}, \mathbf{v}_{\alpha 0}, \theta_{\alpha 0}$ are prescribed functions on $\Omega_\alpha$ and $\beta_0, \eta_0$ are prescribed on $\Sigma$.

We note that the system is fully coupled, since $\kappa_\beta = \kappa_\beta(\mathbf{u}_2 - \mathbf{u}_1, \theta_\Sigma, \beta, \eta)$ and $D = D(\mathbf{u}_2 - \mathbf{u}_1, \theta_\Sigma, \beta, \eta)$. Indeed, in applications the tractions, moisture and temperature affect the adhesive and so affect the diffusion process of water in it.

Moreover, it is nonlinear in part because of the use of the subdifferential terms that enforce the various constraints.

Finally, if one is interested in a thermoelastic system, one has to set all the viscosity coefficients in $B_\alpha$ to be zero.

## 2.1 Comments and Open Issues for Model 1

In this short section, we describe some open questions concerning Model 1.

First, if one wishes to consider the case when the glue layer is not thin, the mechanical equations and the diffusion equations for the temperature and the moisture need to be considered in the layer, together with the boundary conditions between the layer and the two solids. Clearly, this makes the problem much more complex and outside of the current work, but it may be of interest to consider it in the future. In particular, the adhesive boundary conditions between the glue layer and the surfaces of the two bodies need to be carefully studied.

Mathematically, it is of fundamental interest to establish the existence of solutions to a weak formulation of the model. Indeed, the model is rather complex, nonlinear and the existence of solutions is not guaranteed. Then, it is of interest to study possible uniqueness under special conditions on the problem data. Questions of regularity of the solutions and their properties are of interest too, especially since the subdifferential terms introduce a ceiling on the regularity.

Can one establish lower and upper bounds, depending on the problem data, on the time for complete debonding of the two solids, that is when $\beta = 0$ on $\Sigma$?

How does partial debonding, i.e., $\beta = 0$ or even $\beta$ small, on a part of the surface $\Sigma$ looks like and how does it affect the system evolution. Moreover, it is very likely in practice that in such a case the process of debonding will change to a much faster one and so the model will have to be modified.

If one adds randomness to some of the problem coefficients, such as $D$, $K_n$, $K_\tau$, $\mathbf{p}$, and especially $\Phi$, which are all hard to measure, there is interest to find the conditions for the measurability of the solutions. This allows one to consider the system expectation and some other statistical properties. Some recent results for related problems can be found in [11, 13].

Computationally, it is of interest to construct convergent numerical schemes for the reliable simulations of the process. In particular, dealing with the subdifferential constraints. Moreover, if one wishes to use a parameter identification scheme, the running time of a computer code must be short enough. Indeed, running the code many times will allow, as a first step, to perform sensitivity analysis to find which are the parameters that the system is most sensitive to, and which parameters can be estimated imprecisely and still provide useful information.

Finally, to make the model useful in predicting the debonding processes of real systems, there is a need to find experimentally the form and dependence of the various coefficient functions, especially $\Phi$, but also $\kappa_\beta$ and $D$, on the process variables. This may entail detailed experiments, collection of the relevant data and running computer simulations. Some of the parameter identification may be done on the simpler 1D systems described next.

## 3 1D Two Rod-Beams Model

We now describe a model for the process of debonding of two thin long bonded slabs that act as rods and beams, that is caused by humidity and mechanical effects. A derivation from the 3D model, Model 1 above, in the limit of two such long thin slabs is described in [20] as $l_{12}, l_{13}, l_{23} \to 0$. The method is that of [19, 24]. However, here the model development is *ad hoc*. We use the rod equations to describe the horizontal motion of the two slabs and the beam equations for the vertical motion, since both contribute to the debonding process. For the sake of simplicity, we assume that the system is kept in uniform ambient temperature $\theta$ that may be a function of time. Here, we present a somewhat different variant of the rod-beam system model in [14], assuming the temperature is that of the environment.

We consider two slabs occupying the intervals $[0, l_2]$ and $[l_1, 1]$, with $0 < l_1 < l_2 < 1$, that are bonded over the interval $[l_1, l_2]$. The left end of the first slab is clamped while its right end and the left end of the second slab are both free. Time dependent, possibly periodic, horizontal traction $p = p(t)$ and vertical shear $q(t)$ are acting at the right end of the second slab. We denote the horizontal displacements of the central axes by $u_1 = u_1(x,t)$ and $u_2 = u_2(x,t)$, while the vertical displacements by $w_1 = w_1(x,t)$ and $w_2 = w_2(x,t)$, respectively. Thus, we treat the horizontal motion of the slabs as that of rods, while the vertical motion as that of two beams. Moreover, the lengths and displacements are scaled so that the systems length is 1, and we note that this scaling is different from the one in Sect. 2. The setting is depicted in Fig. 2.

The slabs are assumed to be either elastic or viscoelastic, thus, the rod stresses are given by

$$\sigma_{r1} = E_1 A_{c1} u_{1x} + \overline{\nu_{r1}} u_{1xt}, \qquad \sigma_{r2} = E_2 A_{c2} u_{2x} + \overline{\nu_{r2}} u_{2xt}, \tag{27}$$

where $E_\alpha$ are the Young moduli, $A_{c\alpha}$ are the cross sections and $\overline{\nu_{r\alpha}}$ are the coefficients of viscosity, where here and below $\alpha = 1, 2$. The vertical (beam) moments and shear stresses are given by

$$M_1 = E_1 B_{a1} w_{1xx} + \overline{\nu_{b1}} w_{1xxt}, \qquad \sigma_{b1} = E_1 B_{a1} w_{1xxx} + \overline{\nu_{b1}} w_{1xxxt}, \tag{28}$$

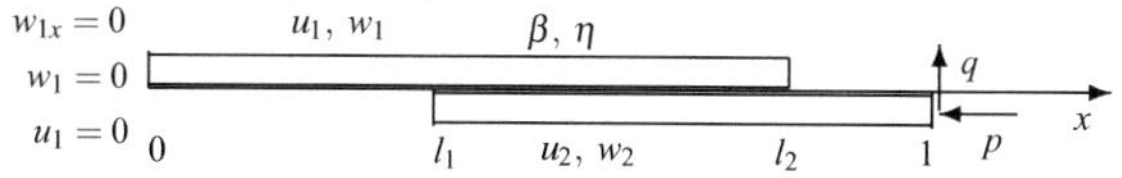


**Fig. 2** The two rod-beam system. The adhesive occupies the interval $l_1 \le x \le l_2$ where the bonding function $\beta$ and moisture function $\eta$ are defined

defined on $[0, l_2]$, where $B_{a\alpha}$ are the area moments, and

$$M_2 = E_2 B_{a2} w_{2xx} + \overline{\nu_{b2}} w_{2xxt} \qquad \sigma_{b2} = E_2 B_{a2} w_{2xxx} + \overline{\nu_{b2}} w_{2xxxt}, \tag{29}$$

defined on $[l_1, 1]$. Here and below, the subscripts $x$ and $t$ denote the partial $x$ and $t$ derivatives, respectively. When the slabs are elastic the $\overline{\nu}$s vanish. The equations of motion of the rods and beams are the usual ones, except for the unilateral condition $w_1 \geq w_2$ that must be imposed and is described below.

Below, we use the notation $\rho_\alpha$ for the density of the materials,

$$c_{r\alpha}^2 = \frac{E_\alpha}{\rho_\alpha}, \quad c_{b\alpha}^2 = \frac{E_\alpha B_\alpha}{(A_\alpha \rho_\alpha)}, \quad \nu_{r\alpha} = \frac{\overline{\nu_{r\alpha}}}{(\rho_\alpha A_\alpha)}, \quad \nu_{b\alpha} = \frac{\overline{\nu_{b\alpha}}}{(\rho_\alpha A_\alpha}.$$

We assume, as above, that the process of debonding is irreversible and denote by $\beta = \beta(x, t)$ the *bonding field* defined on $[l_1, l_2]$ that measure the strength of the bonding, actually, the pointwise fraction of active bonds and so it must satisfy the constraint (2). However, for the sake of convenience, when appropriate, we extend $\beta$ as zero to $[0, l_1]$ and $[l_2, 1]$. The shear force transmission between the slabs is reduced to $\beta K_\tau$ and the normal traction to $\beta K_n$. The full shear force in the adhesive is assumed to depend on $(u_2 - u_1)$ and we let it be $F_\tau = \beta K_\tau (u_2 - u_1)$. The vertical force transmission is assumed to be $F_n = \beta K_n (w_1 - w_2)$, and we note that outside $[l_1, l_2]$ these forces vanish since there is no glue there. To prevent interpenetration, we must impose the Signorini constraint (4) which in this context reads

$$w_2(x, t) \leq w_1(x, t). \tag{30}$$

For the sake of simplicity $K_\tau$ and $K_n$ are assumed to be large positive constants. We note that in applications, it is very likely that the bonding will break down when $\beta$ reaches a small value on a large portion of the interval and we remark on this issue below. The evolution of the bonding field is affected by $\beta K_\tau |u_2 - u_1|$, the vertical force $\beta K_n (w_1 - w_2)$, the humidity or *moisture* $\eta = \eta(x, t)$, in the adhesive, so it is defined on $[l_1, l_2]$ and the ambient temperature $\theta = \theta(t)$. Following [7], we assume that the debonding field is affected by neighboring elements and so we add some diffusion (see also [22, 23]) and describe the process with the debonding rate equation

$$\beta_t - (\kappa_\beta \beta_x)_x \in -\Phi(\beta K_\tau |u_2 - u_1|, \beta K_n (w_1 - w_2), \eta, \theta) - \partial I_{[0,1]}(\beta). \tag{31}$$

Here, $\kappa_\beta$ is the debonding diffusion coefficient function. The subdifferential term on the right-hand side guarantees that (2) holds. The *debonding source function* $\Phi$ is nonnegative and depends on the indicated variables. Moreover, we assume that debonding is irreversible, that is rebonding or mending do not take place, so the rate in nonpositive. Below, we discuss possible forms of $\Phi$, and we write

$$\Phi(u_1, u_2, w_1, w_2, \beta, \eta, \theta) = \Phi(\beta K_\tau |u_2 - u_1|, \beta K_n (w_1 - w_2), \eta, \theta).$$

We allow the debonding process to depend on the ambient temperature $\theta$.

To complete the model, we need to prescribe boundary conditions and the initial bonding field, which usually is assumed to be fully bonded, that is $\beta_0(x) = 1$, but we allow for a more general case with $\beta_0$ with $0 < \beta_0(x) \leq 1$. In practice, this is extremely difficult to find out without interfering with the system.

Next, we describe the diffusion of moisture in the adhesive layer, which is, as was noted in the Introduction, one of the main causes of the adhesive deterioration. We use the humidity or *moisture function* $\eta = \eta(x,t)$, which measures the fractional water content per unit length, and assume that the diffusion is affected by the internal strain and the fraction of the active bonds. Therefore, we model the diffusion as

$$\eta_t - (D\eta_x)_x \in -\partial I_{[0,1]}(\eta), \qquad l_1 < x < l_2. \tag{32}$$

Here, $D = D(u_2 - u_1, w_1 - w_2, \beta, \theta)$ is the humidity diffusion coefficient function assumed to be continuous, and for the sake of simplicity does not depend on $\eta$. Humidity diffusion is driven from the ends where it is that of the ambient air around the system, so we assume that at the ends $\eta(l_1, t) = \eta_L(t)$ and $\eta(l_2, t) = \eta_R(t)$. Although it is usually assumed that there is no initial moisture, for the sake of generality, we allow the initial condition $\eta = \eta_0 \in [0, 1]$.

Finally, to address the constraint (30) we introduce the characteristic function $\chi_{[l_1,l_2]}$ by

$$\chi_{[l_1,l_2]}(x) = \begin{cases} 1 & \text{if } l_1 \leq x \leq l_2, \\ 0 & \text{otherwise,} \end{cases}$$

and the subdifferential of the indicator function $I_{(-\infty,0]}(x)$ given in (8). Then, we add a term with $-\chi_{[l_1,l_2]}(x)\partial I_{(-\infty,0]}(w_2 - w_1)$ to the equations of motion for the beams, actually it is sufficient to add to only one of them, thus changing it into a differential inclusion. This term guarantees that $w_1 \geq w_2$ over $[l_1, l_2]$.

Next, for $T > 0$, we let

$$S_{1T} = (0, l_2) \times (0, T], \qquad S_{2T} = (l_1, 1) \times (0, T], \qquad G_T = (l_1, l_2) \times (0, T].$$

We recall that $\beta$ is extended by zero outside of $[l_1, l_2]$. However, for the sake of clarity, we use the characteristic function $\chi_{[l_1,l_2]}$ on the right-hand sides of the equations of motion.

Collecting the equations and the conditions above and writing the equations of motion in terms of the displacements in the slabs, leads to the following *dynamical model for the debonding of a viscoelastic rod-beam system caused by strain and humidity.*

**Model 2** Find the functions $u_1, w_1 : [0, l_2] \times [0, T] \to \mathbb{R}$, $u_2, w_2 : [l_1, 1] \times [0, T] \to \mathbb{R}$, and $\beta, \eta : [l_1, l_2] \times [0, T] \to \mathbb{R}$, such that,

$$u_{1tt} - c_{r1}^2 u_{1xx} - \nu_{r1} u_{1txx} = \chi_{[l_1,l_2]}(\cdot)\beta K_{\tau 1}(u_2 - u_1) \quad \text{on } S_{1T}, \tag{33}$$

$$u_{2tt} - c_{r2}^2 u_{2xx} - \nu_{r2} u_{2txx} = -\beta K_{\tau 2}(u_2 - u_1) \quad \text{on } S_{2T}, \tag{34}$$

$$w_{1tt} + c_{b1}^2 w_{1xxxx} + \nu_{b1} w_{1txxxx} + \beta K_{n1}(w_1 - w_2) \in \chi_{[l_1,l_2]}(\cdot)\partial I_{(-\infty,0]}(w_2 - w_1) \quad \text{on } S_{1T}, \tag{35}$$

$$w_{2tt} + c_{b2}^2 w_{2xxxx} + \nu_{b2} w_{2txxxx} = \beta K_{n2}(w_1 - w_2) \quad \text{on } S_{2T}, \tag{36}$$

$$\beta_t - (\kappa_\beta \beta_x)_x + \Phi \in -\partial I_{[0,1]}(\beta) \quad \text{on } G_T, \tag{37}$$

$$\eta_t - (D\,\eta_x)_x \in -\partial I_{[0,1]}(\eta) \quad \text{on } G_T, \tag{38}$$

where $\kappa_\beta = \kappa_\beta(u_2 - u_1, (w_1 - w_2), \beta, \eta)$, $D = D(u_2 - u_1, (w_1 - w_2), \beta, \eta)$, and $\Phi = \Phi(u_1, u_2, w_1, w_2, \beta, \eta, \theta)$,

$$u_1(0,t) = 0, \ \ \sigma_{r1}(l_2,t) = 0, \tag{39}$$

$$\sigma_{r2}(l_1,t) = 0, \ \ \sigma_{r2}(1,t) = p(t), \tag{40}$$

$$w_1(0,t) = w_{1x}(0,t) = 0, \ \ M_1(l_2,t) = \sigma_{b1}(l_2,t) = 0, \tag{41}$$

$$M_2(l_1,t) = \sigma_{b2}(l_1,t) = 0, \ \ M_2(1,t) = 0, \ \ \sigma_{b2}(1,t) = q(t), \tag{42}$$

$$\beta_x(l_1,t) = 0, \ \ \beta_x(l_2,t) = 0, \tag{43}$$

$$\eta(l_1,t) = \eta_L(t), \ \ \eta(l_2,t) = \eta_R(t), \tag{44}$$

$$u_1(\cdot,0) = u_{10}, \ \ u_{1t}(\cdot,0) = v_{10}^r, \ \ u_2(\cdot,0) = u_{20}, \ \ u_{2t}(\cdot,0) = v_{20}^r, \tag{45}$$

$$w_1(\cdot,0) = w_{10}, \ \ w_{1t}(\cdot,0) = v_{10}^b, \ \ w_2(\cdot,0) = w_{20}, \ \ w_{2t}(\cdot,0) = v_{20}^b, \tag{46}$$

$$w_{10} \geq w_{20}, \ \ \text{on } [l_1\, l_2], \tag{47}$$

$$\beta(\cdot,0) = \beta_0, \ \ \eta(\cdot,0) = \eta_0. \tag{48}$$

Here, the initial conditions are given in (45)–(48), with $0 < \beta_0(x), \eta_0(x) \leq 1$ on $[l_1, l_2]$. Condition (47) is the compatibility condition for the initial vertical displacements. We assume zero flux of $\beta$ at the end points, (43). The diffusion of moisture is given in (38) and the humidity boundary conditions are given in (44). In practice, $\eta_L(t) = \eta_R(t)$ and both are given functions. However, for the sake of generality, we allow them to be distinct. The subdifferential term in (35) guarantees that $w_1 \geq w_2$.

We note that the system is coupled via $\beta$, $K$, $\Phi$, $D$ and $\kappa$, and we assume that the $K$s are positive constants, although using a function with appropriate properties leads to similar results.

Finally, we note that $\Phi$ needs to be obtained from experimental data.

A similar system can be found in [14] where the slabs were assumed to be thermal insulators and the thermal diffusion in the adhesive layer was taken into account. There, an existence of the weak solutions of the model was established.

### 3.1 Comments and Open Issues for Model 2

In this short section we describe some open questions concerning Model 2, some are similar to those in Sect. 2.1.

First, there is a need to obtain the model, or some version of it, in the limit $l_{12}, l_{13}, l_{23} \to 0$ using an asymptotic approach, such as in [19, 24], which is in progress in [20].

Mathematically, as above, it is of fundamental interest to establish the existence of solutions to a weak formulation of the model. Then, it is of interest to study possible uniqueness under special conditions on the problem data. Questions of regularity of the solutions and their properties are of interest too, especially since the subdifferential terms introduce a ceiling on the regularity.

Moreover, can one establish lower and upper bounds, depending on the problem data, on the time for complete debonding of the two solids, that is when $\beta = 0$ on $\Sigma$?

Adding randomness to some of the problem coefficients, especially $\Phi$, is of interest.

Computationally, it is of interest to construct convergent numerical schemes that are fast. Indeed, if one wishes to use a parameter identification scheme, the running time of a computer code must be short enough. Since the model is essentially 1D, this is likely to be much easier to achieve than in Model 1.

As was noted above, to make the model useful in predicting the debonding processes of real systems, there is a need to find experimentally the form of $\Phi$ and its dependence of the various coefficient functions. Again, most of the parameter identification can be done on the system in Model 2, which is one of the reason a version of it was studied in [14].

## 4 The 1D Two Rods Model

In this short section we describe a model that is obtained from Model 2 when only the horizontal motion and the bonding and humidity are taken into account. Thus, the mechanical system is that of two rods, assumed to be either elastic or viscoelastic, the traction in the glue is shear stress and the restricted model is as follows. The notation is as in Sect. 3 and the setting is depicted in Fig. 3. It seems that this setting and the model may be useful for parameter identification purposes. It is much easier to set up experimentally, and easier to have optical access to the process. As was already mentioned, there is a need to obtain the form of the debonding source function $\Phi$ and its dependence on the system state.

We assume that a time dependent, possibly periodic, traction $p = p(t)$ is acting at the right end. The lengths and displacements are scaled so that the systems length is 1. The setting is depicted in Fig. 3.

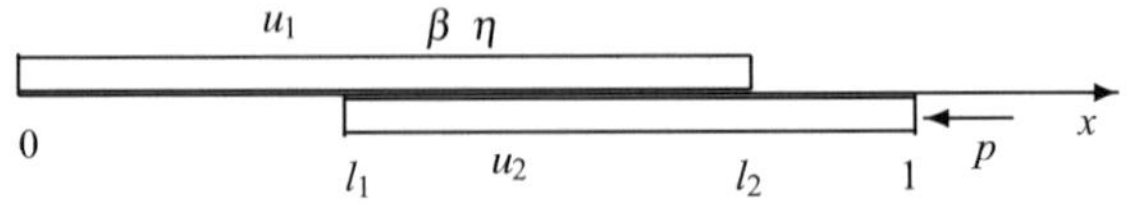


**Fig. 3** The adhesive occupies the interval $l_1 \le x \le l_2$, where the bonding function $\beta$ and the moisture function $\eta$ are defined

Modifying the equations and conditions in Model 2 leads to the following *dynamical model for the debonding of two viscoelastic rods caused by stress and humidity.*

**Model 3** Find the functions $u_1 : [0, l_2] \times [0, T] \to \mathbb{R}$, $u_2 : [l_1, 1] \times [0, T] \to \mathbb{R}$, $\beta, \eta : [l_1, l_2] \times [0, T] \to \mathbb{R}$, such that,

$$u_{1tt} - c_1^2 u_{1xx} - \nu_1 u_{1txx} = \chi_{[l_1,l_2]}(\cdot)\beta K_1(u_2 - u_1) \quad \text{in } S_{1T}, \tag{49}$$

$$u_{2tt} - c_2^2 u_{2xx} - \nu_2 u_{2txx} = -\chi_{[l_1,l_2]}(\cdot)\beta K_2(u_2 - u_1) \quad \text{in } S_{2T}, \tag{50}$$

$$\beta_t - (\kappa_\beta \beta_x)_x + \Phi(u_2 - u_1, \beta, \eta) \in -\partial I_{[0,1]}(\beta) \quad \text{in } G_T, \tag{51}$$

$$\eta_t - (D\eta_x)_x \in -\partial I_{[0,1]}(\eta) \quad \text{in } G_T; \tag{52}$$

$$u_1(0, t) = 0, \quad \sigma_1(l_2, t) = 0, \tag{53}$$

$$\sigma_2(l_1, t) = 0, \quad \sigma_2(1, t) = p(t), \tag{54}$$

$$\eta(l_1, t) = \eta_L(t), \quad \eta(l_2, t) = \eta_R(t), \tag{55}$$

$$u_1 = u_{10},\ u_{1t} = v_{10}, \quad u_2 = u_{20},\ u_{2t} = v_{20}, \tag{56}$$

$$\beta = \beta_0, \quad \eta = \eta_0. \tag{57}$$

Here, we recall that $\beta$ is extended by zero outside of $[l_1, l_2]$, the initial conditions are given in (56) and (57), with $0 < \beta_0(x), \eta_0(x) \le 1$ on $[l_1, l_2]$. The influx of moisture is given by the conditions (54), where $D = D(u_2 - u_1, \beta)$ and $\eta_L(t)$ and $\eta_R(t)$ represent the humidity just outside of the edges. In practice, $\eta_L(t) = \eta_R(t)$ and both given by functions that need to be found experimentally. However, here, for the sake of generality, we allow them to be distinct.

## 4.1 Two Comments on Model 3

Since Model 3 is considerably simplified both mathematically and in terms of its setting, it may be of interest to use it for parameter identification simulations coupled with experiments to obtain a useful form of $\Phi$ and its dependence on the shear stress, on the bonding field, and the humidity in the adhesive layer.

Although this problem is simplified, it is still a nonlinear problem and its analysis, especially existence, and numerical methods for its approximations are of interest. Numerical simulations of a version of this problem are currently underway, [15], and the existence is scheduled to appear there.

**Acknowledgements** I would like to thank the referees for their nice and useful comments. I would also like to express my gratitude to Prof. A. Rodríguez-Arós for his detailed comments that improved the paper considerably.

# References

1. Andrews, K.T., Chapman, L., Fernández, J.R., Fisackerly, M., Shillor, M., Vanerian, L., Van-Houten, T.: A membrane in adhesive contact. SIAM J. Appl. Math. **64**(1), 152–169 (2003)
2. Baldan, A.: Adhesion phenomena in bonded joints. Int. J. Adhes. Adhes. **38**, 95–116 (2012)
3. Chau, O., Fernández, J.R., Shillor, M., Sofonea, M.: Variational and numerical analysis of a quasistatic viscoelastic contact problem with adhesion. J. Comput. Appl. Math. **159**, 431–465 (2003)
4. Chau, O., Shillor, M., Sofonea, M.: Dynamic frictionless contact with adhesion. Z. Angew. Math. Phys. ZAMP **55**, 32–47 (2004)
5. Eck, C., Jarušek, J.: On thermal aspect of dynamic contact problems. Math. Bohem. **126**, 337–352 (2001)
6. Eck, C., Jarušek, J., Krbec, M.: Unilateral contact problems. Pure and Applied Mathematics. Chapman & Hall/CRC Press, Boca Raton (2005)
7. Frémond, M.: Non-Smooth Thermomechanics. Springer, Berlin (2002)
8. Han, W., Kuttler, K.L., Shillor, M., Sofonea, M.: Elastic beam in adhesive contact. Int. J. Solids Struct. **39**, 1145–1164 (2002)
9. Jianu, L., Shillor, M., Sofonea, M.: A viscoelastic bilateral frictionless contact problem with adhesion. Appl. Anal. **80**(1–2), 233–255 (2001)
10. Klarbring, A., Movchan, A.: Asymptotic modeling of adhesively bonded beams. Mech. Mater. **28**, 137–145 (1998)
11. Kuttler, K.L., Shillor, M.: Product measurability with applications to a stochastic contact problem with friction. Electron. J. Differ. Equ. **2014**(258), 1–29 (2014)
12. Kuttler, K.L., Shillor, M., Fernandez, J.R.: Existence and regularity for dynamic viscoelastic adhesive contact with damage. Appl. Math. Optim. **53**, 31–66 (2006)
13. Kuttler, K.L., Li, J., Shillor, M.: A general product measurability theorem with applications to variational inequalities. Electron. J. Differ. Equ. **2016**(90), 1–12 (2016)
14. Kuttler, K.L., Marcinek, P., Shillor, M.: Analysis and simulations of debonding of bonded rods/beams caused by humidity and thermal effects (2016, in preparation)
15. Marcinek, P.: Doctoral Dissertation, Oakland University (expected 2017)
16. Nassar, S.A., Mazhari, E.: A coupled shear stress- diffusion model for adhesively bonded single lap joints. J. Appl. Mech. **83**(10), 101006–1 (2016)
17. Nassar, S.A., Sakai, K.: Effect of cyclic heat, humidity, and joining method on the static and dynamic performance of lightweight multi-material single lap joints. J. Manuf. Sci. Eng.-ASME Trans. 137, 051026-1-051026-11 (2015)
18. Raous, M., Cangémi, L., Cocu, M.: A consistent model coupling adhesion, friction and unilateral contact. Comput. Meth. Appl. Mech. Eng. **177**, 383–399 (1999)
19. Rodriguez-Aros, A., Viano, J.M.: Mathematical justification of viscoelastic beam models by asymptotic methods. J. Math. Anal. Appl. **370**(2), 607–634 (2010)
20. Rodriguez-Aros, A., Kuttler, K.L. , Shillor, M.: Derivation of a 1D system for humidity caused debonding of rods and beams (2016, in preparation)
21. Sakai, K., Nassar, S.: Failure analysis of adhesively bonded GFRP single lap joints after cyclic environmental loading. In: Proceedings of the American Society of Composites, East Lansing, Michigan September 28–30, Paper 1820
22. Shillor, M., Sofonea, M., Telega, J.J.: Models and Analysis of Quasistatic Contact Variational Approach. Lecture Notes in Physics. Springer, Berlin (2004)

23. Sofonea, M., Han, W., Shillor, M.: Analysis and Approximations of Contact Problems with Adhesion or Damage. Pure and Applied Mathematics, vol. 276. Chapman & Hall/CRC Press, Boca Raton (2006)
24. Trabucho, L., Viano, J.M.: Mathematical modelling of rods. Handb. Numer. Anal. **4**, 487–974 (1996)
25. Tsai, M.Y., D.W., Oplinger, Morton, J.: Improved theoretical solutions for adhesive lap joints. Int. J. Solids Struct. **35**(12), 1163–1185 (1998)

# A Variational-Hemivariational Inequality in Contact Mechanics

**Mircea Sofonea, Weimin Han and Mikaël Barboteu**

**Abstract** This chapter deals with a new mathematical model for the frictional contact between an elastic body and a rigid foundation covered by a deformable layer made of soft material. We study the model in the form of a variational-hemivariational inequality for the displacement field. We review a unique solvability result of the problem under certain assumptions on the data. Then we turn to the numerical solution of the problem, based on the finite element method. We derive an optimal order error estimate for the linear finite element solution. Finally, we present numerical simulation results in the study of a two-dimentional academic example. The theoretically predicted optimal convergence order is observed numerically. Moreover, we provide mechanical interpretations of the numerical results for our contact model.

## 1 Introduction

Phenomena of contact involving deformable bodies abound in industry and daily life. Due to their inherent complexity, they lead to mathematical models expressed in terms of nonlinear boundary value problems which, in variational formulation, give rise to challenging inequality problems. Analysis of these problems is based on arguments of nonlinear functional analysis through the theory of variational and hemivariational inequalities.

The theory of variational inequalities started in early sixties and has gone through substantial development since then, see for instance [1, 5, 6, 14] and the references therein. It was built on arguments of monotonicity and convexity, including properties

M. Sofonea (✉) · M. Barboteu
Laboratoire de Mathématiques et Physique, Université de Perpignan Via Domitia, 52 Avenue Paul Alduy, 66100 Perpignan, France
e-mail: sofonea@univ-perp.fr

M. Barboteu
e-mail: barboteu@univ-perp.fr

W. Han
Department of Mathematics, University of Iowa, Iowa City, IA 52242, USA
e-mail: weimin-han@uiowa.edu



F. dell'Isola et al. (eds.), *Mathematical Modelling in Solid Mechanics*,
Advanced Structured Materials 69, DOI 10.1007/978-981-10-3764-1_16

of the subdifferential of a convex function. In contrast, the theory of hemivariational inequalities is based on properties of the subdifferential in the sense of Clarke, defined for locally Lipschitz functions which may be nonconvex. Analysis of hemivariational inequalities, including existence and uniqueness results, can be found in [12, 17, 20, 23]. Both variational and hemivariational inequalities have been extensively used in the study of various problems in Mechanics, Physics and Engineering Sciences and, in particular, in Contact Mechanics. References on this matter include [4, 7, 8, 13, 15, 17, 22–24, 26], among others. Variational-hemivariational inequalities are inequality problems where both convex and nonconvex functions are involved. They have been introduced in the pioneering work [21] and were further studied in [20, 23].

Recently, a new variational-hemivariational inequality is studied in [9]. The inequality involves two nonlinear operators and two nondifferentiable functionals, of which at least one is convex. There, solution existence, uniqueness and data continuous dependence are shown. Moreover, the finite element method is studied for solving the inequality problem. For the first time in the literature, an optimal order error estimate is derived for the linear element solution of a hemivariational inequality under appropriate solution regularity assumptions. A more general variational-hemivariational inequality is analyzed in [19]. Solution existence and uniqueness are proved, together with a result on the continuous dependence of the solution on the data. This study was continuated in [10, 11] where numerical analysis of variational-hemivariational inequalities was performed.

The purpose of this chapter is to illustrate the use of variational-hemivariational inequalities in the analysis and numerical approximations of an elastic contact problem. We use an abstract result to prove the unique solvability of the problem. For the finite element method of the problem, we derive error estimates, which are of optimal order for the linear elements. We provide numerical simulation results to illustrate the performance of the numerical method, including numerical convergence order.

The rest of the chapter is organized as follows. In Sect. 2 we introduce the contact problem in which the material's behavior is modeled with a nonlinear elastic constitutive law and the contact conditions are in a subdifferential form and are associated with unilateral constraints. In Sect. 3, we list the assumptions on the data and state a unique solvability result on the problem. The proof of the unique solvability statement is based on a recent abstract result obtained in [19]. In Sect. 4, we provide numerical analysis of the contact model, including convergence and error estimation results. Finally, in Sect. 5, we report numerical simulation results which provide numerical evidence of our optimal order error estimate and give rise to interesting mechanical interpretations.

## 2 The Contact Model

Let $\Omega$ be the reference configuration of the elastic body, assumed to be an open, bounded and connected set in $\mathbb{R}^d$ $(d = 2, 3)$. The boundary $\Gamma = \partial\Omega$ is assumed Lipschitz continuous and is partitioned into three disjoint and measurable parts $\Gamma_1$,

$\Gamma_2$ and $\Gamma_3$ such that meas $(\Gamma_1) > 0$. The body is in equilibrium under the action of a body force of density $\mathbf{f}_0$ in $\Omega$ and a surface traction of density $\mathbf{f}_2$ on $\Gamma_2$, is fixed on $\Gamma_1$, and is in frictional contact on $\Gamma_3$ with a foundation. We use $\mathbb{S}^d$ for the space of second order symmetric tensors on $\mathbb{R}^d$. Also, "$\cdot$" and "$\|\cdot\|$" will represent the canonical inner product and the Euclidean norm on the spaces $\mathbb{R}^d$ and $\mathbb{S}^d$. We denote by $\mathbf{u}\colon \Omega \to \mathbb{R}^d$ and $\boldsymbol{\sigma}: \Omega \to \mathbb{S}^d$ the displacement field and the stress field, respectively. In addition, we use $\boldsymbol{\varepsilon}(\mathbf{u})$ to denote the linearized strain tensor. Let $\nu$ be the unit outward normal vector, defined a.e. on $\Gamma$. For a vector field $\mathbf{v}$, we use $v_\nu := \mathbf{v}\cdot\nu$ and $\mathbf{v}_\tau := \mathbf{v} - v_\nu\nu$ for the normal and tangential components of $\mathbf{v}$ on $\Gamma$. Similarly, for the stress field $\boldsymbol{\sigma}$, its normal and tangential components on the boundary are defined as $\sigma_\nu := (\boldsymbol{\sigma}\nu)\cdot\nu$ and $\boldsymbol{\sigma}_\tau := \boldsymbol{\sigma}\nu - \sigma_\nu\nu$, respectively.

With the above notation, the contact model to be studied is the following.

PROBLEM $P$. *Find a displacement field* $\mathbf{u}\colon \Omega \to \mathbb{R}^d$, *a stress field* $\boldsymbol{\sigma}: \Omega \to \mathbb{S}^d$ *and an interface force* $\xi_\nu\colon \Gamma_3 \to \mathbb{R}$ *such that*

$$\boldsymbol{\sigma} = \mathscr{F}(\boldsymbol{\varepsilon}(\mathbf{u})) \quad \text{in } \Omega, \tag{1}$$

$$\operatorname{Div}\boldsymbol{\sigma} + \mathbf{f}_0 = \mathbf{0} \quad \text{in } \Omega, \tag{2}$$

$$\mathbf{u} = \mathbf{0} \quad \text{on } \Gamma_1, \tag{3}$$

$$\boldsymbol{\sigma}\nu = \mathbf{f}_2 \quad \text{on } \Gamma_2, \tag{4}$$

$$u_\nu \le g,\ \sigma_\nu + \xi_\nu \le 0,\ (u_\nu - g)(\sigma_\nu + \xi_\nu) = 0,\ \xi_\nu \in \partial j_\nu(u_\nu) \quad \text{on } \Gamma_3, \tag{5}$$

$$\|\boldsymbol{\sigma}_\tau\| \le F_b(u_\nu), \quad -\boldsymbol{\sigma}_\tau = F_b(u_\nu)\frac{\mathbf{u}_\tau}{\|\mathbf{u}_\tau\|} \ \text{ if } \mathbf{u}_\tau \ne \mathbf{0} \quad \text{on } \Gamma_3. \tag{6}$$

In (1)–(6) and sometimes below, we do not indicate explicitly the dependence of various functions on the spatial variable $\mathbf{x} \in \Omega \cup \Gamma$. We now present a short description of the equations and conditions in Problem $P$ and we refer the reader to the books [17, 26] for more details on the modelling of contact problems. First, Eq. (1) is the constitutive law for elastic materials in which $\mathscr{F}$ represents the elasticity operator, allowed to be nonlinear. Equation (2) is the equilibrium equation and is used here since the process is assumed to be static. Condition (3) represents the displacement condition and condition (4) is the traction condition. Relations (5) and (6) represent the contact condition and the friction law, respectively. Here $g \ge 0$, $\partial j_\nu$ denotes the Clarke subdifferential of the given function $j_\nu$, and $F_b$ denotes a positive function, the friction bound.

Note that condition (5) models the contact with a foundation made of a rigid body covered by a layer of soft material, say asperities. It is obtained through the following considerations:

(a) The penetration is restricted by the rigid body, i.e.

$$u_\nu \le g, \tag{7}$$

where $g \geq 0$ represents the thickness of the soft layer. We consider the non-homogeneous case, i.e., $g$ is allowed to be a function of the spatial variable $\mathbf{x} \in \Gamma_3$.

(b) The normal stress has an additive decomposition of the form

$$\sigma_\nu = \sigma_\nu^D + \sigma_\nu^R, \tag{8}$$

where the term $\sigma_\nu^D$ describes the reaction of the soft layer and $\sigma_\nu^R$ describes the reaction of the rigid body.

(c) The component $\sigma_\nu^D$ satisfies a multivalued normal compliance condition of the form

$$-\sigma_\nu^D \in \partial j_\nu(u_\nu). \tag{9}$$

Examples of contact conditions of the form (9) can be found in [17], for instance.

(d) The component $\sigma_\nu^R$ satisfies the Signorini unilateral condition in a form with the gap $g$, i.e.

$$\sigma_\nu^R \leq 0, \qquad \sigma_\nu^R(u_\nu - g) = 0. \tag{10}$$

Comments and mechanical interpretation on the contact condition (10) can be found in [24] and the references therein.

Denote $-\sigma_\nu^D = \xi_\nu$. Then, it is easy to see that the contact condition (5) is a direct consequence of relations (7)–(9).

The friction law (6) was used in [25], associated with a multivalued normal compliance contact condition without unilateral constraint. Here the friction bound $F_b$ may depend on the normal displacement $u_\nu$, which is reasonable from the physical point of view, as explained in [25].

Note that, due to the strong nolinearities involved, in general Problem $P$ does not have classical solution. Therefore, as usual in Contact Mechanics, its study is made by using a weak formulation, the so-called variational formulation. The formulation will allow one to prove the unique solvability of the problem and to construct numerical schemes for the approximation of the weak solution.

## 3 Variational Analysis

In the study of Problem $P$ we use standard notation for Lebesgue and Sobolev spaces. For the stress and strain fields, we use the space $Q = L^2(\Omega; \mathbb{S}^d)$, which is a Hilbert space with the canonical inner product

$$(\boldsymbol{\sigma}, \boldsymbol{\tau})_Q := \int_\Omega \sigma_{ij}(\mathbf{x})\, \tau_{ij}(\mathbf{x})\, dx, \quad \boldsymbol{\sigma}, \boldsymbol{\tau} \in Q$$

and the associated norm $\|\cdot\|_Q$. The displacement fields will be sought in a subset of the space

$$V = \left\{ \mathbf{v} = (v_i) \in H^1(\Omega;\mathbb{R}^d) \mid \mathbf{v} = \mathbf{0} \ \text{ on } \Gamma_1 \right\}.$$

Since meas $(\Gamma_1) > 0$, it is known that $V$ is a Hilbert space with the inner product

$$(\mathbf{u},\mathbf{v})_V := \int_\Omega \boldsymbol{\varepsilon}(\mathbf{u}) \cdot \boldsymbol{\varepsilon}(\mathbf{v})\,dx, \quad \mathbf{u},\mathbf{v} \in V$$

and the associated norm $\|\cdot\|_V$. We denote by $V^*$ the topological dual of $V$, and by $\langle\cdot,\cdot\rangle_{V^*\times V}$ the duality pairing of $V$ and $V^*$. When no confusion may arise, we simply write $\langle\cdot,\cdot\rangle$ instead of $\langle\cdot,\cdot\rangle_{V^*\times V}$. For $\mathbf{v} \in H^1(\Omega;\mathbb{R}^d)$ we use the same symbol $\mathbf{v}$ for the trace of $\mathbf{v}$ on $\Gamma$. By the Sobolev trace theorem we have

$$\|\mathbf{v}\|_{L^2(\Gamma_3;\mathbb{R}^d)} \le \|\Gamma\|\,\|\mathbf{v}\|_V \quad \forall\,\mathbf{v} \in V, \tag{11}$$

$\|\gamma\|$ being the norm of the trace operator $\gamma\colon V \to L^2(\Gamma_3;\mathbb{R}^d)$.

We now turn to the assumptions on the data. First, the elasticity operator $\mathscr{F}\colon \Omega \times \mathbb{S}^d \to \mathbb{S}^d$ and the potential function $j_\nu\colon \Gamma_3 \times \mathbb{R} \to \mathbb{R}$, are assumed to have the following properties:

$$\left\{\begin{array}{l}
\text{(a) there exists } L_{\mathscr{F}} > 0 \text{ such that for all } \boldsymbol{\varepsilon}_1, \boldsymbol{\varepsilon}_2 \in \mathbb{S}^d, \ \text{a.e. } \mathbf{x} \in \Omega, \\
\qquad \|\mathscr{F}(\mathbf{x},\boldsymbol{\varepsilon}_1) - \mathscr{F}(\mathbf{x},\boldsymbol{\varepsilon}_2)\| \le L_{\mathscr{F}} \|\boldsymbol{\varepsilon}_1 - \boldsymbol{\varepsilon}_2\|; \\
\text{(b) there exists } m_{\mathscr{F}} > 0 \text{ such that for all } \boldsymbol{\varepsilon}_1, \boldsymbol{\varepsilon}_2 \in \mathbb{S}^d, \ \text{a.e. } \mathbf{x} \in \Omega, \\
\qquad (\mathscr{F}(\mathbf{x},\boldsymbol{\varepsilon}_1) - \mathscr{F}(\mathbf{x},\boldsymbol{\varepsilon}_2)) \cdot (\boldsymbol{\varepsilon}_1 - \boldsymbol{\varepsilon}_2) \ge m_{\mathscr{F}}\, \|\boldsymbol{\varepsilon}_1 - \boldsymbol{\varepsilon}_2\|^2; \\
\text{(c) } \mathscr{F}(\cdot,\boldsymbol{\varepsilon}) \text{ is measurable on } \Omega \text{ for all } \boldsymbol{\varepsilon} \in \mathbb{S}^d; \\
\text{(d) } \mathscr{F}(\mathbf{x},\mathbf{0}) = \mathbf{0} \text{ for a.e. } \mathbf{x} \in \Omega.
\end{array}\right. \tag{12}$$

$$\left\{\begin{array}{l}
\text{(a) } j_\nu(\cdot,r) \text{ is measurable on } \Gamma_3 \text{ for all } r \in \mathbb{R} \text{ and there} \\
\qquad \text{exists } \bar{e} \in L^2(\Gamma_3) \text{ such that } j_\nu(\cdot,\bar{e}(\cdot)) \in L^1(\Gamma_3); \\
\text{(b) } j_\nu(\mathbf{x},\cdot) \text{ is locally Lipschitz on } \mathbb{R} \text{ for a.e. } \mathbf{x} \in \Gamma_3; \\
\text{(c) } |\partial j_\nu(\mathbf{x},r)| \le \bar{c}_0 + \bar{c}_1 |r| \text{ for a.e. } \mathbf{x} \in \Gamma_3, \\
\qquad \text{for all } r \in \mathbb{R} \text{ with } \bar{c}_0,\ \bar{c}_1 \ge 0; \\
\text{(d) } j_\nu^0(\mathbf{x},r_1;r_2 - r_1) + j_\nu^0(\mathbf{x},r_2;r_1 - r_2) \le \alpha_{j_\nu} |r_1 - r_2|^2 \\
\qquad \text{for a.e. } \mathbf{x} \in \Gamma_3, \ \text{all } r_1, r_2 \in \mathbb{R} \text{ with } \alpha_{j_\nu} \ge 0.
\end{array}\right. \tag{13}$$

On the penetration bound $g\colon \Gamma_3 \to \mathbb{R}$ and the friction bound $F_b\colon \Gamma_3 \times \mathbb{R} \to \mathbb{R}_+$, we assume

$$g \in L^2(\Gamma_3), \quad g(\mathbf{x}) \ge 0 \ \text{ a.e. on } \Gamma_3, \tag{14}$$

$$\left\{\begin{array}{l}
\text{(a) there exists } L_{F_b} > 0 \text{ such that} \\
\qquad |F_b(\mathbf{x},r_1) - F_b(\mathbf{x},r_2)| \le L_{F_b} |r_1 - r_2| \ \ \forall\, r_1, r_2 \in \mathbb{R}, \ \text{a.e. } \mathbf{x} \in \Gamma_3; \\
\text{(b) } F_b(\cdot,r) \text{ is measurable on } \Gamma_3, \ \text{for all } r \in \mathbb{R}; \\
\text{(c) } F_b(\mathbf{x},r) = 0 \text{ for } r \le 0, \ F_b(\mathbf{x},r) \ge 0 \text{ for } r \ge 0, \ \text{a.e. } \mathbf{x} \in \Gamma_3.
\end{array}\right. \tag{15}$$

Finally, on the densities of the body force and the surface traction, we assume

$$\mathbf{f}_0 \in L^2(\Omega; \mathbb{R}^d), \quad \mathbf{f}_2 \in L^2(\Gamma_2; \mathbb{R}^d). \tag{16}$$

Define $\mathbf{f} \in V^*$ by

$$\langle \mathbf{f}, \mathbf{v} \rangle_{V^* \times V} = (\mathbf{f}_0, \mathbf{v})_{L^2(\Omega;\mathbb{R}^d)} + (\mathbf{f}_2, \mathbf{v})_{L^2(\Gamma_2;\mathbb{R}^d)} \quad \forall\, \mathbf{v} \in V. \tag{17}$$

Corresponding to the constraint $u_\nu \le g$ on $\Gamma_3$ in (5), we introduce the following subset of the space $V$:

$$U := \{\mathbf{v} \in V \mid v_\nu \le g \text{ on } \Gamma_3\}. \tag{18}$$

Also, we use the notation $j_\nu^0(u, v)$ for the generalized directional derivative of $j_\nu$ at $u \in \mathbb{R}$ in the direction $v \in \mathbb{R}$, defined by

$$j_\nu^0(u; v) := \limsup_{y \to u,\, \lambda \downarrow 0} \frac{j_\nu(y + \lambda v) - j_\nu(y)}{\lambda}.$$

Then, from the definition of Clarke subdifferential the following implication holds:

$$\xi_\nu \in \partial j_\nu(u_\nu) \text{ a.e. on } \Gamma_3 \quad \Longrightarrow \quad j_\nu^0(u_\nu; v_\nu) \ge \xi_\nu v_\nu \text{ a.e. on } \Gamma_3, \ \forall\, \mathbf{v} \in V. \tag{19}$$

By a standard approach, based on integration by parts and the inequality (19), the following weak formulation of the contact problem $P$ can be derived.
PROBLEM $P_V$. *Find a displacement field* $\mathbf{u} \in U$ *such that*

$$\begin{aligned}(\mathscr{F}(\boldsymbol{\varepsilon}(\mathbf{u})), \boldsymbol{\varepsilon}(\mathbf{v} - \mathbf{u}))_Q &+ \int_{\Gamma_3} F_b(u_\nu)\,(\|\mathbf{v}_\tau\| - \|\mathbf{u}_\tau\|)\, d\Gamma \\ &+ \int_{\Gamma_3} j_\nu^0(u_\nu; v_\nu - u_\nu)\, d\Gamma \ge \langle \mathbf{f}, \mathbf{v} - \mathbf{u} \rangle_{V^* \times V} \quad \forall\, \mathbf{v} \in U.\end{aligned} \tag{20}$$

Note that the inequality (20) has both a convex and nonconvex structure. Its convex structure is given by the subset of the admisible displacement fields $U$, which is convex, and the function

$$\mathbf{v} \mapsto \int_{\Gamma_3} F_b(u_\nu)\|\mathbf{v}_\tau\|\, d\Gamma,$$

which is a convex function on $V$. The nonconvex structure of the inequality (20) follows from the term

$$\int_{\Gamma_3} j_\nu^0(u_\nu; v_\nu - u_\nu)\, d\Gamma$$

which involves a possibly nonconvex locally Lipschitz functions $j_\nu$. We conclude from here that the inequality (20) represents a variational-hemivariational inequality.

The analysis of inequalities of the form (20) has been carried out in [11, 19], in an abstract functional framework. There, a general existence and uniqueness result for inequalities with pseudomonotone operators was provided, under a smallness assumption on the data. The use of this abstract result in the study of (20) is straighforward and, therefore, we skip it. The main point is the use of smallness assumption, that we describe in what follows.

Let $\lambda_{1,V} > 0$ be the smallest eigenvalue of the eigenvalue problem

$$\mathbf{u} \in V, \quad \int_{\Omega} \boldsymbol{\varepsilon}(\mathbf{u})\cdot\boldsymbol{\varepsilon}(\mathbf{v})\, dx = \lambda \int_{\Gamma_3} \mathbf{u}\cdot\mathbf{v}\, d\Gamma \quad \forall\, \mathbf{v} \in V,$$

and let $\lambda_{1\nu,V} > 0$ be the smallest eigenvalue of the eigenvalue problem

$$\mathbf{u} \in V, \quad \int_{\Omega} \boldsymbol{\varepsilon}(\mathbf{u})\cdot\boldsymbol{\varepsilon}(\mathbf{v})\, dx = \lambda \int_{\Gamma_3} u_\nu v_\nu d\Gamma \quad \forall\, \mathbf{v} \in V.$$

Assume also that

$$L_{F_b}\lambda_{1,V}^{-1} + \alpha_{j_\nu}\lambda_{1\nu,V}^{-1} < m_{\mathscr{F}}, \tag{21}$$

Then, using the abstract result in [11] it follows that, under the assumptions (12), (12)–(16) and (21), Problem $P_V$ has a unique solution $\mathbf{u} \in U$.

Let $\mathbf{u} \in U$ be the solution of Problem $P_V$ and denote by $\boldsymbol{\sigma} \in Q$ the function given by $\boldsymbol{\sigma} = \mathscr{F}\boldsymbol{\varepsilon}(\nu)$. The couple $(\mathbf{u}, \boldsymbol{\sigma})$ is called a weak solution to the contact problem $P$. We conclude from the above discussion that the latter has a unique weak solution.

## 4 Numerical Analysis

We now consider the finite element method of solving Problem $P_V$. For simplicity, assume $\Omega$ is a polygonal/polyhedral domain and express the three parts of the boundary, $\Gamma_k$, $1 \le k \le 3$, as unions of closed flat components with disjoint interiors:

$$\overline{\Gamma_k} = \cup_{i=1}^{i_k} \Gamma_{k,i}, \quad 1 \le k \le 3.$$

Let $\{\mathscr{T}^h\}$ be a regular family of partitions of $\overline{\Omega}$ into triangles/tetrahedrons that are compatible with the partition of the boundary $\partial\Omega$ into $\Gamma_{k,i}$, $1 \le i \le i_k$, $1 \le k \le 3$, in the sense that if the intersection of one side/face of an element with one set $\Gamma_{k,i}$ has a positive measure with respect to $\Gamma_{k,i}$, then the side/face lies entirely in $\Gamma_{k,i}$. Construct the linear element space corresponding to $\mathscr{T}^h$:

$$V^h = \left\{\mathbf{v}^h \in C(\overline{\Omega})^d \mid \mathbf{v}^h|_T \in \mathbb{P}_1(T)^d,\ T \in \mathscr{T}^h,\ \mathbf{v}^h = \mathbf{0} \text{ on } \Gamma_1\right\},$$

and the related finite element subset $U^h = V^h \cap U$. Assume $g$ is a concave function. Then

$$U^h = \left\{ \mathbf{v}^h \in V^h \mid v_\nu^h \le g \text{ at node points on } \Gamma_3 \right\}.$$

Note that $\mathbf{0} \in U^h$. Define the following numerical method for Problem $P_V$.

PROBLEM $P_V^h$. *Find a displacement field* $\mathbf{u}^h \in U^h$ *such that*

$$(\mathscr{F}(\boldsymbol{\varepsilon}(\mathbf{u}^h)), \boldsymbol{\varepsilon}(\mathbf{v}^h - \mathbf{u}^h))_Q + \int_{\Gamma_3} F_b(u_\nu^h) \left( \|\mathbf{v}_\tau^h\| - \|\mathbf{u}_\tau^h\| \right) d\Gamma$$
$$+ \int_{\Gamma_3} j_\nu^0(u_\nu^h; v_\nu^h - u_\nu^h)\, d\Gamma \ge \langle \mathbf{f}, \mathbf{v}^h - \mathbf{u}^h \rangle_{V^* \times V} \quad \forall\, \mathbf{v}^h \in U^h. \tag{22}$$

For an error analysis, we assume

$$\mathbf{u} \in H^2(\Omega)^d, \quad \boldsymbol{\sigma}\nu \in L^2(\Gamma_3)^d. \tag{23}$$

Note that for many application problems, $\boldsymbol{\sigma}\nu \in L^2(\Gamma_3)^d$ follows from $\mathbf{u} \in H^2(\Omega)^d$; e.g., this is the case where the material is linearized elastic with suitably smooth coefficients, or where the elasticity operator $\mathscr{F}$ depends on $\mathbf{x}$ smoothly.

The starting point for obtaining error estimates is the inequality

$$\|\mathbf{u} - \mathbf{u}^h\|_V^2 \le c \left[ \|\mathbf{u} - \mathbf{v}^h\|_V^2 + \|\mathbf{u} - \mathbf{v}^h\|_{L^2(\Gamma_3)^d} + R(\mathbf{v}^h) \right] \quad \forall\, \mathbf{v}^h \in U^h. \tag{24}$$

This inequality is based on the properties of the operators $\mathscr{F}$, the function $F_b$, the potential $j_\nu$ and the trace inequality (11). Its proof follows from an abstract error estimation result in the study of elliptic variational-hemivariational inequalities which can be found in [11]. In (24) and below, $c$ represents a positive constant which does not depend on $h$ and whose value may change from line to line and $R(\mathbf{v}^h)$ is a residual term defined by

$$R(\mathbf{v}^h) = (\mathscr{F}(\boldsymbol{\varepsilon}(\mathbf{u})), \boldsymbol{\varepsilon}(\mathbf{v}^h - \mathbf{u}))_Q + \int_{\Gamma_3} F_b(u_\nu) \left( \|\mathbf{v}_\tau^h\| - \|\mathbf{u}_\tau\| \right) d\Gamma$$
$$+ \int_{\Gamma_3} j_\nu^0(u_\nu; v_\nu^h - u_\nu)\, d\Gamma - \langle \mathbf{f}, \mathbf{v}^h - \mathbf{u} \rangle_{V^* \times V}.$$

We now derive a bound for this residual term and, to this end, we follow the procedure found in [7]. Take $\mathbf{v} = \mathbf{u} \pm \mathbf{w}$ with $\mathbf{w}$ in the subset $\tilde{U}$ of $U$ defined by

$$\tilde{U} := \left\{ \mathbf{w} \in C^\infty(\overline{\Omega})^d \mid \mathbf{w} = \mathbf{0} \text{ on } \Gamma_1 \cup \Gamma_3 \right\},$$

and derive from (20) that

$$(\mathscr{F}(\boldsymbol{\varepsilon}(\mathbf{u})), \boldsymbol{\varepsilon}(\mathbf{w}))_Q = \langle \mathbf{f}, \mathbf{w} \rangle_{V^* \times V} \quad \forall\, \mathbf{w} \in \tilde{U}.$$

Therefore,

$$\mathrm{Div}\mathscr{F}(\boldsymbol{\varepsilon}(\mathbf{u})) + \mathbf{f}_0 = \mathbf{0} \quad \text{in } \Omega, \tag{25}$$

$$\boldsymbol{\sigma}\nu = \mathbf{f}_2 \quad \text{on } \Gamma_2. \tag{26}$$

Then multiply (25) by $\mathbf{v} - \mathbf{u}$ with $\mathbf{v} \in U$, integrate over $\Omega$, and integrate by parts,

$$\int_{\partial\Omega} \boldsymbol{\sigma}\nu\cdot(\mathbf{v} - \mathbf{u})\, d\Gamma - \int_{\Omega} \mathscr{F}(\boldsymbol{\varepsilon}(\mathbf{u}))\cdot\boldsymbol{\varepsilon}(\mathbf{v} - \mathbf{u})\, dx + \int_{\Omega} \mathbf{f}_0\cdot(\mathbf{v} - \mathbf{u})\, dx = 0,$$

i.e.,

$$\int_{\Omega} \mathscr{F}(\boldsymbol{\varepsilon}(\mathbf{u}))\cdot\boldsymbol{\varepsilon}(\mathbf{v} - \mathbf{u})\, dx = \langle \mathbf{f}, \mathbf{v} - \mathbf{u}\rangle_{V^*\times V} + \int_{\Gamma_3} \boldsymbol{\sigma}\nu\cdot(\mathbf{v} - \mathbf{u})\, d\Gamma. \tag{27}$$

Thus,

$$R(\mathbf{v}^h) = \int_{\Gamma_3} \left[\boldsymbol{\sigma}\nu\cdot(\mathbf{v}^h - \mathbf{u}) + F_b(u_\nu)\left(\|\mathbf{v}_\tau^h\| - \|\mathbf{u}_\tau\|\right) + j_\nu^0(u_\nu; v_\nu^h - u_\nu)\right] d\Gamma,$$

and then,

$$\left|R(\mathbf{v}^h)\right| \le c\, \|\mathbf{u} - \mathbf{v}^h\|_{L^2(\Gamma_3)^d}. \tag{28}$$

Finally, from (24), we derive the inequality

$$\|\mathbf{u} - \mathbf{u}^h\|_V^2 \le c\left(\|\mathbf{u} - \mathbf{v}^h\|_V^2 + \|\mathbf{u} - \mathbf{v}^h\|_{L^2(\Gamma_3)^d}\right) \quad \forall\, \mathbf{v}^h \in U^h. \tag{29}$$

Under additional solution regularity assumption

$$\mathbf{u}|_{\Gamma_{3,i}} \in H^2(\Gamma_{3,i}; \mathbb{R}^d), \quad 1 \le i \le i_3, \tag{30}$$

we have the optimal order error bound

$$\|\mathbf{u} - \mathbf{u}^h\|_V \le c\, h. \tag{31}$$

We comment that similar results hold for the frictionless version of the model, i.e., where the friction condition (6) is replaced by

$$\boldsymbol{\sigma}_\tau = \mathbf{0} \quad \text{on } \Gamma_3.$$

Then the problem is to solve the inequality (20) without the term

$$\int_{\Gamma_3} F_b(u_\nu)\,(\|\mathbf{v}_\tau\| - \|\mathbf{u}_\tau\|)\, d\Gamma.$$

The condition (21) reduces to $\alpha_{j_\nu}\lambda_{1\nu,V}^{-1} < m_{\mathscr{F}}$. The inequality (29) and the error bound (31) still hold for the linear finite element solution.

# 5 Numerical Simulations

This section is devoted to some numerical simulation results in order to illustrate the solution of the frictional contact Problem $P_V^h$ and to provide a numerical evidence of the theoretical error bound obtained in Sect. 4. We comment that the solution of Problem $P_V^h$ is based on numerical methods presented in detail in [2, 3]. Numerous standard numerical methods for contact mechanics can be found for instance in [16, 27].

**Numerical example.** The physical setting of the numerical example related to Problem $P_V^h$ is depicted in Fig. 1. There, the unit square body $\Omega = (0,1) \times (0,1) \subset \mathbb{R}^2$ is considered and

$$\Gamma_1 = [0,1] \times \{1\},\ \ \Gamma_2 = (\{0\} \times (0,1)) \cup (\{1\} \times (0,1)),\ \ \Gamma_3 = [0,1] \times \{0\}.$$

The domain $\Omega$ represents the cross section of a three-dimensional linearly elastic body subjected to the action of tractions in such a way that a plane stress hypothesis is assumed. On the part $\Gamma_1$ the body is clamped and, therefore, the displacement field vanishes there. Horizontal compressions act on the part $(\{0\} \times [0.5, 1)) \cup (\{1\} \times [0.5, 1))$ of the boundary $\Gamma_2$ and the part $(\{0\} \times (0, 0.5)) \cup (\{1\} \times (0, 0.5))$ is traction free. Constant vertical body forces are assumed to act on the elastic body. We consider that the deformable body is in frictional contact with an obstacle on the subset $\Gamma_3 = [0,1] \times \{0\}$ of its boundary.

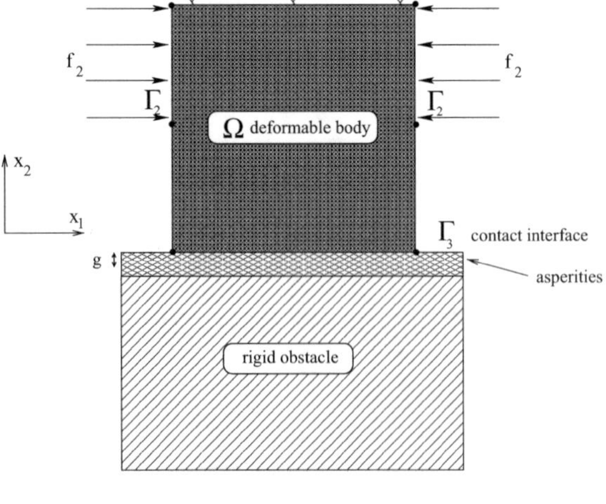


**Fig. 1** Reference configuration of the two-dimensional example

Let $0 < r_\nu^1 < r_\nu^2$ be given, and let $p_\nu : \mathbb{R} \to \mathbb{R}$, $j_\nu : \mathbb{R} \to \mathbb{R}$ be the functions defined by

$$p_\nu(r) = \begin{cases} 0 \ \text{ if } \ r \le 0, \\ c_\nu^1 r \ \text{ if } \ r \in (0, r_\nu^1], \\ c_\nu^1 r_\nu^1 + c_\nu^2 (r - r_\nu^1) \ \text{ if } \ r \in (r_\nu^1, r_\nu^2), \\ c_\nu^1 r_\nu^1 + c_\nu^2 (r_\nu^2 - r_\nu^1) + c_\nu^3 (r - r_\nu^2) \ \text{ if } \ r \ge r_\nu^2, \end{cases} \tag{32}$$

$$j_\nu(r) = \int_0^r p_\nu(s)\, ds \quad \forall\, s \in \mathbb{R}. \tag{33}$$

In the numerical example, we consder the frictional contact conditions (5) and (6) in which the function $j_\nu$ is given by (32), (33) and

$$F_b(r) = \mu p_\nu(r) \quad \forall\, r \in \mathbb{R} \tag{34}$$

where $\mu \ge 0$ represents a given coefficient of friction. Note that the fuction $p_\nu$ is continuous but is not monotone and, therefore, $j_\nu$ is a locally Lipschitz nonconvex function. With this choice, the frictional contact condition we use on $\Gamma_3$ takes the following form:

$$u_\nu \le g, \quad \sigma_\nu + \xi_\nu \le 0, \quad (u_\nu - g)(\sigma_\nu + \xi_\nu) = 0,$$

$$\xi_\nu = \begin{cases} 0 \ \text{ if } \ u_\nu \le 0, \\ c_\nu^1 u_\nu \ \text{ if } \ u_\nu \in (0, r_\nu^1], \\ c_\nu^1 r_\nu^1 + c_\nu^2 (u_\nu - r_\nu^1) \ \text{ if } \ u_\nu \in (r_\nu^1, r_\nu^2), \\ c_\nu^1 r_\nu^1 + c_\nu^2 (r_\nu^2 - r_\nu^1) + c_\nu^3 (u_\nu - r_\nu^2) \ \text{ if } \ u_\nu \ge r_\nu^2, \end{cases}$$

$$\|\boldsymbol{\sigma}_\tau\| \le \mu \xi_\nu, \qquad -\boldsymbol{\sigma}_\tau = \mu \xi_\nu \frac{\mathbf{u}_\tau}{\|\mathbf{u}_\tau\|} \quad \text{if } \ \mathbf{u}_\tau \ne \mathbf{0}.$$

The compressible material response, considered here, is governed by a linear elastic constitutive law defined by the elasticity tensor $\mathscr{F}$ given by

$$(\mathscr{F}\boldsymbol{\tau})_{\alpha\beta} = \frac{E\kappa}{1-\kappa^2}(\tau_{11} + \tau_{22})\delta_{\alpha\beta} + \frac{E}{1+\kappa}\tau_{\alpha\beta}, \quad 1 \le \alpha, \beta \le 2, \ \forall \boldsymbol{\tau} \in \mathbb{S}^2,$$

where $E$ and $\kappa$ are Young's modulus and Poisson's ratio of the material and $\delta_{\alpha\beta}$ denotes the Kronecker symbol.

For the numerical simulations, the following data are used:

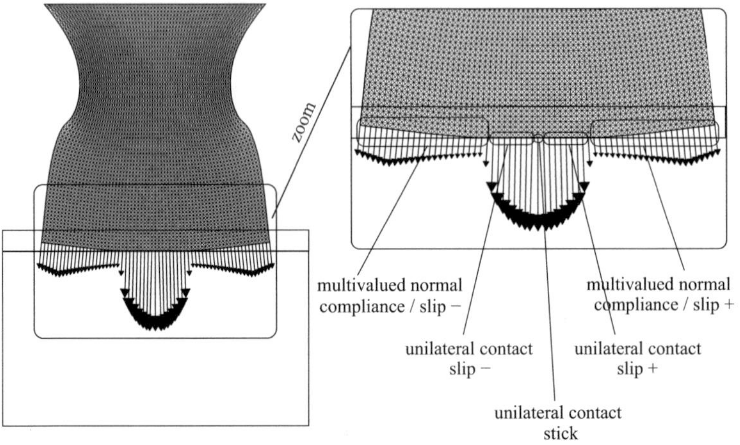


**Fig. 2** Deformed meshes and interface forces on $\Gamma_3$ corresponding to the Problem $P_V^h$

$$
\begin{aligned}
&E = 2000 N/m^2, \quad \kappa = 0.4,\\
&\mathbf{f}_0 = (0, -0.5 \times 10^{-3}) N/m^2,\\
&\mathbf{f}_2 = \begin{cases} (8 \times 10^{-3}, 0)\, N/m & \text{on } \{0\} \times [0.5, 1),\\ (-8 \times 10^{-3}, 0)\, N/m & \text{on } \{1\} \times [0.5, 1), \end{cases}\\
&c_\nu^1 = 100, \quad c_\nu^2 = -100, \quad c_\nu^3 = 400, \quad r_\nu^1 = 0.1\, m, \quad r_\nu^2 = 0.15\, m,\\
&g = 0.15\, m \quad \mu = 0.2.
\end{aligned}
$$

In Fig. 2, we plotted the deformed mesh and the interface forces on $\Gamma_3$. We observe that the contact nodes on the extremities of the boundary $\Gamma_3$ are in multivalued normal compliance with either backward slip (slip-) or forward slip (slip+); there, the normal displacement $u_\nu$ does not reach the penetration bound, that is $u_\nu < g$. All the remaining nodes of $\Gamma_3$ are in unilateral contact; there, the penetration bound is reached, that is $u_\nu = g$. Most of these nodes are in the slip status, except the node in the center of the boundary $\Gamma_3$ which is in stick status.

**Numerical convergence orders**. The aim of this part is to illustrate the convergence of the discrete solutions and to provide numerical evidence of the optimal error estimate obtained in Sect. 4. To this end, we computed a sequence of numerical solutions by using uniform discretization of the Problem $P_V^h$ according to the spatial discretization parameter $h$. For instance, for $h = 1/64$, we obtained the deformed configurations and the interface forces plotted in Fig. 2.

The numerical errors $\|\mathbf{u} - \mathbf{u}^h\|_E$ are computed by using the energy norm $\|\cdot\|_E$ for several discretization parameters of $h$. The energy norm $\|\cdot\|_E$ is equivalent to the canonical norm $\|\cdot\|_V$. Since it is not possible to calculate the exact solution $\mathbf{u}$ in an analytical way, we consider a "reference" solution $\mathbf{u}_{\text{ref}}$ corresponding to a fine discretization of $\Omega$, instead of the exact solution. Here, each line segment component of the boundary $\Gamma$ of $\Omega$ is divided into $1/h$ equal parts. We start with $h = 1/4$ which is successively halved. The numerical solution $\mathbf{u}_{\text{ref}}$ corresponding to

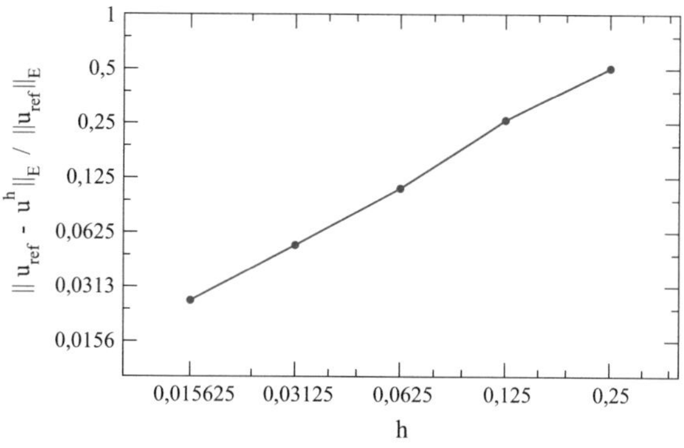


**Fig. 3** Relative numerical errors in the energy norm for Problem $P_V^h$

$h = 1/256$ was taken as the "reference" solution. This fine discretization corresponds to a problem with 132612 degrees of freedom and 131329 finite elements. The numerical results are presented in Fig. 3 where the dependence of the relative error $\|\mathbf{u}_{\mathrm{ref}} - \mathbf{u}^h\|_E / \|\mathbf{u}_{\mathrm{ref}}\|_E$ with respect to $h$ is plotted for the Problem $P_V^h$. Note that these results provide a numerical evidence of the theoretically predicted optimal order estimate obtained in Sect. 4 and highlight the linear asymptotic convergence of the numerical solutions.

**Acknowledgements** The work of W.H. was partially supported by NSF under grant DMS-1521684.

# References


1. Baiocchi, C., Capelo, A.: Variational and Quasivariational Inequalities: Applications to Free-Boundary Problems. John Wiley, Chichester (1984)
2. Barboteu, M., Bartosz, K., Kalita, P.: An analytical and numerical approach to a bilateral contact problem with nonmonotone friction. Int. J. Appl. Math. Comput. Sci. **23**, 263–276 (2013)
3. Barboteu, M., Bartosz, K., Kalita, p, Ramadan, A.: Analysis of a contact problem with normal compliance, finite penetration and nonmonotone slip dependent friction. Commun. Contemp. Math. **15**, 1350016 (2013). doi:10.1142/S0219199713500168
4. Eck, C., Jarušek, J., Krbec, M.: Unilateral Contact Problems: Variational Methods and Existence Theorems, Pure and Applied Mathematics 270. Chapman/CRC Press, New York (2005)
5. Glowinski, R.: Numerical Methods for Nonlinear Variational Problems. Springer-Verlag, New York (1984)
6. Glowinski, R., Lions, J.-L., Trémolières, R.: Numerical Analysis of Variational Inequalities. North-Holland, Amsterdam (1981)
7. Han, W., Sofonea, M.: Quasistatic Contact Problems in Viscoelasticity and Viscoplasticity, Studies in Advanced Mathematics, vol. 30. American Mathematical Society, Providence, RI-International Press, Somerville, MA (2002)
8. Han, W., Reddy, B.D.: Plasticity: Mathematical Theory and Numerical Analysis, 2nd edn. Springer-Verlag, New York (2013)
9. Han, W., Migórski, S., Sofonea, M.: A class of variational-hemivariational inequalities with applications to frictional contact problems. SIAM J. Math. Anal. **46**, 3891–3912 (2014)
10. Han, W., Sofonea, M., Barboteu, M.: Numerical Analysis of Elliptic Hemivariational Inequalities, to appear in SIAM J. Numer. Anal

11. Han, W., Sofonea, M., Danan, D.: Numerical Analysis of Stationary Variational-Hemivariational Inequalities (submitted)
12. Haslinger, J., Miettinen, M., Panagiotopoulos, P.D.: Finite Element Method for Hemivariational Inequalities. Theory, Methods and Applications. Kluwer Academic Publishers, Boston, Dordrecht, London (1999)
13. Hlaváček, I., Haslinger, J., Necǎs, J., Lovíšek, J.: Solution of Variational Inequalities in Mechanics. Springer-Verlag, New York (1988)
14. Kinderlehrer, D., Stampacchia, G.: An Introduction to Variational Inequalities and their Applications, Classics in Applied Mathematics, vol. 31. SIAM, Philadelphia (2000)
15. Kikuchi, N., Oden, J.T.: Contact Problems in Elasticity: A Study of Variational Inequalities and Finite Element Methods. SIAM, Philadelphia (1988)
16. Laursen, T.: Computational Contact and Impact Mechanics. Springer, Berlin (2002)
17. Migórski, S., Ochal, A., Sofonea,: Nonlinear Inclusions and Hemivariational Inequalities. Models and Analysis of Contact Problems. Advances in Mechanics and Mathematics, vol. 26. Springer, New York (2013)
18. Migórski, S., Ochal, A., Sofonea, M.: History-dependent variational-hemivariational inequalities in contact mechanics. Nonlinear Anal. Real World Appl. **22**, 604–618 (2015)
19. Migórski, S., Ochal, A., Sofonea, M.: A Class of Variational-Hemivariational Inequalities in Reflexive Banach Spaces, to appear in Jounral of Elasticity
20. Naniewicz, Z., Panagiotopoulos, P.D.: Mathematical Theory of Hemivariational Inequalities and Applications. Marcel Dekker Inc., New York, Basel, Hong Kong (1995)
21. Panagiotopoulos, P.D.: Nonconvex problems of semipermeable media and related topics. ZAMM Z. Angew. Math. Mech. **65**, 29–36 (1985)
22. Panagiotopoulos, P.D.: Inequality Problems in Mechanics and Applications. Birkhäuser, Boston (1985)
23. Panagiotopoulos, P.D.: Hemivariational Inequalities, Applications in Mechanics and Engineering. Springer-Verlag, Berlin (1993)
24. Shillor, M., Sofonea, M., Telega, J.J.: Models and Analysis of Quasistatic Contact. Lecture Notes in Physics, vol. 655. Springer, Berlin, Heidelberg (2004)
25. Sofonea, M., Han, W., Migórski, S.: Numerical analysis of history-dependent variational inequalities with applications to contact problems. Eur. J. Appl. Math. **26**, 427–452 (2015)
26. Sofonea, M., Matei, A.: Mathematical Models in Contact Mechanics, London Mathematical Society Lecture Note Series, vol. 398. Cambridge University Press, Cambridge (2012)
27. Wriggers, P.: Computational Contact Mechanics. Wiley, Chichester (2002)

# Truncation and Indirect Incremental Methods in Hencky's Perfect Plasticity

**Stanislav Sysala and Jaroslav Haslinger**

**Abstract** The contribution is concerned with reliable and computable bounds of the limit (or safety) load in the deformation theory of perfect plasticity. We consider truncation and indirect incremental methods of limit analysis which can be interpreted as penalization techniques. Further, convergence for higher order finite elements is shown. The efficiency of the proposed approaches is illustrated on numerical experiments with the von Mises and Drucker–Prager yield criteria.

## 1 Introduction

The paper is focused on reliable and easily computable bounds of limit loads in elastic-perfectly plastic problems. We summarize and slightly extend the results presented in [2, 7, 8, 12]. In particular, we extend the finite element analysis to higher order elements to reduce the observed locking effect.

The paper is organized as follows. Section 2 contains preliminaries from the generalized Hencky plasticity and the related limit analysis. The von Mises and Drucker–Prager yield criteria are mentioned as particular cases. Section 3 is devoted to the truncation method where unbounded yield surfaces are approximated by bounded ones. The indirect incremental method is introduced in Sect. 4. Both methods are firstly defined for the continuous setting of the problem in order to demonstrate their independency of the problem discretization. The discretized problem is analyzed in Sect. 5. Section 6 summarizes our strategy how to find computable and reliable lower and upper bounds of $\lambda^*$. The lower and upper bounds of the limit load $\lambda^*$ for two model examples with the above mentioned yield criteria are established. Unlike [8], we consider $P_2$-elements and different meshes, geometries or elastic parameters in order to improve the bounds.

S. Sysala (✉) J. Haslinger
Institute of Geonics of the Czech Academy of Sciences, Ostrava, Czech Republic
e-mail: stanislav.sysala@ugn.cas.cz

J. Haslinger
Charles University, Prague, Czech Republic
e-mail: hasling@karlin.mff.cuni.cz



F. dell'Isola et al. (eds.), *Mathematical Modelling in Solid Mechanics*,
Advanced Structured Materials 69, DOI 10.1007/978-981-10-3764-1_17

# 2 Generalized Hencky Plasticity Problem

## 2.1 Basic Definitions and Properties

The classical Hencky plasticity model of the deformation plasticity theory is based on the von Mises yield law. Since an abstract yield criterion is used we rather write "generalized" Hencky plasticity model in order to stress this fact. This static model is usually completed by a parametric study in order to be close to the incremental (quasistatic) theory of elasto-plasticity. The parametrized model is sufficient to treat the limit load analysis. For more details, we refer to, e.g., [6, 7, 14].

The space of admissible displacement fields has the form

$$\mathbb{V} = \{v \in H^1(\Omega; \mathbb{R}^3) \mid v|_{\Gamma_D} = 0\},$$

where $\Omega$ is a *bounded* domain with the Lipschitz continuous boundary $\partial\Omega$ and $\Gamma_D$, $\Gamma_N$ are open and nonempty parts of $\partial\Omega$ such that $\Gamma_D \cap \Gamma_N = \emptyset$ and $\bar{\Gamma}_D \cup \bar{\Gamma}_N = \partial\Omega$. Further, $f \in L^2(\Gamma_N; \mathbb{R}^3)$, $F \in L^2(\Omega; \mathbb{R}^3)$ denote the density of surface and volume forces, respectively, and

$$L(v) = \int_\Omega F \cdot v\,dx + \int_{\Gamma_N} f \cdot v\,ds, \quad v \in \mathbb{V}, \qquad \|F\|_{L^2(\Omega;\mathbb{R}^3)} + \|f\|_{L^2(\Gamma_N;\mathbb{R}^3)} \neq 0. \tag{1}$$

Stress and strain tensors are represented locally by symmetric matrices, i.e., elements of $\mathbb{R}^{3\times3}_{sym}$. In particular, we consider the infinitesimal small strain tensor represented by a symmetric part of the displacement gradient:

$$\varepsilon(v) = \frac{1}{2}(\nabla v + (\nabla v)^T).$$

The biscalar product and the corresponding norm in $\mathbb{R}^{3\times3}_{sym}$ will be denoted by $e : \eta = e_{ij}\eta_{ij}$ and $\|e\|^2 = e : e$ for any $e, \eta \in \mathbb{R}^{3\times3}_{sym}$, respectively.

Let $B$ be a closed, convex subset of $\mathbb{R}^{3\times3}_{sym}$ containing a vicinity of the origin. This set represents *plastically admissible stresses* and it is defined by a plastic criterion, see Sects. 2.2 or 2.3. To formulate the constitutive stress-strain relation we introduce the function $\Pi_B$ which is a *generalized* projection of $\mathbb{R}^{3\times3}_{sym}$ onto $B$ (in the sense of [11]):

$$\Pi_B : e \mapsto \Pi_B(e), \quad \|\mathbb{C}e - \Pi_B(e)\|_{\mathbb{C}^{-1}} = \min_{\tau\in B} \|\mathbb{C}e - \tau\|_{\mathbb{C}^{-1}}, \quad e \in \mathbb{R}^{3\times3}_{sym},$$

where $\mathbb{C} : \mathbb{R}^{3\times3}_{sym} \to \mathbb{R}^{3\times3}_{sym}$ is a linear, positive definite, fourth order elasticity tensor characterizing the elastic material response, $\mathbb{C}^{-1}$ is the corresponding inverse and $\|\tau\|^2_{\mathbb{C}^{-1}} := \mathbb{C}^{-1}\tau : \tau$ for any $\tau \in \mathbb{R}^{3\times3}_{sym}$. The potential $j : \mathbb{R}^{3\times3}_{sym} \to \mathbb{R}_+$ of $\Pi_B$ is defined by

$$j(e) = \sup_{\tau \in B} \left\{ \tau : e - \frac{1}{2} \|\tau\|^2_{\mathbb{C}^{-1}} \right\}, \quad e \in \mathbb{R}^{3\times 3}_{sym}. \tag{2}$$

It is a convex, continuously Fréchet differentiable function, and

$$\frac{\varepsilon}{2}\|e\|_{\mathbb{C}} - \frac{\varepsilon^2}{8} \le j(e) \le \frac{1}{2}\|e\|^2_{\mathbb{C}}, \quad \|e\|^2_{\mathbb{C}} := \mathbb{C}e : e, \quad \forall e \in \mathbb{R}^{3\times 3}_{sym}, \tag{3}$$

where $\varepsilon > 0$ is such that the ball $\{\tau \in \mathbb{R}^{3\times 3}_{sym} \mid \|\tau\|_{\mathbb{C}^{-1}} \le \varepsilon\}$ belongs to $B$. Thus, only a linear growth of $j$ at infinity is guaranteed.

The generalized Hencky plasticity problem (in terms of displacements) for a given value of the load parameter $\lambda \ge 0$ reads as follows:

$$(\mathscr{P})_\lambda \qquad \inf_{v \in V} J_\lambda(v), \quad J_\lambda(v) = \int_\Omega j(\varepsilon(v))\, dx - \lambda L(v).$$

Notice that $J_\lambda$ need not be bounded from below for all $\lambda > 0$ due to the lower bound of $j$ in (3). Moreover, even if $J_\lambda$ is bounded from below, problem $(\mathscr{P})_\lambda$ need not have a minimizer belonging to $\mathbb{V}$. For the existence analysis, it is necessary to use the relaxation of the problem including the extension of $\mathbb{V}$ to the BD-space of functions with bounded deformations, see, e.g., [10, 14]. This space allows discontinuities of displacements along surfaces in 3D and thus the model is capable to predict possible failure zones in the investigated body. This fact will be illustrated in Sect. 6.

In order to decide whether $J_\lambda$ is bounded from below in $\mathbb{V}$ or not, it is natural to introduce the limit load parameter

$$\lambda^* = \sup \left\{ \lambda \ge 0 \mid \inf_{v \in \mathbb{V}} J_\lambda(v) > -\infty \right\}. \tag{4}$$

This definition also admits the value $\lambda^* = +\infty$, however $\lambda^*$ is usually finite in meaningful settings of the problem. One can easily check that the function $\phi(\lambda) = \inf_{v\in\mathbb{V}} J_\lambda(v)$ is decreasing on $(0, \lambda^*)$ and thus $J_\lambda$ is bounded from below for any $\lambda < \lambda^*$. Other straightforward but useful consequences of (4) are introduced in the following lemma.

**Lemma 1** (Basic bounds of $\lambda^*$.)

*(i)* *Let $j_1, j_2 : \mathbb{R}^{3\times 3}_{sym} \to \mathbb{R}_+ \cup \{+\infty\}$, $0 \le j_1 \le j_2$, be two convex and proper functions. Then the corresponding limit load parameters $\lambda_1^*$, $\lambda_2^*$ defined by (4) satisfy $\lambda_1^* \le \lambda_2^*$.*

*(ii)* *Let $K$ be a subset of $\mathbb{V}$ and $\lambda_K^* := \sup\{\lambda \ge 0 \mid \inf_{v\in K} J_\lambda(v) > -\infty\}$. Then $\lambda^* \le \lambda_K^*$.*

For the limit analysis, it is very useful to introduce a special minimization problem. Following [8], we firstly consider the function

$$j_\alpha(e) = \frac{1}{\alpha} j(\alpha e) \quad \forall e \in \mathbb{R}^{3\times 3}_{sym}, \quad \alpha > 0. \tag{5}$$

As for $j$, one can define the corresponding limit parameter $\lambda^*_\alpha$. By substitution, we have:

$$\inf_{v\in V}\left[\int_\Omega j_\alpha(\varepsilon(v))\,dx - \lambda L(v)\right] = \frac{1}{\alpha}\inf_{v\in V}\left[\int_\Omega j(\varepsilon(v))\,dx - \lambda L(v)\right] \quad \forall\alpha > 0.$$

Hence, $\lambda^*_\alpha = \lambda^*$ for any $\alpha > 0$. Next, define the function

$$\begin{aligned} j_\infty &: \mathbb{R}^{3\times 3}_{sym} \to \overline{\mathbb{R}}_+, \quad \overline{\mathbb{R}}_+ := \mathbb{R}_+ \cup \{+\infty\}, \\ j_\infty(e) &= \lim_{\alpha\to+\infty} j_\alpha(e) = \sup_{\tau\in B} \tau : e, \quad e \in \mathbb{R}^{3\times 3}_{sym}. \end{aligned} \tag{6}$$

Clearly, $j_\infty(0) = 0$ and $j_\infty$ is a proper, convex function in $\mathbb{R}^{3\times 3}_{sym}$ which is also positively 1 - homogeneous. Further, it holds:

$$j \le j_\alpha \le j_\infty \quad \forall\alpha \ge 1. \tag{7}$$

The limit load parameter associated with $j_\infty$ is defined as follows:

$$\zeta^* = \sup\left\{\lambda \ge 0 \mid \inf_{v\in\mathbb{V}}\left[\int_\Omega j_\infty(\varepsilon(v))\,dx - \lambda L(v)\right] > -\infty\right\}. \tag{8}$$

It is readily seen that

$$\lambda^* \le \zeta^*, \tag{9}$$

making use of (7) and Lemma 1, i.e., $\zeta^*$ is in an upper bound of the limit load parameter $\lambda^*$. The properties of $j_\infty$ enable us to derive a more convenient definition of $\zeta^*$ than (8), see, e.g., [8].

**Lemma 2** *It holds:*

$$\zeta^* = \inf_{\substack{v\in\mathbb{V}\\ L(v)=1}} J_\infty(v), \qquad J_\infty(v) = \int_\Omega j_\infty(\varepsilon(v))\,dx, \quad v \in \mathbb{V}. \tag{10}$$

*Remark 1* The inf-problem (10) is termed the *problem of limit analysis* using the terminology of perfect plasticity [4, 14]. This minimization problem is important from several reasons:

- It enables us to estimate $\lambda^*$ by a straightforward manner (minimization), see e.g., in [1, 4]. Moreover, the values $J_\infty(v)$, where $v \in \mathbb{V}$, $L(v) = 1$ and $v \in \operatorname{dom} J_\infty$, are the guaranteed upper bounds of $\lambda^*$.
- The inf-problem (10) is in a certain sense dual to the sup-problem (4) defining $\lambda^*$. Using the duality approach, one can prove that $\lambda^* = \zeta^*$ for some sets $B$.

- It is well-known that the additional constraint $v \in \mathrm{dom} J_\infty = \{w \in \mathbb{V} \mid J_\infty(w) < +\infty\}$ may cause locking effect in perfect plasticity. For example, if $B$ represents the von Mises yield criterion then the divergence free constraint appears, see Sect. 2.2.
- The limit load parameters $\lambda^*$ and $\zeta^*$ are independent of the elasticity tensor $\mathbb{C}$. For $\zeta^*$ this fact follows from the sup-definition of $j_\infty$ in (6) and for $\lambda^*$ from the duality approach. This simple observation is useful mainly for soil materials with the Poisson ratio close to the critical value 0.5 when significant rounding errors of numerical solutions arise.
- The formation of failure zones producing discontinuities of displacements is typical for the limit load. From (10), it seems to be natural that values of $j_\infty$ vanish far from the expected failure during the minimization. On such subdomains, one can expect rigid body displacements. This will be illustrated on numerical examples in Sect. 6. In particular, in Sect. 6.3, we will study the slope stability benchmark where the failure is localized only in a vicinity of the slope. Moreover, the expected rigid body displacements far from the slope vanish due to prescribed boundary conditions. This leads to a simple observation that the limit parameter remains unchanged when we use a much smaller domain than in [5, 8, 13].

In the subsequent parts of this section, we introduce the von Mises and Drucker–Prager yield criteria as particular examples of $B$.

## 2.2 The Von Mises Yield Criterion

The set $B$ defined by the von Mises yield criterion has the form

$$B = \left\{\tau \in \mathbb{R}^{3\times 3}_{sym} \mid \|\tau^D\| \leq \gamma\right\}, \tag{11}$$

where $\|\tau^D\|^2 := \tau^D : \tau^D$, $\tau^D = \tau - \frac{1}{3}(tr\,\tau)\iota$ is the deviatoric part of $\tau$, $tr\,\tau = \tau_{ii}$ is the trace of $\tau$, $\iota = \mathrm{diag}(1, 1, 1)$ is the unit matrix, and $\gamma > 0$ represents an initial yield stress. Notice that $B$ is the unbounded cylinder with the (hydrostatic) axis $\{\tau \in \mathbb{R}^{3\times 3}_{sym} \mid \tau = a\iota,\ \ a \in \mathbb{R}\}$. If the elastic stress-strain relation is isotropic and expressed in terms of the bulk ($K > 0$) and shear ($G > 0$) moduli, i.e.,

$$\tau = \mathbb{C}e = K(tr\,e)\iota + 2Ge^D \quad \forall e \in \mathbb{R}^{3\times 3}_{sym}, \tag{12}$$

then $j$ defined by (2) can be written as

$$j(e) = \begin{cases} \frac{1}{2}K(tr\,e)^2 + G\|e^D\|^2, & \text{if } 2G\|e^D\| \leq \gamma \\ \frac{1}{2}K(tr\,e)^2 + \gamma\|e^D\| - \frac{\gamma^2}{4G}, & \text{if } 2G\|e^D\| > \gamma \end{cases}, \quad \forall e \in \mathbb{R}^{3\times 3}_{sym},$$

see, e.g., [14]. It is readily seen that

$$j_\infty(e) = \lim_{\alpha\to+\infty} \frac{1}{\alpha} j(\alpha e) = \begin{cases} \gamma\|e^D\|, & \text{if } \operatorname{tr} e = 0 \\ +\infty, & \text{if } \operatorname{tr} e \neq 0 \end{cases}, \quad \forall e \in \mathbb{R}^{3\times 3}_{sym}$$

and the corresponding problem of the limit analysis (10) becomes:

$$\zeta^* = \inf_{\substack{v\in\mathbb{V},\ div\, v=0\\ L(v)=1}} \int_\Omega \gamma\|\varepsilon(v)\|\, dx. \tag{13}$$

This is a non-smooth optimization problem involving the divergence-free constraint. Further, it is known that $\lambda^* = \zeta^*$ (see [14]).

## 2.3 The Drucker–Prager Yield Criterion

The set $B$ of the admissible stresses for the Drucker–Prager yield criterion reads as follows:

$$B = \left\{\tau \in \mathbb{R}^{3\times 3}_{sym} \mid \frac{a}{3}\operatorname{tr}\tau + \|\tau^D\| \leq \gamma\right\}, \quad a, \gamma > 0. \tag{14}$$

$B$ is an unbounded cone with the hydrostatic axis and the apex $\tau = \frac{\gamma}{\alpha}\iota$. For the shape of the yield surface in the Haigh-Westergaard coordinates we refer to [5]. Assume that $\mathbb{C}$ is the same as in (12) and denote

$$q_s(e) := Ka(\operatorname{tr} e) + 2G\|e^D\| - \gamma, \quad q_a(e) := Ka(\operatorname{tr} e) - Ka^2\|e^D\| - \gamma, \quad e \in \mathbb{R}^{3\times 3}_{sym}.$$

Notice that $q_s \geq q_a$. Then

$$\begin{aligned} j(e) &= \frac{K}{2}(\operatorname{tr} e)^2 + G\|e^D\|^2 - \frac{1}{2(Ka^2+2G)}\left\{\left[(q_s(e))^+\right]^2 + \frac{2G}{Ka^2}\left[(q_a(e))^+\right]^2\right\} \\ &= \begin{cases} \frac{K}{2}(\operatorname{tr} e)^2 + G\|e^D\|^2, & \text{if } q_s(e) \leq 0, \\ -\frac{\gamma^2}{2Ka^2} + \frac{\gamma}{a}\operatorname{tr} e + \frac{G}{Ka^2(Ka^2+2G)} q_a(e)^2, & \text{if } q_s(e) \geq 0 \geq q_a(e), \\ -\frac{\gamma^2}{2Ka^2} + \frac{\gamma}{a}\operatorname{tr} e, & \text{if } q_a(e) \geq 0, \end{cases} \end{aligned}$$

where $g^+$ denotes the positive part of $g$. Another form of $j$ can be found in [9] as well as the proof of the equality $\lambda^* = \zeta^*$ which holds for sufficiently small values of the parameter $a$ and under appropriate assumptions. Further,

$$j_\infty(e) = \begin{cases} \frac{\gamma}{a}\operatorname{tr} e, & \text{if } \operatorname{tr} e \geq a\|e^D\| \\ +\infty, & \text{if } \operatorname{tr} e \leq a\|e^D\| \end{cases}$$

and

$$\zeta^* = \inf_{\substack{v\in\mathbb{V},\ L(v)=1\\ div\, v\geq a\|\varepsilon^D(v)\|}} \int_\Omega \frac{\gamma}{a}\operatorname{div} v\, dx, \qquad \operatorname{div} v = \operatorname{tr}\varepsilon(v).$$

Unlike the von Mises yield criterion, the problem of limit analysis leads to minimization of a linear functional but subject also to the inequality constraint.

## 3 Truncation Method

The von Mises and Drucker–Prager yield criteria lead to the unbounded sets $B$. The same holds also for the Tresca or Mohr-Coulomb yield criteria. For the Cam-Clay or capped Drucker–Prager criteria, the set $B$ is bounded. On the other hand, models with bounded yield surfaces are usually accompanied by internal variables like hardening/softening or damage and so they are not perfectly plastic. The mentioned yield criteria and many others are presented, e.g., in [5].

The aim of this section is to emphasize that the limit analysis is much simpler for bounded than for unbounded $B$. Using this fact, it is quite natural to consider the truncation method for unbounded $B$. This section summarizes the results presented in [8].

### *3.1 Limit Analysis for Bounded B*

Assume that $B$ is *bounded*. Owing to this fact, one can derive the following additional results:

- From (6), it is readily seen that $j_\infty$ is everywhere real-valued:

$$j_\infty(e) = \sup_{\tau \in B} \tau : e < +\infty \quad \forall e \in \mathbb{R}^{3\times 3}_{sym} \tag{15}$$

- The load assumption (1) yields $\zeta^* < +\infty$.
- From the definitions of $j$ and $j_\infty$, i.e., (2) and (6), we have:

$$j_\infty(e) - c \le j(e) \le j_\infty(e) \quad \forall e \in \mathbb{R}^{3\times 3}_{sym}, \quad c := \sup_{\tau \in B} \frac{1}{2}\mathbb{C}^{-1}\tau : \tau. \tag{16}$$

- It holds:

$$\lambda^* = \zeta^*, \qquad \inf_{v \in \mathbb{V}} J_{\lambda^*}(v) < +\infty. \tag{17}$$

- The following criterion for $\lambda$ to be admissible or not holds:

$$\lambda > \lambda^* \quad \Longleftrightarrow \quad \exists v \in \mathbb{V}: \; J_\lambda(v) < -c|\Omega|, \quad c := \sup_{\tau \in B} \frac{1}{2}\mathbb{C}^{-1}\tau : \tau. \tag{18}$$

Notice that the constant $c$ is usually a priori known. This criterion leads to a *guaranteed and easily computable upper bound* of $\lambda^*$. Indeed, one can easily construct a minimization sequence $\{u_n\}$ of $J_\lambda$ in $\mathbb{V}$ or in its subspace since $J_\lambda$ is convex and differentiable. If $J_\lambda(u_n) < -c|\Omega|$ for some $n$ then $\lambda$ is an upper bound of $\lambda^*$. We use this criterion to verify numerical results in Sect. 6.

## 3.2 Truncation Method for Unbounded B

For unbounded $B$, the assertions (15)–(18) do not hold, in general. For this reason, we consider truncations of $B$ using an appropriate system $\{B_k\}$, $\overline{\bigcup_{k>0} B_k} = B$ of bounded subsets of $B$. With any $B_k$, we associate the functions $j_k, j_{k,\infty}$ and the limit load parameters $\lambda_k^*, \zeta_k^*$ analogously to $j, j_\infty$, and $\lambda^*, \zeta^*$ for unbounded $B$, respectively. From Lemma 1, (2), and (6), it follows that

$$\left.\begin{array}{l} j_k \le j, \quad j_{k,\infty} \le j_\infty, \\ \zeta_k^* = \lambda_k^* \le \lambda^* \le \zeta^*, \\ \lim\limits_{k\to+\infty} j_{k,\infty}(e) = j_\infty(e)\ \forall e \in \mathbb{R}^{3\times 3}_{sym}. \end{array}\right\} \tag{19}$$

Therefore, $\lambda_k^*$ is a lower bound of $\lambda^*$ for any $k > 0$. Knowledge of a reliable lower bound of $\lambda^*$ is important since it presents a safety parameter. The truncation method can be also interpreted as a penalty approach in the problem of limit analysis making use of $(19)_3$. Sufficient conditions ensuring $\lambda_k^* \to \lambda^*$ as $k \to +\infty$ are presented in [8].

The truncated $B$ for the von Mises yield criterion can be defined as follows:

$$B_k = \left\{\tau \in \mathbb{R}^{3\times 3}_{sym} \mid \frac{1}{3}|tr\,\tau| \le k\gamma, \ \ \|\tau^D\| \le \gamma\right\}, \quad k > 0. \tag{20}$$

The functions $j_k$ and $j_{k,\infty}$ associated with such $B_k$ are derived in [8] and the criterion (18) reads:

$$\lambda > \lambda^* \iff \exists v \in \mathbb{V}: \int_\Omega j_k(\varepsilon(v))\,dx - \lambda L(v) < -\frac{\gamma^2}{2}\left(\frac{k^2}{K} + \frac{1}{2G}\right)|\Omega|. \tag{21}$$

A similar truncation can be also used for the Drucker–Prager yield criterion:

$$B_k = \left\{\tau \in \mathbb{R}^{3\times 3}_{sym} \mid \frac{a}{3}\mathrm{tr}\,\tau \ge -k\gamma, \ \ \frac{a}{3}\mathrm{tr}\,\tau + \|\tau^D\| \le \gamma\right\}, \quad k \ge 1. \tag{22}$$

The functions $j_k$ and $j_{k,\infty}$ associated with this $B_k$ are derived in [8] and the criterion (18) reads:

$$\lambda > \lambda^* \iff \exists v \in \mathbb{V}: \int_\Omega j_k(\varepsilon(v))\,dx - \lambda L(v) < -\frac{\gamma^2}{2}\left(\frac{k^2}{Ka^2} + \frac{(1+k)^2}{2G}\right)|\Omega|. \tag{23}$$

# 4 Indirect Incremental Method

By enlarging $\lambda$ up to its limit value $\lambda^*$, one can define the direct incremental method of the limit analysis. Notice that $\lambda$ must be enlarged adaptively since $\lambda^*$ is unknown. In [2, 12], another parameter $\alpha \to +\infty$ has been introduced together with an auxiliary minimization problem enabling an indirect control of the loading process for $\lambda \to \lambda^*$ in a discretized version of the problem. This technique has been extended in [7] for the continuous setting of the problem using the duality approach in terms of stresses. Now, we present a more straightforward derivation.

To this end, we use the sequence $\{j_\alpha\}$, $\alpha > 0$, of functions defined by (5) which pointwisely converges to $j_\infty$ as follows from (6). Define the function $\bar{\psi} : \mathbb{R}_+ \to \mathbb{R}_+$ by the following penalization of the limit analysis problem:

$$(\mathscr{P})^\alpha \qquad \bar{\psi}(\alpha) := \inf_{\substack{v\in\mathbb{V}\\ L(v)=1}} \int_\Omega j_\alpha(\varepsilon(v))\,dx, \quad \alpha > 0. \tag{24}$$

Problem $(\mathscr{P})^\alpha$ is a smooth convex program with just one linear constraint. Numerically, it is not difficult to solve it, however, it is worth mentioning that minimizers need not belong to $\mathbb{V}$ similarly as for problem $(\mathscr{P})_\lambda$. Further, it holds:

$$\begin{aligned}
\bar{\psi}(\alpha) &\overset{(17,5)}{=} \frac{1}{\alpha} \inf_{\substack{v\in\mathbb{V}\\ L(v)=\alpha}} \int_\Omega j(\varepsilon(v))\,dx = \frac{1}{\alpha} \inf_{v\in\mathbb{V}} \sup_{\lambda\in\mathbb{R}} \left\{ \int_\Omega j(\varepsilon(v))\,dx - \lambda(L(v)-\alpha) \right\} \\
&= \frac{1}{\alpha} \sup_{\lambda\in\mathbb{R}} \inf_{v\in\mathbb{V}} \left\{ \int_\Omega j(\varepsilon(v))\,dx - \lambda(L(v)-\alpha) \right\} \\
&= \sup_{\lambda\in\mathbb{R}} \left[ \frac{1}{\alpha} \inf_{v\in\mathbb{V}} \left\{ \int_\Omega j(\varepsilon(v))\,dx - \lambda L(v) \right\} + \lambda \right] \\
&= \sup_{\lambda\in\mathbb{R}_+} \left[ \frac{1}{\alpha} \inf_{v\in\mathbb{V}} J_\lambda(v) + \lambda \right] = \sup_{\lambda\in\mathbb{R}_+} \left[ \frac{1}{\alpha}\phi(\lambda) + \lambda \right] \quad \forall \alpha > 0,
\end{aligned} \tag{25}$$

where $\phi(\lambda) = \inf_{v\in\mathbb{V}} J_\lambda(v)$. The properties of $\phi$ has been derived in [7]. Under the assumption (1), it holds that $\phi$ is negative, strictly concave, decreasing and continuous in $(0, \lambda^*)$. Further, $\phi$ has at least quadratic decrease at infinity when $\lambda^* = +\infty$. Otherwise, $\phi(\lambda) = -\infty$ for any $\lambda > \lambda^*$. These properties of $\phi$ ensure that the function $\lambda \mapsto \frac{1}{\alpha}\phi(\lambda) + \lambda$ has a unique maximizer in $\mathbb{R}_+$ for any $\alpha > 0$. This enables us to introduce the function

$$\psi(\alpha) = \arg\max_{\lambda\in\mathbb{R}_+} \left[ \frac{1}{\alpha}\phi(\lambda) + \lambda \right], \quad \alpha > 0. \tag{26}$$

From [7], we know:

(i) $\psi$ is nondecreasing, continuous, $\lim_{\alpha\to 0_+}\psi(\alpha)=0$, and $\lim_{\alpha\to+\infty}\psi(\alpha)=\lambda^*$;
(ii) if there exists a minimizer $u_\alpha\in\mathbb{V}$ in problem $(\mathscr{P})^\alpha$ then $\alpha u_\alpha$ solves $(\mathscr{P})_\lambda$ for

$$\lambda=\psi(\alpha)=\int_\Omega \Pi_B(\varepsilon(\alpha u_\alpha)):\varepsilon(u_\alpha)\,dx. \tag{27}$$

Moreover, the right hand side in (27) does not depend on the choice of $u_\alpha$ with the above mentioned properties;
(iii) conversely, if $u_\lambda$ is a solution to $(\mathscr{P})_\lambda$ then $\frac{u_\lambda}{L(u_\lambda)}$ solves $(\mathscr{P})^\alpha$ for $\alpha=L(u_\lambda)$.

From $(i)$, we see that knowledge of $\psi$ enables us to introduce the indirect incremental method of limit analysis which corresponds to $\alpha\to+\infty$. Moreover, the values $\psi(\alpha)$ approximate $\lambda^*$ from below. From $(ii)$, we see that the values $\psi(\alpha)$ can be computed solving problem $(\mathscr{P})^\alpha$. Formula (27) is more convenient from the numerical point of view than (26). Due to $(iii)$, one can interpret the parameter $\alpha$ as the complience or the work of external forces. Notice that the inverse mapping $\psi^{-1}:\lambda\mapsto\alpha$ need not be singlevalued unlike $\psi$.

*Remark 2* Using (24) and (25) it is not difficult to show that the function $\bar{\psi}$ has the same properties as $\psi$, i.e. $\bar{\psi}$ is nondecreasing, continuous, $\lim_{\alpha\to 0_+}\bar{\psi}(\alpha)=0$, and $\lim_{\alpha\to+\infty}\bar{\psi}(\alpha)=\lambda^*$. Moreover, $\bar{\psi}(\alpha)=\frac{1}{\alpha}\phi(\psi(\alpha))+\psi(\alpha)<\psi(\alpha)\quad\forall\alpha>0$.

## 5 Finite Element Approximation

In this section, classical finite element approximations are considered for computing bounds of the limit load in the generalized Hencky plasticity. Let $\{\mathbb{V}_h\}$ be a system of finite element subspaces of $\mathbb{V}$ which is limit dense in $\mathbb{V}$. For the sake of simplicity, we will not consider influences of a domain approximation and numerical integration. Due to this simplification, most of the results from [7, 8] proven for the linear simplicial elements ($P_1$-elements) can be straightforwardly extended for higher-order elements. We summarize them.

The discrete forms of $(\mathscr{P})_\lambda$ and $(\mathscr{P})^\alpha$ read as follows:

$$(\mathscr{P}_h)_\lambda \qquad \inf_{v_h\in\mathbb{V}_h} J_\lambda(v_h),\quad J_\lambda(v_h)=\int_\Omega j(\varepsilon(v_h))\,dx-\lambda L(v_h),\quad \lambda>0,$$

$$(\mathscr{P}_h)^\alpha \qquad \bar{\psi}_h(\alpha)=\inf_{\substack{v_h\in\mathbb{V}_h\\ L(v_h)=1}}\int_\Omega j_\alpha(\varepsilon(v_h))\,dx,\quad j_\alpha(e)=\frac{1}{\alpha}j(\alpha e),\quad \alpha>0.$$

Unlike the continuous setting, one can find minimizers in $\mathbb{V}_h$ of these problems for any $\alpha>0$ and any $\lambda<\lambda_h^*$, where

$$\lambda_h^*=\sup\{\lambda\geq 0\mid \inf_{v_h\in\mathbb{V}_h} J_\lambda(v_h)>-\infty\} \tag{28}$$

is the discrete limit load parameter. Solutions to $(\mathscr{P}_h)_\lambda$ and $(\mathscr{P}_h)^\alpha$ are related each other as in Sect. 4. Therefore, the discrete counterpart $\psi_h$ of $\psi$ can be defined as follows:

$$\psi_h(\alpha) = \int_\Omega \Pi_B(\varepsilon(\alpha u_{h,\alpha})) : \varepsilon(u_{h,\alpha})\, dx = \arg\max_{\lambda \in \mathbb{R}_+} \left[ \frac{1}{\alpha}\phi_h(\lambda) + \lambda \right], \quad \alpha > 0, \quad (29)$$

where $u_{h,\alpha} \in \mathbb{V}_h$ is a solution to $(\mathscr{P}_h)^\alpha$, $\phi_h(\lambda) = \inf_{v_h \in \mathbb{V}_h} J_\lambda(v_h)$, again using that the value $\psi_h(\alpha)$ is independent of the choice of the solution to $(\mathscr{P}_h)^\alpha$. The function $\psi_h$ is continuous, nondecreasing, and $\psi_h(\alpha) \to \lambda_h^*$ as $\alpha \to +\infty$.

The discrete problem of limit analysis reads: determine $\zeta_h^*$ such that

$$\zeta_h^* = \inf_{\substack{v_h \in \mathbb{V}_h, \\ L(v_h)=1}} J_\infty(v_h), \quad J_\infty(v_h) = \int_\Omega j_\infty(\varepsilon(v_h))\, dx. \quad (30)$$

If there exists $v_h \in \mathbb{V}_h$, $L(v_h) = 1$, such that $J_\infty(v_h) < +\infty$ then problem (30) has a minimizer $u_{h,\infty}$ and any sequence $\{u_{h,\alpha}\}_\alpha$ of the solutions to $(\mathscr{P}_h)^\alpha$ is bounded in $\mathbb{V}_h$ for any $h > 0$. It is possible to show (see [8]) that any accumulation point of $\{u_{h,\alpha}\}_\alpha$ minimizes (30), and it holds that $\lambda_h^* = \zeta_h^*$. If such $v_h$ does not exist, then $\lambda_h^* = \zeta_h^* = +\infty$.

From Lemma 1 we see that $\lambda_h^* \geq \lambda^*$. Further, it is known from [7] that if $B$ is bounded then

$$\lambda_h^* \to \lambda^*, \quad h \to 0_+. \quad (31)$$

If $B$ is unbounded, then (31) does not hold, in general. For unbounded $B$, one can apply the truncation technique from Sect. 3 to the discretized problems. Let $\{B_k\}$ be a system of bounded, closed and convex subsets of $B$. As in Sect. 3 we associate with any $B_k$ the functions $j_k$, $J_{k,\lambda}$ and the limit values $\lambda_k^*$, $\zeta_k^*$. The discrete limit load parameters associated with $B_k$ and $\mathbb{V}_h$ are denoted as $\lambda_{k,h}^*$ and $\zeta_{k,h}^*$. Then $\lambda_{k,h}^* = \zeta_{k,h}^*$ and the following criterion holds:

$$\lambda > \lambda_{k,h}^* (\geq \lambda_k^*) \iff \exists v_h \in \mathbb{V}_h: \quad J_{k,\lambda}(v_h) < -c_k|\Omega|, \quad c_k = \frac{1}{2}\sup_{\tau \in B_k} \|\tau\|^2_{\mathbb{C}^{-1}}. \quad (32)$$

In [7], pointwise convergence $\psi_h \to \psi$ has been established for $\mathbb{V}_h$ constructed by $P_1$-elements. However, just this result cannot be straightforwardly extended to higher order elements as the ones mentioned above. Therefore, we sketch another proof based on the function $\bar{\psi}$ and its discretization.

**Lemma 3** *It holds:*

$$\lim_{h\to 0_+} \bar{\psi}_h(\alpha) = \bar{\psi}(\alpha) \quad \forall \alpha > 0. \tag{33}$$

*Proof* For any $v \in \mathbb{V}$, $L(v) = 1$, there is a sequence $\{v_h\}$, $v_h \in \mathbb{V}_h$, $L(v_h) = 1$, such that $v_h \to v$ in $\mathbb{V}$. Hence,

$$\bar{\psi}_h(\alpha) \leq \int_\Omega j_\alpha(\varepsilon(v_h))\, dx \to \int_\Omega j_\alpha(\varepsilon(v))\, dx \quad \forall \alpha > 0.$$

At the same time, from the definitions of $\bar{\psi}_h$ and $\bar{\psi}$ it follows that $\bar{\psi}_h(\alpha) \geq \bar{\psi}(\alpha)$ for any $\alpha > 0$. Therefore, (33) holds.

**Theorem 1** *It holds:*

$$\lim_{h\to 0_+} \psi_h(\alpha) = \psi(\alpha) \quad \forall \alpha > 0. \tag{34}$$

*Proof (Sketch).* It is possible to show that the sequence $\{u_{h,\alpha}\}_h$ of solutions to $(\mathscr{P}_h)^\alpha$ is bounded in $\mathbb{V}$ for any $\alpha > 0$. Therefore, (29) implies boundedness of the sequence $\{\psi_h(\alpha)\}_h$ for any $\alpha > 0$. Then there exist: $\bar{\lambda} \in \mathbb{R}_+$ and a subsequence $\{\psi_{h'}(\alpha)\}$ such that $\psi_{h'}(\alpha) \to \bar{\lambda}$ as $h' \to 0_+$. From [7], it follows that $\phi_{h'}(\lambda) \to \phi(\lambda)$ for any $\lambda \in \mathbb{R}_+$. Using the continuity of $\psi_h$, we arrive at $\phi_{h'}(\psi_{h'}(\alpha)) \to \phi(\bar{\lambda})$ and

$$\bar{\psi}_{h'}(\alpha) = \frac{1}{\alpha}\phi_{h'}(\psi_{h'}(\alpha)) + \psi_{h'}(\alpha) \to \frac{1}{\alpha}\phi(\bar{\lambda}) + \bar{\lambda} \overset{(17,33)}{=} \bar{\psi}(\alpha) = \frac{1}{\alpha}\phi(\psi(\alpha)) + \psi(\alpha).$$

From the definition of $\psi(\alpha)$, it follows that $\bar{\lambda} = \psi(\alpha)$ proving (34).

*Remark 3* Observe that the only one assumption on $\{\mathbb{V}_h\}$ is needed: namely that this system is limit dense in $\mathbb{V}$.

## 6 Computable Bounds of $\lambda^*$ and Numerical Experiments

Since $\lambda^*$ is a safety parameter, reliable computable bounds of this quantity are important. Unlike [2, 7, 8, 12], we now use the $P_2$-elements with a 7-point numerical integration formula instead of the $P_1$-elements in order to reduce strong dependence on the mesh density observed just for the $P_1$-elements and unbounded yield surfaces. We use similar benchmarks for the von Mises and Drucker–Prager yield surfaces as in [7, 8]. The computational experiments presented below were implemented in MatLab.

### 6.1 Computable Bounds of $\lambda^*$ and Numerical Methods

Notation $\lambda^*$, $\lambda_h^*$, $\psi$, $\psi_h$, $(\mathscr{P}_h)^\alpha$ will be related to unbounded $B$ while $\lambda_k^*$, $\lambda_{k,h}^*$, $\psi_k$, $\psi_{k,h}$, $(\mathscr{P}_{k,h})^\alpha$, and $(\mathscr{P}_{k,h})_\lambda$ will be associated with a bounded subset $B_k$ of $B$. For unbounded $B$, two kinds of lower bounds were mentioned: $\psi(\alpha), \alpha > 0$, and $\lambda_k^*, k > 0$. Convergence $\psi_h(\alpha) \to \psi(\alpha)$ and $\lambda_{k,h}^* \to \lambda_k^*$ as $h \to 0_+$ hold for the bounds but $\lambda_h^* \to \lambda^*$ need not hold, in general. Values $\lambda_{k,h}^*$ can be approximated from below by $\psi_{k,h}(\alpha)$ where $\alpha$ is sufficiently large. Moreover, the guaranteed upper bound (32) of $\lambda_{k,h}^*$ and $\lambda_k^*$ is at our disposition. This bound should be close to $\psi_{k,h}(\alpha)$ for verification of numerical results. Further, $\lambda^*$ will be estimated from above by $\lambda_h^* \approx \psi_h(\alpha)$, where $\alpha$ is sufficiently large. We will compare numerically the mentioned bounds for several meshes.

The functions $\psi_h$ and $\psi_{k,h}$ can be assessed solving $(\mathscr{P}_h)^\alpha$ and $(\mathscr{P}_{k,h})^\alpha$, respectively. To this end, we use the semismooth Newton method with damping or as a case may be with regularized tangent stiffness matrices. In the latter case the tangent stiffness matrix is replaced by a convex combination of the tangent and elastic stiffness matrices to get positive definiteness. Damped parameters belong to (0, 1] and guarantee a decrease of minimized functions in the Newton direction. In context of the minimization problem, the method can be interpreted as a sequential quadratic programming approach. The load constraint is enforced by the Lagrange multiplier in each Newton's iteration. For convergence analysis and numerical experiments with the variants of the semismooth Newton method we refer to [2]. We use the relative tolerance $1e - 10$ in a termination criterion of the Newton-like method.

This enables us to construct the loading $\lambda - \alpha$ curve to estimate $\lambda_h^*$. Firstly, a constant increment $\delta\alpha > 0$ is considered. If the computed increment $\delta\lambda$ is less than the prescribed threshold (we use 0.001) then we enlarge $\delta\alpha$ twice. So the values of $\alpha$ can growth exponentially in a vicinity of the limit load. The loading process is terminated when either $\alpha > \alpha_{max}$ or $\delta\lambda \ll 0.001$.

Denote $\bar{\alpha}_h$ as the maximal value of $\alpha$ obtained in this way for given $\mathscr{T}_h$ and $B$. For the truncation $B_k$, we write $\bar{\alpha}_{k,h}$ to emphasize also dependence on $k$. We use the values $\psi_h(\bar{\alpha}_h)$ and $\bar{\lambda}_{k,h} = \psi_{k,h}(\bar{\alpha}_{k,h})$ as an approximation of $\lambda_h^*$ and as a lower bound of $\lambda_{k,h}^*$, respectively. In order to find the guaranteed upper bound of $\lambda_{k,h}^*$, we contruct a minimization sequence of problem $(\mathscr{P}_{h,k})_\lambda$ using the damped semismooth Newton method. We observed that the value $\lambda = \bar{\lambda}_{k,h} + \delta\lambda$, where $\delta\lambda = 0.001$, is usually sufficient to satisfy criterion (32). Otherwise, we enlarge $\delta\lambda$.

### 6.2 Numerical Example with the Von Mises Criterion

We consider a benchmark used in [7, 8] and many other papers, namely a plane strain problem with $\Omega$ depicted in Fig. 1: $\Omega$ is a quarter of the $10 \times 10$ (m) square with the circular hole of radius 1 (m) in its center. The constant traction of density $f = (0, 450), (0, 0)$ (MPa) acts on the upper, and the right vertical side, respec-

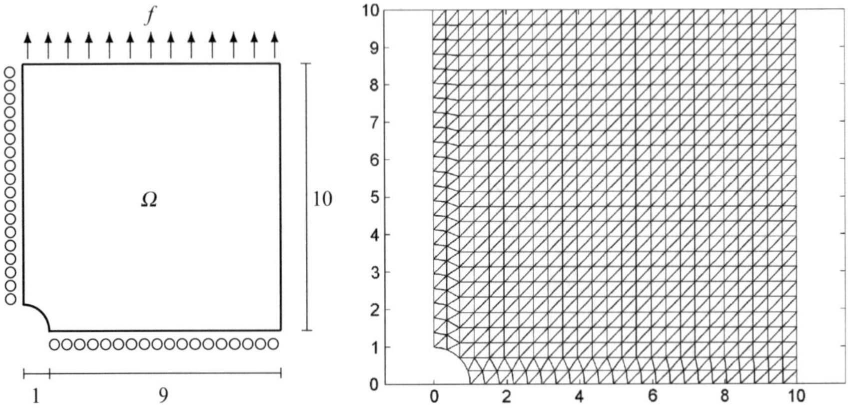


**Fig. 1** Geometry and triangulation

**Fig. 2** Loading paths (zoom) for $B$

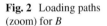

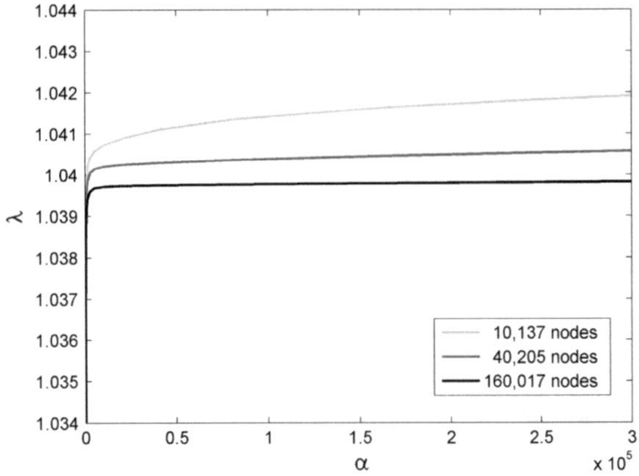


tively. The volume forces are neglected. This load corresponds to $\lambda = 1$. On the rest of $\partial\Omega$ the symmetry boundary conditions are prescribed. The material parameters are set as follows: $E = 206900$ MPa (Young's modulus), $\nu = 0.29$ (Poisson's ratio) and $\gamma = 450\sqrt{2/3}$ MPa. Hence, the values of $K$ and $G$ needed in (12) are $K = \frac{E}{3(1-2\nu)}$ and $G = \frac{E}{2(1+\nu)}$. To obtain more accurate bounds than in [7, 8], we use the $P_2$-elements, a different mesh structure depicted in Fig. 1, and the truncation coefficient $k = 0.7$ for defining $B_k$.

The loading paths represented by the graphs of $\psi_h$ and $\psi_{k,h}$ are computed and compared for three different triangulations $\mathscr{T}_h$ with 10,137, 40,205, and 160,017 nodes (including the midpoints). Notice that the mesh in Fig. 1 has 2,585 nodes. Zoom of the resulting loading paths in a vicinity of the limit load is depicted in Fig. 2 for $B$ and in Fig. 3 for $B_k$.

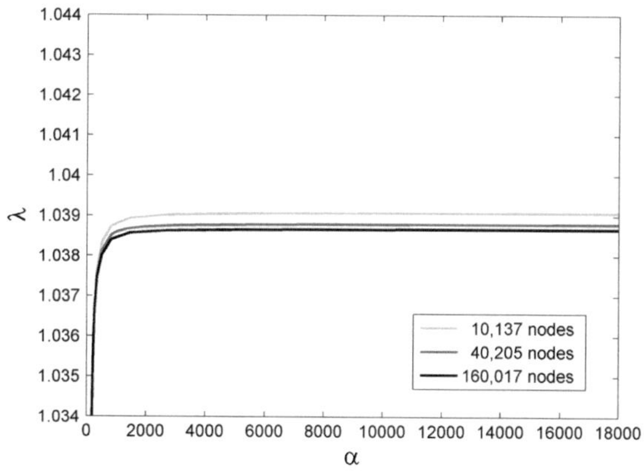


**Fig. 3** Loading paths (zoom) for $B_k$, $k = 0.7$

**Table 1** Lower and upper bounds of $\lambda^*_{k,h}$, $k = 0.7$

| No. of nodes | 10,137 | 40,205 | 160,017 |
|---|---|---|---|
| Lower bound | 1.0391 | 1.0388 | 1.0387 |
| Upper bound | 1.0401 | 1.0398 | 1.0397 |

One can observe that for any $\alpha > 0$ fixed, the sequences $\{\psi_h(\alpha)\}_h$ and $\{\psi_{k,h}(\alpha)\}_h$ are decreasing and converging. Moreover, the curves are almost constant for sufficiently large values of $\alpha$. In Fig. 2, this is visible only for the finest mesh (black color) due to the zoom. From the black curve, we obtain the value close to 1.040 as a reliable upper bound of $\lambda^*$. The curves for $B_k$ are less dependent on the number of nodes of $\mathcal{T}_h$ than for $B$. The computed values $\psi_{k,h}(\bar{\alpha}_{k,h})$, i.e., the lower bounds of $\lambda^*_{k,h}$ are shown in Table 1 and compared with the guaranteed upper bounds of $\lambda^*_{k,h}$. The bounds practically coincide for the used meshes. This confirms that our results are reliable. From Table 1 we see that $\lambda^*_k \approx 1.038$.

We arrive at the following bounds of $\lambda^*$: $1.038 \leq \lambda^* \leq 1.040$. Further, it is useful to have a look at Figs. 4 and 5 where the displacements in the horizontal and vertical directions (solution to $(\mathcal{P}_h)^\alpha$) at the end of the loading process (i.e. $\alpha = \bar{\alpha}_h$) are depicted for $B$ and the finest mesh.

We observe a significant jump of the values along the same thin diagonal band. Far from it, the material is rigid (constant displacements) which is in accordance with Remark 1. Moreover, the vector field depicted in these figures can be simply approximated by the function $v_\delta = (v_{\delta,1}, v_{\delta,2})$, where $\delta > 0$, $\delta << 1$,

$$v_{\delta,1}(x, y) = \begin{cases} 0, & \text{if } y - x + 1 \geq \delta \\ \frac{a}{\delta}(y - x + 1 - \delta), & \text{if } 0 \leq y - x + 1 \leq \delta \\ -a, & \text{if } 0 \geq y - x + 1 \end{cases},$$

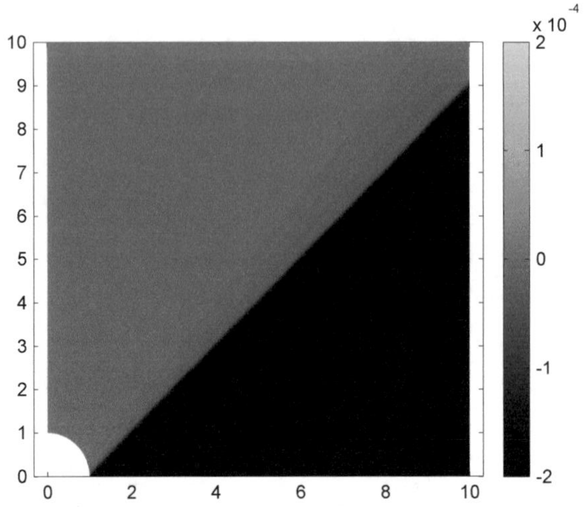


**Fig. 4** Horizontal displacements for $B$

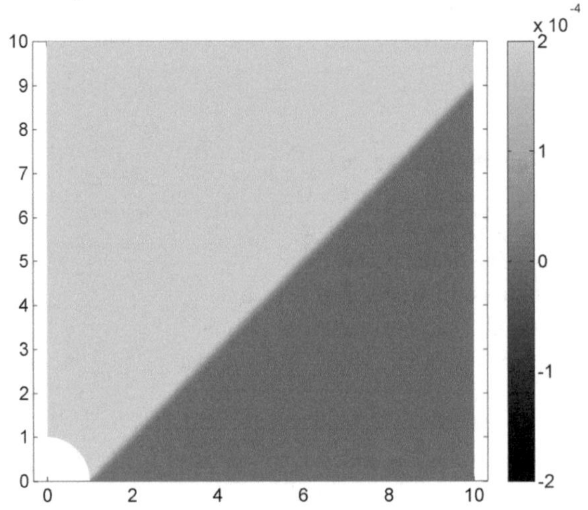


**Fig. 5** Vertical displacements for $B$

$$v_{\delta,2}(x,y) = \begin{cases} a, & \text{if } y-x+1 \geq \delta \\ \frac{a}{\delta}(y-x+1), & \text{if } 0 \leq y-x+1 \leq \delta \\ 0, & \text{if } 0 \geq y-x+1 \end{cases}, \quad a = \frac{1}{4500}.$$

It is easy to see that $v_\delta \in \mathbb{V}$, div $v_\delta = 0$ in $\Omega$, and $L(v_\delta) = 1$ for any $\delta \ll 1$. Therefore, from (13), we have:

$$\lambda^* \leq \int_\Omega \gamma \|\varepsilon(v_\delta)\| \, dx = \gamma \frac{a}{\delta}\sqrt{2}|\Omega_\delta| = \frac{9}{5\sqrt{3}} + O(\delta) \doteq 1.0392 + O(\delta),$$

where $\Omega_\delta = \{(x, y) \in \Omega \mid 0 \le y - x + 1 \le \delta\}$ and $|\Omega_\delta| = 9\delta + O(\delta^2)$. Hence, the value 1.0392 is a guaranteed upper bound of $\lambda^*$ which is fully in accordance with our numerical results. Notice that the sequence $\{v_\delta\}$ is converging to a function which belongs to $BD(\Omega; \mathbb{R}^2) \setminus \mathbb{V}$.

In [7, 8], a much more pessimistic upper bound was obtained for the $P_1$-elements. Moreover, the corresponding curves for $B$ strongly depended on the mesh. On the other hand, for $B_k$, the results for $P_1$ and $P_2$-elements are similar. So, it seems that the truncation limit analysis is useful for $P_2$-elements and necessary for $P_1$- elements.

## 6.3 Numerical Example with the Drucker–Prager Criterion

The second example is a slope stability benchmark considered as a plane strain problem [3, 5, 13]. In comparison to [5, 8, 13], we use much smaller geometry and the Poisson ratio $\nu = 0.25$ instead of $\nu = 0.49$. From Remark 1, it follows that the limit load parameter should be independent of these settings. The shape and sizes of 2D domain $\Omega$ with a uniform triangular mesh are shown in Fig. 6. The slope inclination is 45°. On the bottom we assume that $\Omega$ is fixed and the zero normal displacements are prescribed on both vertical sides. The remaining part of $\partial\Omega$ is free. The load $L$ is represented by the gravity force $F$. We set the specific weight $\rho g = 20\,\text{kN/m}^3$ with $\rho$ being the mass density and $g$ the gravitational acceleration. The Drucker- Prager parameters $a$ and $\gamma$ appearing in (14) are computed from the friction angle $\phi$ and the cohesion $c$ as follows [5]:

$$a = \frac{3\sqrt{2}\tan\phi}{\sqrt{9 + 12(\tan\phi)^2}}, \quad \gamma = \frac{3\sqrt{2}c}{\sqrt{9 + 12(\tan\phi)^2}}.$$

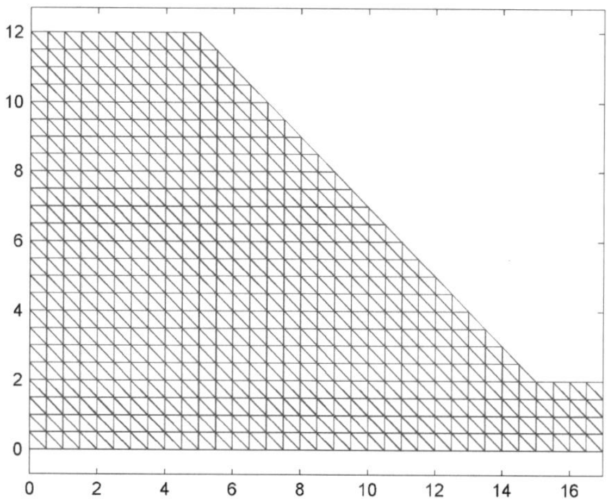

**Fig. 6** Geometry of the problem and the mesh for $h = 0.5\,\text{m}$

Fig. 7 Loading paths for $B$

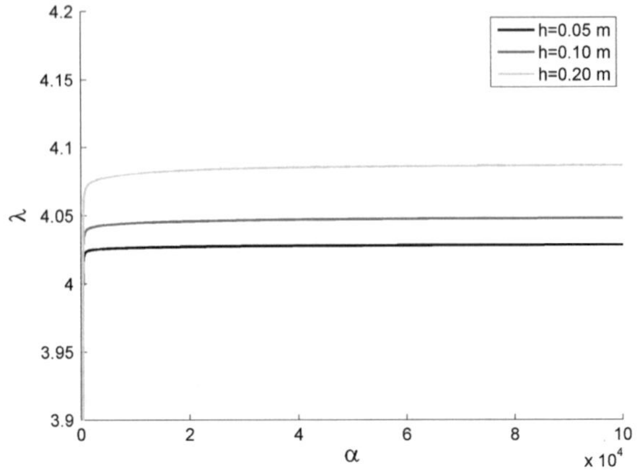


Fig. 8 Loading paths for $B_k$, $k = 3.9$

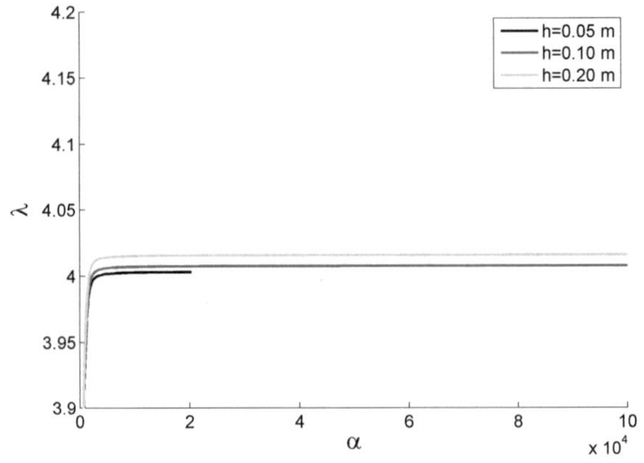


Finally, $E = 20\,000$ kPa, $\phi = 20°$ and $c = 50$ kPa. The bulk and shear moduli are computed as in Sect. 6.2. The discretization of $(\mathscr{P}_h)^\alpha$ is done by $P_2$-elements using three uniform triangulations $\mathscr{T}_h$ of $\bar{\Omega}$ with $h = 0.05,\ 0.1,\ 0.2$ meters, where $h$ stands for the length of the leg of the isosceles right triangles creating $\mathscr{T}_h$.

The loading paths for all these meshes are depicted in Figs. 7 and 8 for $\alpha \in [0, 1e5]$. In Fig. 7 we see the loading paths for the original set $B$ defined by (14). Again, one can observe that the curves converge to some limit curve. Since the paths are almost constant for $\alpha > 2e4$ the respective values of $\psi_h$ at $\alpha = 1e5$ can be considered to be equal to $\lambda_h^*$. Consequently, $\lambda_{h=0.05}^* = 4.03$ is a reliable upper bound of $\lambda^*$.

To get a lower bound of $\lambda^*$ we use the truncation approach with $B_k$ defined by (22). Figure 8 depicts the resulting loading paths for $B_{k=3.9}$. The curves are less dependent on the number of the nodes of $\mathscr{T}_h$ than for $B$. The computed values $\psi_{k,h}(\bar{\alpha}_{k,h})$, i.e., the lower bounds of $\lambda_{k,h}^*$ are displayed in Table 2 and compared with the guaranteed upper bounds of $\lambda_{k,h}^*$. The bounds practically coincide for the used meshes. This confirms the reliability of our results. From Table 2, we see that $\lambda_k^* \approx 4.00$.

**Table 2** Lower and upper bounds of $\lambda^*_{k,h}$, $k = 3.9$

| No. of nodes | $h = 0.20$ | $h = 0.10$ | $h = 0.05$ |
|---|---|---|---|
| Lower bound | 4.016 | 4.008 | 4.003 |
| Upper bound | 4.017 | 4.009 | 4.005 |

Based on this experiment we may conclude that the values 4.00 and 4.03 could serve as reliable lower and upper bounds to $\lambda^*$, respectively. The analytical estimate to this problem for the Mohr-Coulomb yield function presented in [3] gives the value $\lambda^* \approx 4.045$ which is close to the computed bounds.

## 7 Conclusion

The paper completes our research presented in [2, 7, 8, 12]. Unlike these papers, some results are extended to $P_2$-elements. For quadratic elements and unbounded yield surfaces, the loading paths are not so much dependent on the number of mesh nodes as for $P_1$-elements. On the other hand, for bounded yield surfaces, $P_1$ and $P_2$-elements, we obtain more or less the same results. We also illustrated using the slope stability benchmark that the limit load parameter is independent of the elastic parameters and the size of the geometry. These facts enable us to significantly improve estimates of $\lambda^*$ in comparison to [7, 8].

**Acknowledgements** This work was supported by The Ministry of Education, Youth and Sports of the Czech Republic from the National Programme of Sustainability (NPU II), project "IT4 Innovations excellence in science - LQ1602".

## References

1. Caboussat, A., Glowinski, R.: Numerical solution of a variational problem arising in stress analysis: the vector case. Discret. Contin. Dyn. Syst. **27**, 1447–1472 (2010)
2. Cermak, M., Haslinger, J., Kozubek, T., Sysala, S.: Discretization and numerical realization of contact problems for elastic-perfectly plastic bodies. PART II – numerical realization. ZAMM-Journal of Applied Mathematics and Mechanics/Zeitschrift für Angewandte Mathematik und Mechanik **95**, 1348–1371 (2015)
3. Chen, W., Liu, X.L.: Limit analysis in soil mechanics. Elsevier, Amsterdam (1990)
4. Christiansen, E.: Limit analysis of colapse states. In: Ciarlet, P.G., Lions, J.L.: (eds.) Handbook of Numerical Analysis, vol. IV, Part 2, pp. 195–312. North-Holland (1996)
5. de Souza Neto, E.A., Perić, D., Owen, D.R.J.: Computational Methods for Plasticity: Theory and Application. Wiley, New Jersey (2008)
6. Duvaut, G., Lions, J.L.: Inequalities in Mechanics and Physics. Springer, Berlin (1976)
7. Haslinger, J., Repin, S., Sysala, S.: A reliable incremental method of computing the limit load in deformation plasticity based on compliance: Continuous and discrete setting. J. Comput. Appl. Math. **303**, 156–170 (2016)
8. Haslinger, J., Repin, S., Sysala, S.: Guaranteed and computable bounds of the limit load for variational problems with linear growth energy functionals. Appl. Math. **61**, 527–564 (2016)

9. Repin, S., Seregin, G.: Existence of a weak solution of the minimax problem arising in Coulomb-Mohr plasticity. In: Nonlinear Evolution Equations, American Mathematical Society Translations: (2), vol. 164, pp. 189–220. American Mathematical Society, Providence, RI (1995)
10. Suquet, P.: *Existence et régularité des solutions des équations de la plasticité parfaite*, These de 3e Cycle, Université de Paris VI (1978)
11. Sysala, S.: Properties and simplifications of constitutive time-distretized elastoplastic operators. ZAMM-Journal of Applied Mathematics and Mechanics/Zeitschrift für Angewandte Mathematik und Mechanik **94**, 233–255 (2014)
12. Sysala, S., Haslinger, J., Hlaváček, I., Cermak, M.: *Discretization and numerical realization of contact problems for elastic-perfectly plastic bodies. PART I – discretization, limit analysis.* ZAMM-Journal of Applied Mathematics and Mechanics/Zeitschrift für Angewandte Mathematik und Mechanik **95**, 333–353 (2015)
13. Sysala, S., Cermak, M., Koudelka, T., Kruis, J., Zeman, J., Blaheta, R.: *Subdifferential-based implicit return-mapping operators in computational plasticity*. ZAMM-Journal of Applied Mathematics and Mechanics/Zeitschrift für Angewandte Mathematik und Mechanik **96**, 1318–1338 (2016)
14. Temam, R.: Mathematical Problems in Plasticity. Gauthier-Villars, Paris (1985)

# Can a Hencky-Type Model Predict the Mechanical Behaviour of Pantographic Lattices?

**Emilio Turco, Maciej Golaszewski, Ivan Giorgio and Luca Placidi**

**Abstract** Current research in metamaterials design is pushing to fill the gap between mathematical modeling and technological applications. To meet these requirements, predictive and computationally effective numerical tools need to be conceived and applied. In this paper we compare the performances of a discrete model already presented in [1], strongly influenced by Hencky approach [2], versus some interesting experiments on pantographic structures built using the 3D printing technology. The interest in these structures resides in the exotic behavior that they have already shown, see [3, 4], and their study seems promising. In this work, after a brief presentation of the discrete model, we discuss the results of three experiments and compare them with the corresponding predictions obtained by the numerical simulations. An in-depth discussion of the numerical results reveals the robustness of the numerical model but also clearly indicates which are the focal points that strongly influence the accuracy of the numerical simulation.

E. Turco
Department of Architecture, Design and Urban Planning,
via Garibaldi 35, Alghero, Italy
e-mail: emilio.turco@uniss.it

M. Golaszewski
DCEBM, Warsaw University of Technology, Warsaw, Poland
e-mail: golaszewski.maciej07@gmail.com

I. Giorgio
MeMoCS, International Research Center for the Mathematics and
Mechanics of Complex Systems, Università dell'Aquila, L'Aquila, Italy
e-mail: ivan.giorgio@uniroma1.it

L. Placidi (✉)
Faculty of Engineering, International Telematic University Uninettuno,
Rome, Italy
e-mail: luca.placidi@uninettunouniversity.net



F. dell'Isola et al. (eds.), *Mathematical Modelling in Solid Mechanics*,
Advanced Structured Materials 69, DOI 10.1007/978-981-10-3764-1_18

# 1 Introduction

Current research in metamaterials design is pushing to fill the gap between mathematical modeling and technological applications. Although both evolutionary selection in living organism and the past engineering scientifically based research have already promoted the conception of *exotic* metamaterials (the bone tissue is one example while woven fabrics gives another one) it is only a recent issue the systematic research of tailored materials having fixed well-determined a priori uses and applications. To meet all the requirements imposed by determined and well-specified applications it is needed to establish a designing procedure which involve the important step concerning the development of some predictive and computationally effective numerical tools. These tools will be then used to verify experimental measurements output and subsequently to design specifically adapted materials.

In this paper we focus on a specific, but in our opinion relevant, task: to compare the performances of a discrete model for pantographic lattices, sometimes also called pantographic sheets, with some interesting experiments. The interest in these structures resides in the exotic behavior that they have shown [3, 4] and their study seems promising. In particular pantographic structures:

- are the actual realization of a (often disputed) continuum model: i.e. second gradient materials; indeed pantographic sheets are one of the first mechanical structures which have been proven, see [5], to need a second gradient models at a given macroscopic length-scale;
- have been proven to have very promising properties in wave propagation, representing an example of effective wave-guides, see [6];
- have shown promising toughness properties, which suggest that they could be fruitfully embedded in novel composite materials.

The comparison between an effective discrete model and some experiments shows, on one side, its robustness and, on the other side, the focus points which require strong attention when accurate results are desired.

In the following we briefly present the discrete model, Sect. 2, also mentioning the algorithm used to reconstruct the complete equilibrium path of the mechanical problem. Successively, in Sect. 3, we thoroughly discuss some experiments and the comparison with the corresponding numerical simulations. Finally, in Sect. 4, there are some concluding remarks and future perspectives.

# 2 Discrete Model for Pantographic Sheets

In this section we shortly describe the discrete Lagrangian model which we consider here to be possible model for planar pantographic structures. Their predictive performances will be analysed in the following Sect. 3. We limit ourselves to remark here that:

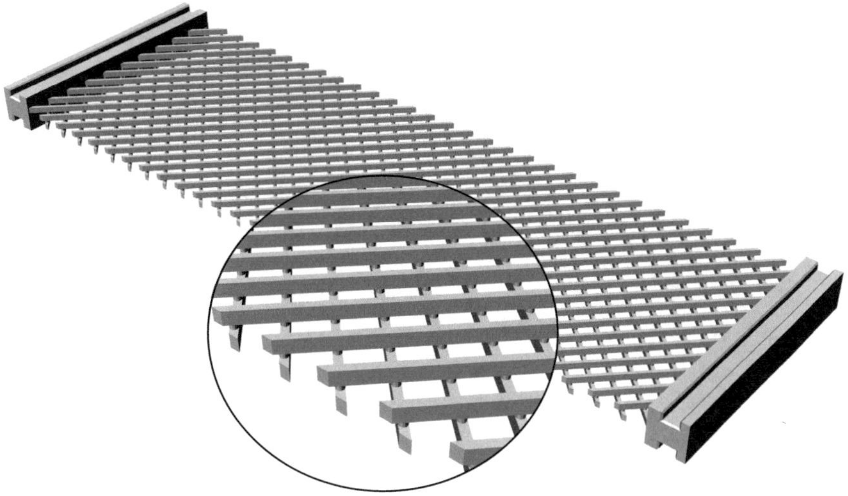

**Fig. 1** Pantographic structure fabric built by using 3D printing technology

- pantographic structures can be nowadays simply built using 3D printing technologies (see, e.g., Fig. 1) based on the concept first proposed in [4] where elastic pivots were realized with small size elastic cylinders interconnecting the two arrays of beams;
- discrete models, see Fig. 2, are conceived by modelling interconnecting pivots as nodes linked each other by means of extensional (pairwise interaction, see Fig. 3a) and rotational springs, i.e. bending springs (triple interaction on a fiber, see Fig. 3b) and shear springs (triple interaction on a fiber of the same array and also on the nearest pivots on the other direction, see Fig. 3c).

The kinematical description of the pantographic lattice model involves a finite configuration space. To be precise, the discrete model involves the introduction of a set of Lagrangian parameters specifying the position of all the material particles modelling the pivots. They are initially located in the nodes of the reference configuration and then they displace in the actual configuration. If we limit ourself to consider planar motions, only a set of $2N$ coordinates is sufficient (if $N$ is the number of considered nodes, the generic of which has referential position given by $P_{i,j}$, such a set of Lagrangian coordinates could be given by the corresponding actual position $p_{i,j}$).

The strain energies of the discrete model are the only kind of energy to be specified in hard devices deformations (in absence of relevant volume forces). The postulated expression for the Lagrangian discrete deformation energy $W_{\text{int}}$ (in terms of the Lagrangian coordinates $p_{i,j}$) is completely defined specifying the contribution of each one kind of spring (see Figs. 2 and 3):

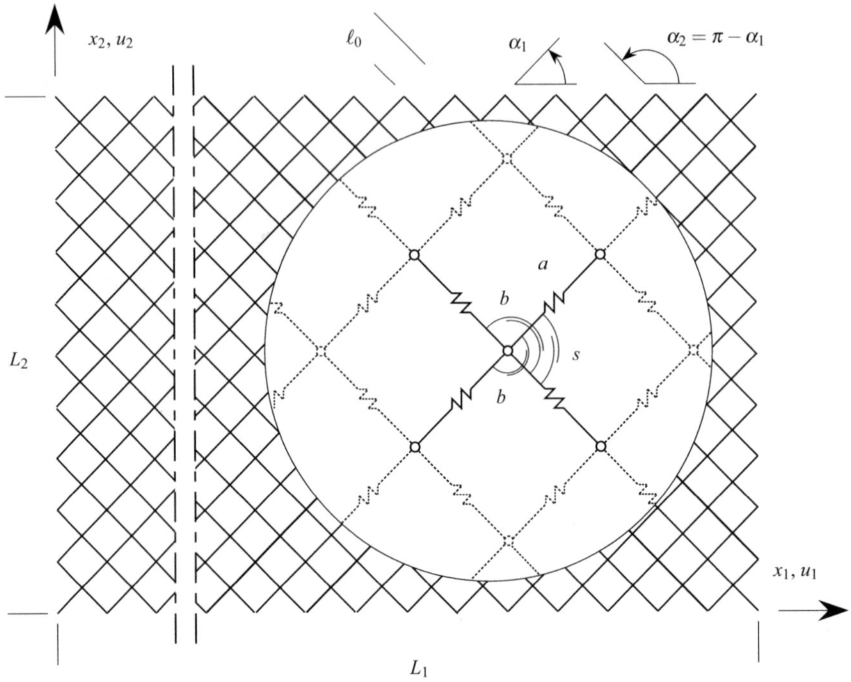


**Fig. 2** Hencky-type mechanical model of a pantographic structure

$$w_a = \frac{1}{2}a(\ell - \ell_0)^2 , \tag{1}$$

$$w_b = b(\cos\beta + 1) , \tag{2}$$

$$w_s = \frac{1}{2}s(\sigma - \sigma_0)^2 , \tag{3}$$

where $\ell$ and $\ell_0$ are the actual and the reference length, respectively, of pantographic bar (i.e. the distance between two consecutive pivots along fibers' direction), $\beta$ is the angle between two consecutive pantographic bars (in the reference configuration this angle is $\pi$) and $\sigma$ is the angle between two fibers starting from the same pivot (in the reference configuration this angle is related to $\alpha_1$ or $\alpha_2$ and $\sigma_0$). Furthermore, $a$, $b$ and $s$ are the axial, bending and shear stiffnesses of each one type of spring, respectively.

Some remarks:

i. The shear springs used for the discrete model, and depicted in Figs. 2 and 3, are actually four, having the same stiffness, for each node or pivot, one for each quadrant and having origin, in the reference configuration, at $P_{i,j}$.
ii. The bending deformation energy is expressed by means of the $\cos\beta$ instead of the corresponding angle $\beta$, these two possibilities are equivalent, at least in principle,

**Fig. 3** Kinematics of axial (**a**), bending (**b**) and shear (**c**) springs: reference (*dashed line*) and actual (*continuous line*) configurations

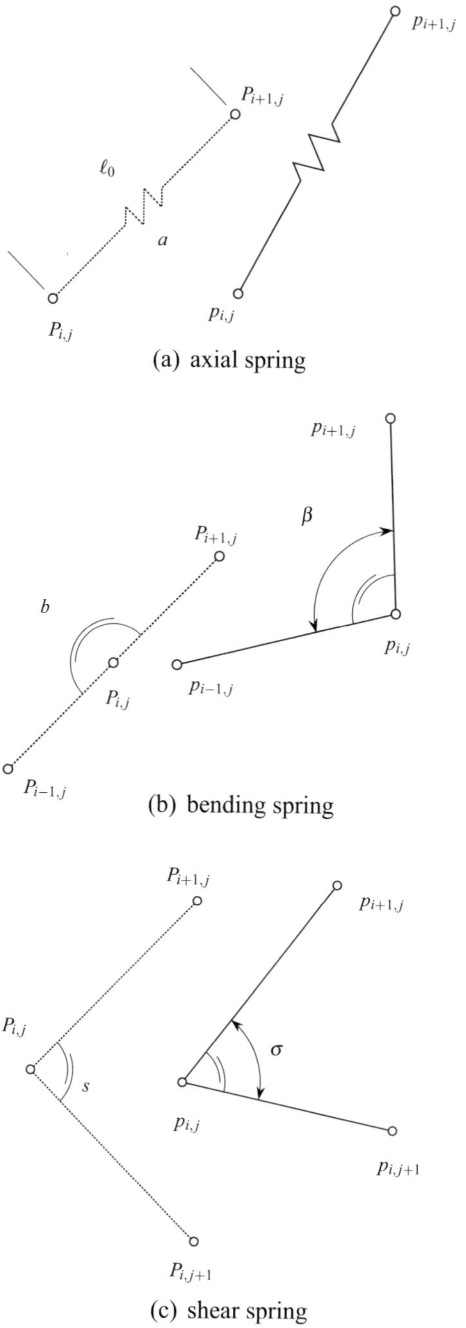

but the first, avoiding the use of arccos$(\cdot)$ function, results more convenient from the computational point of view since it produces a more compact and effective code.

In order to have a complete solution of the considered equilibrium problem, i.e. the displacements (from which can be easily evaluated the structural reaction and the forces or couples exerted by each spring) a step-by-step procedure was implemented to reconstruct the complete equilibrium path of the pantographic sheet.

We briefly sketch here the basic ingredients of the procedure. The total energy of the pantographic structure can be computed in a straightforward manner simply adding the strain contribution of each spring. Formally we can write:

$$W(\mathbf{d}) = W_{\text{int}} - W_{\text{ext}} = \sum_{e} (w_a + w_b + w_s) - W_{\text{ext}}\,, \tag{4}$$

where $e$ ranges on all the springs, extensional, bending and shear, and $W_{\text{ext}}$ is the work of the external loads and all quantities on the *RHS* depend on the vector $\mathbf{d}$ which collects the nodal displacements of the pantographic lattice.

The equilibrium problem which we want to consider is a mixed one: we assume that the displacements of some particles are imposed and that some externally conservative forces are applied to the remaining particles. Let us therefore decompose $\mathbf{d}$ into the pair composed by two vectors: the assigned displacements $\mathbf{u}_a$ and the free displacements $\mathbf{u}$. For notational aims, we will reorder $\mathbf{d}$ to get the decomposition

$$\mathbf{d} = (\mathbf{u}, \mathbf{u}_a)\ .$$

Because of our assumption we have that $W_{\text{ext}}$ depends only on $\mathbf{u}$.

The nonlinear system of equilibrium equations is obtained by imposing that the first variation of $W$ vanish:

$$\mathbf{s}(\mathbf{u}) - \mathbf{p}(\mathbf{u}) = \mathbf{0}\,, \tag{5}$$

where $\mathbf{p}(\mathbf{u})$ is the vector which collects the Lagrangian components of external forces (which may be assumed to be dead loads, for instance, so that $\mathbf{p}$ becomes independent of $\mathbf{u}$) and $\mathbf{s}(\mathbf{u})$ is the vector of the internal forces (called also, in the context of structural mechanics, *structural reaction*), as defined by:

$$\mathbf{s}(\mathbf{u}) = \frac{\mathrm{d}W_{\text{int}}}{\mathrm{d}\mathbf{u}}\,, \qquad \mathbf{p}(\mathbf{u}) = \frac{\mathrm{d}W_{\text{ext}}}{\mathrm{d}\mathbf{u}}\,. \tag{6}$$

The tangent stiffness matrix is defined as the derivative of the structural reaction $\mathbf{s}(\mathbf{u})$ with respect to the displacement vector $\mathbf{u}$, in formulas:

$$\mathbf{K}_T(\mathbf{u}) = \frac{\mathrm{d}\mathbf{s}(\mathbf{u})}{\mathrm{d}\mathbf{u}} = \frac{\mathrm{d}^2 W_{\text{int}}}{\mathrm{d}\mathbf{u}^2}\,, \tag{7}$$

The solution of the nonlinear equilibrium system of Eq. (5) can be found by means of an incremental-iterative procedure based on the Newton–Raphson scheme.

**Table 1** Scheme of a basic algorithm to compute a new point of the equilibrium path $(\lambda_{j+1}, \mathbf{u}_{j+1})$ given the previous $(\lambda_j, \mathbf{u}_j)$

set

  exit := false

  $\mathbf{K}_T := \mathbf{K}_T(\mathbf{u}_j)$

  $\Delta\mathbf{u} := \Delta\lambda\bar{\mathbf{u}}$   ($\bar{u} = 0$ for free nodes and $\bar{u} = u_a$ for the assigned ones)

while (loop < maxloop) and (exit=false)

  $\mathbf{s} := \mathbf{s}(\mathbf{u}_j + \Delta\mathbf{u})$

  $\dot{\mathbf{u}} := \mathbf{K}_T^{-1}\mathbf{s}$

  if $\|\dot{\mathbf{u}}\| > \eta$

    $\Delta\mathbf{u} := \Delta\mathbf{u} - \dot{\mathbf{u}}$

    $\mathbf{K}_T := \mathbf{K}_T(\mathbf{u}_j + \Delta\mathbf{u})$

  else

    exit := true

  end

save

  $\lambda_{j+1} := \lambda_j + \Delta\lambda$

  $\mathbf{u}_{j+1} := \mathbf{u}_j + \Delta\mathbf{u}$

We will limit ourselves to the case of equilibrium paths depending only by the single parameter $\lambda$. Starting from an estimated point of the equilibrium path $(\lambda_j, \mathbf{u}_j)$ verifying that the residue $\mathbf{r}$ of Eq. (5) is

$$\left\|\mathbf{r}(\mathbf{u}_j, \lambda_j)\right\| \leq \eta\,, \tag{8}$$

i.e.with a pair being an $\eta$-approximate solution of the equilibrium condition (5), the iterative scheme, once the step $\Delta\lambda$ is fixed, is obtained by constructing the $\eta$-approximate solution $(\mathbf{u}_{j+1} =: \mathbf{u}_j + \Delta\mathbf{u}_j\,, \lambda_{j+1} := \lambda_j + \Delta\lambda)$ by using the iteration scheme reported in Table 1 to compute $\Delta\mathbf{u}_j$.

Further details on the Hencky-type model and on the strategy used to compute the complete equilibrium path are contained in [1], in addition [7–9] report a comparison of numerical simulations, also with second gradient numerical model, with experimental results in the case of fiber push-out and extensional and bending cases.

## 3 Comparison with Experiments

The reader will remark that very few parameters are postulated to characterize the discrete model. On the contrary a wealth of experimental data are nearly perfectly fitted using these few parameters. In [3] an identification of the parameters of the continuum model in terms of the discrete model was proposed, see [5, 10, 11].

Below we consider three experiments performed on specimen built by using 3D printing technologies. Each one test is characterized by a different orientation of the fibers. By referring to Fig. 2, the first one considers the case $\alpha_1 = \pi/4$, the second one $\alpha_1 = \pi/6$ and the last one $\alpha_1 = \pi/3$. We remark that in the first case the fibers are orthogonal, contrarily to the second and third case. Another remarkable difference regards the boundary conditions. More precisely, in the first test the same extensional displacements have been assigned only on two fibers of the small side (using a small bridge, see subsection 3.1) whereas in the other cases, see subsections 3.2 and 3.3, the displacements are assigned on all the fibers which intersect the small side.

## 3.1 Fiber Push-Out Test

On the basis of the technical drawing reported in Fig. 4, a specimen was built, by using the 3D printing technology, in polyamide (PA 2200) by a SLS Formiga P100. The Young's modulus for this material was estimated between 1.5 and 1.7 GPa following the rules of EN ISO 527 and EN ISO 178.

This specimen was tested by clamping the entire left side and assigning an increasing displacement $u$ (parallel to the larger sides) to two fibers of the right side until

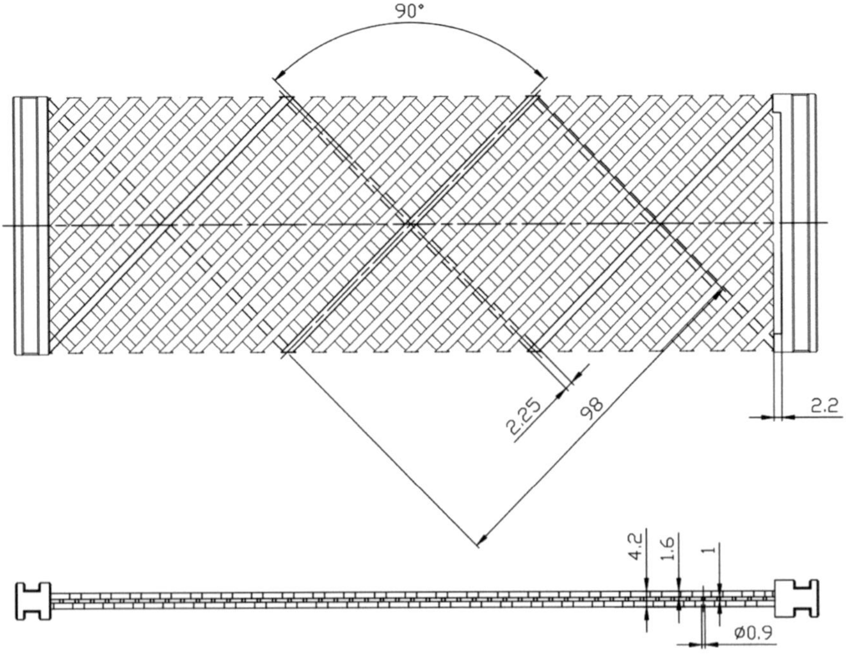


**Fig. 4** Drawing of the specimen for the fiber push-out test

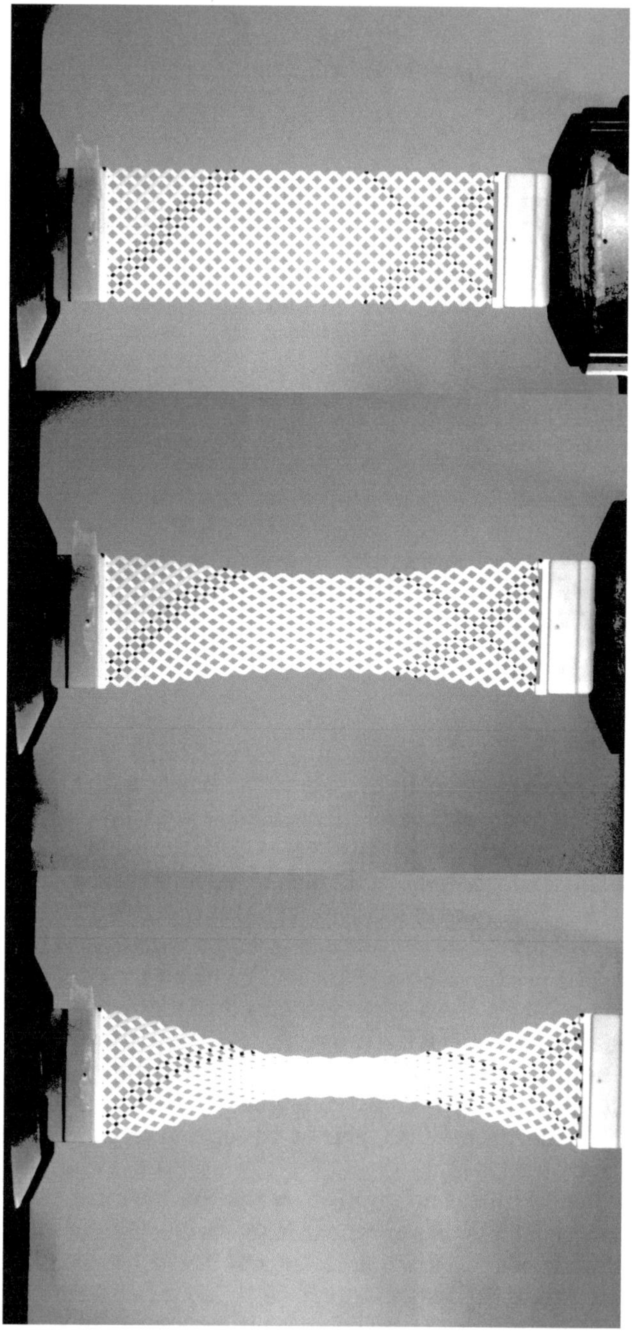

**Fig. 5** Push-out test: sequence of deformations for $\lambda = 0, 0.5$ and 1

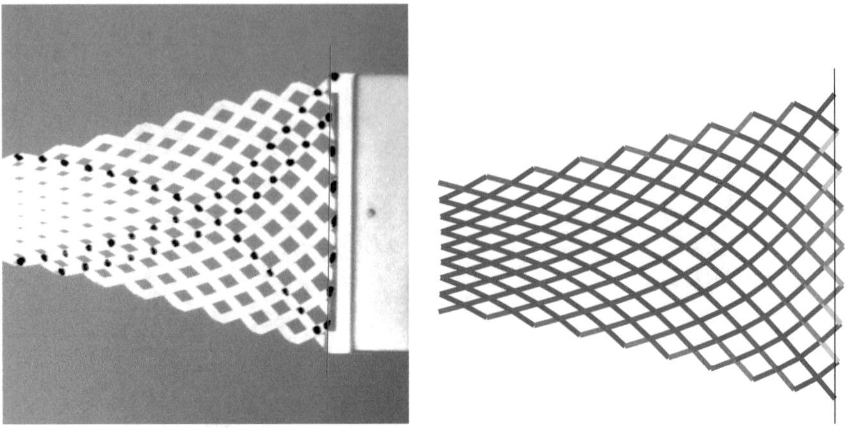

(a) experiments (b) numerical simulation

**Fig. 6** Push-out test: particular of the deformation near the push-out fibers

**Table 2** Spring stiffnesses consequent to different values of the Young's modulus

| $E$ (GPa) | $a$ (N/mm) | $b_1$ (Nmm) | $b_2$ (Nmm) | $s$ (Nmm) |
|---|---|---|---|---|
| 1.6 | 265.0 | 238.2 | 238.2 | 0.9739 |
| 1.2 | 198.8 | 178.7 | 148.9 | 0.7304 |

its maximum $u_{max}$ was reached by using the MTS Bionix system strength machine selecting a velocity of about 5 mm/min. From the drawing, it is clear that the assigned displacement engages only on two fibers thanks to the small bridge on the right side. We remark the there is a small gap, i.e. 2.2 mm, between the fibers and the small bridge on the right side. The reasons of this gap will be better clarified in the following.

Figure 5 reports three pictures taken during the test execution and distinct for the displacement parameter $\lambda = u/u_{max}$. It has to be highlighted the particular effect reported in Fig. 6a which makes clear the necessity of the gap on the specimen.

Numerical simulation of this test was performed by assuming for the spring stiffnesses the same values already used in [7] for similar specimens. These values are reported in the first row of Table 2 and correspond to the value of the Young's modulus $E = 1.6$ GPa (an intermediate value of the declared range) estimated by using the suggestion reported in [12].

In Fig. 7a is reported the comparison between the structural reaction on the left side of the specimen both for the experiment, in black, and for the numerical simulation, in red. The uncertainties on $E$ and the awareness that this parameter could hardly affects all the stiffness parameters, see e.g. [8, 9, 13] for some insights, suggested us for trying to improve the curve fitting simply modifying the Young's modulus, and consequently the stiffness parameters of the springs. A surely better fitting was

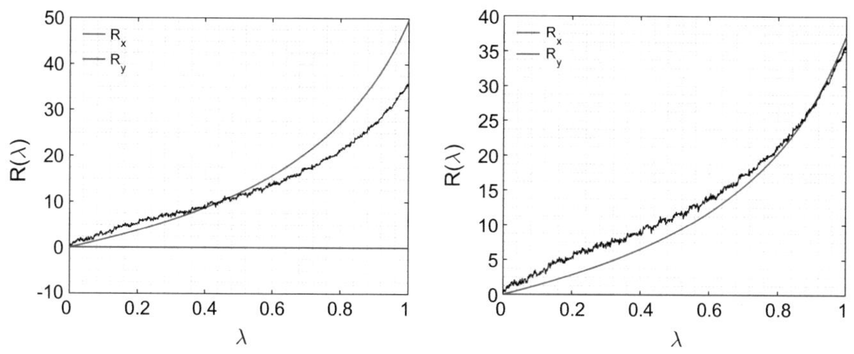


(a) $E = 1.6$ GPa (b) $E = 1.2$ GPa

**Fig. 7** Push-out test: structural reaction for different values of the Young's modulus

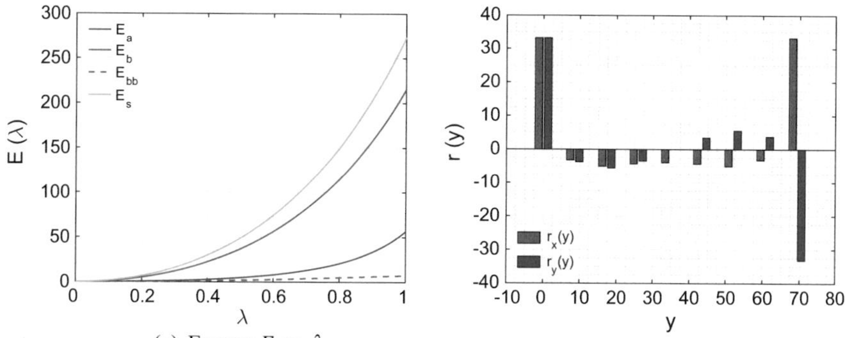


(a) Energy $E$ *vs.* $\lambda$

(b) Density of structural reaction $r$ on the side $x_1 = 0$ for $\lambda = 1$

**Fig. 8** Push-out test: energy, structural reaction and its density computed by using the Hencky-type numerical model

obtained using $E = 1.2$ GPa, see Fig. 7b, and the consequent stiffness parameters of the springs, see the second row of Table 2. For this value of $E$, and for the consequent springs stiffnesses, we also report in Fig. 8 both the strain energy, distinct for axial, bending and shear, as $\lambda$ increases and the density of the structural reaction on the left side. Particularly remarkable is the presence of negative (red) values which indicate compressions on the most of the center of the side.

Figure 9 reports the deformations as $\lambda$ increases, in particular for the cases $\lambda = 0.25, 0.5, 0.75$ and 1. In grey is reported the reference configuration whereas colors shows the density of the achieved energy level.

Finally, we observe that the numerical simulation reproduces, see Fig. 6b, the same remarkable effect reported in Fig. 6a.

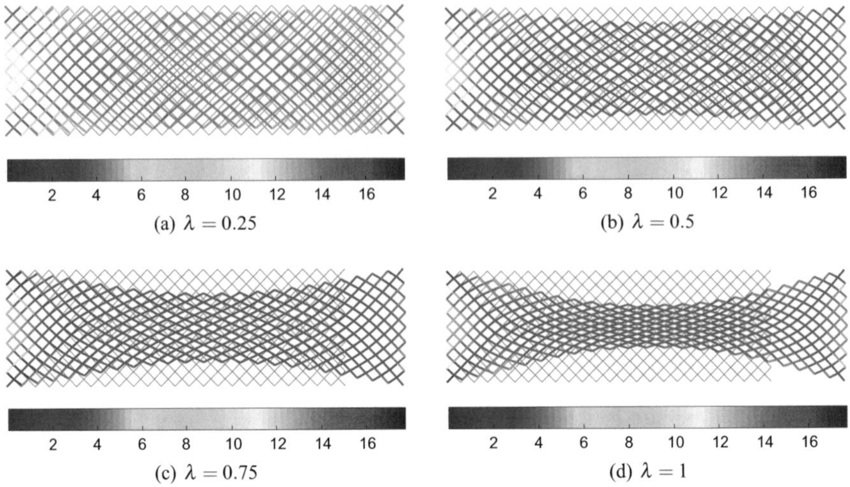


**Fig. 9** Push-out test: deformations by using the Hencky-type numerical model

It has to be remarked that the used springs stiffnesses give quite satisfactory results for what concerns the structural reaction fitting. Conversely, looking at the contraction in the central part of the specimen, the numerical simulation appears more thick than that detected during in the experiment. The reason of this difference is probably due to a non precise choice of the constitutive parameters used in the numerical simulations.

## 3.2 Pantograph with Non-orthogonal Fibers: $\alpha_1 = \pi/6$

This test concerns the specimen depicted in Fig. 10. In this case the most remarkable thing is the lack of orthogonality between intersecting fibers.

Also in this case the specimen was built using the polyamide in the 3D printing process. In this case both the left and right side are clamped and an assigned displacement $u$, parallel to the larger side of the specimen, was imposed on the right side until the value $u_{\max} = 23.7$ mm. Using the same strength machine and the same assigned velocity (5 mm/min), the three pictures, distinct by the displacement parameter $\lambda = u/u_{\max}$, were taken, see Fig. 11.

The experience of the previous test suggested us to choose the springs stiffnesses using, as a fitting rule, the agreement between the deformations at the final stage both for the experiments and for the numerical simulations. This choice leads to the springs stiffnesses reported in Table 3.

Using these values for springs stiffnesses we obtained the results reported in Fig. 12 where we reported the energies and the structural reactions vs $\lambda$ and the density of the structural reaction, for $\lambda = 1$, on the left side of the specimen. In

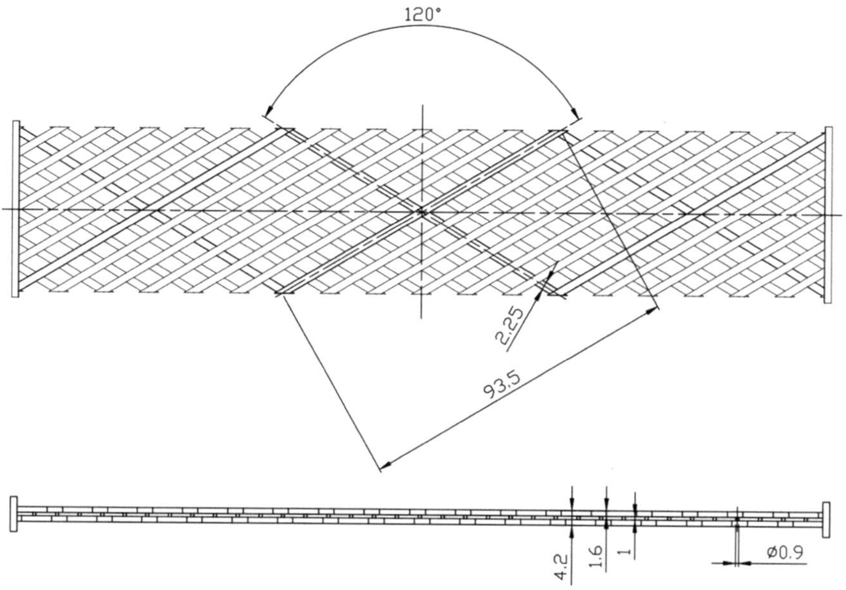


**Fig. 10** Drawing of a pantographic structure with non-orthogonal fibers: $\alpha_1 = \pi/6$

**Table 3** Stiffnesses of the springs

| $a$ (N/mm) | $b_1$ (Nmm) | $b_2$ (Nmm) | $s$ (Nmm) |
|---|---|---|---|
| 165.6 | 148.9 | 148.9 | 0.977 |

Fig. 12b there is the comparison between the structural reaction evaluated during the experiment, in black, and that computed by the numerical simulation, in red. The closeness of the two curves it is remarkable.

The deformation history computed by using the numerical model is reported in Fig. 13 for $\lambda = 0.25, 0.5, 0.75$ and 1.

Finally we reported in Fig. 14 an overlapping of the picture taken at the final deformation of the experiment and that computed numerically which clearly shows the quality of the numerical model when an accurate fitting of the spring stiffnesses is preventively performed.

### 3.3 *Pantograph with Non-orthogonal Fibers: $\alpha_1 = \pi/3$*

The last test concerns again a specimen made by the same polyamide of the previous tests and with non-orthogonal fibers but this time with a different orientation, see the drawing reported in Fig. 15.

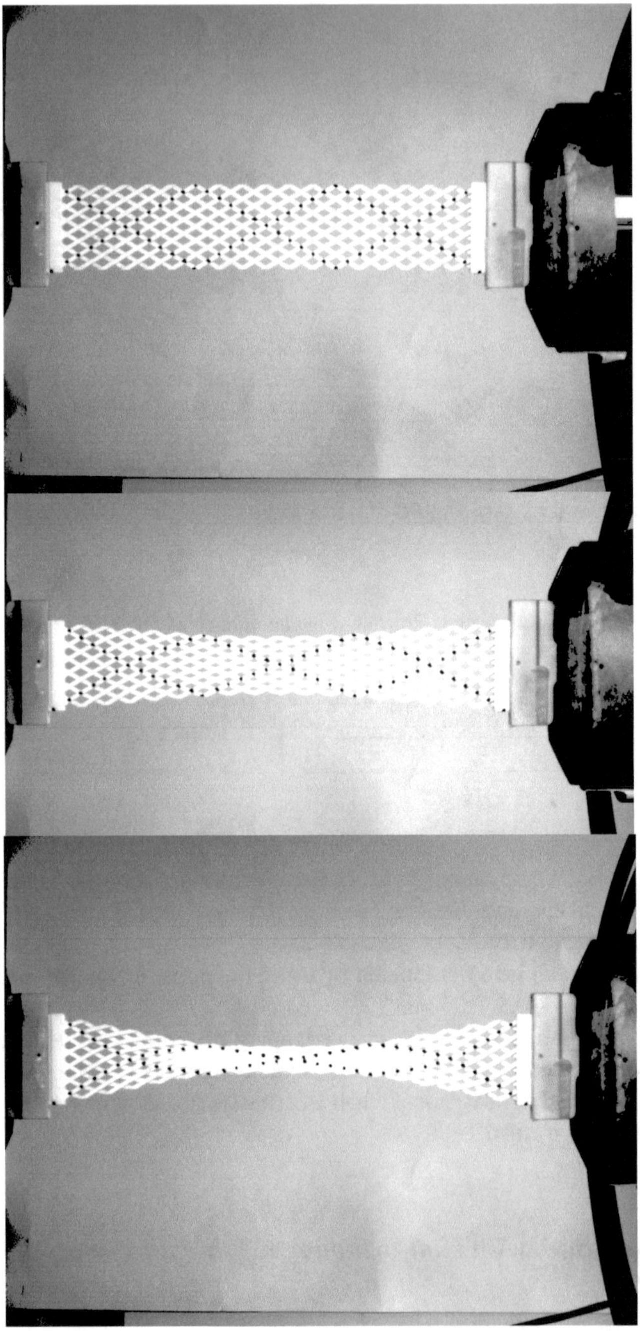

**Fig. 11** Pantograph with non-orthogonal fibers ($\alpha_1 = \pi/6$): sequence of deformations for $\lambda = 0$, 0.5, 1

**Fig. 12** Pantograph with non-orthogonal fibers ($\alpha_1 = \pi/6$): energy, structural reaction and its density computed by using the Hencky-type numerical model

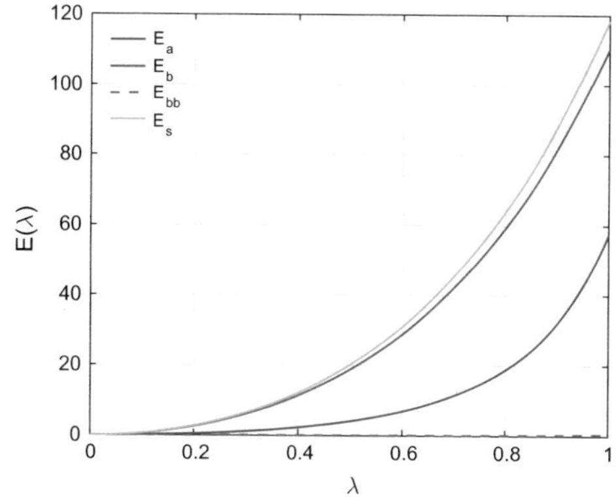


(a) Energy $E$ *vs.* $\lambda$

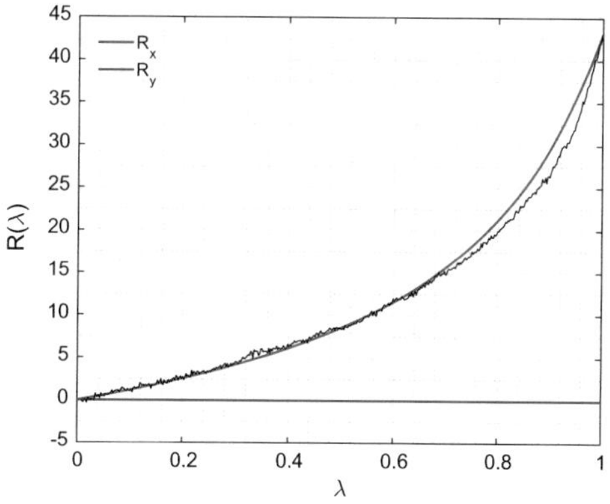


(b) Structural reaction $R$ *vs.* $\lambda$
(in black is reported the experimental result)

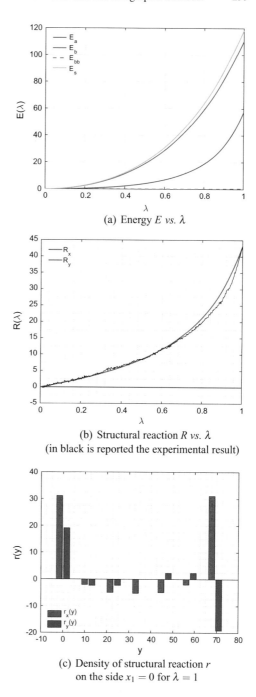

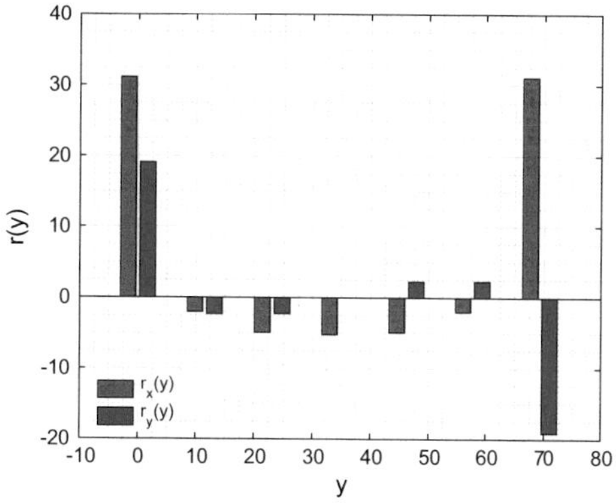


(c) Density of structural reaction $r$ on the side $x_1 = 0$ for $\lambda = 1$

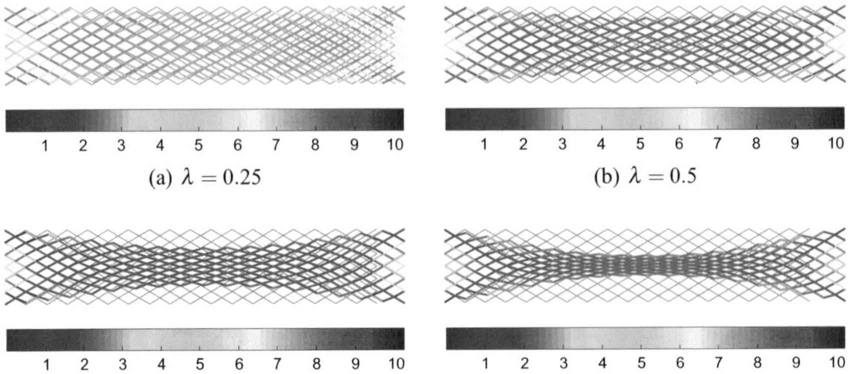


(a) $\lambda = 0.25$ (b) $\lambda = 0.5$

(c) $\lambda = 0.75$ (d) $\lambda = 1$

**Fig. 13** Pantograph with non-orthogonal fibers ($\alpha_1 = \pi/6$): deformations by using the Hencky-type numerical model

Also in this case on the specimen was applied an assigned displacement on the right smaller side in the direction parallel to the larger sides until the value $u_{\text{max}} = 74.7$ mm was reached. Three pictures were taken in the loading process (the strength machine and the velocity of the test are unchanged), see Fig. 16, corresponding to $\lambda = 0$, 0.5 and 1.

Numerical simulation of this experiment was performed using as springs stiffnesses those reported in Table 3. The results of the simulation are reported in Fig. 17, energies, structural reaction and density of structural reaction, and in Fig. 18, deformation history. We have to remark that in this case although there is a lack of closeness between the structural reaction given from the experiment and that computed numerically, mostly for values of $\lambda > 0.7$ (see Fig. 17b), the agreement on the whole set of displacement is again remarkable as can be observed by the overlapping of the two configurations at the final stage, see Fig. 19.

In our opinion this difference, unexpected if we consider the test with $\alpha_1 = \pi/6$, between the structural reaction evaluated in the experiment and computed numerically can be explained if we consider that in the $\alpha_1 = \pi/6$ test the ratio between the maximum assigned displacement and the long side, macro-strain, of the specimen is about 10% whereas the same quantity for the $\alpha_1 = \pi/3$ test is 53%. Looking again at Fig. 17b, if we consider only the first part of the curves (those corresponding approximatively to $\lambda \leq 0.4$) then there is a remarkable agreement between them. We highlight that $\lambda = 0.4$ corresponds to a macro-strain of about 21% less than half of that imposed on the specimen (53%).

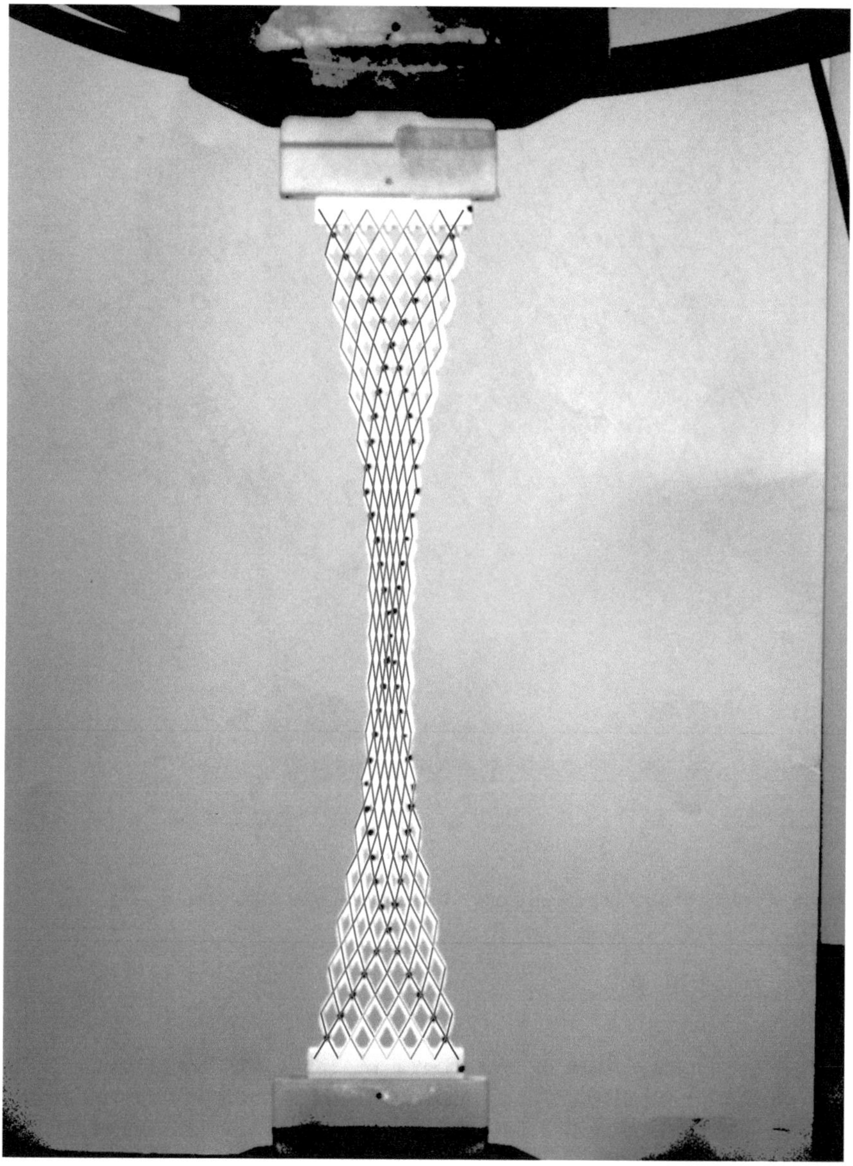

**Fig. 14** Pantograph with non-orthogonal fibers ($\alpha_1 = \pi/6$): deformations by using the Hencky-type numerical model

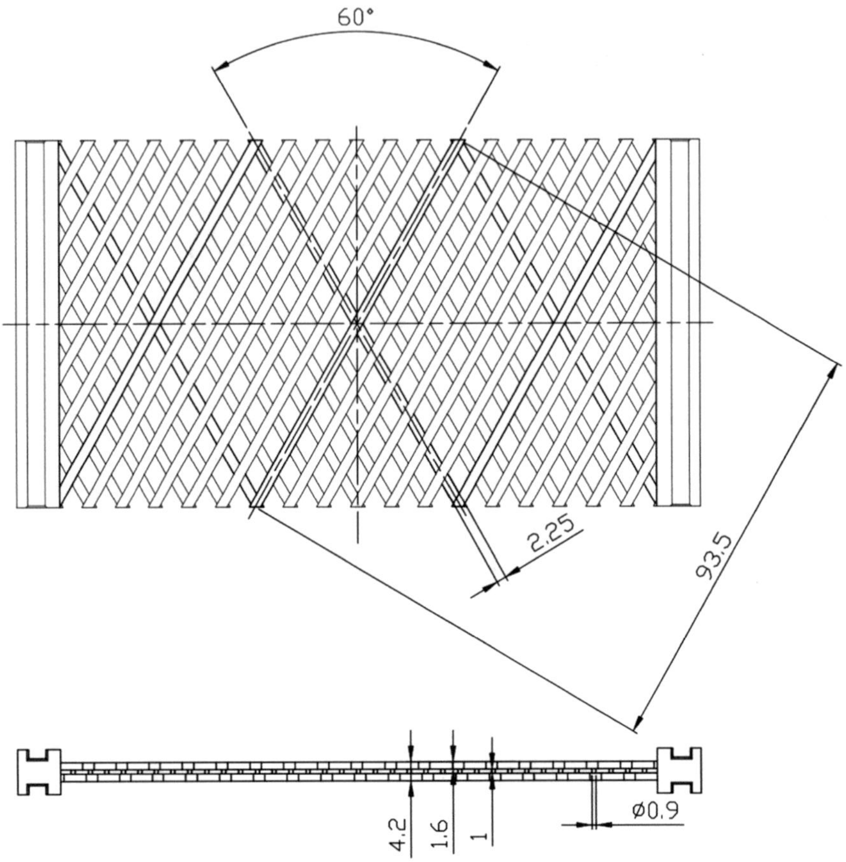


**Fig. 15** Drawing of a pantographic structure with non-orthogonal fibers ($\alpha_1 = \pi/3$)

## 4 Concluding Remarks

In [5, 10, 14] more or less rigorous homogenization results are presented, in the framework of linear elasticity: i.e. small deformations and quadratic deformation energies. The model presented here, instead, tries to model the behavior of real pantographic structures undergoing large displacements. In the proposed experiments, while the majority of the beams constituting the pantographic lattice are in the small deformation regime, we can however distinguish some boundary layers in which the involved beam elements undergo very large deformations (more that 5% of elongations, for instance, as remarked in [3]). These experimental evidence compelled us to introduce strongly nonlinear models in order to be able to design a priori pantographic sheet having tailored properties.

While the numerical simulations show a surprising agreement with experimental evidence, we feel that a rigorous basis on the homogenization results presented in

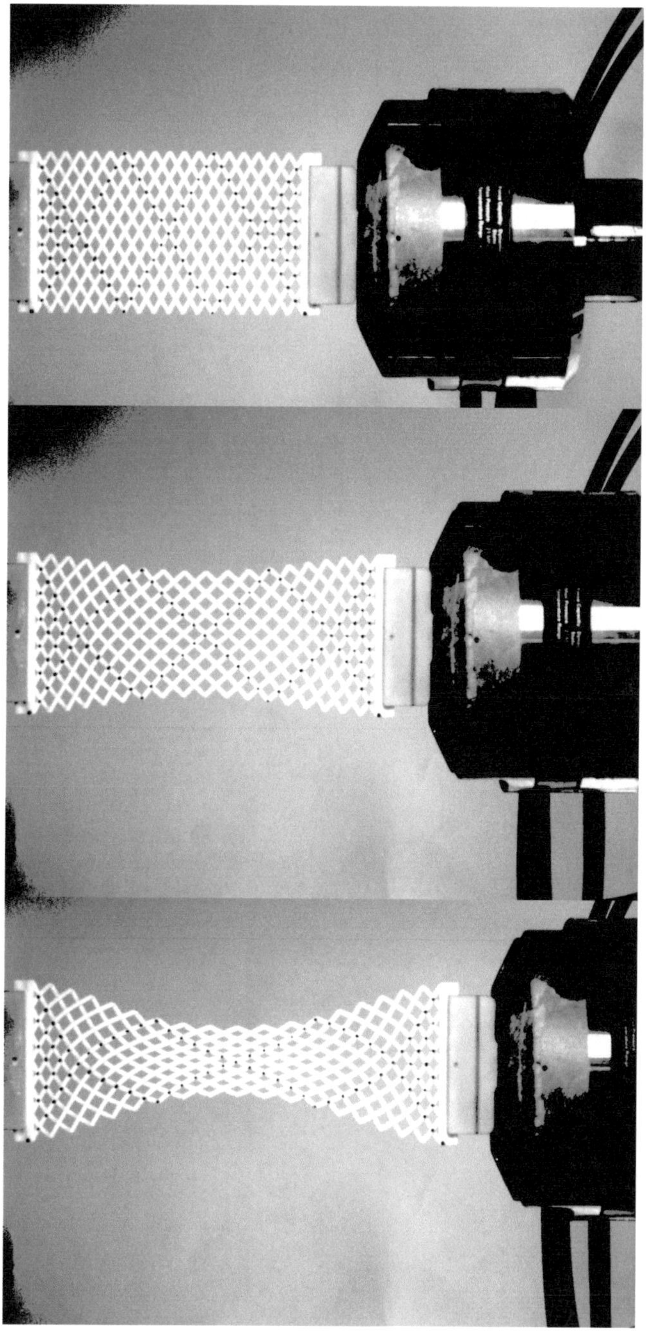

**Fig. 16** Pantograph with non-orthogonal fibers ($\alpha_1 = \pi/3$): sequence of deformations for $\lambda = 0$, 0.5 and 1

**Fig. 17** Pantograph with non-orthogonal fibers ($\alpha_1 = \pi/3$): energy, structural reaction and its density computed by using the Hencky-type numerical model

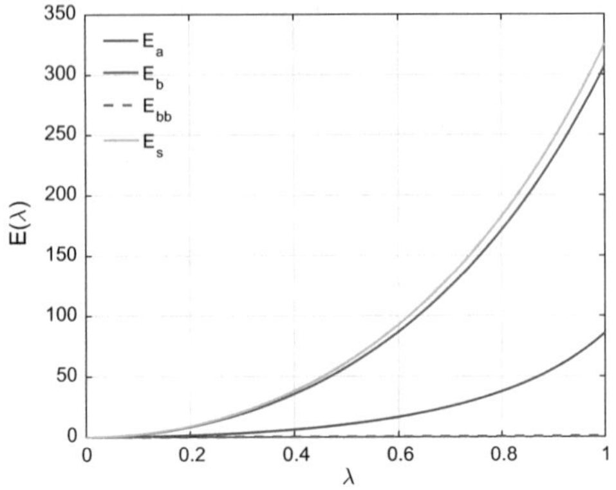


(a) Energy $E$ *vs.* $\lambda$

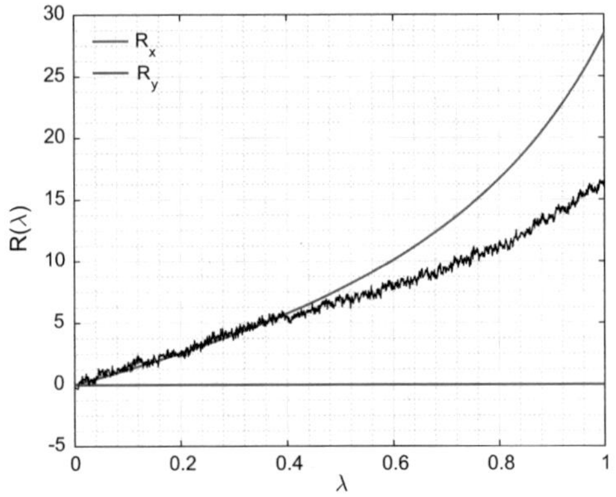


(b) Structural reaction $R$ *vs.* $\lambda$
(in black is reported the experimental result)

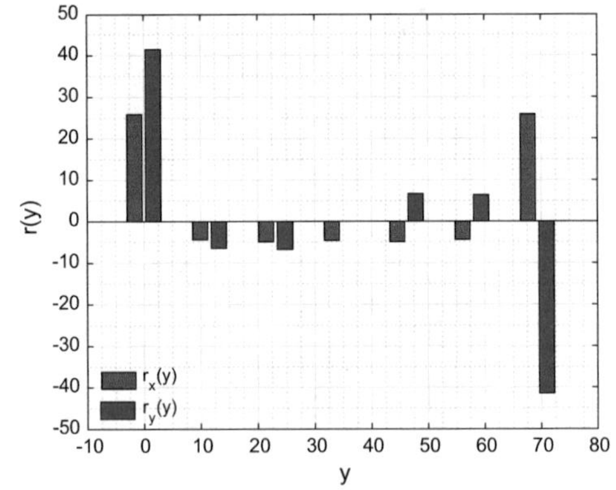


(c) Density of structural reaction $r$
on the side $x_1 = 0$ for $\lambda = 1$

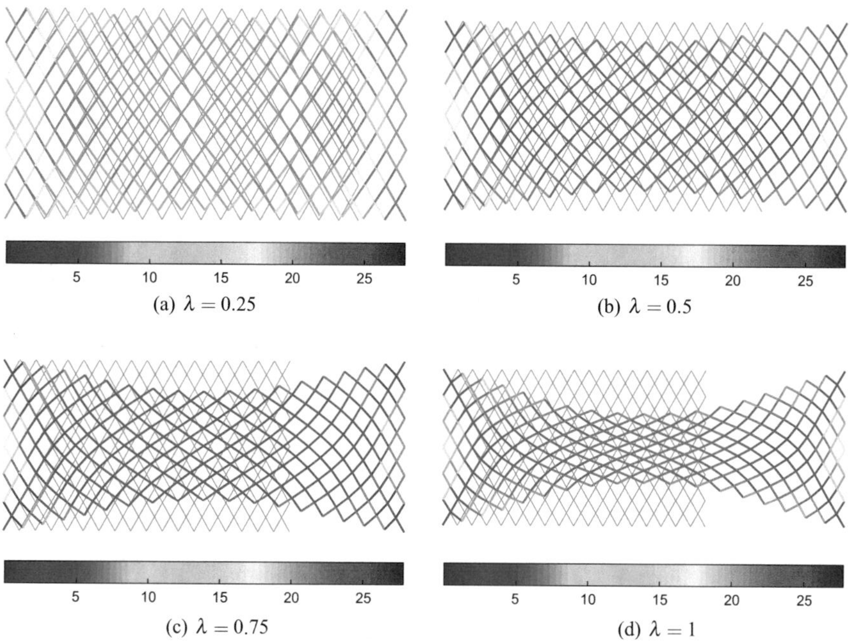


**Fig. 18** Pantograph with non-orthogonal fibers ($\alpha_1 = \pi/3$): deformations by using the Hencky-type numerical model

[3] needs to be firmly established. We expect that $\Gamma$-convergence results can be now confidently formulated and conjectured, see [15]. Moreover we expect that the methods exploited in [16] could be adapted to get also a priori error estimates in the replacement process involved when passing from discrete to continuum models.

A final remark is needed: many cases of out of plane buckling of *exotic* pantographic sheets were observed. A phenomenological model proposed in [17, 18] has been successfully used to get qualitative predictions. However to get more general quantitative predictions an identification procedure involving discrete Lagrangian models with concentrated springs is needed, which applies to three-dimensional motion of two-dimensional pantographic sheets.

Future challenges concern:

i. Although pantographic structures were conceived to give an example of second gradient metamaterial, see e.g. [19–29], the development of 3D printing technology allowed for the *practical* synthesis of such metamaterials. It deserves to be investigated how to improve the design of 3D printed fabrics in order to fully exploit the exotic behaviour of higher gradient metamaterials. We remark that, as seen in [3, 30], the behaviour of higher gradient continua shows many peculiarities which deserve a deeper experimental investigation.
ii. The discrete nature of suitably designed beam lattices may be modelled also by means of more refined tools, see e.g. [31–38] for an in-depth description of

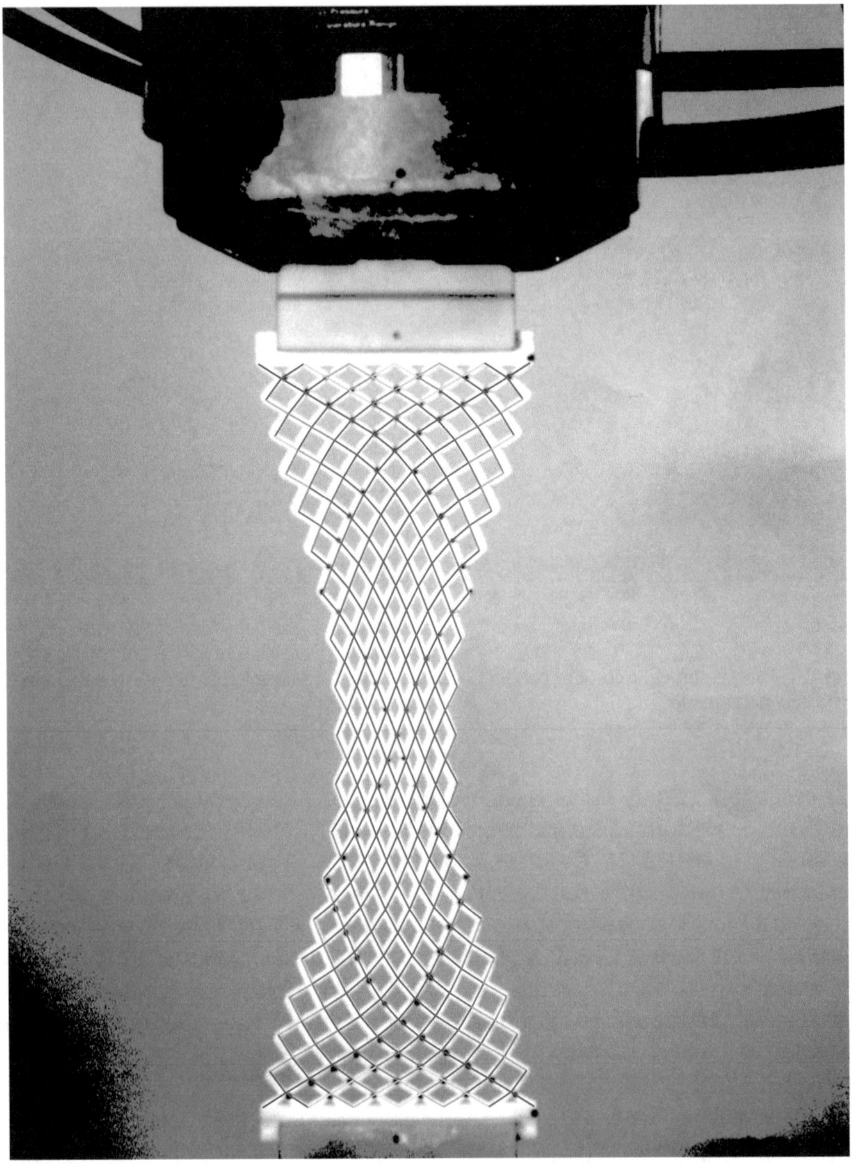

**Fig. 19** Pantograph with non-orthogonal fibers ($\alpha_1 = \pi/3$): deformations by using the Hencky-type numerical model

NURBS interpolation or using the generalized beam theory, see [39, 40], this in order to design even more complex metamaterials also in the 3D case where could be efficiently used the Pipkin model described in [41] and in the review paper [42].

iii. Another crucial point concerns the modelling of the breakdown evolution of pantographic sheets. Indeed some evidence has been already gathered about the onset and the evolution of failure. It is rather evident that have to be considered ruptures concerning both fibers and pivots. A first modelling effort to model such rupture phenomena was presented in [43] when the attention was limited to the rupture mechanism initiated by the rupture of a fiber, see also [13] for an insight on the modelling of fiber defects. In this context, surely deserve models able to consider the out-of-plane deformations and the related buckling phenomena, see [44–48] for a quick insight on this argument.
iv. The experimental identification of the parameters of the discrete model, i.e. the stiffnesses of the springs, require a specific investigation (see [12]). In particular methods of best fitting must be coupled to those used in extended sensitivity analysis by adapting, for example, the tools described in [49] and exploited in [50–56], see also [57, 58] for a more specific application to the description of huge and innovative structures.
v. Experimental evidence shows the onset of some vibration phenomena in some specific experimental conditions. Therefore, it is relevant the extension of modelling to dynamic regimes, which can be obtained following the methods presented in see [59, 60] and also in [61, 62].
vi. The discrete Hencky-type model and the related numerical discretization technique could also be used to model granular media interactions, see [63], or generalized and micro-structured continua, see [19, 64–67] and, in particular, [68, 69] for applications in civil engineering and [70] in biomechanics.
vii. In various experiments the contact between fibers was observed; if this kind of phenomenon has to be considered they could be interesting the guidelines reported in [71, 72].

## References


1. Turco, E., dell'Isola, F., Cazzani, A., Rizzi, N.L.: Hencky-type discrete model for pantographic structures: numerical comparison with second gradient continuum models. Zeitschrift für Angewandte Mathematik und Physik **67**(4), 1–28 (2016)
2. Hencky, H.: Über die angenäherte Lösung von Stabilitätsproblemen im Raum mittels der elastischen Gelenkkette. Ph.D. thesis, Engelmann (1921)
3. dell'Isola, F., Giorgio, I., Pawlikowski, M., Rizzi, N.L.: Large deformations of planar extensible beams and pantographic lattices: heuristic homogenisation, experimental and numerical examples of equilibrium. Proc. R. Soc. Lond. A: Math. Phys. Eng. Sci. **472**(2185) (2016)
4. dell'Isola, F., Lekszycki, T., Pawlikowski, M., Grygoruk, R., Greco, L.: Designing a light fabric metamaterial being highly macroscopically tough under directional extension: first experimental evidence. Zeitschrift für angewandte Mathematik und Physik **66**(6), 3473–3498 (2015)
5. Alibert, J.-J., Seppecher, P., dell'Isola, F.: Truss modular beams with deformation energy depending on higher displacement gradients. Math. Mech. Solids **8**(1), 51–73 (2003)
6. Madeo, A., Della Corte, A., Greco, L., Neff, P.: Wave propagation in pantographic 2D lattices with internal discontinuities. Proc. Est. Acad. Sci. **64**(3S), 325–330 (2015)

7. Turco, E., Golaszewski, M., Cazzani, A., Rizzi, N.L.: Large deformations induced in planar pantographic sheets by loads applied on fibers: experimental validation of a discrete Lagrangian model. Mech. Res. Commun. **76**, 51–56 (2016)
8. Turco, E., Barcz, K., Pawlikowski, M., Rizzi, N.L.: Non-standard coupled extensional and bending bias tests for planar pantographic lattices. Part I: numerical simulations. Zeitschrift für Angewandte Mathematik und Physik **67**(122), 1–16 (2016)
9. Turco, E., Barcz, K., Rizzi, N.L.: Non-standard coupled extensional and bending bias tests for planar pantographic lattices. Part II: comparison with experimental evidence. Zeitschrift für Angewandte Mathematik und Physik **67**(123), 1–16 (2016)
10. Alibert, J.-J., Della Corte, A.: Second-gradient continua as homogenized limit of pantographic microstructured plates: a rigorous proof. Zeitschrift für Angewandte Mathematik und Physik **66**(5), 2855–2870 (2015)
11. Pideri, C., Seppecher, P.: A second gradient material resulting from the homogenization of an heterogeneous linear elastic medium. Contin. Mech. Thermodyn. **9**(5), 241–257 (1997)
12. Placidi, L., Andreaus, U., Della Corte, A., Lekszycki, T.: Gedanken experiments for the determination of two-dimensional linear second gradient elasticity coefficients. Zeitschrift für Angewandte Mathematik und Physik (ZAMP) **66**(6), 3699–3725 (2015)
13. Turco, E., Rizzi, N.L.: Pantographic structures presenting statistically distributed defects: numerical investigations of the effects on deformation fields. Mech. Res. Commun. **77**, 65–69 (2016)
14. Boutin, C., dell'Isola, F., Giorgio, I., Placidi, L.: Linear pantographic sheets. Part I: asymptotic micro-macro models identification. Mathematics and Mechanics of Complex Systems (in press)
15. Braides, A., Solci, M.: Asymptotic analysis of Lennard-Jones systems beyond the nearest-neighbour setting: a one-dimensional prototypical case. Math. Mech. Solids **21**(8), 915–930 (2016)
16. Carcaterra, A., dell'Isola, F., Esposito, R., Pulvirenti, M.: Macroscopic description of microscopically strongly inhomogenous systems: a mathematical basis for the synthesis of higher gradients metamaterials. Arch. Rational Mech. Anal. (2015). doi:10.1007/s00205-015-0879-5
17. Steigmann, D.J., dell'Isola, F.: Mechanical response of fabric sheets to three-dimensional bending, twisting, and stretching. Acta Mech. Sin. **31**(3), 373–382 (2015)
18. Scerrato, D., Giorgio, I., Rizzi, N.L.: Three-dimensional instabilities of pantographic sheets with parabolic lattices: numerical investigations. Zeitschrift für Angewandte Mathematik und Physik **67**(3), 1–19 (2016)
19. dell'Isola, F., Steigmann, D., Della Corte, A.: Synthesis of fibrous complex structures: designing microstructure to deliver targeted macroscale response. Appl. Mech. Rev. **67**(6), 060804 (2015)
20. Giorgio, I., Grygoruk, R., dell'Isola, F., Steigmann, D.J.: Pattern formation in the three-dimensional deformations of fibered sheets. Mech. Res. Commun. **69**, 164–171 (2015)
21. Scerrato, D., Giorgio, I., Della Corte, A., Madeo, A., Limam, A.: A micro-structural model for dissipation phenomena in the concrete. Int. J. Numer. Anal. Methods Geomech. **39**(18), 2037–2052 (2015)
22. Scerrato, D., Zhurba Eremeeva, I.A., Lekszycki, T., Rizzi, N.L.: On the effect of shear stiffness on the plane deformation of linear second gradient pantographic sheets. Zeitschrift für Angewandte Mathematik und Mechanik (2016). doi:10.1002/zamm.201600066
23. D'Agostino, M.V., Giorgio, I., Greco, L., Madeo, A., Boisse, P.: Continuum and discrete models for structures including (quasi-) inextensible elasticae with a view to the design and modeling of composite reinforcements. Int. J. Solids Struct. **59**, 1–17 (2015)
24. Placidi, L., Andreaus, U., Giorgio, I.: Identification of two-dimensional pantographic structure via a linear d4 orthotropic second gradient elastic model. J. Eng. Math. (2017). doi:10.1007/s10665-016-9856-8
25. Placidi, L., Dhaba, A.E.: Semi-inverse method à la Saint-Venant for two-dimensional linear isotropic homogeneous second-gradient elasticity. Math. Mech. Solids (2017). doi:10.1177/1081286515616043

26. dell'Isola, F., Giorgio, I., Andreaus, U.: Elastic pantographic 2d lattices: a numerical analysis on static response and wave propagation. Proc. Est. Acad. Sci. **64**(3), 219–225 (2015)
27. dell'Isola, F., Della Corte, A., Greco, L., Luongo, A.: Plane bias extension test for a continuum with two inextensible families of fibers: a variational treatment with Lagrange multipliers and a perturbation solution. Int. J. Solids Struct. **81**, 1–12 (2016)
28. Cuomo, M., dell'Isola, F., Greco, L.: Simplified analysis of a generalized bias test for fabrics with two families of inextensible fibres. Zeitschrift für angewandte Mathematik und Physik **67**(3), 1–23 (2016)
29. dell'Isola, F., Cuomo, M., Greco, L., Della Corte, A.: Bias extension test for pantographic sheets: numerical simulations based on second gradient shear energies. J. Eng. Math. (2017). doi:10.1007/s10665-016-9865-7
30. dell'Isola, F., Andreaus, U., Placidi, L.: At the origins and in the vanguard of peridynamics, non-local and higher-gradient continuum mechanics: an underestimated and still topical contribution of Gabrio Piola. Math. Mech. Solids **20**(8), 887–928 (2015)
31. Cazzani, A., Malagù, M., Turco, E.: Isogeometric analysis of plane curved beams. Math. Mech. Solids **21**(5), 562–577 (2016)
32. Cazzani, A., Malagù, M., Turco, E.: Isogeometric analysis: a powerful numerical tool for the elastic analysis of historical masonry arches. Contin. Mech. Thermodyn. **28**(1), 139–156 (2016)
33. Cazzani, A., Malagù, M., Turco, E., Stochino, F.: Constitutive models for strongly curved beams in the frame of isogeometric analysis. Math. Mech. Solids **21**(2), 182–209 (2016)
34. Bilotta, A., Formica, G., Turco, E.: Performance of a high-continuity finite element in three-dimensional elasticity. Int. J. Numer. Methods Biomed. Eng. **26**, 1155–1175 (2010)
35. Greco, L., Cuomo, M.: B-spline interpolation of Kirchhoff-Love space rods. Comput. Methods Appl. Mech. Eng. **256**, 251–269 (2013)
36. Greco, L., Cuomo, M.: An implicit $G^1$ multi patch B-spline interpolation for Kirchhoff-Love space rod. Comput. Methods Appl. Mech. Eng. **269**, 173–197 (2014)
37. Greco, L., Cuomo, M.: An isogeometric implicit G1 mixed finite element for Kirchhoff space rods. Comput. Methods Appl. Mech. Eng. **298**, 325–349 (2016)
38. Cazzani, A., Stochino, F., Turco, E.: An analytical assessment of finite elements and isogeometric analysis of the whole spectrum of Timoshenko beams. Zeitschrift für Angewandte Mathematik und Mechanik **96**(10), 1220–1244 (2016)
39. Piccardo, G., Ranzi, G., Luongo, A.: A complete dynamic approach to the generalized beam theory cross-section analysis including extension and shear modes. Math. Mech. Solids **19**(8), 900–924 (2014)
40. Piccardo, G., Ranzi, G., Luongo, A.: A direct approach for the evaluation of the conventional modes within the gbt formulation. Thin-Walled Struct. **74**, 133–145 (2014)
41. Placidi, L., Greco, L., Bucci, S., Turco, E., Rizzi, N.L.: A second gradient formulation for a 2D fabric sheet with inextensible fibres. Zeitschrift für angewandte Mathematik und Physik **67**(114), 1–24 (2016)
42. Placidi, L., Barchiesi, E., Turco, E., Rizzi, N.L.: A review on 2D models for the description of pantographic fabrics. Zeitschrift für angewandte Mathematik und Physik **67**(121), 1–20 (2016)
43. Turco, E., dell'Isola, F., Rizzi, N.L., Grygoruk, R., Müller, W.H., Liebold, C.: Fiber rupture in sheared planar pantographic sheets: numerical and experimental evidence. Mech. Res. Commun. **76**, 86–90 (2016)
44. D'Annibale, F., Rosi, G., Luongo, A.: Linear stability of piezoelectric-controlled discrete mechanical systems under nonconservative positional forces. Meccanica **50**(3), 825–839 (2015)
45. Rizzi, N., Varano, V., Gabriele, S.: Initial postbuckling behavior of thin-walled frames under mode interaction. Thin-Walled Struct. **68**, 124–134 (2013)
46. Gabriele, S., Rizzi, N., Varano, V.: A 1D higher gradient model derived from Koiter's shell theory. Math. Mech. Solids **21**(6), 737–746 (2016)
47. AminPour, H., Rizzi, N.: A one-dimensional continuum with microstructure for single-wall carbon nanotubes bifurcation analysis. Math. Mech. Solids **21**(2), 168–181 (2016)

48. Gabriele, S., Rizzi, N.L., Varano, V.: A 1D nonlinear TWB model accounting for in plane cross-section deformation. Int. J. Solids Struct. **94–95**, 170–178 (2016)
49. Turco, E.: Tools for the numerical solution of inverse problems in structural mechanics: review and research perspectives. Eur. J. Environ. Civil Eng. 1–46 (2017). doi:10.1080/19648189.2015.1134673
50. Lekszycki, T., Olhoff, N., Pedersen, J.J.: Modelling and identification of viscoelastic properties of vibrating sandwich beams. Compos. Struct. **22**(1), 15–31 (1992)
51. Bilotta, A., Turco, E.: A numerical study on the solution of the Cauchy problem in elasticity. Int. J. Solids Struct. **46**, 4451–4477 (2009)
52. Bilotta, A., Morassi, A., Turco, E.: Reconstructing blockages in a symmetric duct via quasi-isospectral horn operators. J. Sound Vib. **366**, 149–172 (2016)
53. Bilotta, A., Turco, E.: Numerical sensitivity analysis of corrosion detection. Math. Mech. Solids. **22**(1), 72–88 (2017). doi:10.1177/1081286514560093
54. Alessandrini, G., Bilotta, A., Formica, G., Morassi, A., Rosset, E., Turco, E.: Evaluating the volume of a hidden inclusion in an elastic body. J. Comput. Appl. Math. **198**(2), 288–306 (2007)
55. Alessandrini, G., Bilotta, A., Morassi, A., Turco, E.: Computing volume bounds of inclusions by EIT measurements. J. Sci. Comput. **33**(3), 293–312 (2007)
56. Turco, E.: Identification of axial forces on statically indeterminate pin-jointed trusses by a nondestructive mechanical test. Open Civ. Eng. J. **7**, 50–57 (2013)
57. Buffa, F., Cazzani, A., Causin, A., Poppi, S., Sanna, G.M., Solci, M., Stochino, F., Turco, E.: The Sardinia radio telescope: a comparison between close range photogrammetry and FE models. Math. Mech. Solids 1–22 (2015). doi:10.1177/1081286515616227
58. Stochino, F., Cazzani, A., Poppi, S., Turco, E.: Sardinia radio telescope finite element model updating by means of photogrammetric measurements. Math. Mech. Solids 1–17 (2015). doi:10.1177/1081286515616046
59. Del Vescovo, D., Giorgio, I.: Dynamic problems for metamaterials: review of existing models and ideas for further research. Int. J. Eng. Sci. **80**, 153–172 (2014)
60. Battista, A., Cardillo, C., Del Vescovo, D., Rizzi, N.L., Turco, E.: Frequency shifts induced by large deformations in planar pantographic continua. Nanomech. Sci. Technol.: Int J. **6**(2), 161–178 (2015)
61. Cazzani, A., Stochino, F., Turco, E.: On the whole spectrum of Timoshenko beams. Part I: a theoretical revisitation. Zeitschrift für Angewandte Mathematik und Physik **67**(24), 1–30 (2016)
62. Cazzani, A., Stochino, F., Turco, E.: On the whole spectrum of Timoshenko beams. Part II: further applications. Zeitschrift für Angewandte Mathematik und Physik **67**(25), 1–21 (2016)
63. Misra, A., Poorsolhjouy, P.: Granular micromechanics based micromorphic model predicts frequency band gaps. Contin. Mech. Thermodyn. **28**(1), 215–234 (2016)
64. Altenbach, J., Altenbach, H., Eremeyev, V.A.: On generalized Cosserat-type theories of plates and shells: a short review and bibliography. Arch. Appl. Mech. **80**(1), 73–92 (2010)
65. Eremeyev, V.A., Pietraszkiewicz, W.: Material symmetry group and constitutive equations of micropolar anisotropic elastic solids. Math. Mech. Solids. **21**(2), 210–221 (2016). doi:10.1177/1081286515582862
66. Dos Reis, F., Ganghoffer, J.F.: Construction of micropolar continua from the asymptotic homogenization of beam lattices. Comput. Struct. **112–113**, 354–363 (2012)
67. Elnady, K., Dos Reis, F., Ganghoffer, J.-F.: Construction of second order gradient continuous media by the discrete asymptotic homogenization method. Int. J. Appl. Mech. (2014)
68. Caggegi, C., Pensée, V., Fagone, M., Cuomo, M., Chevalier, L.: Experimental global analysis of the efficiency of carbon fiber anchors applied over CFRP strengthened bricks. Constr. Build. Mater. **53**, 203–212 (2014)
69. Tedesco, F., Bilotta, A., Turco, E.: Multiscale 3D mixed FEM analysis of historical masonry constructions. Eur. J. Environ. Civ. Eng. (2017). doi:10.1080/19648189.2015.1134676
70. Tomic, A., Grillo, A., Federico, S.: Poroelastic materials reinforced by statistically oriented fibres - numerical implementation and application to articular cartilage. IMA J. Appl. Math. **79**, 1027–1059 (2014)

71. Andreaus, U., Chiaia, B., Placidi, L.: Soft-impact dynamics of deformable bodies. Contin. Mech. Thermodyn. **25**, 375–398 (2013)
72. Andreaus, U., Baragatti, P., Placidi, L.: Experimental and numerical investigations of the responses of a cantilever beam possibly contacting a deformable and dissipative obstacle under harmonic excitation. Int. J. Non-Linear Mech. **80**, 96–106 (2016)

# Index




F. dell'Isola et al. (eds.), *Mathematical Modelling in Solid Mechanics*,
Advanced Structured Materials 69, DOI 10.1007/978-981-10-3764-1

Printed in the United States
By Bookmasters